# Seguin's
# COMPUTER
## Concepts & Applications

### for Microsoft® Office 365
#### 2019 Edition

**Denise Seguin**
Fanshawe College, London, Ontario

PARADIGM
EDUCATION SOLUTIONS

St. Paul

**Vice President, Content and Digital Solutions:** Christine Hurney
**Director of Content Development, Computer Technology:** Cheryl Drivdahl
**Developmental Editor:** Jennifer Gehlhar
**Director of Production:** Timothy W. Larson
**Production Editor:** Carrie Rogers
**Design and Production Specialist:** Jack Ross
**Interior Design:** Jack Ross
**Cover Design:** Jack Ross
**Page Layout:** Jack Ross
**Photo Researcher and Illustrator:** Melora Pappas
**Copy Editor:** Penny Stuart
**Proofreader:** Eric Braem
**Tester:** Julia Basham and Becky Jones
**Indexer:** Terry Casey
**Vice President, Director of Digital Products:** Chuck Bratton
**Digital Projects Manager:** Tom Modl
**Digital Solutions Manager:** Gerry Yumul
**Senior Director of Digital Products and Onboarding:** Christopher Johnson
**Supervisor of Digital Products and Onboarding:** Ryan Isdahl
**Vice President, Marketing:** Lara Weber McLellan
**Marketing and Communications Manager:** Selena Hicks

ISBN 978-0-76388-749-0 (print)
ISBN 978-0-76388-748-3 (digital)

© 2020 by Paradigm Publishing, LLC
875 Montreal Way
St. Paul, MN 55102
Email: CustomerService@ParadigmEducation.com
Website: ParadigmEducation.com

# Contents

# Seguin's COMPUTER Applications with Microsoft® Office 365

## Chapter 6
## Creating, Editing, and Formatting Word Documents .................................125

## Chapter 7
## Enhancing a Word Document with Special Features.........................147

## Chapter 8
## Creating, Editing, and Formatting Excel Worksheets...............................175

# Preface

## Course Overview

Many of today's students have grown up with technology all around them and are more connected to the internet with more devices than any generation before them. They have always had the internet as a source of information, entertainment, and communication. Students in the classroom today likely started using the internet and a word processor before entering grade school. Some students never turn off their smartphones, even when they go to bed at night. For many, the cell phone or smartphone is integral to daily life. Being in constant touch with friends, family, and others (including classmates, teammates, colleagues, and social media followers) is a normal way of life.

So what value can a course on introductory computer concepts provide for these technology-driven students? Knowing how to type an essay in a word processor and look up information on the internet is not all there is to this digital world in which we live. To be successful in any career, we need a deeper understanding of technology and how the tools at our disposal can be used effectively. In the pages that follow, you will learn about all the equipment and programs that combine to provide you with technology applications for communicating and performing tasks in your personal and professional life. You will also learn about tools and practices for using technology safely and securely and for being mindful of privacy issues.

Along with well-designed pedagogy, practice and problem solving will help you learn and apply computer concepts and skills. Technology provides opportunities for interactive learning as well as excellent ways to quickly and accurately assess student performance. To this end, this course is supported by Cirrus, Paradigm Education Solution's web-based training and assessment learning management system.

## Course Goals

*Seguin's COMPUTER Concepts*, Third Edition, provides instruction in achieving a basic understanding of the components that comprise computer hardware, system software, application software, internet connectivity and resources, social media, and the security and privacy issues related to technology. No prior experience with computer concepts is required. Even those with some technological knowledge can benefit from completing the course by digging deeper into topics of interest with the instructor's guidance. After completing the courseware, you will be able to:

- identify various types of computers; explain how data is converted into information; and understand technological convergence, cloud computing, green computing, and ergonomics.
- identify hardware and software needed to connect to the internet; explain broadband connectivity options; recognize popular web browsers; search the web effectively; and distinguish various online services.
- recognize and explain the components in a system unit, input devices, output devices, and network adapters; understand how data is represented on a computer; and distinguish various options for storage and storage capacity.
- list the major functions of operating system software and recognize popular operating systems used for computing devices; explain the purpose of embedded and cloud operating systems; describe commonly used utility programs for maintaining a computer; and use troubleshooting tools in system software for solving computer issues.
- identify productivity and multimedia applications used in workplaces and by individuals; differentiate web-based and open source applications; explain the process to acquire, install, uninstall, and upgrade software; and provide examples of mobile apps.

- describe popular social networking, media sharing, blogging, and wiki websites; and provide examples of how businesses are using social media to connect and communicate with consumers.
- explain various risks to security and privacy encountered by connected individuals, and describe strategies for protecting against security and privacy intrusions from unwanted sources.

# Course Organization and Methodology

This course is divided into seven chapters that can be done in any order. Some instructors will follow the chapter sequence as written; however, others may assign the chapters in a different order. For example, your instructor may prefer that you study the social media content before you learn about hardware.

Each chapter opens with a brief overview, including a list of the objectives covered by the chapter. All the chapter content is presented in short topics with many visual aids to minimize reading time; you will learn just what you need to know to succeed. You want to spend your time *doing*—not just reading. With that in mind, this course includes several hands-on activities. Completing these activities will allow you to dig deeper into specific topics. Some activities involve getting out and asking classmates, friends, or family what they do with technology; other activities have you try out the technology yourself; and some activities require investigative research on your own. Learning by discovery can be the most rewarding learning of all. At the end of each chapter, you will have a chance to review a summary of the topics covered in the chapter.

## Chapter Features

The following elements provide clear information to help you master the key computer concepts taught in this course.

**Chapter introductions** briefly introduce topics and key terms presented in the chapter.

**Learning objectives** are numbered and align with the topics presented in the chapter.

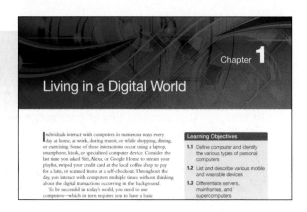

**Green IT** boxes identify environmentally conscious products and solutions.

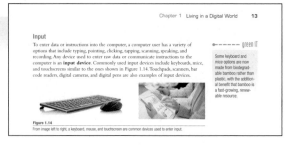

**Secure IT** boxes provide ways to protect your computer and files from security breaches.

**Go Online** features provide links to websites where you can explore and learn more.

**In the Know** features highlight inter-esting facts or trivia related to chapter content.

**Photos** show you the latest in technology while **illustrations** and **tables** organize information in a streamlined format to minimize your reading load.

A **Topics Review** chart at the end of each chapter summarizes the main chapter content and key terms learned.

## Review and Assessment

At the end of each chapter, seven to ten **assessments** of varying complexity provide opportunities to apply your knowledge. Work individually, in pairs, or in teams to produce documents, presentations, blogs, videos, and podcasts that demonstrate your understanding of topics and how they apply to the workplace or an individual application. Special assessments focused on ethics, green computing, job readiness, and mobile computing are covered throughout the course.

An **Ethics assessment** gives you the opportunity to explore real-world ethical decisions you will encounter when using computers in your personal life and career.

A **Green assessment** prompts exploration of ways to protect the environment and make good use of its resources.

---

**Living in a Digital World**

Chapter **1**
Review and Assessment

The following assessments offer opportunities to apply what you have learned in relevant, real-world situations. Save your solution files and URLs, and submit them for evaluation as directed by your instructor.

**Assessment 1.1  A Day in Your Life with the IoT**
**Type:** Individual or Pairs
**Deliverable:** Document or Presentation

Imagine a day in your life five years from now, in which all your household appliances and gadgets as well as your car are connected to the internet. You interact with these devices using voice commands, and some of the devices communicate directly with each other automatically, without your intervention or command.
 1. Write a script or prepare a presentation that describes a typical school or work day in this future IoT world.
 2. Save the document or presentation as **MyIoTWorld-YourName**.
 3. Submit the assessment to your instructor in the manner requested.

**Assessment 1.2  The Information Processing Cycle**
**Type:** Individual
**Deliverable:** Document or Presentation

Take a task you completed today using technology and break the task down into each of the four operations in the information processing cycle. Consider the hardware and software that would be used during each operation. Review Figure 1.9 on page 12 for assistance, if necessary.
 1. Create a document or presentation in which you list the operation and then describe the task at each stage in the information processing cycle. Include at the beginning of the document or presentation a brief description of the task you analyzed, including a list of the hardware and software you determined was used.
 2. Save the document or presentation as **IPCycle-YourName**.
 3. Submit the assessment to your instructor in the manner requested.

- Glossary
- Infographic
- Crossword Puzzle
- Multiple Choice
- Completion
- Matching
- Assessments
- Chapter Exam

---

**Ethics**  **Assessment 1.7  Ethics Discussion—Is It OK to Use Work Technology for Personal Reasons?**
**Type:** Team
**Deliverable:** Document, Blog Entry, or Presentation

A friend at your workplace uses her computer for personal reasons several times throughout the day when she should be working. You have observed her updating her Facebook page, scrolling Instagram photos, tweeting, watching YouTube videos, shopping online, and looking at travel websites. At lunch you casually mentioned her use of the web for personal reasons, and she told you she likes the higher speed access at work over her home setup.
 1. Within your team, discuss how you should handle this situation. Should you ignore the situation? Would your answer be different if the employee was not your friend? In what instances, if any, is it OK to use employer equipment for personal use? How should management handle employees' personal use of company-provided technology?

---

**Green**  **Assessment 1.4  Greener Computing—Benefits of Green Computing in a Newspaper Office**
**Type:** Individual, Pairs, or Team
**Deliverable:** Blog Entry

You are involved with the school newspaper and decide to write a blog entry that will help educate students on the benefits to the environment of embracing green computing practices.
 1. Write and post a blog entry that will educate the readers about green computing and encourage all of them to participate with you in adopting green computing practices.
 2. Submit the assessment to your instructor in the manner requested.

**Optional**
Read the blog entries of at least two classmates and post a comment to each. Submit the URLs of your classmates' blogs in Step 2.

A **Jobs assessment** specifically challenges you to apply your skills to real-world job experiences in order to prepare you for work opportunities.

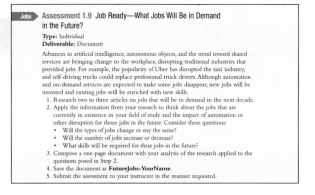

A **Mobile assessment** allows you to explore your new skills using mobile devices.

# The Cirrus Solution
*Elevating student success and instructor efficiency*

Powered by Paradigm, Cirrus is the next-generation learning solution for developing skills in Microsoft Office and computer concepts. Cirrus seamlessly delivers complete course content in a cloud-based learning environment that puts students on the fast-track to success. Students can access their content from any device anywhere, through a live internet connection. Cirrus is platform independent, ensuring that students get the same learning experience whether they are using PCs, Macs, or Chromebook computers.

## Dynamic Training

Cirrus online courses for *Seguin's COMPUTER Concepts* include interactive assignments to guide and enrich student learning.

Cirrus provides access to all student resources, delivered in a series of **scheduled assignments** that report to a grade book to track progress and achievement.

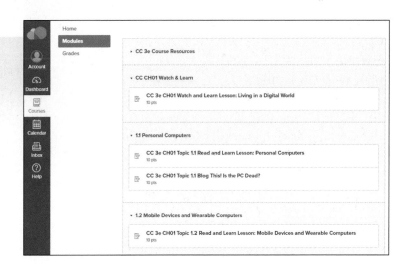

**Watch and Learn Lessons** provide chapter overviews in a video format, and include a short quiz that allows you to check your understanding of the content.

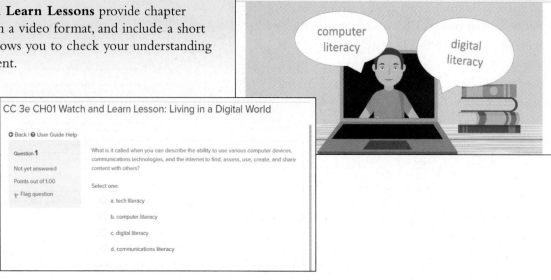

**Read and Learn Lessons** (with topic-based multiple-choice, true/false, matching, and completion quizzes) provide immediate feedback and an entry into the Cirrus grade book.

**Videos** expand on interesting tech topics. Each video ends with a brief quiz.

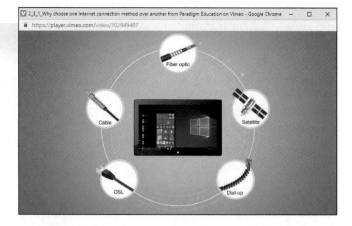

Essay quizzes provide response boxes with a text editor, and in some cases file upload options, to collect your work for instructor review and evaluation. **Blog This!** activities offer a chance for you to write about your experience or opinions on the topic. **Let's Explore!** activities invite you to dig deeper into the topic by interviewing others, trying out the technology yourself and researching related topics. **Job Connect!** activities give job information to spark interest and invite exploration of computer-related careers.

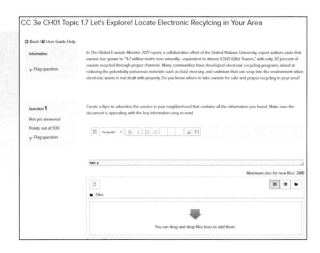

## End-of-Chapter Review

Interactive chapter review and assessment activities in the Cirrus environment offer students with all types of learning styles ample opportunity to reinforce learning and check understanding.

- Topics Review
- Glossary
- Infographics
- Crossword Puzzle
- Multiple Choice
- Completion
- Matching
- Assessments
- Chapter Exam

## Student eBook

The Student eBook makes *Seguin's COMPUTER Concepts* content available from any device (desktop, laptop, tablet, or smartphone) anywhere. The ebook is accessed through the Cirrus online course.

## Instructor eResources

All instructor resources are available through Cirrus and visible only to instructors. The instructor materials include the following items:

- Grading rubrics for evaluating responses to chapter assessments
- Lesson blueprints with teaching hints, lecture tips, and discussion questions
- Syllabus suggestions and course planning resources
- Topic-based quizzes
- Chapter-based exams

# Acknowledgments

The author and editors would like to thank the following contributors and reviewers for their involvement in this project: Julia Basham, Becky Jones; and Patti Ann Reynolds, Toni McBride, and Nicole Oke of Fanshawe College. A special thank you is extended to Michael Seguin for his contributions.

## About the Author

Denise Seguin served on the Faculty of Business at Fanshawe College of Applied Arts and Technology in London, Ontario, from 1986 until her retirement from full-time teaching in December 2012. She developed curriculum and taught a variety of office technology, software applications, and accounting courses to students in postsecondary Information Technology diploma programs and Continuing Education courses. Seguin served as Program Coordinator for Computer Systems Technician, Computer Systems Technology, Office Administration, and Law Clerk programs and was acting Chair of the School of Information Technology in 2001. Along with authoring *Seguin's COMPUTER Concepts,* First and Second Editions, and *Seguin's COMPUTER Applications with Microsoft® Office,* First and Second Editions, Seguin has also authored Paradigm Education Solution's *Microsoft Outlook* and co-authored *Our Digital World, Benchmark Series Microsoft® Excel®, Benchmark Series Microsoft® Access®, Marquee Series Microsoft® Office,* and *Using Computers in the Medical Office.*

In 2007, Seguin earned her Masters in Business Administration specializing in Technology Management, choosing to take her degree at an online university. She has an appreciation for those who are juggling work and life responsibilities while furthering their education, and she has taken her online student experiences into account when designing instruction and assessment activities for this course.

Seguin's

# COMPUTER
## Concepts

Third Edition

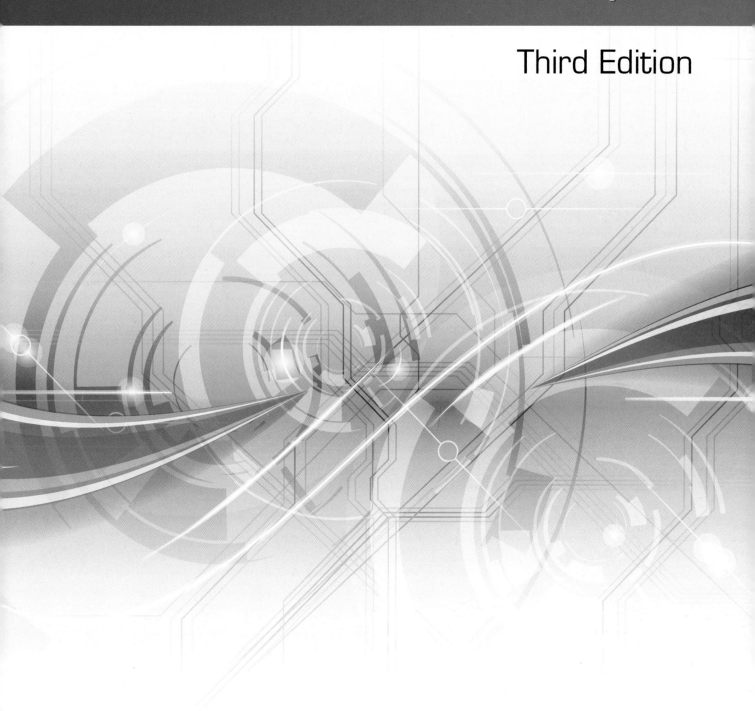

# Living in a Digital World

Individuals interact with computers in numerous ways every day at home, at work, during transit, or while shopping, dining, or exercising. Some of these interactions occur using a laptop, smartphone, kiosk, or specialized computer device. Consider the last time you asked Siri, Alexa, or Google Home to stream your playlist, swiped your credit card at the local coffee shop to pay for a latte, or scanned items at a self-checkout. Throughout the day, you interact with computers multiple times without thinking about the digital transactions occurring in the background.

To be successful in today's world, you need to use computers—which in turn requires you to have a basic understanding of computers, computer terminology, and how computers work. These essential computer skills are often referred to as **computer literacy**. Once you can use various computer devices, communications technologies, and the internet to find, assess, use, create, and share content with others, then you will have achieved **digital literacy**.

In this chapter, you will begin to build computer literacy and gain a better understanding of digital literacy. You will distinguish the various types of digital devices around you and discover how they impact your life, explain how data is transformed into useful information, and explore new information related to three topics unique to our digital world: cloud computing, green computing, and ergonomics. This chapter is also an introduction to the terminology and concepts you will explore in more detail in the chapters that follow. Some of the objectives may sound very basic, yet you may be surprised by what you learn—even when it comes to defining *computer*, our first topic.

## Learning Objectives

**1.1** Define *computer* and identify the various types of personal computers

**1.2** List and describe various mobile and wearable devices

**1.3** Differentiate servers, mainframes, and supercomputers

**1.4** Describe embedded computers, the Internet of Things, and explain an autonomous object

**1.5** Define *information technology* and explain the four operations of information processing: input, processing, output, and storage

**1.6** Explain cloud computing and list its advantages and disadvantages

**1.7** Describe the goals of green computing and list environmentally sustainable computing practices

**1.8** Recognize good ergonomic design for digital device use

### Watch & Learn

Living in a Digital World

### Content Online

The online course includes additional training and assessment resources.

# 1.1 Personal Computers

A **computer** is an electronic device that has been programmed to process, store, and output data. Data consists of the characters the computer has accepted as input. Computers come in many sizes and shapes to meet a variety of computing needs. Most people are familiar with desktops, laptops, and tablets and recognize them as computers. Other electronic devices that contain programmed chips—such as digital cameras, game consoles, GPSs, and even some children's toys—also have computer capabilities designed to perform specific functions.

A computer needs both **hardware** (physical components and devices) and **software** (instructions that tell the computer what to do, also called *applications* or *programs* to be useful. You will learn about hardware in Chapter 3, operating system programs in Chapter 4, and productivity software applications in Chapter 5.

The personal computer continues to become lighter in weight and more mobile. The tablet shown is a wireless, portable personal computer with a touchscreen. A tablet is smaller than a laptop and larger than a smartphone. Its compact design and touchscreen make it a popular device when on the go.

## Personal Computers

A **personal computer (PC)** is a computer in which the input, processing, storage, and output are self-contained. A PC is generally used by one person at a time, although PCs can be connected to a network where other people can access PC resources, such as files on storage media. A PC is sometimes referred to as a *microcomputer*.

**Read & Learn**
Personal Computers

**History of the PC**  The personal computer evolved from second-generation, bulky, large, expensive computers used mainly by government and corporations to small, custom-made computers made by individual hobbyists in the 1970s. The Computer History Museum credits John Blankenbaker's Kenbak-1 as the first commercially available PC advertised for sale in 1971 at $750. Other PCs emerged in the 1970s, such as the Apple I and the Commodore PET, but it was the release of IBM's Personal Computer in 1981 (Figure 1.1) that started the dramatic change in the workplace with widespread use of PCs as business tools. As companies innovated processing, speed, and storage capabilities, PCs became household items that we have come to rely on for many daily activities.

**Figure 1.1**

The introduction of the IBM PC in 1981 made owning a PC affordable. Other companies soon made their own versions from IBM's design, but at a lower cost, advertising their "clones" as IBM-compatible.

**PC Platform**  A computing **platform** describes the combination of hardware architecture and operating system (OS) software that determines the applications that can be used on the PC. The PC is generally identified as a PC-compatible computer or a Mac computer. PC-compatible computers are made by several technology companies, such as Acer, Lenovo, Dell, and HP. Most use the Windows operating system, but some may operate another operating system such as Chrome OS. Mac is the exclusive brand made by Apple and runs its own operating system called macOS.

## Desktop Computers

A **desktop computer** is a PC designed to sit on a desk or table. Traditional desktop PCs house the processor, memory, and main storage device within a horizontal or vertical case called the **system unit**. A separate monitor, keyboard, and mouse connect to the system unit. A vertical case is sometimes referred to as a *tower*.

Modern desktops integrate the system unit and monitor into one case as an **all-in-one PC**. Figure 1.2a shows a traditional desktop computer with a tower. Figure 1.2b shows a modern all-in-one desktop PC.

**Figure 1.2**

(a) A desktop computer may have a vertical system unit, which is sometimes called a *tower*. (b) An all-in-one PC requires less desk space because the monitor, processor, memory, and storage media are housed in one unit.

## Mobile Computers

A **mobile computer** is a PC that can be moved from place to place. Mobile computers come in a number of configurations that vary in size, shape, and weight. A **laptop** computer, also referred to as a **notebook**, fits comfortably on a person's lap. These PCs often replace desktops for workers and for home users who want the ability to move easily from room to room with their PCs. A typical laptop has a clamshell-style design (Figure 1.3a). When the laptop is opened, the monitor swivels up to reveal a keyboard. The processor, memory, storage media, and battery are housed below the keyboard.

A **tablet** (also called a *tablet PC*) is a small, lightweight laptop with a touchscreen that you interact with using touch gestures or with a special pen called a *stylus*. Slate tablets have an on-screen keyboard only (Figure 1.3b), while other tablets come with a keyboard that pivots or plugs in so that the tablet acts more like a traditional laptop.

Technology manufacturers developed a 2-in-1 laptop to combine the laptop and tablet experience in a single device. A **2-in-1 laptop** (Figure 1.3c) has a touchscreen that swivels or can be folded down for use as a tablet and has the computing power and software capability of a laptop. An older version of a 2-in-1 laptop was called a **convertible laptop** or **hybrid laptop**.

**Blog This!**

Is the PC Dead?

**Figure 1.3**

(a) A laptop, or *notebook*, has a clamshell design in which the monitor opens up to reveal the system unit. (b) A tablet has a touchscreen that you tap or use touch gestures to navigate (a keyboard is not necessary). A slate tablet has an on-screen keyboard only. (c) A 2-in-1 laptop can be used as either a tablet when the screen is folded down or as a laptop for more intensive tasks such as word processing or data input.

## 1.2 Mobile Devices and Wearable Computers

A handheld computing device smaller than a laptop, such as a smartphone or ebook reader, is referred to as a **mobile device**. A mobile device stores programs and data within the unit or on a memory card. Some mobile devices rely entirely on touch-based input, while others have a built-in or slide-out keyboard or use voice-recognition technology. A **wearable computer** is a device that can be worn on the body, often attaching to an arm, a belt, a helmet, eyeglasses, or clothing. These computers perform special functions such as health monitoring and assist with other hands-free work. Wearable computers are integrated into clothing and accessories, called *smart textiles*, to enhance functionality. These devices constantly monitor the body and environment and send data wirelessly to cloud-based programs or smartphone apps, where wearers can then review the data and interact with the apps.

Technology and fashion companies are joining to design smart jackets that use cuff gestures to get directions, answer a call or text message, and play music.

### Mobile Devices

Mobile devices expand computer use outside our homes and offices. Examples include the following:

- A **smartphone** is a cell phone with a built-in camera (Figure 1.4a). Software applications (called *apps*), web browsing, and messaging capabilities are available via cellular or other wireless networks. A smartphone that has a screen size of 5.5 inches or larger but is not as large as a typical tablet is called a **phablet** (from the combination of *phone* and *tablet*). Advances in mobile processors have brought **augmented reality (AR)** to smartphones, allowing these devices to overlay computer images on the real world displayed in games or apps. For example, one app lets you measure a room with your camera and then digitally place furniture.
- An AR headset for mixed reality blends the physical world with the digital world. This is a self-contained computer or mobile device that lets you interact with computer imagery interspersed with real-world objects (Figure 1.4b).
- A **portable media player**, or **ebook reader** (also called an *ereader*), is a device used to play music, watch videos, view photos, play games, and/or read electronic books (called *ebooks*), newspapers, magazines, or other digital media (Figure 1.4c).

**Read & Learn**

Mobile Devices and Wearable Computers

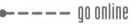

●————— go online

**https://CC3**
**.ParadigmEducation**
**.com/Hootsuite**
In the Hootsuite 2018 digital report, smartphones accounted for a greater share of web traffic than all other devices combined.

a            b            c

**Figure 1.4**

(a) A smartphone is an example of a mobile device that allows you to stay connected, track activities, and access digital resources like music and email anytime and from anywhere. (b) The Microsoft HoloLens® is a mixed-reality computer that lets you interact with holograms. (c) An ebook reader allows you to receive digital magazine subscriptions.

### Wearable Computers

A computing device with the primary function to be worn on the body is called a *wearable computer*. These devices are usually always on, include sensors, and provide for hands-free use with communications capability. They may allow input from voice commands, hand movements, joysticks, buttons, touch, or video. Wearable computers are used to assist or monitor individuals with health conditions, track items, provide real-time information, access documentation, or engage in virtual reality simulations. For example, the US Department of Defense helped develop wearable computers for soldiers who remotely monitor battlefields.

Google started the development of *smart glasses* (computing technology integrated into eyeglasses; Figure 1.5), and is now experiencing success with second-generation eyeglasses in enterprise applications. For example, warehouse workers wearing the eyeglasses record inventory by looking at the item code. Doctors wearing the eyeglasses consult with patients while the glasses record the conversation and take notes.

A digital wristwatch called a *smartwatch* connects to your smartphone and lets you manage calls, messages, appointments, alerts, and other data from your wrist. Smartwatch apps are available to track your fitness activities, monitor your health, browse the web, or show you where you are on a map.

Today, fitness trackers and wearable bands do more than count your steps. Some models monitor your heart rate, automatically sense and track sports activity, and keep track of your sleep statistics. A *wearable video camera* provides a hands-free way to record and share experiences. It can be fastened to an object (Figure 1.6a) or worn on the body (Figure 1.6b).

An advanced intelligence level of wearable technology called *invisibles* are sensor-based devices attached to the skin, woven into clothing or accessories, or implanted into the person (under the skin or inside the body). For example, insertable cardiac monitors provide continuous real-time data. Some of these wearable technologies include the following:

- A thin, waterproof device sticks to skin and tracks heart rate, activity, skin temperature, and resting heart rate. Data syncs to an app on your smartphone.
- A smart insole inside a shoe sends data on the wearer's gait to a smartphone app. You can monitor your feet while running to improve performance or avoid injury.
- Smart fabric measures the body's heat, movement, or other body conditions, and lets you interact with the clothing to send or receive data.

**Figure 1.5**

Smart glasses made by Magic Leap have sensors and six cameras that provide the wearer with a 3-D mixed reality experience for web browsing, shopping, or gaming. The software interface allows for voice, gesture, head pose, and eye tracking input modes.

a

b

**Figure 1.6**

The GoPro HERO digital video camera fastens to objects like (a) a paddle or (b) a helmet, to capture stunning action images and videos. The videos can be streamed live or posted online to share with family and friends.

# 1.3 Computers for Connecting Multiple Users and Specialized Processing

In your home, all the PCs and devices share internet connectivity, including printers and media-streaming players. A specialized computer for sharing these resources is not needed. Typical-sized organizations and government agencies do need specialized computers to connect many people to the organization's resources and to store and process large amounts of data. These specialized computers are called *servers*. Individuals at the organizations use PCs to connect via a network to these specialized computers so they can then access company programs and storage resources. Small organizations usually fall in between home computer users and large organizations in that they can use a high-end computer with multiple processors, large amounts of memory, and a large storage capacity to connect PCs to their system. When organizations with hundreds or thousands of users need to process transactions simultaneously, a midrange server or mainframe is used (Figure 1.7). Finally, when massive computing power is needed to perform advanced calculations, supercomputers are used.

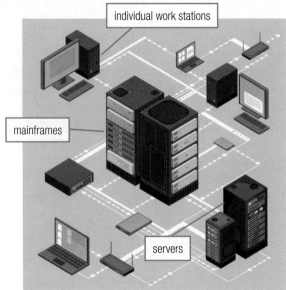

**Figure 1.7**

Businesses, government agencies, schools, not-for-profits, and other large organizations need specialized computers that connect and process data for hundreds or thousands of workers at the same time.

## Servers

A **server** is a computer with hardware and software that allows individual computers to access shared services (such as internet access) and resources (such as applications, data, printers, and storage). A computer that connects to a server is called a **client**. Servers are high-performance computers often mounted in racks that are then stacked in a cabinet (Figure 1.8) and stored in a specialized **server room** (a room that controls the heat, humidity, and dust for optimal performance; often this room is secure to prevent data or equipment theft/damage).

A **midrange server** is used in small- and medium-sized organizations that need to connect hundreds of client computers at the same time. A midrange server is more powerful than a PC but not as powerful as a mainframe.

A computer connected to a network that does not store any data or software on the local PC is called a **thin client**. Some companies use thin clients to reduce costs and/or maintain tighter security on software licensing and data.

A type of thin client called a *virtual desktop infrastructure (VDI)* allows a computer user to access a virtual desktop from any device connected to the internet, including a smartphone or tablet. VDI software is installed on each device, and the user's operating system, applications, and files are all stored on a central server that can be accessed from anywhere in the world.

**Figure 1.8**

Servers mounted in racks and stacked in a cabinet like this one are stored in rooms protected from excessive heat, humidity, and dust.

 **Read & Learn**

Computers for Connecting Multiple Users and Specialized Processing

 in the know

Midrange servers used to be known as *mini-computers*.

## Mainframes

A large, powerful, and expensive computer used by governments and large organizations—such as banks, insurance companies, and corporations—to connect hundreds or thousands of users simultaneously is called a **mainframe** (Figure 1.9). These computers have specialized hardware and software capable of processing millions of transactions and retaining massive volumes of data.

**Figure 1.9**

A mainframe simultaneously processes and stores millions of transactions for thousands of connected workers (Shown: Google data center.).

## Supercomputers

The fastest, most expensive, and most powerful of all computers is a **supercomputer**. A supercomputer is capable of performing trillions of calculations per second. These massive computers are usually designed to perform complex mathematical calculations needed for specific tasks, such as weather forecasting, nuclear research, oil exploration, and scientific research analysis.

Supercomputing speed is expressed in petaflops. A **petaflop** (a measurement used in scientific calculations) represents a quadrillion floating point operations per second. Figure 1.10 shows Summit, the most powerful supercomputer in the United States as of June 2018. Summit is installed at the Oak Ridge National Laboratory (ORNL) in Tennessee, which is sponsored by the US Department of Energy. The supercomputer is used for scientific research in sustainable energy solutions, safe nuclear energy systems, sustainable transportation, climate change, and detailed knowledge of advanced materials, artificial intelligence (AI), and human health research.

Summit, developed by IBM, delivers more than eight times the computational performance of Titan, the previous top-ranked supercomputer at ORNL. Titan's speed was described by National Geographic as being equivalent to seven billion people solving three million math problems per second.

● — — — — **go online**

**https://CC3 .ParadigmEducation .com/Top500**

Go here to read about the location of super-computers around the world. A list of the top 500 sites for the most powerful computers is released twice per year.

● — — — — **in the know**

The United States has set 2023 as the target date for the next leap in supercomputing devel-opment. The Department of Energy announced plans to develop an exascale system with IBM. One exaflop in this system equals one quin-tillion (1 followed by 18 zeros) calculations per second! Top research-ers say the technical challenges involved in developing software for such a system may require international cooperation.

**Figure 1.10**

Summit, a supercomputer installed at the Oak Ridge National Laboratory, can perform 200,000 trillion calculations per second (200 petaflops). 2018 Summit projects include the analysis of exploding stars, advanced materials used for transmitting electricity, and AI used in machine learning (to identify patterns in human health and diseases).

## 1.4 Embedded Computers, the Internet of Things, and Autonomous Objects

Embedded computers are being added to everyday objects that we use throughout the course of our days. Connecting these objects to the internet for sending and receiving data is called the **Internet of Things (IoT)**. This trend is making computers ubiquitous and leading an evolution toward autonomous objects such as self-driving cars, drones, and robots. These objects interact with us and other objects.

### Embedded Computers

Numerous appliances and consumer electronics incorporate an **embedded computer** that has a processor programmed to perform a particular task. An embedded object with wireless connectivity is referred to as a **smart device**. Examples include a refrigerator that displays a recipe based on ingredients stored inside and a smart lock that opens the door as you reach for the handle. Digital cameras, electronic thermostats, security systems, ATMs, and car navigation consoles are a few other examples of devices with embedded systems.

### The Internet of Things

The IoT describes a world where everyday objects containing embedded processors are connected to networks, allowing devices to interact with the owner and/or other connected "things" (Figure 1.11). This world is here now, where products—such as alarm clocks, coffeemakers, refrigerators, stoves, dishwashers, washing machines, clothes dryers, door locks, thermostats, entertainment systems, window shades, and cars— exchange data with each other and via the internet to apps. In this connected world, imagine your day begins with the alarm clock awakening you with current weather

Embedded computers added to everyday objects are making computers ubiquitous and leading to the development of smart cities that leverage the technology to improve city life.

**Read & Learn**
Embedded Computers, the Internet of Things, and Autonomous Objects

**Video**
What Is the Internet of Things?

**Figure 1.11**
Electronic devices with embedded systems are each designed to perform a specific task. Adding the ability to connect these devices to the internet and share data is part of the Internet of Things.

**in the know**
Market research firm International Data Corporation (IDC) forecasted the global market for IoT devices and services to be over $7 trillion in 2020 and beyond.

and traffic conditions, reminding you about a meeting, adjusting the room temperature, and starting the coffeemaker. This is just how your day begins! The possibilities for other interactions throughout the day are numerous. Some of these interactions between connected devices occur without ongoing human directive. This process is referred to as **machine-to-machine communications** (or *M2M communications*).

The emergence of the IoT is allowing municipalities to use and manage data from connected sensors in roads, lights, meters, and other public infrastructure objects to manage resources and systems more efficiently and sustainably. *Smart cities* can also use technology to improve emergency response and increase safety for citizens.

## Autonomous Objects

An area of technology that is experiencing a surge in innovation is autonomous objects. An **autonomous object** uses computers and other technologies to bring self-directed objects into the world. These objects can operate without human intervention and have the ability to interact with humans and other objects.

For several years, manufacturers have been adding autonomous technology to vehicles, such as wireless connectivity, cameras, collison avoidance warnings, and self-parking features. More recently, manufacturers have partnered with technology companies Google, Uber, and others to harness innovations for the next-generation autonomous vehicles (Figure 1.12a). Self-driving cars use sensors, cameras, radar, GPS, laser light detection and measurement, AI, and machine learning. Predictions vary for when fully autonomous vehicles will be mainstream, with some as early as 2025. They are currently regularly test-driven in Cupertino, California, and in other places around the world.

a

b

Autonomous drones that make deliveries or offer security monitoring are being refined and flight tested. In 2017 a fully autonomous drone taxi developed by German company Volocopter (Figure 1.12b) completed an eight-minute test run over the streets of Dubai.

Robots used for repetitive assembly work continue to evolve in the manufacturing sector. Today, autonomous robots are also being used in offices, hotels, and warehouses to deliver supplies, check inventory, and provide security monitoring. For consumers, robot vacuum cleaners and robot lawn mowers are now affordable. Improvements in AI are leading to interactive robot assistants and companions that can read and react to human emotion. Figure 1.12c shows Pepper, a robot assistant that talks to you and can interpret your facial expressions and body language.

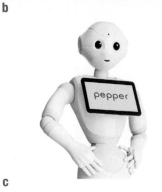

c

**Figure 1.12**

A (a) self-driving car, (b) autonomous drone taxi, and (c) robot assistant are three examples of autonomous object innovations made possible by embedded computers, the IoT, and AI.

● – – – – – **go online**

**https://CC3
.ParadigmEducation
.com/Retail2020**

Go here to watch a short video about five technologies that will change the way you shop in the next 10 years.

● – – – – **in the know**

Intel cofounder Gordon E. Moore described a trend in 1965 in which the number of transistors that could be placed on a circuit doubled every two years. He predicted the trend would continue, and this became Moore's law, which is a guiding principle for the computer industry.

☁ **Video**

The Future Is Now: Robots, Drones, and Driverless Cars

☁ **Let's Explore!**

What New Technology is Occurring in Your Field of Study?

# 1.5 Information Technology and the Information Processing Cycle

**Information technology (IT)** refers to the use of computers, software, networks, and other communications systems and processes to store, retrieve, send, process, and protect information. The IT department in an organization manages the hardware, software, networks, and other systems that form the information infrastructure. IT specialists design, develop, implement, and maintain information systems.

## Data and Information

Characters (text, numbers, or symbols) that are typed or otherwise entered into a computer are **raw data** or **data**. Data entered into the computer can also be in the form of an image, an audio recording, or a video recording. At the entry stage, data is not meaningful, since it is just a string of characters—not yet processed. Software programs provide the instructions for the computer to process (calculate or otherwise manipulate the data) and organize the results into meaningful and useful **information**.

Another way to understand the difference between data and information is to relate it to your grades in a course. Each individual grade on a test, assignment, project, or other activity is data. When you organize the individual grades into their weighted categories and perform the calculations to arrive at the final grade at the end of the term, you now have information. The final grade gives context to how well you performed in the course.

The **information processing cycle** includes the operations that transform data into information: input, processing, output, and storage, as shown in Figure 1.13 and described next. The terminology and hardware introduced in this topic will be explored in more detail in Chapter 3.

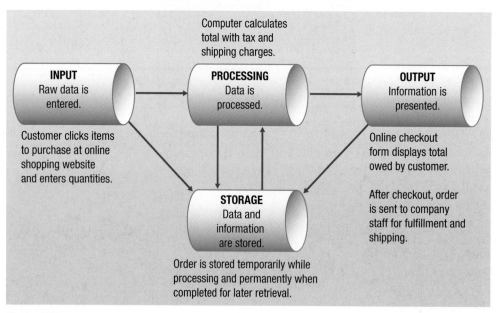

**Figure 1.13**

The information processing cycle includes four operations to transform raw data into useful information.

## Input

To enter data or instructions into the computer, a computer user has a variety of options that include typing, pointing, clicking, tapping, scanning, speaking, and recording. Any device used to enter raw data or communicate instructions to the computer is an **input device**. Commonly used input devices include keyboards, mice, and touchscreens similar to the ones shown in Figure 1.14. Touchpads, scanners, bar code readers, digital cameras, and digital pens are also examples of input devices.

**Figure 1.14**

From image left to right, a keyboard, mouse, and touchscreen are common devices used to enter input.

When you make a handsfree call from a car, your voice gives the instructions. Similarly, your voice is input when you use a digital assistant like Siri or Cortana or a smart speaker such as Amazon's Alexa. When you place a video call on a smartphone or tablet via Skype, FaceTime, or another video-calling app, the video feed sent over the internet is input. With voice input, you use the microphone in your car, smartphone, or other computing device for input (Figure 1.15a). When placing a video call or videoconferencing via the internet, you use a webcam for input (Figure 1.15b).

a          b

**Figure 1.15**

(a) Voice input via an embedded microphone is used when speaking into a device to make a call, texting by voice (known as *speech to text*), or asking a digital assistant for help. (b) Video input uses an embedded web camera during a video call or videoconference to capture the image seen on the screen by the receiving person.

## Processing

A computer chip called the **central processing unit (CPU)** carries out the instructions given by the software to perform the processing cycle. Think of the CPU as the brain of the computer. A CPU is a silicon chip referred to as the **microprocessor** (Figure 1.16). It is housed on the computer's *motherboard* (printed circuit board that contains the main computing components). Microprocessors have multiple cores, which are independent units on the same chip that process data simultaneously. A **multicore processor** significantly increases speed and performance.

**Figure 1.16**

Intel®, a leading manufacturer of microprocessors, developed the eighth generation family of Core™ processors for high performance, enhanced security, and stability. The processors, with up to six cores, are designed for better virtual reality, gaming, and ultra high definition experiences.

## Output

Anything used to view the information processed and organized by the computer is an **output device**. Typically, you view information on a display monitor or embedded screen (such as in a mobile device), and you print information onto paper using a printer. Other output devices include the screens in smart devices or appliances, such as those in smart TVs or smart home systems. Output may also be audio that you listen to on speakers, headphones, earbuds, or smart speakers. Figure 1.17 shows some common devices used for viewing or listening to output.

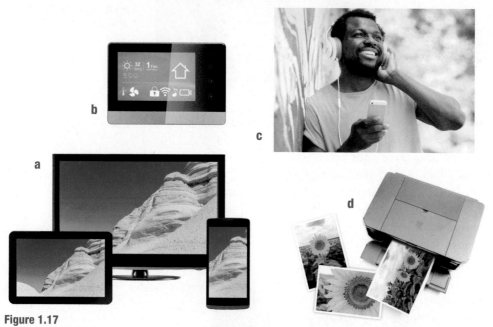

**Figure 1.17**

A variety of (a) display monitors, embedded screens, and (b) smart devices are commonly used to view processed information. You might also listen to output on (c) headphones or view output on a (d) printed copy.

## Storage

Two types of storage are in a computing system. During data entry and processing, storage occurs in a temporary storage location. When the data has been organized and processed, the information is copied to a permanent storage device, where it remains until removed by the user.

**Temporary Storage**  While input and processing take place, instructions and data are temporarily stored in a holding area called **random access memory (RAM)**. RAM is housed on the computer motherboard. RAM's temporary storage is deleted whenever you turn off the computer or the power is lost due to a power failure.

    **Cache memory**, located either on the CPU chip or on a separate chip between the CPU and RAM, is used for storing frequently used data. Cache memory increases speed and performance because the CPU checks cache memory before checking RAM. The faster retrieval of frequently used data saves processor time.

**Permanent Storage**  Permanent storage, called a **storage medium**, is any device that saves information for later use. Several storage media, some portable and some housed inside system units, are available in various capacities. An internal hard disk drive is housed inside the system unit, while universal serial bus (USB) flash drives (also called *memory sticks*) and external hard disk drives are portable (Figure 1.18). You might also access storage by connecting via a network to a hard disk drive on a server. Optical discs such as CDs, DVDs, and Blu-ray discs are still available but are not widely used due to their lower storage capacity, greater instability (for example, a scratch on a DVD could make information on the DVD inaccessible), and lack of availability (most new computers no longer include an optical drive).

a                                        b                                c

**Figure 1.18**

Frequently used permanent storage devices include: (a) an internal hard disk drive, (b) a USB flash memory stick, and (c) a portable, external hard drive.

Another place to store information permanently is **cloud storage**, where you copy documents, photos, or videos to a server at a cloud service provider's website. Cloud service providers, such as Dropbox and MediaFire, offer some free storage. For a subscription, they offer larger capacity storage. Information stored in the cloud is easily shared with others and accessible anywhere there is internet connectivity. You will learn more about cloud computing in the next topic.

## Communications

In our connected world, a fifth operation called *communications* is often added to the information processing cycle. Computers accept input from and send output to other computers and smart devices via wired (physical cables) and wireless communications channels (wireless signals). A **communications device** is any component used to facilitate the transmission of data, such as a cable modem, digital subscriber line (DSL) modem, router, or combination modem/router (Figure 1.19).

An integrated modem and wireless router connects the computer and smart devices on the home network to the internet.

**Figure 1.19**

Computers and smart devices connect to other computers or devices to receive input and send output using a communications apparatus such as the integrated modem and wireless router shown.

# 1.6 Cloud Computing

Individuals and businesses are turning to providers of software and computing services that are accessed entirely on the internet. This delivery model of software and services is called **cloud computing**. With cloud computing, all you need to get your work done is a computer with a web browser and active internet connection, since all software and files are stored online.

Cloud computing places the processing and storage operations of the information processing cycle on an online service provider's server and data center, rather than on the hardware and software of your own PC or mobile device (Figure 1.20).

Individuals can use some cloud computing services for free. For example, 15 GB of cloud storage is free at Google Drive, and 5 GB of cloud storage is free at Microsoft OneDrive. Both providers offer more storage with paid subscription plans. Some cloud service providers, such as Flickr, are used for storing photos and videos only. If you open your web browser to create documents or upload photos, music, videos, or other files to store and share them with others, then you are already using cloud computing services.

Businesses are using cloud computing services to save money on software updates and equipment. With cloud computing, a business always has access to the latest technology and software for a monthly or annual fee. A business may also have reduced inhouse IT staffing costs by using a cloud service provider. Lastly, during a period of rapid growth, additional computing resources can be quickly brought into operations with a cloud computing model.

**Figure 1.20**

With cloud computing, your software applications and data are launched from a web browser and stored on the service provider's equipment. The software and data are accessible from any PC or mobile device with an active internet connection.

Free productivity suites in the cloud are ideal options for any person or business with a need to create only basic documents, presentations, or worksheets. For example, Microsoft offers a suite of free web-based applications called Office Online from which you can open Word Online, Excel Online, PowerPoint Online, and OneNote Online. To start any of these applications requires only that you sign up for a Microsoft account, which can be any email address you want to use as your username. Any Hotmail.com or Live.com email address is a Microsoft account, so you do not need to sign up to use the Office Online applications if you have one of these accounts. A Google account gives you access to the free, web-based Google Docs, Google Sheets, and Google Slides applications. Delivery of software via the internet is referred to as **Software as a Service (SaaS)**. Advantages and disadvantages of cloud computing are listed in Table 1.1.

**Figure 1.21**

Chromebooks were designed for computer users who wanted a cloud computing experience. Google's Pixelbook (shown) is a browser-based laptop with most document storage in the cloud.

Cloud computing was the basis for a laptop design called the **Chromebook**. A Chromebook is a laptop that runs the Chrome operating system (an OS designed by Google) with most of the software and documents accessed from the cloud. These devices were designed to be used primarily while connected to the internet, but many apps will still work when you are offline. Several technology manufacturers, such as Acer, Dell, HP, and Samsung, are now offering Chromebooks. In 2017, Google released their own Chromebook called the Pixelbook (Figure 1.21).

●----- **go online**

**https://CC3
.ParadigmEducation
.com/OfficeOnline**
Go here to start a document, presentation, worksheet, or laptop using Microsoft's web-based Office Online suite. You will need to sign in with a valid Microsoft, work, or school email address to use the free software.

**Table 1.1** Advantages and Disadvantages of Cloud Computing

| Advantages | Disadvantages |
|---|---|
| Applications and data are available at any time from any place, since all you need is an internet connection. | You are totally dependent on your internet connection; if the service goes down, you cannot access your programs or files. |
| Updates to software are automatic and immediately available. | You do not have control over changes made to software. Upgrading is not at your discretion. |
| The service provider makes continual investment in hardware and software. You pay only the subscription fees after you have purchased your computer and internet connectivity. | Over a long period, monthly subscription fees could end up costing more than the initial outlay of cash for hardware and software, and ongoing internet connectivity may increase your costs. |
| Web applications are generally easy to use and navigate. | Web applications may be too limiting to meet your needs. |
| Web-based applications are compatible with PCs, Macs, and most mobile devices. | Some businesses may have compatibility issues exchanging data with legacy systems or infrastructure. |
| Documents, photos, music, videos, and other files are stored off-site. If you experience equipment malfunction, theft, or some other disaster, all that is lost is the physical equipment, which is easily replaced. | Large cloud computing vendors have reliable and secure storage; however, there is a risk that personal and/or sensitive data could be accessed by unauthorized users. Also be mindful of who actually owns the data you are storing online. Is it you or the provider? |

 **Video**

How Do I Access My Software Using Cloud Computing?

## 1.7 Green Computing

**Green computing** refers to the use of computers and other electronic devices in an environmentally responsible manner. Green computing in the workplace encompasses new or modified computing practices, policies, and procedures. The green computing trend is growing, with more individuals and businesses adopting green strategies every year.

Green computing strategies that can be used by any individual or business include reducing the energy consumption of computers and other devices; reducing the use of paper, ink, and toner; purchasing only certified sustainable and/or recycled products (such as paper); and reusing, recycling, or properly disposing of electronic waste (referred to as **ewaste** or *etrash*). Governments and industries are making available more recycling programs to reuse or properly dispose of electronics.

In addition to changes in computing practices, manufacturers are working to produce eco-friendly electronics (products that have reduced negative impact on the environment). When considering a new computing device or other gadget, look for and support a manufacturer that is making products that are energy efficient, use less toxic chemicals in the manufacturing and/or components, and incorporate recyclable/sustainable materials in both the product and the packaging.

The EPEAT program is a global ecolabel registry that allows purchasers to search for green electronics. EPEAT-registered products are verified by independent testing and certification organizations around the world for adherence to sustainability criteria that includes the full life cycle for each product. The EPEAT program is managed by the Green Electronics Council, a nonprofit group with a mission to have only sustainable IT products worldwide. It is estimated that from 2006 to 2018, use of the registry reduced greenhouse gases by over 38 million metric tons, eliminated over 670,000 tons of hazardous waste, and reduced landfill waste by the equivalent of almost 300,000 US households' annual waste.

With the IoT expected to exponentially increase the number of connected electronic gadgets worldwide, environmentally sustainable computing is becoming more important than ever before. That means reusing, reducing, and recycling as often as possible.

 **Read & Learn**
Green Computing

● ─ ─ ─ ─ ─ **go online**

**https://CC3 .ParadigmEducation .com/EPEAT**
Go here to use the EPEAT registry to search for a device by category and by country.

● ─ ─ ─ ─ **in the know**

Research labs around the world are engineering dissolvable electronics to help solve the ewaste blight. Examples of inventions that have been reported include a wearable electronic film that dissolves in vinegar, self-destructing batteries that dissolve in water, biodegradable packaging that doubles as fish food, and an electronic device that dissolves in a controlled timeframe process when exposed to water in the atmosphere.

### Five Ways to Go Green

The five strategies described next are ways that every person can start today to practice green computing at home or at work. These strategies are listed in order by ease of adoption.

1. Turn off devices when you are not using them, and unplug them if you will not be using them for an extended period of time. Even better—use smart power strips that automatically reduce the power to devices when not in use.

   Devices that consume power when turned off or in standby mode are called *energy vampires*. The Department of Energy estimates the average home energy bill is 10 percent higher due to energy vampires.

   The Department of Energy set new standards for microwave ovens manufactured on or after June 17, 2016, that reduce standby energy use. The department estimates that over a 30-year period, this change will be equivalent to removing 12 million new cars from the roads for one year! A bonus for consumers will be lower energy bills and cost savings over time.

2. Reduce printing to only those documents, messages, or photos that absolutely must be in paper copy, and make sure you efficiently use paper space.

   Make it a policy to distribute PDF copies of documents and messages and share photos online with friends and relatives.

When shopping, opt for emailed receipts; you can always unsubscribe from promotional email messages from the retailer if you don't want them.

When you must print, use paper that is certified recycled and/or certified as a sustainable forest product by the Forest Stewardship Council® (FSC®). Also, recycle all ink and toner cartridges. Consider printing on both sides of the paper and/or using narrower margins so you can fit more on a page; the cost savings from reduced paper consumption, toner, and ink cartridges is significant. Consider also that fewer print cartridges will end up in landfills.

Reduce printing—only print what you absolutely must have in paper copy.

3. Modify the default power options on your computer to conserve more energy. Most Windows-based desktop and laptop computers default to a balanced power plan, which aims to balance performance with energy conservation. You can adjust the sleep and display settings to turn off power sooner than the default times. Search for *power and sleep* in the Settings app on a Windows PC. On Mac computers, adjust the energy features using the Energy Saver pane in System Preferences.

4. Consider reselling or donating equipment you no longer need. Computers, smartphones, and electronics that are still in working order can be sold or donated to schools, churches, or nonprofit groups. If you must dispose of electronics, try to drop off the ewaste at an electronics recycling facility.

   "The Global Ewaste Monitor 2017," by the United Nations University, estimated that by 2016, the world recycled only 20 percent of ewaste through appropriate channels. The authors also estimated that about 4 percent more ewaste is being generated each year.

5. Make sure you buy Energy Star–certified devices, or use the EPEAT search tool to choose EPEAT-registered products. See Table 1.2 for the measured benefits to the environment of one organization's use of the registry for new purchases.

   The Energy Star label on a device means the manufacturer has reached or exceeded the minimum federal standards for reduced energy consumption. In 2017 performance standards for five product categories were updated due to technology changes: commerical ice machines, clothes washers, ceiling fans, uninterruptible power supplies, and commercial water heaters. In addition to benefiting the environment, you could save up to 30 percent on your energy bill, according to government estimates.

**Table 1.2** Environmental Benefits of EPEAT-Registered Devices

| A leading nationwide healthcare provider's purchase of 162,653 EPEAT-registry computers and displays instead of devices that were not on the registry saved the environment . . . |
| --- |
| enough toxic mercury to fill 1,574 fever thermometers |
| the disposal of hazardous waste equivalent to the weight of 979 refrigerators |
| the equivalent of 301 US households' annual solid waste |

*(source: greenelectronicscouncil.org/epeat)*

- - - - - - - green IT

The FSC logo on paper products means that purchasing the paper supports responsible management of the world's forests. The Forest Stewardship Council sets standards and independent third parties verify that certified companies have met the standards. Choosing products with an FSC label protects water and air quality, habitat, and respects the rights of people and communities that live in and depend on the forests.

 **Blog This!**
Are You Green?

 **Let's Explore!**
Locate Electronic Recycling Services in Your Area

# 1.8 Computers and Your Health

Frequent use of computing devices can adversely affect your health if you don't use proper care and preventive strategies. Physical ailments that can occur include fatigue, eye strain, blurred vision, backaches, wrist and forearm pain, finger numbness or pain, and neck and shoulder pain. A leading job-related illness and injury in North America is **repetitive-strain injury (RSI)**. An RSI is an injury or disorder of the joints, nerves, muscles, ligaments, or tendons.

Two common computer-related RSIs are **tendonitis** and **carpal tunnel syndrome (CTS)**. Tendonitis occurs when a tendon in your wrist becomes inflamed. CTS occurs when the nerve that connects the forearm to the palm of the hand becomes inflamed (Figure 1.22). Both conditions are caused by excessive typing, mouse scrolling, mouse clicking, and thumb typing/movements on mobile devices. Symptoms of tendonitis and CTS are listed in Table 1.3.

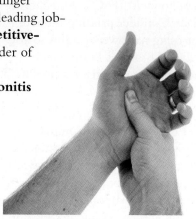

**Figure 1.22**

Carpal tunnel syndrome is a common RSI from computer use that causes pain in the hand and wrist.

**Table 1.3** Symptoms Associated with RSI Injuries from Computer Use

| Symptoms of Tendonitis | Symptoms of CTS |
|---|---|
| Tingling or numbness in the fingers | Numbness and tingling in the thumb and the first two fingers, especially at night (can cause sleep disruptions) |
| Pain in the forearm and wrist | Burning sensation when the nerve is compressed |
| Decreased mobility of the wrist or fingers | Decreased grip strength leading the user to drop objects |
| Loss of strength in the hand | Loss of strength in the hand |

## Computer Vision Syndrome

**Computer vision syndrome (CVS)**, also called *digital eye strain*, is a temporary condition caused by prolonged computer, tablet, and smartphone use that involves eye strain; weak, itchy, burning, or dry eyes; blurred or double vision; difficulty with focus; sensitivity to light; headaches; and neck pain. According to the American Optometric Association, 50 to 90 percent of computer workers will experience some CVS symptoms.

The American Optometric Association recommends all digital users practice the 20-20-20 rule to alleviate problems associated with CVS: take a 20-second break to view something 20 feet away every 20 minutes.

## Posture-Related Injuries

Pain in the back, neck, shoulders, and arms are common complaints from computer users. Sitting for long periods of time in the same position causes fatigue and reduces circulation to muscles and tendons, which can lead to stiffness and soreness.

**Read & Learn**

Computers and Your Health

----- in the know

CVS affects 50 to 90 percent of computer users at some point in their lives.

Neck pain is a common complaint among computer workers.

## Prevention Strategies to Reduce Risk of RSI, CVS, and Muscular Pain

Knowing the three primary risk factors for developing health issues when using a computer is key to identifying strategies to reduce risk of injury. These factors include: poor posture, poor technique, and excessive use. Prevention strategies are listed in Table 1.4. Many of these strategies involve the use of **ergonomics** (the study of the way equipment and a person's workspace are designed to promote safe and comfortable operation of the equipment). In other words, good ergonomic design fits the computer equipment to the worker by adjusting components to the optimal height, distance, and angles (Figure 1.23).

●————— **go online**

**https://CC3
.ParadigmEducation
.com/Workstation**

Go here for an interactive etool to learn more about proper workstation setup and to view setup checklists.

●———— **in the know**

According to the Mayo Clinic, CTS affects more women than men, which may be because the wrist's carpal tunnel is relatively smaller in women than in men.

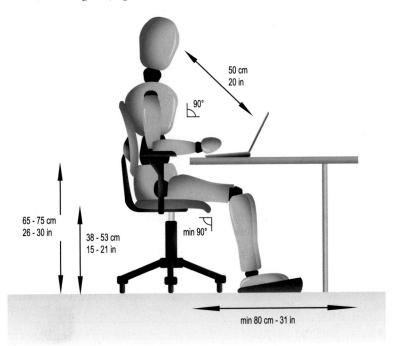

**Figure 1.23**
An ergonomically correct computer workspace helps prevent RSI.

**Table 1.4** Preventive Strategies for Avoiding Computer-Related Illness

| Preventive Strategies |
| --- |
| Take frequent breaks. Walk around, stretch, and do another activity for a few moments. |
| Maintain good posture by sitting up straight with feet flat on the floor, shoulders relaxed, head and neck balanced and in line with torso, and your lower back supported by your chair or a rolled towel. If the desk height is not adjustable, use a footrest to allow the angle of your knees to be at 90 degrees as shown in Figure 1.23. |
| Sit in a chair that has five legs for stability and allows adjustments to height and angle of backrest, seat, and armrests. |
| When typing, the keyboard should be at elbow level, with the elbows close to your body and supported by your chair. Wrists and hands should be positioned in line with forearms. If using a laptop, consider plugging in an external keyboard so that you can adjust the height and position. |
| The top of the computer screen should be at or slightly below your eye level. If using a laptop computer, consider buying an adjustable laptop stand to raise the monitor up, preventing neck strain. Optimal viewing distance is 20 to 40 inches. Increase font size in documents and browsers, if necessary, to avoid eye strain. |
| Remember to blink—studies have shown computer users tend to blink about five times less than normal. Minimize glare and use ambient lighting rather than overhead fluorescents, if possible. |
| Look away from the screen once in a while to focus on something in the distance. Practice the 20-20-20 rule: take a 20-second break to view something 20 feet away every 20 minutes. |

 **Let's Explore!**

Laptop, Texting, and Smartphone Ergonomics

## Topics Review

| Topic | Key Concepts and Key Terms |
|---|---|
| **1.1 Personal Computers** | An understanding of computers, computer terminology, and how computers work is **computer literacy**. |
| | **Digital literacy** is the ability to use various computer devices, communications technologies, and the internet to find, assess, use, create, and share content with others. |
| | A **computer** is an electronic device programmed to process, store, and output data that has been accepted as input. |
| | A computer needs both **hardware** (physical devices and components) and **software** (instructions that tell the computer what to do) to work. |
| | A **personal computer (PC)** is a self-contained computer with input, processing, storage, and output resources. |
| | A computing **platform** is the combination of hardware technology and operating system (OS) software that specifies the types of applications that can be performed on the PC. PC-compatible and Mac are two platforms used for a personal computing device. |
| | A **desktop computer** includes a **system unit** (a case that houses the memory, processing, and storage resources) and a separate monitor, keyboard, and mouse designed to sit on a desk or table. |
| | A **mobile computer** is a computer that can be moved around. |
| | An **all-in-one PC** is a desktop computer that has the system unit and monitor integrated into one case. |
| | A typical **laptop** (also called a **notebook**) has a monitor that when swiveled up reveals a keyboard with the remaining components housed below. |
| | A **tablet** is a lightweight laptop with a smaller screen that you interact with using touch or a digital pen. |
| | A laptop that can be converted into a tablet is called a **2-in-1 laptop**, a **convertible laptop**, or a **hybrid laptop**. |
| **1.2 Mobile Devices and Wearable Computers** | A **mobile device** is a handheld computing device with a small screen and with programs and data stored within the unit. |
| | A **wearable computer** is a device that can be worn on the body, often attaching to an arm, a belt, a helmet, eyeglasses, or clothing. |
| | A **smartphone** is a cell phone with a built-in camera, apps, web-browsing, and messaging capabilities via cellular or wireless networks. |
| | A **phablet** is a mobile device that is larger than a smartphone but smaller than a tablet. |
| | Advances in mobile processors have enabled smartphone users to overlay computer images on real-world content, thus experiencing **augmented reality (AR)**. |
| | A **portable media player** is used to play games or other media, while an **ebook reader** is used primarily to read ebooks. |

*continued...*

| Topic | Key Concepts and Key Terms |
|---|---|
| **1.3 Computers for Connecting Multiple Users and Specialized Processing** | A **server** is a computer with hardware and software that allow individual computers to access shared services and resources. |
| | A computer that connects to a server is called a **client**. |
| | A **server room** stores servers in a room that controls the heat, humidity, and dust for optimal performance of the servers stored therein. |
| | A **midrange server** connects hundreds of clients at the same time. |
| | A computer that relies on the server for all programs and data is called a **thin client**. |
| | A **mainframe** is a large, powerful, and expensive computer capable of handling hundreds or thousands of users. |
| | A **supercomputer** is the fastest, most expensive computer, capable of processing trillions of calculations per second. |
| | Supercomputing speed is measured in petaflops; one **petaflop** is a quadrillion floating point operations per second. |
| **1.4 Embedded Computers, the Internet of Things, and Autonomous Objects** | An **embedded computer** is an appliance or other gadget that has a built-in processor and is designed to perform a specific task; an embedded computer is sometimes referred to as a **smart device**. |
| | The **Internet of Things (IoT)** describes a trend where everyday objects are connected to a network and can interact with the owner or other connected "things." |
| | Some of these interactions between connected devices occur without ongoing human directive, referred to as **machine-to-machine communications** (or *M2M communications*). |
| | An **autonomous object** can interact with humans and other objects without human intervention. Examples of autonomous objects include self-driving cars, autonomous drones, and robot assistants or robot companions. |
| **1.5 Information Technology and the Information Processing Cycle** | **Information technology (IT)** involves the use of technology to store, retrieve, send, process, and protect information. IT specialists design, develop, implement, and maintain information systems. |
| | **Raw data** (or **data**) consists of characters entered into a computer that software turns into meaningful **information** (useful data that has been processed by the computer). |
| | The **information processing cycle** includes four operations: input, processing, output, and storage. |
| | A device used to enter raw data is an **input device**. |
| | An operation that carries out instructions to manipulate and organize data is called *processing* and is carried out by the **central processing unit (CPU)**, also called a **microprocessor**. |
| | A **multicore processor** has an independent unit on the same chip that allows multiple simultaneous processes. |
| | A device used to view processed and organized information is an **output device**. |
| | **Random access memory (RAM)** is temporary storage used to hold data while processing. |
| | **Cache memory** is where frequently used data is stored. |
| | A permanent **storage medium** such as a hard disk drive, USB flash drive, external hard disk drive, optical disc, or **cloud storage** (storage on web servers) is where data is saved for later use. |
| | A **communications device** such as a cable modem or router transmits data to/from computers. |

*continued...*

| Topic | Key Concepts and Key Terms |
|---|---|
| **1.6 Cloud Computing** | **Cloud computing** delivers software and computing services, including processing and storage, entirely on the web.<br><br>Software delivered via the internet is called **Software as a Service (SaaS)**.<br><br>A laptop that accesses most of its software and documents from the cloud is called a **Chromebook**. |
| **1.7 Green Computing** | Using computers and other electronic devices in an environmentally responsible manner is known as **green computing**.<br><br>Green computing practices, policies, and procedures strive to reduce energy consumption; reduce use of paper and paper supplies; and reuse, recycle, or properly dispose of **ewaste** (electronic waste).<br><br>Five ways to practice green computing are to turn off or unplug devices when not in use, reduce printing of documents, modify power options on your computers, resell or donate old equipment, and buy Energy Star-rated or EPEAT-registered products. |
| **1.8 Computers and Your Health** | **Repetitive-strain injury (RSI)** is the leading cause of job-related injury in North America and involves a disorder of the joints, nerves, muscles, ligaments, or tendons.<br><br>**Tendonitis** occurs when a tendon in your wrist becomes inflamed from overuse.<br><br>**Carpal tunnel syndrome (CTS)** occurs when the nerve that connects the forearm to the palm becomes inflamed from overuse. Symptoms of tendonitis and CTS include pain, numbness, tingling, decreased mobility, and decreased strength in the hand.<br><br>**Computer vision syndrome (CVS)**, or *digital eye strain*, is a temporary problem with the eyes caused by prolonged computer use.<br><br>**Ergonomics** involves the design of equipment and a person's computer workspace to promote safe and comfortable operation of the equipment by the individual. |

## Topics Review

# Exploring the World Using the Internet

Chances are you have used the internet today at home, at school, or at work to search for information, access study notes, play games, watch videos or movies, update your status, upload photos, or post a comment on a social media site. For many people reading this book, the internet has always been available, and connecting and sharing information online with anyone in the world is taken for granted. The majority of people use a smartphone to access the internet in order to update information, stream content, and communicate with others.

The internet can also be accessed from a watch on your wrist or a console in your car. The IoT has also connected smart devices in our homes to the internet. Such devices include security systems, appliances, speakers, lights, and thermostats. In the future, smart clothing that communicates with the internet may also be common.

Consumers and workers need to be network and web savvy. Knowing how to find the right information quickly, connect with others, and communicate professionally are essential skills for personal and professional success. In this chapter, you will learn the difference between the internet and the World Wide Web, study the equipment options used for internet connectivity, review the most popular web browsers, search and evaluate web content, and identify the various services available to you in the online world.

Following this introduction to the internet, you will learn in Chapter 6 to use social media and web publishing options to professionally communicate and connect with others.

### Watch & Learn
Exploring the World Using the Internet

## Learning Objectives

**2.1** Distinguish a network, the internet, and the World Wide Web

**2.2** Identify the hardware and various types of connectivity options for internet service

**2.3** Recognize the popular web browsers used to view online content and browse on a mobile device

**2.4** Describe parts of a URL and browse to web pages via web addresses and hyperlinks

**2.5** Locate content using a search engine, narrow a search using search tools, and describe content aggregators

**2.6** Evaluate the accuracy and timeliness of web content

**2.7** List and describe various online services for shopping, communicating, and education

### Content Online
The online course includes additional training and assessment resources.

## 2.1 Networks, the Internet, and the World Wide Web

A **network** is two or more computers or other devices (such as a printer) that are linked together to share resources and communicate with each other. Computers are linked by a communication medium, such as a wireless signal or a cable that physically connects each device. You may have a small home network to share internet access among a desktop, laptop, tablet, or smartphone. Often, home networks also share a printer.

Networks in business, government, and other organizations exist in a variety of sizes and types for sharing and communicating among workers. In addition to internet access, these larger networks share software, storage space, printers, copiers, and other devices. Businesses typically use the network for centralized data storage and for controlling access to computers and other equipment.

### The Internet

The **internet** (or **net**) is a global network that links together networks, such as government departments, businesses, nonprofit organizations, educational and research institutions, and individuals. Think of the internet as the physical pathway made up of thousands of other connected networks to make a worldwide network. For this network to transmit data around the world, special high-speed communications and networking equipment is needed to provide the pathway on which the data travels. For example, a text message you send using WhatsApp to your friend across the world travels through several networks, including cellular, telephone, cable, or satellite networks to reach its destination (Figure 2.1). This collection of networks is the internet.

You connect your personal computer (PC) or mobile device to the internet by subscribing to internet service through an **internet service provider (ISP)**, a company that provides access to the internet infrastructure for a fee. When you use your smartphone to access the internet, your cellular provider (such as Verizon) is your ISP. You will learn about connecting to the internet in the next topic.

> ☁ **Read & Learn**
>
> Networks, the Internet, and the World Wide Web

> ●–––– **in the know**
>
> The internet had its beginning in 1969 when the US Department of Defense Advanced Research Projects Agency (ARPA) connected four computers together, creating a network for scientists to share data and communicate with each other. The network was called ARPANET. As the network expanded and hundreds of public and private networks were connected to it, the collection of networks evolved to become today's internet.

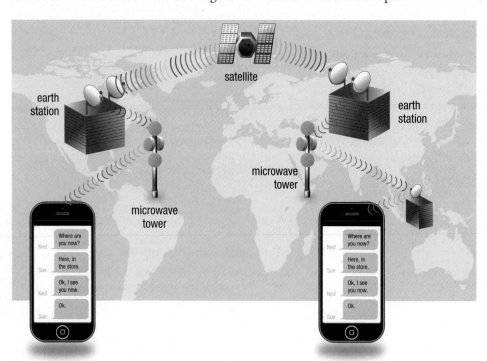

**Figure 2.1**

A smartphone uses radio signals to communicate with high-speed equipment such as microwave towers, earth stations, and satellites to send and receive data across the world.

## The World Wide Web

The global collection of electronic documents circulated on the internet in the form of web pages makes up the **World Wide Web** (or **web**). A **web page** is a document that contains text and multimedia content, such as images, video, sound, and animation. Web pages are usually linked to other pages so that a person can start at one page and click hyperlinks to several other related pages. Web pages are stored in a format that is read and interpreted for display within a **web browser** (a software program used to view web pages). A **website** is a collection of related web pages for one organization or individual. For example, all the web pages about your school that are available for viewing on the internet make up your school's website. All the web pages and resources—such as photos, videos, sounds, and animations—that make the website work are stored on a special server called a **web server**. Web servers are connected to the internet continuously so that anyone can access the web pages and linked resources at any time from any place.

While the internet provides the path on which a web page is transmitted, web pages are just one type of data that uses the internet. Other services that use the same network include email messages, instant messages, Voice over Internet Protocol (VoIP) services (telephone services via the internet), and file transfer services. Figure 2.2 illustrates how the internet delivers the web content to you on your PC or mobile device, and Figure 2.3 on the next page provides an example of web pages with multimedia.

●----- **go online**

**https://CC3 .ParadigmEducation .com/UsageStatistics**
Go here for internet usage statistics. Check the latest internet usage and population by country or region. For example, in June 2018, 95 percent of North Americans were internet users, compared to 55 percent for the world.

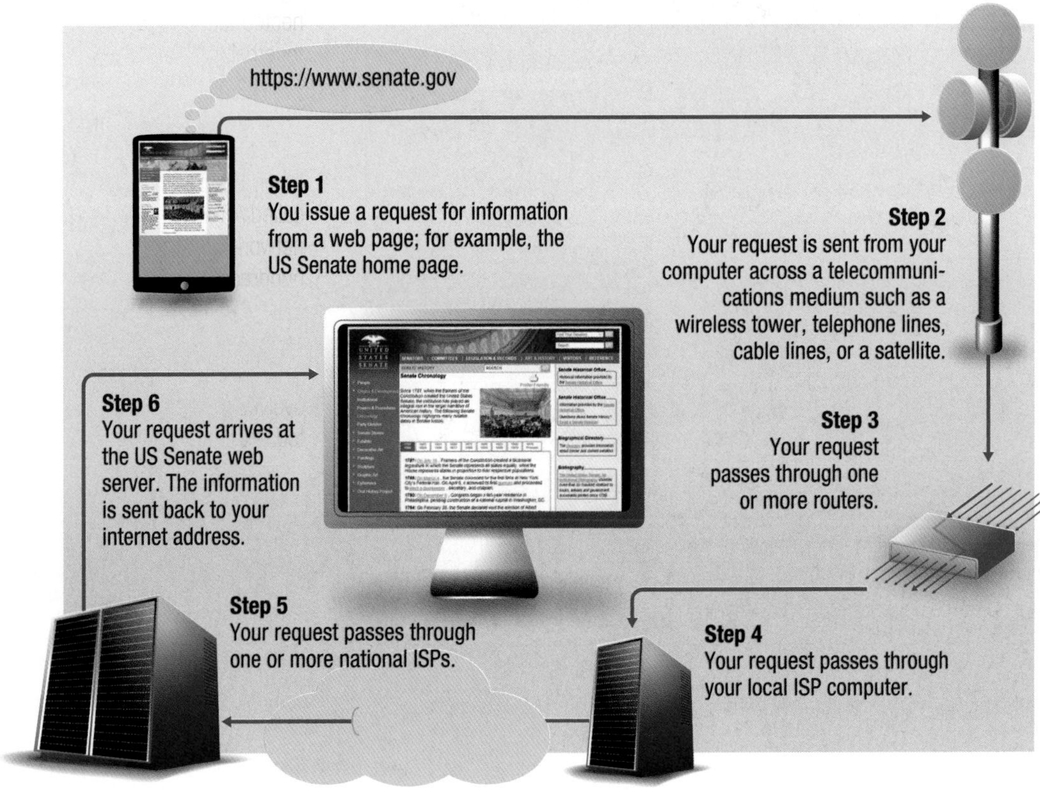

**Step 1**
You issue a request for information from a web page; for example, the US Senate home page.

https://www.senate.gov

**Step 2**
Your request is sent from your computer across a telecommunications medium such as a wireless tower, telephone lines, cable lines, or a satellite.

**Step 3**
Your request passes through one or more routers.

**Step 4**
Your request passes through your local ISP computer.

**Step 5**
Your request passes through one or more national ISPs.

**Step 6**
Your request arrives at the US Senate web server. The information is sent back to your internet address.

**Figure 2.2**
A web page travels along a collection of internet networking equipment and telecommunications systems to your PC or mobile device.

> The starting page for a website is called the *home page*.

> A website includes a navigation area with the main headings or topics for the site. The navigation area is typically a horizontal bar along the top of the website or vertically along the left side.

> Web pages contain text and multimedia content. Related web pages are linked together.

**Figure 2.3**

A website is a collection of related web pages such as these from the US Navy. Web pages can contain text, images, video, sound, and animation.

**Web 2.0** Initially, web pages mostly were a one-way communication medium where the organization or individual controlled the content and used the web simply as a means to provide information. Over time, a second generation of two-way communication web pages became available that allowed people to interact with the organization or with one another by sharing ideas, feedback, content, and multimedia. These websites that encourage and facilitate collaboration and sharing and allow users to add or edit content became known as **Web 2.0**.

**Web 3.0**  The next generation of the web happening now is known as **Web 3.0**. In this generation, meaningful connections between data and web pages exist, allowing you to find personalized complex information easily and quickly. Web 3.0 uses automated or intelligent agents (essentially, machines that adapt to the type of content you want based on previous searches and browsing you have done). For example, you might start a search with *Where is a good place to eat out tonight?* and the pages shown in the browser display a list of restaurants for the type of food you have searched for in the past within a few miles of your current location. Many of these searches use voice interaction with digital assistants such as Siri, Cortana, Alexa, and Google Assistant.

Web 3.0 focuses on making online experiences more intuitive and customized to each user. Artificial intelligence is continually improving the digital assistant's ability to interpret natural language to find what the user wants. As the Internet of Things increases the presence of machine-to-machine communications and autonomous objects, some experts predict that Web 3.0 will blend the physical world with the digital world, so viewing web content will no longer be restricted to screens but will also operate in virtual "web spaces." Experts differ on the exact meaning and capabilities of Web 3.0, and some even disagree with the name, preferring to call this evolution the "Semantic Web."

**Internet2**  In 1996, a group of US research and educational organizations formed **Internet2**, a not-for-profit organization geared toward solving society's challenges through innovations using advanced technology. Internet2 operates the largest and fastest research and education network in the US. Projects at Internet2 focus on advanced networking, information security, high-performance computing, and cloud computing. In 2017, a collaboration between Internet2 members at three universities established a customized cloud-based electronic medical records system for supporting patient care and research in dentistry schools. The electronic record solution integrates with hospitals, drug databases, and other medical information support systems, providing a model for the use of health information technology in dental practices. This example illustrates the type of innovation Internet2 collaborative work seeks to achieve—practical technology solutions that can have an impact on improving society.

**Net Neutrality**  In 2015, a set of rules governing **net neutrality** was passed. These rules prohibited ISPs from restricting access to the internet or from providing priority access or additional capacity to promote paid content. The rules were overseen by the Federal Communications Commission (a government agency that regulates communications). Supporters argued that the rules were necessary to keep the internet open and equal for all users. However, in 2017, the rules were repealed. This means that it is now possible for ISPs to set up "fast lanes" to push content to users who pay a fee for premium service. A person viewing a website that is not paying for premium service may experience slower page loading. In May 2018, the US Senate voted to reinstate net neutrality rules; however, for net neutrality rules to return, the US House of Representatives needs to vote to reinstate the rules and the president has to sign off on the final language (Figure 2.4).

**Video**

How Will Web 3.0 Change Our World?

**Figure 2.4**

In January 2018, protestors held a rally outside the Comcast headquarters in Philadephia to advocate for a return to net neutrality rules.

## 2.2 Connecting to the Internet

☁ **Read & Learn**
Connecting to the Internet

At work or school, you connect to the internet using the high-speed service installed on the organization's network. After signing in to a PC or other device by providing your user name and password, you can immediately launch a web browser and start searching the web. The networks you use in these contexts are designed to handle more functions beyond sharing internet access, such as providing access to shared software and centralized management of other resources. In this topic, you will learn about internet connectivity options for a small home network.

### Components Needed for Internet Access

Computers and mobile devices sold today are already equipped with the hardware needed to connect to a network. For example, a laptop may be equipped with a **network interface card (NIC)**, often referred to in computer ads as an **Ethernet port**. With this type of hardware, you plug one end of an Ethernet cable (a twisted-pair cable similar to a telephone cable but larger) into the laptop and the other end into a device that connects to the internet. Most PCs, all mobile devices, and all smart devices (such as smart TVs) also come equipped with a built-in **wireless interface card** to connect to a network using wireless technology. Wireless technology can be radio signals to connect to Wi-Fi equipment, or it can be **Bluetooth** technology, which pairs one Bluetooth device to another Bluetooth-enabled device, such as your smartphone or your car. You will learn more about wireless adapters in Chapter 3.

In addition to your network-enabled PC, laptop, mobile device, or smart device, you also need to have the following:
- an account with an ISP
- networking equipment (usually provided by your ISP), such as a cable modem, digital subscriber line (DSL) modem, wireless modem, or satellite modem
- a router if you will be connecting multiple devices to the service (newer modems integrate a wireless router within the modem)
- a web browser, such as Microsoft Edge, Firefox, Chrome, Safari, or mobile web browser (you will learn about these software programs in the next topic)

### ISPs

ISPs are companies that sell internet access by providing the equipment and servers that allow your PC or mobile device to connect to the internet backbone (main data routes). Typically, you contract for service with a telephone or cellular provider, cable company, or satellite company.

Fees for internet access vary and usually are based on the connection speed. High-speed access costs more than a slower connection and is usually priced at various speed levels. Also be aware of download limits (or *data caps*) in your contract. Some plans attach a data cap to a service, and if you exceed the cap during the billing cycle, additional fees are charged. An unlimited plan means you don't have to worry about extra fees, but these plans sometimes cost more than those with data limits. A marketing scheme by some companies is to offer "unlimited data," yet once you've used a certain amount, your connection speed slows down (sometimes to the point in which you aren't able to access the internet).

Your choice of connection speed is usually based on your anticipated usage of the internet. If you primarily only use email and browse web pages, you can get by with a slower speed than someone who wants to stream movies/music or play games online.

Along with internet access, you also receive multiple email accounts and security tools to keep your PC or mobile device virus- and spam-free.

## High-Speed Internet Connectivity

High-speed internet access is called **broadband** and includes any always-on connection capable of carrying a large amount of data at a fast speed. Speed for broadband connectivity is expressed as **megabits per second (Mbps)**, meaning data transfers at the rate of 1 million bits per second. As providers upgrade networks and equipment, high-speed connectivity may be described in **gigabits per second (Gbps)**, which is 1 billion bits per second. See Table 2.1 for a list of typical connection speeds by type of connection. When you evaluate ISP contracts, a higher Mbps or Gbps value means a higher subscription fee. Speeds are faster for downloading content than for uploading content. Upload speeds are slower, and some ISPs restrict upload speeds. Figure 2.5 puts meaningful context to connection speed on a cellular network by comparing downloading a movie at three speed levels. Typical broadband options are described next.

**Table 2.1** Average Internet Connection Speeds (Speeds Vary by Location)

| Type of Internet Connectivity | Average Speed (varies by type of equipment in use) |
|---|---|
| Cable | 20 to 250 Mbps |
| Cellular network | 3 to 7 Mbps for 3G and 20 to 100 Mbps for 4G |
| Digital subscriber line (DSL) | 5 to 100 Mbps |
| Fiber-Optic (FiOS) | up to 2,000 Mbps (2 Gbps) |
| Satellite | 12 to 25 Mbps |
| Wireless media | 10 to 1,000 Mbps (1 Gbps) |

● ---- **in the know**

Actual connection speed will vary due to factors such as distance from ISP, time of day (congestion slows down networks), and the web activities of others sharing the connection.

● ----- **go online**

https://CC3
.ParadigmEducation
.com/ProviderStats

Go here, type in your ZIP code, and click Search to view broadband providers and their advertised speeds in your area.

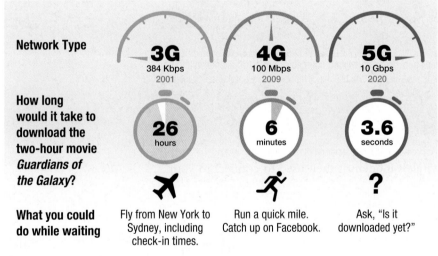

| Network Type | 3G 384 Kbps 2001 | 4G 100 Mbps 2009 | 5G 10 Gbps 2020 |
|---|---|---|---|
| How long would it take to download the two-hour movie *Guardians of the Galaxy*? | 26 hours | 6 minutes | 3.6 seconds |
| What you could do while waiting | Fly from New York to Sydney, including check-in times. | Run a quick mile. Catch up on Facebook. | Ask, "Is it downloaded yet?" |

**Figure 2.5**

The speed of your internet connection on your smartphone is important if you want to use it for downloading multimedia content.

## Cable Internet Access

**Cable internet access** is provided by a cable company such as Comcast or Spectrum, depending on the companies in your area. Plans can be standalone or bundled with television and voice calling plans. A cable modem connects to the cable network outlet using a coaxial cable (Figure 2.6). You connect a device to a cable modem via twisted-pair cables that plug into a port on the modem and your device (such as a PC or smart device) or via Wi-Fi signals. Cable connections are very fast; however, performance can vary because you share the connection with others.

**Figure 2.6**

A cable modem is connected to the cable company network using coaxial cable.

## Digital Subscriber Line Internet Access

A **digital subscriber line (DSL)** provided by your telephone company connects using telephone lines. A DSL modem uses twisted-pair cable (Figure 2.7a) to plug into your telephone jack and your computer or wireless router. With DSL, your telephone line is not tied up while you are using the internet.

## Fiber-Optic Internet Access

With **fiber to the premises (FTTP)**, also referred to as *fiber to the home (FTTH)* or *fiber-optic service (FiOS)*, the ISP runs a fiber-optic cable directly to your home or business (Figure 2.7b). This is the most expensive option and is not available in all areas. A piece of networking equipment is installed to convert optical signals transmitted over fiber-optic cable to and from the electrical signals transmitted by your computer devices.

## Satellite Internet Access

**Satellite internet access** is an option in rural areas where cable or DSL is not available (Figure 2.7c). Satellite requires the installation of a satellite dish and satellite modem. Access is typically slower than other broadband options and is more expensive. Some newer services offer higher speeds using geostationary satellites.

a          b          c

**Figure 2.7**

(a) A DSL modem connects to the telephone company network using twisted-pair cable. (b) Fiber-optic networks transmit beams of light along lines capable of speeds measuring in the billions of bits per second. (c) Satellite internet is used in rural areas where cable and DSL are not available.

## Fixed Wireless Internet Access

**Fixed wireless internet access** involves installing an antenna outside your home to send and receive radio signals. A wireless modem is then used to connect to your computer or router.

## Wi-Fi Hotspots

In a **Wi-Fi network**, wireless access points and radio signals transmit data (Figure 2.8). Wi-Fi networks are common in home networks, in public spaces such as coffee shops, in schools, and in workplaces. The area in which a Wi-Fi network is within range is called a **hotspot** and is a short distance. Homeowners connect their mobile devices via a wireless router or access point integrated into their cable or DSL modem.

Connection speeds vary depending on the equipment in use. Older equipment can be less than 10 Mbps, midrange equipment is in the range of 25 to 30 Mbps, and newer devices are capable of transmitting up to 1 Gbps.

**Figure 2.8**

A wireless access point is used in a Wi-Fi network to provide connectivity to wireless users.

☁ **Video**

5 Choices for Internet Connection

●————— secure IT

When you access the internet using free public Wi-Fi, be aware that anything you send over a public network can be intercepted by hackers or cybercriminals. Avoid entering personal information. When using free public Wi-Fi, do not check your bank or investment accounts or shop online.

## Cellular Communications

**4G** (fourth generation) smartphone devices provide internet access through cell towers. Average speed for a 4G device is in the range between 20 and 100 Mbps. Internet access via 4G is sometimes called *mobile broadband*. The network connecting a 4G smartphone is a 4G **Long-Term Evolution (LTE) network**, LTE-Advanced network, or LTE-U network, each of which is a variant of the original LTE technology used to bring high-speed data to mobile devices. 4G technology evolved from the earlier 3G (third generation) wireless standard. 3G, while still in use, is expected to be phased out within the next few years.

In 2018, wireless carriers including Verizon and AT&T rolled out the first 5G (fifth generation) networks. **5G** involves a new industry standard for radio signals that will increase mobile broadband speed by up to 30 to 50 times compared to 4G. Soon, 5G smartphones are expected to be widely available. 5G networks are being built to provide enough capacity to accommodate IoT and autonomous devices that send and receive data continuously.

## Mobile Hotspots

If you are not always within range of a free public Wi-Fi hotspot, a **mobile hotspot** can provide internet access from either your smartphone or a standalone external device. Most smartphones have a built-in mobile hotspot feature. By turning the feature on, you can share the internet connectivity on your smartphone with other devices, such as laptops or tablets. On some smartphones, this feature is called *tethering*. You will learn more about this feature in Chapter 3. Devices connected to the internet from your smartphone use the mobile data in your smartphone plan.

Another option is to buy an external mobile hotspot device. There are many to choose from; some are small enough to fit in the palm of your hand. Prices and data plans vary, with some requiring that you commit to a contract for a specified term.

An older technology that is still in use is a **mobile broadband stick** (also called an *internet dongle*). It looks like a USB flash drive but is a portable modem that plugs into a USB port on your laptop or tablet and provides internet access via a cellular network. The stick is purchased, usually for a monthly fee, through a wireless carrier or phone provider.

## Slow Internet Connectivity

In some communities, broadband internet access may not be available, or the cost may be too high for some people. A low-cost way of connecting to the internet is to use **dial-up**, in which you connect your PC to your telephone system via a modem built into your PC. The ISP provides you with a local telephone number that the computer dials to get on the internet. The disadvantages to dial-up are the slow speed and the inability to use your telephone while online.

## Setting Up Your Internet Connection

The ISP provider gives you instructions to connect to the ISP network or may have someone install it for you. Generally, the installation requires that you plug in the modem and run a software program. See Appendix B for help setting up a wireless network in your home. Once you've established connectivity to your ISP service, you are ready to explore the web using a web browser.

●————— go online

**https://CC3 .ParadigmEducation .com/SpeedTest**
Go here to learn the actual speed of your connection. Perform a speed test at different times of the day until you find the time when speed is fastest.

●———— in the know

In 2018, approximately 24 million US residents living in rural areas did not have access to high-speed internet, according to the Federal Communications Commission (FCC), creating a digital divide between those with easy access to the internet and those without.

## 2.3 Popular Web Browsers

Once you are connected to the internet, you are ready to start browsing, shopping, watching videos, playing games, or checking out what your friends and family are posting on social media websites. Web pages are viewed using browser software. A web browser is a program that locates a web page on the internet and interprets the code in which the web page has been stored to compose the page as text and multimedia. Many web pages are created using a markup language called **HTML**, which stands for **Hypertext Markup Language**. A markup language uses tags to describe page content.

In this topic, you will review the most popular web browsers for PCs: Microsoft Edge, Firefox, Chrome, and Safari. You will also be introduced to the unique needs of mobile web browsers.

### Microsoft Edge

With the release of Windows 10, Microsoft introduced a new web browser called **Microsoft Edge** (Figure 2.9). Microsoft Edge provides fast searches with a clean and clear reading experience. Pages can be saved to a reading list. You can highlight text, draw, or write notes on a web page, and those notations can be saved and shared. Cortana, a digital assistant, searches and helps you do things online faster. For example, while viewing a restaurant page, Cortana will pop up with the phone number, directions, and hours of operation. The Favorites pane that you open at the right side of the screen, collects your favorites, downloads, browsing history, and reading list in one place accessible across all devices. Web browsers have similar standard features such as navigation buttons, an Address bar, and tabs for opening multiple pages.

**Read & Learn**
Popular Web Browsers

**Video**
How is HTML 5 Transforming the Web?

— — — in the know

Browsers use *extensions* to extend the functionality of the browser. For example, an extension might be used to play media content or block ads. Extensions are also known as *add-ons* and *plug-ins*. Older websites required that users download and install a plug-in or player to display media content. You may still encounter a website that uses an external player like Adobe Flash. In that case, you will be prompted to download the player to view the content.

**Figure 2.9**
Microsoft Edge is the browser included with Microsoft Windows 10.

### Chrome

Google **Chrome** is a free web browser that runs on PCs and mobile devices, including Apple desktops, tablets, and smartphones. The browser was initially released in 2008 with features that included fast page loading. Chrome was the first browser to provide the ability to launch a search request from the Address bar. Signing in with a Google account while using Chrome allows you to sync open tabs, bookmarks, history, and saved passwords with your other devices. In Figure 2.10a, a partial view of Chrome with the navigation buttons and Address bar is shown with the web page for Grand Canyon National Park.

StatCounter, a company that measures internet usage, reported Chrome was the browser preferred by 50 percent of web users in the United States in October 2018. Web surfers report a preference for Chrome's fast page loading, minimalist design, and synchronized searches across all devices.

## Firefox

**Firefox** is a free, open-source (source code is freely available to use or modify) web browser offered by the Mozilla Foundation. The browser runs on PCs, Macs, and mobile devices. Since the initial release of Firefox in 2004, the browser has enjoyed a steadfast fan base. Figure 2.10b shows a partial Firefox view of navigation buttons and the Address bar on the Library of Congress website. Firefox and Chrome have similar placements of the navigation buttons, Address bar, and tabs for browsing multiple pages. Fans of Firefox indicate they favor the browser's speed in loading web pages and believe the security and privacy features are superior.

**a**

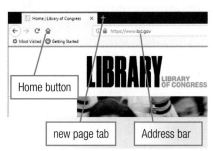

**b**

**Figure 2.10**

(a) Google Chrome and (b) Firefox have similar navigation buttons and an Address bar. Each browser also provides tabs for browsing multiple pages within the same window.

---- in the know

The Mozilla Foundation is a nonprofit organization dedicated to promoting openness on the web. Over 6,000 volunteers (called Mozillians) worldwide contribute technical expertise to the development of Firefox and other open-source projects.

## Safari

Apple's **Safari** web browser is installed on Macs. A mobile version of Safari is included on Apple mobile devices, such as iPads and iPhones (Figure 2.11). The browser is also available as a free download for a PC. Safari stores passwords, bookmarks, history, and the reading list in Apple's cloud storage service so that the information is available across all devices. StatCounter reports that Safari was the second most popular web browser in the United States in October 2018 when desktop and mobile applications are combined; however, Safari moved to first place for users browsing from their mobile devices within the same period.

## Mobile Browsers

A **mobile browser** is designed to quickly display web pages optimized for smaller screens. Since many people now access websites from their mobile devices, more companies provide a mobile app and/or responsive website. A **responsive website** automatically queries the device requesting the page for the screen size to deliver the page layout optimized for the screen.

Mobile versions of Microsoft Edge, Chrome, Firefox, and Safari are available in addition to other browser apps for Android and Apple smartphones and tablets.

**Figure 2.11**

Apple's Safari browser is the favorite browser for mobile users in the United States.

----- go online

**https://CC3 .ParadigmEducation .com/Opera**

Go here to download Opera if you want to try an alternative web browser. The free browser was started in Norway in 1995 as part of a research project. The browser runs on any device and advertises a more secure, ad-free experience with a built-in ad blocker.

 **Let's Explore!**

Add-on Toolbars for Your Browser

## 2.4 Understanding Internet and Web Addresses, and Navigating Web Pages

All the popular web browsers share similar features and navigation options designed to assist with finding and viewing web pages. Regardless of the browser you use, web addressing and search techniques (discussed in the next topic) are universal.

If you are invited to a friend's house, you need to know the address. Similarly, on the internet, you locate a connected device (PC, smartwatch, tablet, smart appliance, and so on) by using an address. A computing device connected to the internet is assigned a unique address called an **internet protocol (IP) address** so that the device can send and receive data. Two addressing systems are in use for the internet. In Figure 2.12, the two addresses for a web server installed for the National Oceanic and Atmospheric Administration (NOAA) are shown. The **Internet Corporation for Assigned Names and Numbers (ICANN)** is a nonprofit organization that is in charge of keeping track of the internet addresses and names all around the world.

A system of addressing that uses four groups of numbers from 0 to 255 separated by periods is referred to as IPv4. A system called IPv6 was launched in 2012 because the number of unique IPv4 addresses was running out. IPv6 uses eight groups of characters separated by colons, where a character can be a number 0 through 9 or a letter *a* through *f*. Methods are available to abbreviate the full IPv6 address. With IPv6 there will be enough IP addresses for years to come. For now, both systems coexist, since some equipment and software connected to the internet cannot handle IPv6 addressing.

**Read & Learn**

Understanding Internet and Web Addresses and Navigating Web Pages

● – – – – **in the know**

ICANN allocates IP address blocks to five Regional Internet Registries (RIRs) around the world. The RIRs then allocate smaller IP address blocks to ISPs and other network operators.

● – – – – **in the know**

IPv6 provides over 340 trillion trillion trillion addresses! That's enough for every person on the planet to have about 4,000 addresses.

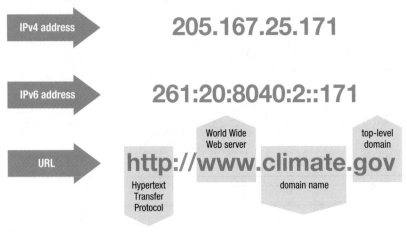

**Figure 2.12**

A URL contains a domain name that allows you to locate a website using a text-based name rather than the numeric IP address.

### Web Addresses

Using a complicated numeric address to find a website is not feasible. Instead, we use a **web address**, which is a text-based address. A web address is also called a **uniform resource locator (URL)**. Figure 2.12 shows that the URL for NOAA is http://www.climate.gov. The four parts of this URL are explained in Table 2.2 on the next page. The parts of a URL are separated by punctuation consisting of a colon (:), one slash (/) or two slashes (//), and a period (.), referred to as a *dot*. When telling someone the NOAA URL, you would say "climate dot gov."

When you type a URL, a server locates the IP address for the domain name to transmit data to the correct computer. A server that holds this directory is called a **Domain Name System (DNS) server** and is owned by a business or an ISP.

## Navigating Web Pages

Most people browse the web in one of two ways: by typing the web address of a company or organization to go directly to a website, or by typing a search phrase into a search engine website such as Google or Bing to search for information (Figure 2.13).

A **hyperlink (link)** is any item on a web page that when clicked displays another page. For example, a title, word, phrase, photo, video, icon, button, or audio object may be a hyperlink. As you move a mouse pointer around a page, the white arrow pointer changes to a white hand with the index finger pointing upward when you are pointing at text or an object that is a hyperlink.

### Video
What Are IP Addresses and Domain Names?

### Let's Explore!
How Do I Register a Domain Name?

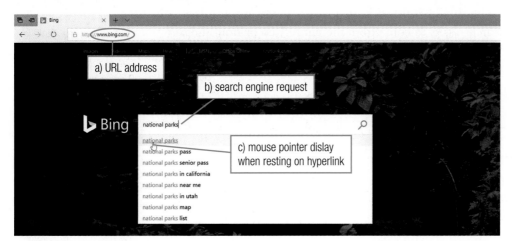

a) URL address

b) search engine request

c) mouse pointer dislay when resting on hyperlink

**Figure 2.13**

(a) If you know a web address, navigate to the website by typing the address in the Address bar. (b) Another way to navigate to a website is to use a search engine and type a search request to find a website. (c) The mouse pointer changes shape when resting on a hyperlink; clicking moves to the linked page.

The first page of a website that displays when the web address is requested is called the **home page**. Main headings appear along the top or edges and are organized by topic to guide you to subpages within the site.

**Table 2.2** The Parts of a URL

| Part of URL | What It Represents | Examples | |
|---|---|---|---|
| http | The protocol for the page. A protocol is a set of rules for transmitting data. | http<br>ftp<br>https | Hypertext Transfer Protocol<br>File Transfer Protocol<br>Hypertext Transfer Protocol Secure |
| www | World Wide Web server | | |
| climate | The organization's **domain name** is also referred to as a *second-level domain name*. Domain names are the text-based version of the IP address and are usually the owner's name, an abbreviation of an entity, or an alias. | nytimes<br>navy<br>loc | *The New York Times* domain name<br>US Navy domain name<br>Library of Congress domain name |
| gov | The part of the domain name that identifies the **top-level domain (TLD)**. A TLD identifies the type of organization associated with the domain name. Several TLDs with three or more characters are known as a **generic top-level domain (gTLD)**. ICANN is expanding the number of approved gTLDs. Watch for websites with domain names such as myroadsideinn.*hotel* or mylawfirm.*legal*. | com<br>edu<br>gov<br>net<br>org<br>biz | Commercial organization<br>Educational institution<br>Government website<br>Network provider such as an ISP<br>Nonprofit organization<br>Business |
| | A two-character TLD is a country code called a **ccTLD**. The country code for the United States is .*us* and the country code for Canada is .*ca*, for example. | ca | Canada |

## 2.5 Searching the Web

A **search engine** is a company that searches web pages and indexes the pages by keywords or by subject. When you type a search phrase at a search engine website, a results list displays hyperlinks to the web pages associated with the keywords used in the search phrase. To generate indexes, a search engine uses a program called a **spider** (or **crawler**) that reads web pages and other information to generate index entries. Generally, website owners submit information to search engines to make sure they are included in search results.

Some search engines display a subject directory that provides hyperlinks to categories of information, such as *Music*, *News*, *Shopping*, or *Travel*. Clicking a category brings you to another page with subtopics for that category. In most cases, you can specify whether you want search results to come from the entire web or from indexes associated with images, maps, news, shopping, or videos. Table 2.3 lists five popular search engines; however, be aware that many other search engines exist.

**Table 2.3**  Popular Search Engines

| Search Engine | URL |
|---|---|
| Google | https://www.google.com/ |
| Bing | https://www.bing.com/ |
| Yahoo! | https://www.yahoo.com/ |
| Ask | https://www.ask.com/ |
| Dogpile | https://www.dogpile.com/ |

You will get different results from various search engines. Figure 2.14 displays the search results from the same search phrase, *augmented reality*, entered at two search engines, Bing and Dogpile. Differences in results can occur for a variety of reasons. For example, spider and index program parameters differ by search engine, some search engines locate and index new or updated pages at different speeds or times, and page relevance to the search phrase may be ranked differently.

The same search phrase returns different search results at two search engines.

**Figure 2.14**

Search engines provide different results for the same search phrase for many reasons, including the way in which spiders and indexes find and rank pages and keywords.

**Read & Learn**
Searching the Web

● ─ ─ ─ ─ in the know

Most browsers allow you to type a search phrase directly into the Address bar to search using a default search engine. Some browsers also place an Instant Search text box near the Address bar in which you can type a search phrase.

● ─ ─ ─ ─ in the know

Web analytics companies consistently rank Google as the leading search engine worldwide that people use for finding information, followed by Bing.

## Fine-Tuning a Search

Regardless of the search engine you like to use, knowing how to fine-tune a search request will make your searching quicker and easier. With practice, you will develop a search technique that works best for you. Following are some guidelines for searching:

 **Video**

How and Why Is the Deep Web Hidden from Us?

- A search phrase in quotation marks returns pages where the keywords are in the same sequence; otherwise, the search results may have the keywords in any order (Figure 2.15). For example, "endangered species" returns pages with the word *endangered* immediately followed by the word *species*.
- Type a minus symbol in front of a keyword to *exclude* the keyword from the search results; for example, "endangered species" –africa returns pages about endangered species in continents other than Africa.
- Rather than searching the entire web, consider using search engine categories, such as News, Images, or Videos, to restrict the search results to a specific type of page or object.
- Look for filter options or advanced search tools to further refine your search terms, criteria, or region.
- Most search engines allow you to restrict searches to a specific timeframe; for example, in the Bing search tools, you can filter results by *Past 24 hours*, *Past week*, *Past month*, or *Past year*. Google has a *Custom range* date option.
- Look for Help at the search engine website to learn more about how the search engine indexes pages and the recommendations for searching at its site.

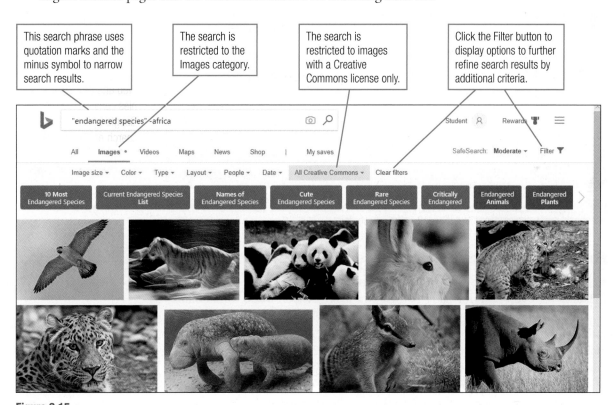

This search phrase uses quotation marks and the minus symbol to narrow search results.

The search is restricted to the Images category.

The search is restricted to images with a Creative Commons license only.

Click the Filter button to display options to further refine search results by additional criteria.

**Figure 2.15**

Using the search tools offered by a search engine can help you refine a search.

## Using a Specialized Search Engine

Some search engines are metasearch search engines. A **metasearch search engine** sends your search phrase to other search engines and then compiles the results in one list. Using a metasearch search engine allows you to type your search phrase once and access results from a wider group of search engines. Dogpile is an example of a metasearch search engine that provides search results in one place from Google, Bing, and Yahoo!, among others.

Some metasearch search engines specialize in one type of search service. For example, KAYAK (https://www.kayak.com) specializes in searching travel websites to provide you with a comparison of flights, hotels, car rentals, travel packages, and restaurants from many websites all in one place (Figure 2.16).

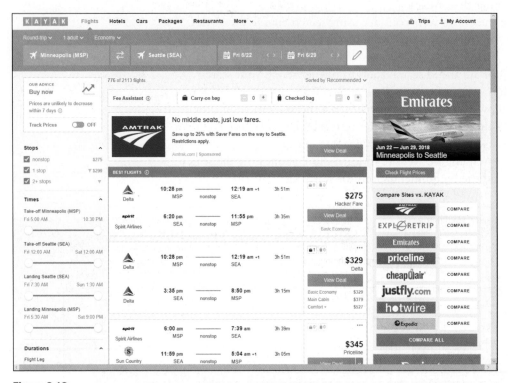

**Figure 2.16**

KAYAK, a metasearch search engine, specializes in searching travel websites.

## Educational Search Engines

Even with filters and advanced search tools, it can be time consuming to find scholarly articles or reputable information when you have research to do for a course or need to find resources to help you do your homework. When you need to find reliable information for academic purposes, consider using an educational search engine such as one of the six search engines listed in Table 2.4 on the next page.

You can also check with your school librarian or search your school website for information about research guides, databases, or eJournals that are available to registered students. Many schools subscribe to databases and journals that are not available on public web pages. Access to information on these resources usually requires you to sign in using your school user name and password on a web page that the library maintains.

**Table 2.4** Educational Search Engines

| Educational Search Engine | URL | Description |
|---|---|---|
| Google Scholar | https://scholar.google.com | broad search of scholarly literature |
| Microsoft Academic | https://academic.microsoft.com | broad search of scholarly publications indexed by Bing |
| Infotopia | http://www.infotopia.info | search results curated by librarians, teachers, and other educational experts |
| Lexis Web | https://www.lexisweb.com | legal research topics |
| Science.gov | https://www.science.gov | science information from 13 federal agencies in the US government |
| BASE (Bielefeld Academic Search Engine) | https://www.base-search.net | index of more than 1 million academic documents |

●- - - - - go online

**https://CC3 .ParadigmEducation .com/TimeSaving**

Go here for a list of 100 search engines, directories, databases, and other resources for finding information suitable for academic purposes.

## Content Aggregators

A **content aggregator** is an organization that searches the web for new content and collects and organizes the content in one place. Subscribers receive an update, sometimes referred to as a **web feed** (or *feed*) based on selections made for the type of information in which the subscriber is interested. For example, you can elect to receive new items related to news or music. Some content aggregators provide this service for free, while others charge a fee to send out updates to subscribers. The advantage to subscribers is that the content aggregator does the work of searching the web and organizing related information. Flipboard (https://flipboard.com) is a news aggregator that lets you pick the topics within themes (such as technology, news, or entertainment) for which you want to receive stories (Figure 2.17). Another aggregator, Techmeme (https://techmeme.com), tracks changes to technology news and presents a summary each day.

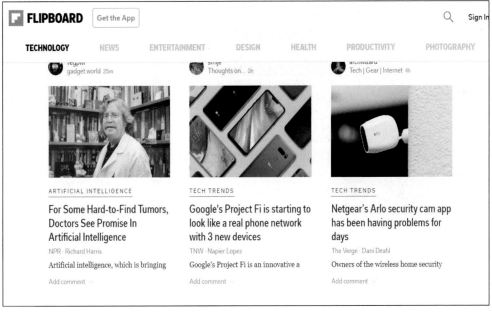

**Figure 2.17**

Flipboard is a news aggregator that sends you news feeds to content personalized to themes and topics you select.

## 2.6 Evaluating Web Content

Anyone with an internet connection and space allocated on a web server can publish a page on the web. When Web 2.0 websites became popular, users began generating their own content on others' websites. With thousands of new websites coming online every year, making sure information presented at a site is true, accurate, and timely is important if you are going to cite the content in a research paper or reuse it in some other manner. In this topic, you will learn how to perform four checks to evalute web content.

### Start at the Web Address

Begin with the domain name in the web address. Is the site hosted by an organization you recognize and trust? For example, an article published at http://www.nytimes.com is associated with a highly recognizable media organization, *The New York Times.* Look for the domain names of print-based publishers with which you are familiar. If you are familiar with the US Defense Department Science Blog, then you will be comfortable reading content at http://science.dodlive.mil (Figure 2.18). If you do not recognize a domain, further investigate the site's credibility by searching the organization's name to find out whether the organization is reputable.

Next look at the TLD in the web address (the three characters after the period in the domain name). A *.gov* or *.edu* TLD is a government department or agency or an educational institution—both are trusted sources.

**Read & Learn**
Evaluating Web Content

**in the know**

A *.org* TLD indicates a nonprofit organization. While these websites can have a wealth of high-quality information, statistics, and/or advice, be aware that some are also advocacy groups that may have an inherent bias or slant in the information presented.

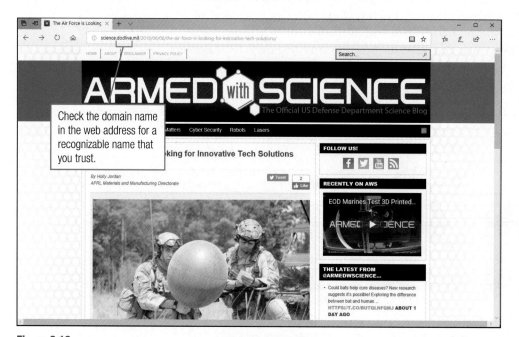

**Figure 2.18**
Begin evaluating web content by looking in the web address for a domain name that is one you recognize and trust.

### Look for an Author and the Author's Affiliation

Check for an author's name on the page (Figure 2.18). Is the author affiliated with the organization in the web address? Expert authors are often identified as such in the introduction, in other biographical information, or by credentials after their names. If no author name is shown, use other means to try to determine whether the information is credible. For example, check for a hyperlink that describes the organization presenting the content. Most websites have an <u>About Us</u> or a <u>Contact Us</u> hyperlink. Check these sources to determine whether a bias might exist for the information you are reading.

## Check the Publication Date

In most cases, you want to find the most recent information about a topic. Look for the date of publication on the page (Figure 2.19). Use filter or search tools to refine a search by date. Sometimes the original publication date may be older, but a notation (usually near the top or bottom of the page) can provide a date the content was updated. Finally, a clue to the date of publication may appear in the web address. For example, a page may have an address like http://www.companyname.com/articles/2020/March/pagetitle.html.

If no date can be found, consider whether the content seems dated. For example, look for something in the article that may give a clue to the timeframe in which it was published, such as a reference to an event.

**Figure 2.19**

This page has an author's name, a recent publication date, and is from a government source—all indicators the content can be trusted.

## Purpose and Design of the Site

Finally, consider the purpose and design of the page you are reading. Is the publisher of the web page a business that markets products or services? If yes, the content may be biased toward convincing you to buy the company's goods. Check the source of funding grants to make sure a study that is providing scientific or medical results is not subject to bias by the funding agency. Check peer-reviewed journals at an educational website to corroborate findings for a study that you want to verify.

The design of a web page and correctness of spelling and grammar will also indicate the credibility of the content. A professionally designed, well-written page invites trust. Some reliable websites may have pages that do not include all the qualities described in this topic; however, verifying sources and spot-checking facts for accuracy will help you confirm credibility of the information found on a web page.

 **Blog It!**

Is Wikipedia a Valid Source of Information for a Research Paper?

# 2.7 E-Commerce, Communications, and Online Learning

The internet is not just for connecting with others at social media websites or looking up information you need for work, school, or your personal life. Many people turn to the internet for e-commerce (shopping), electronic mail (email), text and instant messaging, telephone calls, conferencing services, and online learning.

## E-Commerce

**E-commerce** is the abbreviation for *electronic commerce*, which involves buying or selling over the internet (Figure 2.20). Most software is now purchased directly via the internet, where the hyperlink to download the program becomes available once the license fee is paid. Shopping online for music, videos, books, clothing, and other merchandise may require a credit card for payment; some merchants allow customers to pay directly from their bank accounts or through third-party payment services, such as PayPal or Google Pay.

In 2017, over two billion smartphone or tablet users completed an online transaction (called *mobile commerce* or **m-commerce**). A survey funded by delivery company UPS predicts that m-commerce will represent almost one-half of all e-commerce sales by 2020, the majority of which will originate on smartphones, with millennials expected to be doing most of the spending. Millenials represented almost 70 percent of m-commerce sales in the UPS study. Table 2.5 on the next page describes three categories of e-commerce.

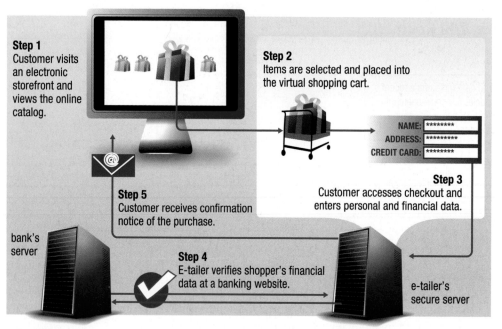

**Step 1**
Customer visits an electronic storefront and views the online catalog.

**Step 2**
Items are selected and placed into the virtual shopping cart.

NAME: ********
ADDRESS: *********
CREDIT CARD: ********

**Step 3**
Customer accesses checkout and enters personal and financial data.

**Step 5**
Customer receives confirmation notice of the purchase.

bank's server

**Step 4**
E-tailer verifies shopper's financial data at a banking website.

e-tailer's secure server

**Figure 2.20**

The online shopping process for business-to-consumer (B2C) transactions involves the five steps illustrated here.

**Read & Learn**
E-Commerce, Communications, and Online Learning

**Job Connect!**
Careers for the Web

●————— secure IT

Never enter personal or financial information online unless the website URL begins with *https*, which means the website is transferring data using encryption (when data is transformed into unreadable text by anyone other than intended recipient).

**Table 2.5** Types of E-Commerce Transactions

| E-Commerce Activity | Description |
|---|---|
| **Business-to-consumer (B2C)** | B2C is a familiar category of e-commerce if you have ever bought music, software, or other merchandise on the web. A business, such as Amazon, sets up a website that allows any consumer to purchase merchandise or services. A business that sells online to consumers is often referred to as an **e-tailer** (short for *electronic retailer*). Figure 2.20 illustrates the typical steps in a B2C transaction. |
| **Business-to-business (B2B)** | In B2B a business sells directly to other businesses using the web. In some cases, B2B transactions occur at websites used by consumers. For example, if you are on a website that shows ads and you click the ad, a payment is charged to the business that posted the ad. B2B website ads allow website owners to provide content free of charge to consumers. |
| **Consumer-to-consumer (C2C)** | C2C activity occurs at websites such as Craigslist or eBay, where transactions occur directly between two consumers. C2C websites generally also involve B2C or B2B transactions, since the website owner that provides the service charges fees to advertisers and/or sellers to post ads. |

## Email

Electronic mail (**email**) is the sending and receiving of digital messages using the internet, sometimes with documents or photos attached (Figure 2.21). Businesses were using the internet for email long before consumers embraced the service when PCs became mainstream. In the workplace, email is the standard communication medium and often is preferred over voice conversation, since email provides a written record of what has been agreed upon.

When you sign up for an account with an ISP, one or more email accounts are included with your internet service. An **email address** is used to connect to your email account at the ISP server and is generated in the format *username@mailserver .com* or *username@mailserver.net*. Your ISP will provide the means for you to create your own user name to include before the @ symbol. The text after the @ symbol is the ISP server name. Once your email account is created, you use a program called an **email client**, such as Microsoft Outlook or Apple Mail, to manage messages.

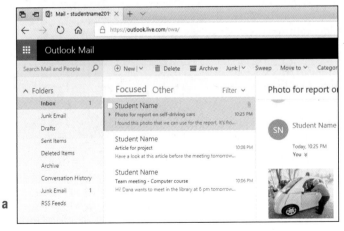

**Figure 2.21**

Many people set up their email accounts on multiple devices. (a) An email account on https://outlook.live.com is viewed in a web browser. (b) The same email message is shown displayed in the Outlook mobile app on a smartphone.

**Video**

What Are the Differences among the Three Kinds of E-commerce?

**- - - - in the know**

In 2021, email celebrates 50 years. In 1971 the first message was sent across a network, and that was the start of email that we know of today. Ray Tomlinson, a contractor on the ARPANET project, is credited with inventing email. He picked the @ symbol from the computer keyboard to separate the sender's name from the computer name.

Web-based email is popular with many users because all you need to send and receive email messages is a web browser. You can create an account at the website where you get to choose the email address you want to use. For example, to use the web-based version of Outlook, navigate to https://outlook.live.com and click the Create free account button to set up a new email address. Another popular web-based email service is Gmail from Google. Once you have an account created and are signed in, all messages are created, sent, stored, and otherwise managed from the browser window. Free web-based email services are supported by ads.

## Text and Instant Messaging

Messaging between mobile devices is often preferred over email, since messages are sent more quickly without a special email program. A **text message** is a short text exchange, referred to as **texting**, sent from one mobile device (Figure 2.22a) to another mobile device using a service called **Short Message Service (SMS)**.

Smartphone apps such as WhatsApp are popular for **instant messaging** (exchanging text messages in real time) family and friends around the world. WhatsApp lets users message each other via the device's internet connection. Users with SMS limits prefer instant messaging apps to avoid SMS fees, making it easier to stay in touch with people around the world.

Messaging app Facebook Messenger lets you conduct a private chat with a Facebook friend or do a group chat with multiple Facebook friends. Many other messaging apps are available, with WeChat and Skype two other options in the top five instant messaging apps of 2018 according to Statista (a market and consumer data company).

## Voice or Video Conversations

The ability to engage in a voice conversation using the internet is called **Voice over Internet Protocol (VoIP)**. Skype is one VoIP service that includes the ability to make a video call using the device's web camera to see one another while chatting (Figure 2.22b). FaceTime is a similar service used on Apple devices. Software and hardware on your PC or mobile device convert your voice and picture into digital signals that travel the internet. Most instant messaging apps such as WhatsApp, Facebook Messenger, and WeChat include voice or video calling.

## Web Conferencing

**Web conferencing** programs allow individuals to connect online to engage in a meeting. Web conferencing software similar to that shown in Figure 2.22c allows each participant to talk and share ideas and information using presentations or online whiteboards. A participant's screen can be "shared" so that other participants can watch a demonstration from their own devices. Online meetings allow businesses to save on travel expenses.

a

b

c

**Figure 2.22**

a) According to the Pew Research Center, texting is the number one way teens contact their friends. b) Video calls using the internet are often used to stay in touch with long-distance friends or relatives. c) Web conferencing software is used to connect and collaborate with a group using the internet.

## Online Learning

Online learning, also referred to as *elearning*, is a way to use the internet to achieve academic credits or even an entire degree or diploma online. In 2018 the US Department of Education's list of accredited online schools and colleges had over 28,000 fully online degrees offered at an associate, bachelor, master, doctoral, or certificate level.

Learners like the convenience and accessibility offered in online courses. In many courses, students can progress at their own pace. Typically, an online course is delivered through a school's learning management system (LMS), which is software used for organizing and delivering the course syllabus, assignments, discussion forums, curriculum content, chat rooms, grades, and group work spaces, and may also include an email system or other messaging feature restricted to registered students and faculty.

In the workplace, more than 75 percent of US employers offer online professional development opportunities to their employees. A type of elearning that is popular in business is called a *webinar*. A **webinar** is a presentation conducted via the internet in real time. In a webinar, participants can ask the presenter questions or make comments by interacting with the presenter during the session. Features in the software used to run a webinar allow for polling, sharing a desktop, and recording the presentation for later viewing.

Initially, many online courses were mostly text-based content; however, in some of today's online classes, multimedia-based content provides a richer, dynamic, and more meaningful experience. Instructional designers are incorporating techniques such as:

- adaptive learning paths using a pre-test and a post-test to map to individual learning needs from a library of learning objects;
- interactive videos that use questions or decision trees to let learners control the progression of the content; and
- mixed reality learning experiences that blend computer-generated content with the learner's physical world to help the learner master concepts or skills in a three-dimensional view (Figure 2.23).

In Chapter 6, you will learn more ways the internet can be used to connect and communicate with others using social media applications.

● – – – – **in the know**

The online education market is expected to reach \$240 billion by 2023.

**Video**

How Does VoIP Work?

**Let's Explore!**

Netiquette Guidelines

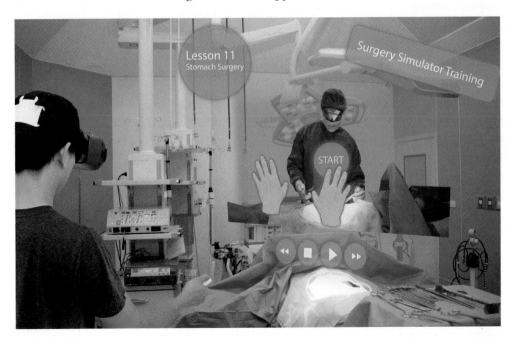

**Figure 2.23**

Augmented learning experiences added to online learning content can improve outcomes for the student. For example, a medical student can practice surgery in a simulation of the operating room.

## Topics Review

| Topic | Key Concepts and Key Terms |
|---|---|
| **2.1 Networks, the Internet, and the World Wide Web** | A **network** links computers together to share resources.<br><br>The **internet** (or **net**) is the physical structure that represents a global network linking other networks from around the world.<br><br>Connection to the internet is provided by an **internet service provider (ISP)** for a fee.<br><br>Web pages circulated on the internet make up the **World Wide Web**, sometimes referred to simply as the *web*.<br><br>A **web page** stored on a **web sever** is viewed using a **web browser**.<br><br>A **website** is a collection of related web pages.<br><br>**Web 2.0** refers to websites that have two-way communication, sharing, and collaboration with users.<br><br>**Web 3.0** incorporates automated agents that adapt and learn the type of content you want to see on the web. This evolution of online experiences is sometimes called the Semantic Web.<br><br>**Internet2** is a group of US research and educational organizations formed to solve current challenges using technology innovations.<br><br>**Net neutrality** is a set of rules governing internet service that aims to prevent access restrictions and to avoid preferential treatment or faster bandwidth to websites that pay for promotions. |
| **2.2 Connecting to the Internet** | Computers have a **network interface card (NIC)** (also called an **Ethernet port**) and/or a **wireless interface card** to facilitate connection to the internet.<br><br>Laptops and mobile devices also integrate wireless **Bluetooth** connectivity.<br><br>ISPs provide the equipment and servers to connect to the internet backbone with fees based on connection speed.<br><br>High-speed internet access is referred to as **broadband** and is measured in **megabits per second (Mbps)**, which is 1 million bits per second, or **gigabits per second (Gbps)**, which is 1 billion bits per second.<br><br>**Cable internet access** requires a cable modem that connects via coaxial cable to the cable company's network.<br><br>With a **digital subscriber line (DSL)**, your telephone company supplies a DSL modem that uses twisted-pair cable to connect into a telephone jack in your home.<br><br>**Fiber to the premises (FTTP)** means a fiber-optic cable is installed directly to your home, and a piece of equipment converts optical signals to electrical signals for your computer.<br><br>**Satellite internet access** is found in rural areas where cable and DSL are not available and requires the installation of a satellite dish and satellite modem.<br><br>**Fixed wireless internet access** requires an antenna outside the home and a wireless modem to send/receive radio signals.<br><br>A **Wi-Fi network**, often used in public spaces and homes, uses wireless access points and radio signals to transmit data. A **hotspot** is an area in range of a Wi-Fi network.<br><br>Smartphones and tablets connect using a **4G** (fourth generation) network to cellular towers, with an **LTE network** the fastest option for mobile devices.<br><br>**5G** projects that use a new industry standard for radio signals are expected to increase mobile broadband speed 30 to 50 times faster than 4G.<br><br>A **mobile hotspot** provides internet access to devices from a smartphone or a standalone device.<br><br>A **mobile broadband stick** is a portable modem that connects to the internet via a cellular network.<br><br>**Dial-up** uses a modem to connect via a telephone line. |

*continued...*

| Topic | Key Concepts and Key Terms |
|---|---|
| **2.3 Popular Web Browsers** | Web pages are viewed using web browser software that interprets the code the page is stored in as text and multimedia content. |
| | **Hypertext Markup Language (HTML)** is used in many web pages and describes page content using tags. |
| | Windows 10 includes a new browser named **Microsoft Edge** with faster page loads, a streamlined design, drawing ability, and Cortana (a digital assistant). |
| | Display the Favorites pane at the right side of the window to navigate to pages saved to the Favorites list, Reading list, History list, or view the Downloads list. |
| | **Chrome**, provided free by Google, runs on PCs, Macs, and mobile devices. Chrome is popular due to its fast page loading, minimalist design, and synchronized searches. |
| | **Firefox** is a free, open-source browser for PCs, Macs, or mobile devices available from the Mozilla Foundation. Firefox fans prefer its speed and security. |
| | **Safari** is used on Mac and Apple mobile devices. |
| | A **mobile browser** is designed to quickly display web pages optimized for much smaller screens. |
| | A **responsive website** optimizes the layout of a website to the screen size in use. |
| **2.4 Understanding Internet and Web Addresses, and Navigating Web Pages** | Every computing device connected to the internet is assigned an **internet protocol (IP) address**. |
| | **Internet Corporation for Assigned Names and Numbers (ICANN)** is a nonprofit organization in charge of keeping track of IP addresses around the world. |
| | An IP address with four groups of numbers from 0 to 255 separated by periods is known as IPv4. |
| | IPv6, developed when the number of unique IPv4 addresses was running out, uses eight sets of characters 0 through 9 or letters *a* through *f* separated by colons. |
| | A **web address**, or **uniform resource locator (URL)**, is a text-based address used to navigate to a website. |
| | A **Domain Name System (DNS) server** holds the directory that associates an internet address with a web address. |
| | A web page contains a **hyperlink (link)** that takes you to a related web page when clicked. |
| | The first page you see when you visit a website is the **home page**. |
| | The *http* in a URL refers to the Hypertext Transfer Protocol used for displaying pages. |
| | A **domain name** is the text-based name for an organization within the URL. |
| | The three- or four-character extension in a domain name is the **top-level domain (TLD)** and identifies the type of organization. For example, *.gov* means "government." |
| | ICANN is expanding the number of approved **generic top-level domain (gTLD)** names, which means you may see a domain name such as mylawfirm.*legal.* |
| | A **ccTLD** is a two-character extension in a domain name that is a country code. For example, *.ca* means "Canada." |

| Topic | Key Concepts and Key Terms |
|---|---|
| **2.5 Searching the Web** | A **search engine** reads web pages and creates indexes using keywords associated with the pages. |
| | A **spider** (or **crawler**) is a program used by search engines to find and index web pages. |
| | Some search engines provide categories of topics and subtopics that you can use to narrow your search. |
| | You will get different results from different search engines using the same search phrase because spider and index programs vary, pages may be updated at different times, and rankings may differ. |
| | Use the filter or advanced tools at a search engine website to fine-tune a search. |
| | A **metasearch search engine** compiles results from other search engines in one place. |
| | To find information suitable for academic papers, use an educational search engine to find scholarly literature or publications. |
| | A **content aggregator** is a website that collects and organizes web content. Subscribers can sign up for an update (called a ***web feed***) that alerts the subscriber when new content is added. |
| **2.6 Evaluating Web Content** | Domain names in the web address can provide clues to the authenticity or trustworthiness of a web page. |
| | Check a web page for an author's name and affiliation; if no name is present, try reading the <u>About Us</u> or <u>Contact Us</u> hyperlinks. |
| | Look for dates to make sure the information you are reading is the most recent; if no date exists, look for clues in the content or web address. |
| | Evaluate the purpose of a website to help decide whether a bias may exist in the information presented. |
| | A poorly designed website with errors in spelling and grammar should have its content corroborated by another source. |

*continued…*

| Topic | Key Concepts and Key Terms |
|-------|----------------------------|
| **2.7 E-Commerce, Communications, and Online Learning** | **E-commerce** involves online transactions, and **m-commerce** involves transactions from mobile devices. |
| | Shopping online generally requires a credit card or account with a third-party payment service such as PayPal or Google Wallet. |
| | Types of e-commerce transactions include between businesses and consumers **business-to-consumer (B2C)**, businesses and other businesses **business-to-business (B2B)**, and consumers and other consumers **consumer-to-consumer (C2C)**. |
| | A business that sells online to consumers is an **e-tailer**. |
| | **Email** is the sending and receiving of messages via the internet. |
| | An **email address** connects you to the ISP mail server to send, receive, and store messages. |
| | An **email client** is a program installed on a PC that is used to create, send, receive, and manage email messages. |
| | **Text messages** (or **texting**) are short messages sent via a **Short Message Service (SMS)** on a smartphone or tablet. |
| | **Instant messaging** is exchanging messages with someone else in real time using an app on a device that transmits using an internet connection. |
| | **Voice over Internet Protocol (VoIP)** is technology that allows users to make telephone and video calls via the internet. |
| | **Web conferencing** software allows a group of individuals to engage in meetings online with shared documents. |
| | Online learning, called *elearning*, uses the internet to earn academic credits or an entire degree. |
| | A **webinar** is a presentation delivered via the internet in real time. |

 **Topics Review**

## Chapter 3

# Computer Hardware

A s you learned in Chapter 1, computers come in all shapes and sizes. Whether you are working with a desktop or a laptop, tablet, or other mobile device, specific components are found in each one. All computers have a central processing unit (CPU), an input device(s), an output device(s), memory, connectivity adapters, and storage. These electronic and physical components found in a computing device are collectively known as *computer hardware*.

Some hardware is visible to you, such as the screen, keyboard, and mouse, while other hardware is housed inside the system, such as the CPU, memory, network adapter, and some storage. Hardware that is plugged in or wirelessly connected to your computer for input, output, connectivity, or storage is called a **peripheral**. An example of a peripheral is a printer.

If you have looked at ads for computers, you know that a basic understanding of the terminology and hardware components is helpful in making a purchase decision. In Chapter 1, you were introduced to many of the hardware components found in a PC. In Chapter 2, you learned about hardware that provide connectivity to the internet, such as cable modems and digital subscriber line (DSL) modems. In this chapter, you will explore what's inside the system unit, input and output devices, network adapters, digital data representation, and storage options in more detail. Appendix A contains tips for buying a computer or mobile device.

 **Watch & Learn**

Computer Hardware

## Learning Objectives

**3.1** Recognize the major components in a system unit and explain their purposes

**3.2** List and recognize various types of input devices

**3.3** List and recognize various types of output devices

**3.4** Identify wired and wireless network adapters used for connectivity purposes

**3.5** Explain how data is represented on a computer and describe typical speed and storage capacities

**3.6** Describe various options for permanent storage and their corresponding storage capacities

**Content Online**

The online course includes additional training and assessment resources.

## 3.1 The System Unit

**Read & Learn**
The System Unit

Recall from Chapter 1 that the system unit is the horizontal or vertical (tower) case in which the computer microprocessor, memory, and storage are located. In a mobile device, such as a laptop, tablet, or smartphone, these components are inside the unit below the integrated keyboard and/or screen. In an all-in-one system, these components are mounted in the same case in which the monitor is housed. Regardless of the configuration, the main components in the system unit are the motherboard, ports for plugging in peripherals, power supply with or without cooling fan, and storage devices.

In this topic, you will learn about the motherboard components, ports, and the power supply. In Topic 3.6, you will learn about storage.

### Motherboard

The main circuit board in the computer is called the **motherboard**. All of the other devices plug directly into or communicate wirelessly with the motherboard. Figure 3.1 shows a typical motherboard inside a laptop.

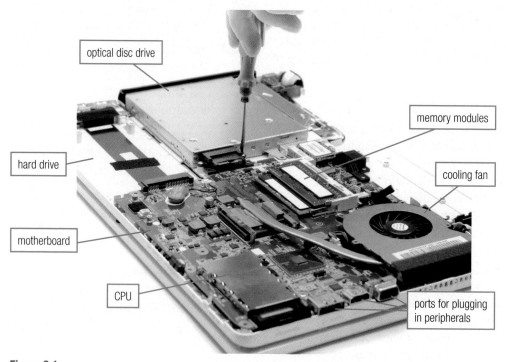

optical disc drive

memory modules

hard drive

cooling fan

motherboard

CPU

ports for plugging in peripherals

**Figure 3.1**

Shown here is the inside view of a laptop. In many laptops sold today, the optical disc drive and a cooling fan is omitted. A design technique called *passive cooling* describes a newer laptop with hardware that uses less power so that a cooling fan is not required to disperse the heat generated by the CPU and other components.

The motherboard includes the CPU, memory, expansion slots, and circuitry attached to ports that are used to plug in external devices. Data travels to the CPU and memory from the other components on the motherboard over wires called a **data bus**. Think of a data bus as the highway that data travels on. The size and speed of a data bus affect performance. A data bus that carries faster memory and more data to the CPU performs better. Similarly, traveling by car on a four-lane highway at high speed means you arrive at a destination faster than if you traveled on a two-lane highway at a slower speed.

Many computers integrate video and sound electronics on the motherboard, while other computers may include one or more of these components as an **expansion card** that plugs into an expansion slot on the motherboard (Figure 3.2a). Expansion slots used to plug in additional circuit boards to your computer either add or improve functionality.

A laptop does not have expansion slots. Most devices plugged into a laptop to extend functionality use a USB connection. However, many laptops have a slot on the side of the case for inserting a secure digital (SD) medium, such as a memory card for a digital camera (Figure 3.2b). You will learn about SD cards in Topic 3.6.

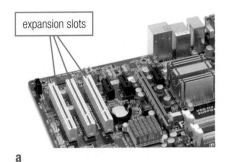

expansion slots

a

SD card slot

b

**Figure 3.2**

(a) Expansion slots on a desktop PC are used to plug in new hardware. (b) The SD card slot on a laptop is used to plug in a memory card.

**CPU** As you learned in Chapter 1, the CPU performs the processing cycle and is often referred to as the brain of the computer. Today, computers offer **parallel processing**, which involves having multiple microprocessor chips (more than one CPU chip on the motherboard) or a multicore processor (one CPU chip separated into independent processor cores). Parallel processing allows for multiple instructions simultaneously. Most laptops are configured with a four-core microprocessor. Parallel processing has vastly improved system performance when combined with software written to take advantage of its capabilities.

A CPU goes through an **instruction cycle** that involves the CPU retrieving, decoding, executing, and storing an instruction. The speed at which a CPU operates is measured in the number of instruction cycles the CPU can process per second, referred to as **clock speed**. Clock speed is typically measured in **gigahertz (GHz)**, which is 1 billion cycles per second. A CPU in most laptops and tablets sold today runs at speeds between 1.5 and 3.6 GHz.

In Chapter 1, you were introduced to the cache memory, which is memory either built into the CPU or located next to the CPU on a separate chip on the motherboard. Cache memory is used for storing frequently used instructions and data. Cache memory located closest to the CPU is called **Level 1 (L1) cache**. L1 cache is memory built on the CPU chip and operates the fastest of all memory. The secondary cache memory may be either built on the CPU chip or located on the motherboard and is called **Level 2 (L2) cache**. L2 cache feeds the L1 cache. The third level of cache memory, called **Level 3 (L3) cache**, is usually located on the motherboard and feeds the L2 cache. The amount of L1, L2, and L3 cache affects performance, since the CPU accesses cache memory faster than other memory.

**Memory** On the motherboard are two types of memory: random access memory (RAM, as described in chapter 1) and **read-only memory (ROM)**. ROM is used for storing instructions that do not change. For example, the programming code with instructions for starting a computer is often stored on a ROM chip. This ROM chip is called the **BIOS (basic input/output system)** or **UEFI (Unified Extensible Firmware Interface)**. UEFI is expected to replace the older BIOS technology with a faster, more secure startup process.

 **Video**

What Are the Various Types of RAM?

RAM is temporary memory that stores data and instructions while the computer is processing. The type of RAM in a PC varies, and the description of the memory in a computer ad can read like alphabet soup. However, the type affects the computer's performance, so knowing what the letters mean will help you understand one of the reasons some PCs operate faster than others.

Today's PCs are built with DDR3 or DDR4 SDRAM. *DDR* stands for "double data rate" and refers to the ability to transfer data twice as often. The number after DDR represents the generation of DDR. For example, *DDR3* means third generation DDR, which is twice as fast as DDR2. DDR4 RAM is engineered to use less power and transfer data at a faster rate than DDR3.

Synchronous dynamic random access memory (SDRAM) refers to a type of RAM that operates faster than older RAM technologies, such as Dynamic RAM that needed to be continually refreshed.

In mobile devices such as tablets and smartphones, you will find LPDDR3 and LPDDR4 RAM (*LP* stands for "low-power" RAM). This RAM is more compact than the RAM used in desktops and laptops. The manufacturers of RAM designed LP memory chips to use less power and therefore extend the time before the battery on a smartphone or tablet has to be recharged.

Installing more RAM into an empty module slot in a desktop PC or a laptop can improve computer performance, extending the life of an older system. Figure 3.3 shows a computer technician inserting a memory module into an empty slot in a desktop PC and in a laptop.

a

b

**Figure 3.3**

(a) On a desktop PC, open the system unit to install a memory module in an empty slot. (b) To install more RAM in a laptop, turn the laptop over, remove the memory module slot cover, and plug in the new memory module.

When adding RAM to a PC, you need to know the correct type, the number of available slots, and the maximum amount of RAM that the system will support. On a Windows PC, you can view the type of RAM, the RAM speed, and the number of available memory slots at the Task Manager window (Figure 3.4). To review the RAM specs on a Mac computer, click the Apple menu, click the *About This Mac* option, and then click

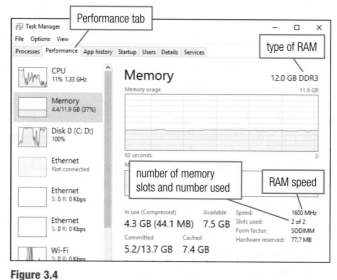

**Figure 3.4**

To review the RAM specs on a PC, open the Task Manager window in the More details view and use the *Memory* option on the Performance tab.

the More Info button. To find out the maximum amount of RAM your device can support, go to the website for your PC manufacturer and navigate to the technical support page. Search the page using the model number of your PC, and then click a hyperlink to the hardware specifications.

## Ports

A **port** is a connector located at the back, front, or side of a computer or mobile device. Ports are used to plug in peripherals. The most common type of port is the **universal serial bus (USB) port**, which is used to connect an external device such as a keyboard, mouse, printer, smartphone, or external storage medium. Most computers today provide several USB ports to accommodate multiple USB devices being used simultaneously. A USB hub can be used to increase the number of USB ports if your computer does not provide enough to meet your needs.

A laptop includes a high-definition multimedia interface (HDMI) port for connecting to a high-definition television. Some laptops also include a port for plugging in an external monitor and a port called an *Ethernet port* for connecting to a wired network. Ports are also found on laptops or other mobile devices to plug in headphones, external speakers, or microphones (Figure 3.5).

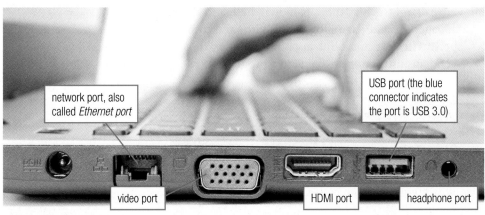

network port, also called *Ethernet port*

USB port (the blue connector indicates the port is USB 3.0)

video port

HDMI port

headphone port

**Figure 3.5**
Laptop ports are at the sides, back, and sometimes the front of the device. Some of the ports shown here may not be available in all laptop models.

## Power Supply

All computers operate with power. In a tower or desktop PC, the power supply is at the back of the system unit where you plug in the power cord. A laptop or tablet is powered by a charged battery or a plugged-in power adapter. A cooling fan near a power supply draws heat away from the CPU and prevents the CPU from overheating. Many laptop users who work with their device sitting on their lap set a cooling pad, also called a *chill mat*, under the laptop to help reduce the heat generated by the PC.

Accessories for portable power and charging batteries are plentiful. Power banks, external battery chargers, charging pads (Figure 3.6), and charging mats are four options that keep your devices powered up. Wireless charging capability is also being integrated into furniture and other surfaces people tend to put their phone when not in use. In the future, public spaces, such as airports and coffee shops, may have charging surfaces.

**Figure 3.6**
You can wirelessly charge some smartphones by placing them on a charging pad, like the IKEA charger shown here.

## 3.2 Input Devices

Any device used to enter raw data or communicate instructions to the computer is an input device. Several input devices exist, with the most common being keyboards and mice. Other devices used for input are also described in this topic.

### Keyboards and Mice

While wired keyboards and mice were the standards for traditional desktop PCs, today's computer users prefer to be untethered. A wireless keyboard and mouse (Figure 3.7) can be used with any computer without limitations set by the wire length. Wireless keyboards and mice use batteries as a source of power. A wireless receiver is built in or plugged into a USB port and connects with the wireless keyboard and mouse. The keyboard device controller interprets electronic signals from the keyboard and sends them to the computer operating system via the wireless receiver, which in turn sends the keystrokes to the active running program for display on the screen.

A **keyboard** is a device used to type data into a computer. In addition to the alphabetic and numeric keys on a keyboard, special-purpose keys (such as Esc, Insert, Delete, function keys labeled F1 through F12, the Start key on a Windows-compatible keyboard, and the Command key on a Mac keyboard) send commands to a program. Directional movement keys (such as the Up, Down, Left, and Right Arrow keys and the Home, End, Page Up, and Page Down keys) are used to move around a screen. Finally, keys labeled Ctrl, Alt, and Shift in combination with a letter, number, or function key send an instruction to the active program.

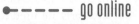
**Read & Learn**
Input Devices

●----- go online

Go here for a list of keyboard shortcuts for use with Windows 10 and Microsoft applications: **https://CC3 .ParadigmEducation .com/Keyboard Shortcuts.**

Go here for a list of Mac keyboard shortcuts: **https://CC3 .ParadigmEducation .com/MacShortcuts.**

**Figure 3.7**

A wireless keyboard, stylus, and mouse are used with traditional desktop PCs and as peripherals for mobile devices.

Laptops include a built-in keyboard. Tablets and smartphones may or may not have an integrated keyboard that is built in or slides out. Some tablets include a keyboard that you can use by plugging the tablet into a port on the top of the keyboard (called a *docking port*). Tablets and smartphones also display a touch-enabled keyboard that you use by tapping with your fingers (Figure 3.8) or with a **stylus** (digital pen or pointer).

A **mouse** is used to point to, select, and manipulate objects on the screen. The mouse works by detecting its motion in relation to the surface beneath it using light technology and by the clicks of its buttons. Some mice use a trackball to move the pointer on the screen when you roll your thumb or fingers over the ball. As you move the mouse, the pointer on the screen moves in the same direction. Buttons on the mouse provide the ability to send a command when the pointer is resting on the target that you wish to manipulate. A wheel on the mouse facilitates scrolling (moving the screen display up, down, left, or right).

**Figure 3.8**

Tablets and smartphones offer an on-screen keyboard, which you tap using your fingers or a stylus.

On a laptop, a **touchpad** is located below the keyboard. This rectangular surface is used in place of a mouse to move the pointer on the screen and manipulate objects. The touchpad senses finger movement and taps much like a mouse senses movement and clicks. Depending on the touchpad device, multi-finger gestures and pinch and zoom movements similar to the gestures you use on your smartphone screen can be used to scroll and zoom a laptop display. Some laptops and smartphones include a track pointer, which senses finger pressure to move the pointer. Figure 3.9 shows a typical track pointer and touchpad.

track pointer

touchpad

**Figure 3.9**

A touchpad and/or a track pointer (red button in middle of keyboard) moves the pointer on the screen using finger movements.

## Touchscreens

Mobile devices, such as tablets, smartphones, and portable media players, are associated with accepting input with a touch-enabled display, called a **touchscreen** (Figure 3.10). Some laptops, desktop monitors, and even TVs are now touch-enabled. A touch-enabled device has a layer of capacitive material below a protective screen. When you touch the screen with your finger or with a stylus, an electrical signal is sensed by the capacitive layer, which sends data about the location you touched and any gesture you used to the computer touch processor and software that interpret the gesture.

Touchscreens used in combination with other input devices, such as keypads, buttons, and magnetic strip readers, are commonly used for input with many other computing devices, such as self-serve kiosks, digital cameras, and ATMs.

**Video**

How Do Touchscreens Work?

a

b

**Figure 3.10**

(a) A tablet and a (b) self-serve kiosk are two of many computing devices that are touch-enabled.

## Scanners and Readers

Sophisticated scanners and readers are becoming more integrated into our day-to-day lives. A **scanner** is a device that uses optical technology to analyze text and images and then convert them into data that can be used by a computer. A scanner can be a separate input device, such as a flatbed scanner (a scanner that sits on a desk) or a handheld scanner. Many printers and photocopiers used today are multipurpose devices that also incorporate scanners.

A **bar code reader** optically scans bar codes to identify products in warehouses or at checkout counters. Smartphones can scan a **QR (Quick Response) code**, a type of bar code that looks like a matrix of black square dots. The smartphone reads the QR code, which usually directs the device to display a website. A **magnetic strip reader** reads data from the magnetic strip on the backs of cards, such as debit cards, credit cards, or key cards (the latter used to open doors or parking gates). A **biometric device** identifies people by a human characteristic, such as a fingerprint, iris, face, or voice. Facial recognition is now used on smartphones as a way to unlock the device and authorize purchases. Facial recognition technology is also being used in some airports for boarding and customs clearances. Finally, a **radio frequency identification (RFID) reader** scans an embedded RFID tag to identify an object. RFID readers and tags use tiny chips with antennae that are readable when the reader and tag are within range of each other. Typically, RFID technology is used to read tags up to 300 feet away. All of these scanners and readers are shown in Figure 3.11.

**go online**

https://CC3 .ParadigmEducation .com/RFIDJournal
Go here to read the latest news on RFID technology innovation in various industries.

**Figure 3.11**
Facial recognition (top left) unlocks a device or authorizes payments. A handheld bar code reader (top right) inputs a product code to receive inventory. A scanner (bottom left) converts a document to text and/or images, and a smartphone camera (bottom right) reads a QR code to display a website. An embedded wireless tag similar to the one shown (center image) identifies an object or provides stored information such as a passport number.

## Digital Cameras, Digital Camcorders, and Webcams

Many media devices today are multipurpose, meaning a single device (like your smartphone or tablet) can snap photos, record audio, and capture videos (Figure 3.12). Digital cameras are mostly used to capture still images, although short live-video segments are also recorded on a digital camera. A digital video camera, called a *camcorder*, captures live video and audio.

**Figure 3.12**
(a) Digital cameras, (b) webcams, and (c) camcorders are used to convert images and live video into a digital format. Many of these devices are now multipurpose. For example, the digital GoPro camera (a) also records video and can stream live footage to a blog, social media site, or website.

A **webcam** is a video camera built into a video display screen (mounted at the top center of the screen edge) or plugged in as a peripheral and is used to provide images to the computer during live web conferencing or chatting online. Webcams are often installed inside or outside homes and workplaces for security monitoring, providing the owners or managers with real-time images accessed on smartphones or PCs. A webcam or other digital camera can also be built into a video doorbell to let the homeowner use a smartphone from anywhere in the world to see and speak to the person at the front door.

## Microphones and Voice Recognition

Microphones are built into laptops and mobile devices. An external microphone or a headset with a microphone attached can also be plugged in as a peripheral. A **microphone** can be used in conjunction with software to create a digital audio file from voice input or other sounds captured by the device. You may also use a microphone to chat with someone online. Some games played online or with game systems accept voice commands. **Voice recognition technology** (also called *speech recognition technology*) is used to recognize voice commands as input for hands-free operation at work, with the navigation and communications system in your car, or with your mobile device. Siri on the Apple iPhone or Alexa on Amazon's smart speaker uses voice recognition input to send messages, schedule meetings, make a call, or search for information on the web.

## Other Input Devices

Gaming systems accept input using a variety of methods, such as wired or wireless game controllers, joysticks, and motion sensors. Some games have specialized controllers, such as guitars, other musical instruments, or tennis rackets. Some exercise games use pressure-sensitive mats or boards.

Other devices, such as keypads and magnetic ink character recognition (MICR) systems, are used in specialized applications to provide input. MICR code consists of the characters at the bottom of your check that are scanned and processed by an MICR reader at your bank to withdraw the funds from your account.

Augmented-reality and mixed-reality headsets that use gestures, eye movements, and voice input with holographic images are no longer considered science fiction. Figure 3.13 shows input using gestures with the Meta 2 AR headset and the Levi's Jacquard smart jacket.

a    b

**Figure 3.13**

(a) With the Meta 2 AR headset, the user manipulates a hologram using hand gestures. (b) Cuff gestures on the Jacquard smart jacket from Levi Strauss & Co. provide input to the smartphone app for accessing music, navigation, texts, or calls.

## 3.3 Output Devices

Devices used to view the information processed and organized by the computer are output devices. The most commonly used output device is a **monitor** (or the screen/display found in mobile devices). Computer output can also be connected to high-definition TVs, projectors, or interactive whiteboards. Other types of output devices explored in this topic include speakers, headsets, printers, and copiers.

### Video Displays

An electronic device called a **video display** presents information from a computer visually. A desktop PC uses a monitor connected to the computer video port. Laptops and other mobile devices have built-in video display screens.

Curved displays are becoming the rage due in part because a curve in the monitor matches the curvature of your eye, offering a field of view that is more natural and comfortable.

Video displays come in many different sizes, which are measured by the diagonal length of the viewing area (Figure 3.14). For a mobile device, the size of the video display is constrained by the size of the laptop, tablet, or smartphone. Small laptops may have screens as small as 11 inches, while large laptop screens can be as big as 18.4 inches. Tablet screens typically range from 7 to 12 inches. Smartphone screens can be as small as 2.5 inches, with newer phones nearing 6 inches.

To take advantage of larger viewing areas or to use two screens at the same time, computers (mainly laptops) are often connected to external video displays, such as computer monitors or high-definition TV screens. Monitors used in this way are typically wide-screen, are available in various sizes upward from 20 inches, and feature flat-panel liquid crystal technology as either **liquid crystal display (LCD)** or **light-emitting diode (LED)**. The difference between LCD and LED lies with the backlighting that allows light to pass through the panels. LCD screens use a type of fluorescent lamp whereas LED screens use smaller, more efficient light-emitting diodes. The technology in LED screens means they consume less energy while displaying brighter, more vibrant colors than LCD screens.

Newer technology in screens made by LG, Panasonic, and Sony known as **organic light-emitting diode (OLED)** does not require backlighting like LCDs and LEDs. OLED technology results in a display that is more energy efficient for any device that runs off battery power. OLED screens are thinner and produce the best picture quality because each organic cell behind the panel creates its light source, meaning no light spills into other areas of the picture. Samsung developed a display marketed as QLED that uses quantum dots (nanocrystal semiconductors) for more accurate and brighter colors on LED screens.

**Read & Learn**

Output Devices

● – – – – **in the know**

In 2018, LG announced a 77-inch OLED display that is transparent and can be rolled up and down as needed to show different content. For example, roll the screen up a short distance to view a weather forecast or to home theater size for watching a movie.

a    b    c

**Figure 3.14**

Display screen sizes vary by device, with (a) smartphone screens measuring 2.5 to 6 inches; (b) tablets typically from 7 to 12 inches; and (c) laptops varying widely, reaching up to 18 inches.

The video display **resolution** setting affects the quality of output on the display screen. Resolution refers to the number of picture elements, called *pixels*, that make up the image shown on the screen. A **pixel** is square with red, green, and blue color values (called *RGB*) that represent each color needed to form the image. Resolution is expressed as the number of horizontal pixels by the number of vertical pixels, such as 1920 × 1080 pixels for a 17-inch laptop display. The more pixels used to render the output, the sharper the image, as shown in Figure 3.15. Displays default to recommended settings, but operating system software can be used to change the resolution.

**Figure 3.15**

A character displayed at a low resolution using fewer pixels to form the character appears fuzzy (left) compared to the same character at a higher resolution using more pixels (right).

Display technology called **4K UHD** (ultra-high-definition) is found on many display screens, such as those in monitors, TVs, and digital cameras. *4K* refers to the resolution of the display. A 4K UHD screen uses more than four times as many pixels as the standard high-definition (HD) screen (8 million pixels for a 4K UHD screen versus 2 million pixels for an HD screen). 5K displays are also available in limited supply, and 8K displays are expected to become widely available in the next few years.

## Interactive Whiteboards and Video Display Projectors

An **interactive whiteboard** displays a computer's output on a large surface similar in size to a whiteboard (Figure 3.16a). Special pens can be used to make notes on the display. In these instances, the whiteboard is both an output device and an input device. The annotations can be saved with the images as electronic files for viewing later. Touch-based tools for interaction with the computer while the display is being viewed are also included.

A **video display projector (VDP)** is used to project computer output on a large screen (Figure 3.16b). VDPs are often ceiling mounted in classrooms and boardrooms, while portable projectors are available for other venues. A laptop or other computer connects to the projector via a video output cable or wireless technology.

## Speakers and Headphones

The audio output from the computer, such as music or sound effects, is heard through internal speakers, external speakers, or headphones. PC monitors, laptops, and other mobile devices have integrated speakers. Plugging headphones or earbuds into the audio port of a computer or mobile device redirects the audio output to your headset. High-quality headphones reduce background noise, improving the audio experience. External speakers can give the listener home theater–quality sound. A **Bluetooth headset** or speaker paired with a smartphone is used to listen to voice conversations or play music. Smart speakers such as the Amazon Echo and Google Home are popular for streaming music, news, or live radio shows.

a

b

**Figure 3.16**

When making presentations to groups, a computer image can be directed to display on (a) an interactive whiteboard or (b) a VDP.

## Printers and Copiers

Video displays provide output that is temporary. When the computer is turned off, the output disappears. Printed copies of computer output are called **hard copy**. A hard copy can be generated by a laser printer, inkjet printer, photo printer, thermal printer, plotter, or digital photocopier.

3-D printers manufacture objects on demand, and while the manufacturing industry was an early adopter of 3-D printers for printing prototypes and parts, the medical field is increasingly using the technology to print implants, external prostheses, and even living organs. NASA is currently studying using 3-D printers to print synthetic food.

A team of volunteer engineers at Limbitless Solutions created this "Iron Man" prosthetic arm for a seven-year-old boy using 3-D printing technology. Robert Downey Jr. surprised the boy when he delivered the arm himself.

**Laser Printers** A **laser printer** is a popular choice for producing a hard copy in offices and homes. A laser beam electrostatically charges a drum with the text and images sent for printing. A dry powder called *toner* sticks to the drum and is transferred to the paper as the paper passes through the printer. Paper feels warm when it comes out of a laser printer because heat is used to adhere the toner to the paper permanently. Color laser printers, once found only in business settings, are becoming popular with home users.

**Inkjet Printers** An **inkjet printer** forms text and images on the page by spraying drops of ink from one or more ink cartridges that move back and forth across the page. Inkjet printers allow home users to print in color relatively inexpensively.

**All-in-One Printers** Many laser and inkjet printer models are multipurpose, meaning that in addition to printing, they can be used for scanning, photocopying, and sending/receiving faxes. Having a multipurpose printer saves desktop space, cuts down on various supply costs, and means you only have to learn to operate one device. The downside is that you cannot perform multiple functions at the same time. For example, you cannot copy a document while printing.

a

**Digital Copiers** A **digital copier** in an office can act as a traditional paper photocopier and, when connected to a network, accept output from computers for printing (Figure 3.17a). Often these copiers are used to share color-printing capabilities in a workplace. Digital copiers usually include other features, such as scanning and saving the output as electronic documents as well as hard copy.

**Plotters** Blueprints, technical drawings, large drawings, and signs are often produced on a **plotter** (Figure 3.17b). A plotter moves one or more pens across the surface of the paper to create the drawing. Plotters are slow, as the movement of the pen or

b

**Figure 3.17**

(a) A digital copier is a multipurpose printer that is connected to a network to print computer output. (b) A plotter prints large format output, such as technical drawings.

pens across the page takes time, but they can print easily on paper, cardboard, plastic, aluminum, plywood, and sheet steel.

**Photo Printers**   A **photo printer** generally connects directly to a digital camera to print high-quality photos on photo paper, usually using inkjet technology. Professional photo-printing service companies use higher-end photo printers that produce a high-quality image on treated paper for long-lasting prints.

**Thermal Printers**   Receipts you receive from retail stores or services are typically printed on a **thermal printer**. Thermal printers produce output by heating coated paper as the paper passes over the printhead (the component inside the printer with nozzles that spray ink onto the paper). These printers produce receipts quickly; however, the image fades away over time.

**3-D Printers**   A **3-D printer** can create anything from a replacement part for a coffeemaker to an entire house (Figure 3.18). A 3-D object is designed using software, and then the object is printed using a technique whereby consecutive layers of material are placed on top of one another precisely cut to build the object. The process is sometimes referred to as *additive manufacturing*. The 3-D printing industry is expected to be worth $32.78 billion by 2023, according to marketsandmarkets.com, a market research firm.

in the know

The average office worker prints 10,000 pages per year!

Video

How Does 3-D Printing Work?

**Figure 3.18**

3-D printers (see inset photo of 3-D printer), print any object, such as a replacement part—or a home! This 650-square-foot house was developed by ICON, a company in Austin, Texas, and is the first 3-D-printed home. It was constructed out of cement and made in 24 hours. ICON's goal is to bring affordable homes to underserved communities.

## Other Output Devices

Other output devices, such as label printers and document cameras (devices that display an image from a hard copy on a video display), are also available. Additionally, virtual reality devices are available to provide output via the images seen through the head-mounted display and the audio heard through the headphones.

## 3.4 Network Adapters

As you learned in Chapter 1, the term *communications device* refers to any component or device used to transmit and receive data. In Chapter 2, the various types of internet connection options were explored, and you learned about hardware used for accessing the internet, such as cable modems, DSL modems, and satellites. A communications device can be thought of as both an input device and an output device since it receives data (input) from and sends data (output) to the computing device, and other computers via a network. In this topic, you will examine further the hardware provided in desktops, laptops, tablets, or mobile devices that enable connectivity to a network.

### Network Adapters

Any device used to connect a computer to a network is called a **network adapter** (also called a *network card*). The network adapter is the interface between your computer and the other networking equipment, such as the modem or router that provides the pathway for data to travel. All pieces of networking equipment communicate with one another using a set of standards called *protocols*. A **protocol** can be thought of as a set of rules that define how data is exchanged between two devices. For each device to understand the other, a network adapter has to use the same protocols as the modem or router. When trying to use an older wireless network adapter with a newer router that uses a faster data transfer protocol, you can purchase a newer adapter that supports the faster protocol.

### Network Adapters for Wired Connectivity

A network adapter for a wired connection is called an *Ethernet port* (sometimes referred to as an *RJ-45 port*). One end of a network cable is plugged into the network port on the desktop or laptop PC, and the other end is plugged into a modem or router (Figure 3.19) to provide the communications channel.

   *Ethernet* refers to the type of cable used to connect two devices and the data transfer speed that the media can support. Typical ads for a computer will state Ethernet 10/100 or 10/100/1000. The numbers after Ethernet refer to the data transfer speed the adapter can support with *10* indicating 10 Mbps, *100* indicating 100 Mbps, and *1000* indicating 1,000 Mbps (or 1 Gbps, called *Gigabit Ethernet*).

1. Plug one end of a network cable into the network port of a computer.

2. Plug the other end into an available port on the back of a modem or router.

**Figure 3.19**

With a wired network adapter, a network cable connects two devices in a network.

### Network Adapters for Wireless Connectivity

Wireless connectivity is preferred by people who do not want their movements constrained by the length of a physical cable. At home, at work, or when you are on the go, a wireless network allows you the freedom to move around. Currently, computers come equipped with a wireless interface card that is integrated into the system unit as either Wi-Fi or Bluetooth. Many devices include both adapters.

**Wi-Fi Adapters**  A **Wi-Fi adapter** transmits data to and from a wireless access point or wireless router using radio frequencies. Wi-Fi is often referred to as the **802.11 protocol**,

**Read & Learn**
Network Adapters

**Video**
What Devices Can Be Built into Your Computer to Enable Communications?

which is the name of the standard developed to facilitate wireless communication among several hardware providers. As the protocol was improved, version letters were added to the 802.11 standards. In 2018, the standards were renamed to simplify identification. The most current Wi-Fi standard is Wi-Fi 6 (also called 802.11ax). A computer ad may state that the Wi-Fi adapter is Wi-Fi 4, Wi-Fi 5, or Wi-Fi 6, which means the adapter can communicate with older Wi-Fi networks as well as the newer technology.

If an integrated wireless adapter does not support a newer wireless network, or you have a desktop PC that does not have a wireless interface, you can purchase an **external wireless adapter**. External wireless adapters have built-in radio transmitters and receivers to connect to a wireless modem or wireless router. The most commonly used external adapter is a **USB wireless adapter**, also called a *USB dongle* or *USB mobile broadband stick* (Figure 3.20).

**Connecting with Bluetooth**   Bluetooth is a communications technology for short-range distances, usually up to 300 feet depending on the device. When the Bluetooth adapter on one of the devices is activated and the other Bluetooth device is turned on, the adapter detects the wireless signal, and the two devices can be *paired* (user has to accept a pair request), meaning they can exchange data. Bluetooth is used to connect a smartphone to a communications system inside a car. The technology is also used to wirelessly connect a PC or smartphone to peripherals, such as Bluetooth speakers, mice, keyboards, printers, or wireless media hubs.

Bluetooth LE (BLE), which means Bluetooth low energy or Bluetooth Smart, is a standard that uses less power to communicate between devices and was developed to conserve battery power on mobile devices. **Bluetooth 5.0** became available in 2017. With Bluetooth 5.0, you can play audio on two connected devices at the same time with faster data transfer speeds over an expanded connectivity range up to 800 feet. Most mobile devices are equipped with an integrated Bluetooth adapter. You can purchase a USB Bluetooth adapter to add connectivity to older devices.

## Sharing Smartphone Network Connectivity

As you learned in Chapter 2, by using a cable, Bluetooth, or Wi-Fi adapter, you may be able to share your smartphone's internet connection with other devices, such as tablets or laptops. On some smartphones, the feature is called *hotspot*, while on others it is called *tethering* (Figure 3.21). Sharing your smartphone's connectivity is useful when access to the internet is not available for your laptop or tablet. In this case, your smartphone becomes a modem that can service other computing devices. To share your smartphone's internet connection, turn on the hotspot or tethering feature in the Settings app. If your smartphone does not have the tethering feature, you can download and install a tethering app to provide the option.

**Figure 3.20**
A USB wireless adapter is used for connectivity in a PC that does not have an integrated adapter or to upgrade an existing Wi-Fi adapter.

---— in the know

The name Bluetooth came from 10th century King Harold Bluetooth, who was instrumental in uniting warring countries in Europe. Bluetooth seemed an appropriate name, as hardware competitors had to collaborate to make devices that work with each other.

**Video**

How Does Bluetooth Work?

**Figure 3.21**
Check the *Networks* category in the Settings app on your smartphone to find the *Tethering* option. On some phones, this may be called *hotspot*.

**Let's Explore!**

What Wireless Technology Is Used by Google Wallet and Apple Pay?

# 3.5 Digital Data

When you type a document on a computing device, you see characters, numbers, and symbols appear on the video display screen. However, the digital data exists as binary digits. First, it is temporarily stored in RAM, then it is processed by the CPU and eventually saved to permanent storage when you save the document. In the **binary system**, only two possible values exist—0 and 1. Every document you save, picture on your camera, or message you send is a series of 0s and 1s according to the computer. Understanding the way digital data is represented is helpful to put context to the speed of an internet connection and the capacity of storage devices and memory.

## Binary Bits and Bytes

The smallest unit for digital data in the binary system is called a **bit**. The name *bit* is derived from *binary digit*. A bit is either 0 or 1. By itself, a single bit of 0 or 1 is not meaningful, because only two possibilities exist. However, by adding additional bits to form a group, more opportunities to represent data are available because different combinations of 0s and 1s can be created to represent something. By grouping 8 bits, 256 possibilities for combinations of 0s and 1s are created. This provides enough combinations to represent each letter of the alphabet in both uppercase and lowercase, each number, and symbols available on a standard keyboard (such as $, ?, and !). Each group of 8 bits is called a **byte** and represents one character. When you type the letter *A* on the keyboard and see *A* appear on the screen, one byte of RAM is being used to store the letter. Table 3.1 illustrates the binary equivalents for the two forms of the letter *A*, the number *4*, and the dollar symbol.

**Table 3.1** Binary Equivalent Examples

| You type . . . | Computer stores in RAM . . . |
| --- | --- |
| a (lowercase) | 01100001 |
| A (uppercase) | 01000001 |
| 4 | 01100100 |
| $ | 00100100 |

## Measuring Internet Speed and Data Caps

Bits are used to measure speed for transmitting data on the internet. Recall in Chapter 2 when discussing broadband, average speeds are measured in mega*bits* per second (Mbps) or giga*bits* per second (Gbps). One megabit is defined as 1 million bits, and one gigabit is 1 billion bits. If it takes eight bits to represent one character, then one megabit transfers 125,000 characters. If your cable modem performs at an average speed of 20 Mbps, that means your cable modem is transmitting 2,500,000 characters per second.

Some internet service providers (ISPs) put a limit on your monthly data transfer capacity. These limits are called *data caps* or *bandwidth caps* and are expressed in gigabytes. For example, a data cap of 100 GB means the ISP is limiting data transfer to 100 gigabytes in the current billing period, which is usually a month in duration. If your provider has a data cap, and your household use during the billing cycle exceeds the cap, you might be charged extra fees by the provider, or your connection speed might slow down.

**Read & Learn**
Digital Data

**Video**
How Does Your Computer Translate Text into Numbers?

●---- in the know

Unicode is a coding system based on binary. It was developed to accommodate every letter in every language. More than 100,000 characters are represented in Unicode, with some requiring two 16-bit groups to represent a character! Unicode is the standard for programming characters in browsers and operating systems.

A data cap may also apply to your smartphone plan. If your plan has a maximum data usage per month, then you need to be mindful when using apps on your phone over the cellular service that stream music or movies over the internet, because these apps will quickly use up your data quota.

Netflix advises users that streaming a movie in standard definition uses 1 GB of data per hour, in high definition 3 GB per hour, and in ultra-high-definition 7 GB per hour. You can adjust options in your Netflix account to change the playback setting to a lower quality that uses less data or choose to allow streaming only when the device is connected to a Wi-Fi network. Customizing your settings is especially helpful if you have a data cap on your smartphone plan.

## Measuring File Size and Storage Capacity

A document, such as an essay that you create in a program like Word, is merely a collection of characters (bytes). When saved, the collection of bytes that comprises the document is called a **file**. The unique name you assign to the document when you save it is called a **file name**, and that name identifies that particular file among others on your storage medium. A file could potentially hold several million characters. Similarly memory, such as RAM, and a permanent storage medium, such as a hard disk drive, can store several billion characters.

Prefixes added to the word *byte* describe progressively larger storage capacities. For example, adding *kilo* in front of *byte* creates the word **kilobyte (KB)**. One KB is 1,024 bytes, but people use the approximated value of 1,000 characters when calculating storage capacity.

The next prefix is *mega*, creating the term **megabyte (MB)**. One MB is approximately 1 million bytes. Table 3.2 summarizes storage units, including a **gigabyte (GB)** and **terabyte (TB)**, with their corresponding abbreviations and capacities for the types of storage media you would encounter at home or at work.

**Table 3.2**  Storage Units and Capacities in Bytes

| Storage Unit | Abbreviation | Popular Reference | Approximate Capacity in Bytes |
|---|---|---|---|
| Kilobyte | KB | K | 1,000 (one thousand) |
| Megabyte | MB | meg | 1,000,000 (one million) |
| Gigabyte | GB | gig | 1,000,000,000 (one billion) |
| Terabyte | TB | TByte | 1,000,000,000,000 (one trillion) |

## Adding Context with Storage Capacity

Understanding that 1 MB is approximately 1,000,000 characters is more helpful to you if you can put this value into some meaningful context. According to eHow, 1 MB equals approximately 500 typed pages of text with no images. However, most documents created today have some images or other graphics added to the text to make the documents more appealing to read. A 25-page report with graphics on each page could easily require 7.5 MB of storage.

Many people like to download and store music and videos. It is not uncommon for one MP3 song to use more than 8 MB of storage and a movie file to require more than 4 GB. If you like to use your computer to store pictures, music, and movies, you need a storage medium with a large capacity. In the next topic, you will learn about the various options for storage and the typical storage capacity of each option.

## 3.6 Storage Options

Saving a document, spreadsheet, or presentation makes a permanent copy of the work in a file that is available after power to the computer is turned off. When saving, a **storage device** for the file is identified. A storage device is a piece of hardware with the means to write new data to and read existing data from storage media. A storage device is often referred to as a **drive**. A unique letter and label identifies each drive on a Windows-compatible computer (Figure 3.22). On a Mac computer, each storage device is assigned a name, such as Macintosh HD, for the hard disk drive.

Several storage options for saving a permanent copy of a file are available, such as internal and external hard disk drives, solid-state drives (SSDs), network storage, USB flash drives, and flash memory cards. Cloud storage is another option for saving files when connected to the internet. Optical discs, while no longer in widespread use, are still available and used by some people. In this topic, you will learn about each of these storage devices and the cloud storage option.

**Read & Learn**
Storage Options

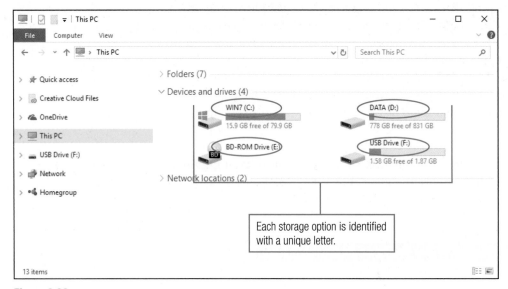

Each storage option is identified with a unique letter.

**Figure 3.22**
On a Windows-compatible computer, each storage drive is identified with a letter.

●----- in the know

Notice the first drive listed in the *Devices and drives* section in Figure 3.22 is the hard disk drive C. Early PCs did not have HDDs and operated with one or two floppy disks (soft plastic disks that you inserted into a drive slot that were) assigned the letters *A* and *B*. When floppy disks became obsolete, the letters *A* and *B* were not reused.

### Hard Disk Drives

Inside the system unit of a PC is an internal **hard disk**, also called a *hard drive* and *hard disk drive (HDD)* (Figure 3.23). The internal HDD is usually the storage device with the largest capacity and is usually assigned drive letter *C* on Windows-compatible computers. This is the storage medium where operating system, applications, and data files are stored. Hard disks are round pieces of metal called *platters* stacked inside a sealed unit with a magnetic coating on their surface. The disks spin, and a read/write head moves across the surface of the disk, magnetizing the data to the surface.

read/write head

platter

**Figure 3.23**
Shown is a hard disk with the top cover plate removed. Each platter can store data on both sides, so each platter requires two read/write heads.

●----- in the know

A hard disk needs to be formatted before data can be stored on the drive. During formatting the disk is partitioned (assigned a logical container drive letter, such as *C*). In many cases, the HDD is split into multiple partitions for organizing programs and data files. For example, you might split a single HDD into two partitions: drive C for software programs, and drive D for data.

An external HDD is a storage device that attaches to the computer using a USB port (Figure 3.24a). External HDDs are popular options for extending the storage capacity of PCs and for backing up documents, photos, videos, music, and other files.

Storage capacities for HDDs vary but are typically in the range of several hundred GB, with newer computers having HDDs in the range of 1 to 3 TB.

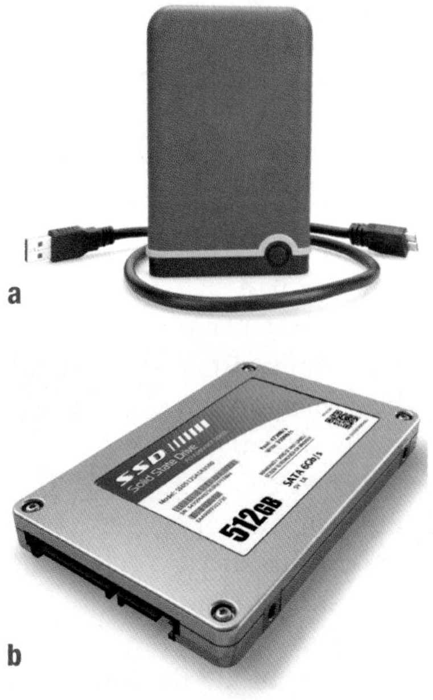

## Solid-State Drives

A type of drive technology called a **solid-state drive (SSD)** uses **flash memory** (Figure 3.24b). Flash memory uses chips to store data electronically instead of magnetically, as on a traditional HDD. With chip-based flash memory, no mechanical parts are required inside the drive, making the device more durable. SSDs also weigh less than traditional HDDs, make no noise when operating, and need less power. Internal and external SDDs are available in storage capacities similar to those of traditional HDDs.

A type of drive called a *hybrid drive* combines solid-state memory into a traditional HDD. The most frequently accessed data is stored in the solid-state area, with the rest on traditional hard disk platters.

## Network Storage

In some workplaces, data is required to be stored on a network drive for security and cost reasons. Network drives are HDDs installed in a server made available to users with access rights to the server storage via their username and password. A network drive often has a letter assigned that is higher in the alphabet, such as *R* or *S*. Networked storage is automatically backed up daily from the server. Centralizing network storage is considered more secure than leaving data on individual HDDs scattered about a workplace.

Home users can connect various storage devices to a single wireless media hub for storing and sharing data with multiple devices and people. Wireless media hubs are also battery-powered Wi-Fi routers and charging stations.

## USB Flash Drives

A **USB flash drive** is a portable storage device with flash memory inside the case. USB flash drives are popular because they are easy to use, small enough to carry comfortably, and inexpensive. The drive is powered through the USB port of the PC. When you plug in a USB flash drive, a drive letter is assigned—usually *E*, *F*, *G*, or a higher letter, depending on the number of ports and other devices.

USB flash drives are made in all kinds of shapes, sizes, and colors (Figure 3.25). Some are disguised inside toys or cartoon characters. Others are available as plastic wristbands. Storage capacities meet all needs; some store as little as 4 to 8 GB, and some store up to 512 GB.

**Figure 3.24**

(a) External HDDs are often used to back up an entire computer. (b) SSDs weigh less and use less power than traditional HDDs.

a

b

**Figure 3.25**

USB flash drives come in all shapes, sizes, colors, and capacities.

USB flash drives with 1 TB of storage are also now available. Many other names are used to refer to a USB flash drive, such as *thumb drive*, *memory stick*, *jump drive*, *key drive*, and *pen drive*.

## Flash Memory Cards

The most common type of storage for a digital camera, smartphone, or other portable device is a **flash memory card** (Figure 3.26). A flash memory card is a small card that contains one or more flash memory chips (similar to SSDs and USB flash drives). Flash memory cards come in a variety of formats, such as Secure Digital (SD), CompactFlash (CF), and proprietary formats made by Sony and Olympus for their digital cameras. Capacities for flash memory cards can range from 1 GB to several hundred GB. The type of flash memory card you would buy for a device depends on the formats the device can read.

**Figure 3.26**

Flash memory cards are used for storage in digital cameras, smartphones, and other portable devices.

Some computers and laptops have a card reader built in for secure transfer of files stored on the flash memory card of a portable device (such as a digital camera) and the PC. Some USB hubs also contain readers.

## Cloud Storage

Recall from Chapter 1 that cloud computing involves accessing software and storage services from the internet (Figure 3.27). Some cloud computing providers offer a limited amount of free storage space to account holders, with subscription plans available for those who need a higher capacity. For example, both Google Drive and Microsoft OneDrive offer free storage for all Microsoft and Google account holders. Other popular cloud storage providers are listed in Table 3.3 on the next page, each of which offers some free storage so that you can try out the service before signing up for the paid options with higher storage limits. Many other cloud storage providers exist. Talk with friends and family about suggestions for a cloud storage provider they use and recommend.

The main advantage with storing files using a cloud service provider is the ability to access the files using a web browser from any location with internet connectivity and from any device. This allows you to share files with others easily. If you are working with a team, storing the project files in the cloud lets all team members have access to the files they need to complete their work.

Transferring files via the internet is fast, and the software interface is generally easy to navigate. Most smartphone settings default to storing files in the cloud. Typical files saved

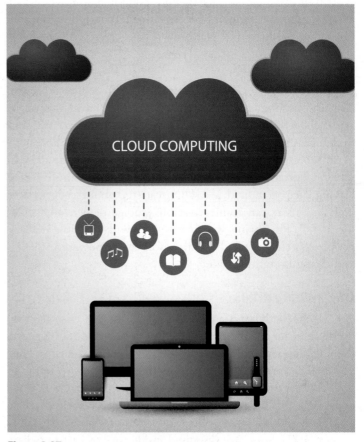

**Figure 3.27**

Being able to access a document from any device and easily share your files or collaborate with others are reasons many people are choosing a cloud storage option for their documents, entertainment, contacts, and other files.

to the cloud include photos taken with the smartphone's camera, contacts, and documents you access from the device. Finally, you no longer need to buy, organize, and store media, such as external HDDs and USB flash drives, if all your files are online.

Make sure you investigate the cloud service provider's terms or rules before signing up for a paid subscription plan. Some providers put a limit on the size of the file that can be uploaded. You may also find that if you have numerous files in your online storage account, syncing those files may take longer. Relying on a cloud service provider is not a viable option for people who do not have consistent access to high-speed internet service.

**Table 3.3** Five Cloud Storage Providers

| Cloud Storage Provider | Free Storage Available |
|---|---|
| Dropbox | 2 GB |
| iCloud | 5 GB |
| Box | 19 GB |
| pCloud | 20 GB |
| IDrive | 5 GB |

## Optical Discs

An **optical disc** is a compact disc (CD) that is recordable or rewritable (CD-R, CD-RW), a digital versatile disc/digital video disc (DVD), recordable or rewritable disc (DVD-R, DVD-RW), or Blu-ray disc (BD). Note that the spelling of this storage medium is *disc* (not *disk*). Two formatting standards were approved for DVD drives: DVD-R and DVD+R. A DVD drive that cannot read and write to both disc standards can only use the +R, +RW, or the -R, -RW media. A drive that can accommodate both standards can use media labeled ±R or ±RW. An optical drive is internally mounted inside the system unit. Although fewer PCs are including optical drives these days, some people still use an optical drive to play music CDs or watch movies on DVD or Blu-ray.

Many laptops omit the optical drive to decrease weight and thickness and extend battery life. Optical drives are expected to eventually become obsolete in part because of their low capacity relative to other storage options, and because they are vulnerable to data loss from a scratch or other mishandling of the media.

Optical drives are usually identified with the letter *E* or *F*, although this varies depending on the number of other installed devices. A laser beam is used to read data from and write data to an optical disc. Recording to an optical disc is referred to as *burning a disc*.

Optical discs are sometimes used to back up important data or to make copies of music, pictures, or movies because the removable media can be used in other devices and players are relatively inexpensive. Optical disc types and storage capacities are listed in Table 3.4.

**Table 3.4** Recordable Optical Discs and Storage Capacities

| Type of Optical Disc | Media Options | Storage Capacity |
|---|---|---|
| CD | CD-R (write once)<br>CD-RW (rewritable up to 1000 times) | 700 MB |
| DVD | DVD +R or –R (write once)<br>DVD +RW or –RW (rewritable up to 1000 times) | 4.7 GB for single-layer<br>8.5 GB for dual-layer |
| BD | BD-R (write once)<br>BD-RE (erase and re-record multiple times) | 25 GB for single-layer<br>50 GB for double-layer |

## Topics Review

| Topic | Key Concepts and Key Terms |
|---|---|
| **3.1 The System Unit** | A **peripheral** device is hardware plugged into the computer used for input, output, connectivity, or storage. |
| | A **motherboard** is the main circuit board into which all other devices connect. |
| | Data travels between components on the motherboard and the CPU via a **data bus**. |
| | An **expansion card** plugs into an expansion slot on the motherboard to provide or improve functionality. |
| | **Parallel processing** means the CPU can execute multiple instructions simultaneously. |
| | The **instruction cycle** is the process the CPU carries out to retrieve, decode, execute, and store an instruction. |
| | **Clock speed** is the number of instruction cycles per second the CPU processes, measured in **gigahertz (GHz)**—1 billion cycles per second. |
| | **Level 1 (L1) cache**, **Level 2 (L2) cache**, and **Level 3 (L3) cache** affect system performance, since the CPU accesses cache memory faster than RAM because it is located closest to the CPU. |
| | **Read-only memory (ROM)** stores instructions that do not change, such as the **BIOS (basic input/output system)** or **UEFI (Unified Extensible Firmware Interface)** instructions used to start up a computer. UEFI instructions operate faster and are more secure. |
| | RAM is configured in modules that plug into the motherboard. The size and type of RAM affect system performance. Adding additional RAM modules is one way to extend the life of an older PC. |
| | A **port** is a connector located at the back, front, or side of a computer or mobile device. Ports are used for plugging in external devices. |
| | A **universal serial bus (USB) port** is the most common type of port. |
| | The power supply is where you plug in the power cord. A cooling fan near the CPU prevents the CPU from overheating. |
| **3.2 Input Devices** | A **keyboard** is a device used to input data by typing. Special-purpose keys on the keyboard are used to send commands to the active program running on the computer. |
| | A **stylus** is a digital pointer (or digital pen) used on touch-enabled screens. |
| | A **mouse** detects movement using light technology and moves the pointer in the same direction you move the mouse. Buttons and a scroll wheel on a mouse are used to manipulate objects or send commands. |
| | On a laptop, a **touchpad** or track pointer senses finger movement and taps or finger pressure in place of a mouse. |
| | A **touchscreen** accepts input from finger gestures or a stylus. |
| | A **scanner** uses optical technology to convert text and images into data. |
| | Other readers used to input objects or data are a **bar code reader**, **QR (Quick Response) code**, **magnetic strip reader**, **biometric device**, and **radio frequency identification (RFID) reader**. |
| | Digital cameras, camcorders, and webcams are input devices used to provide pictures or live video to the computer or mobile device. A **webcam** is a video camera built into a video display screen or plugged into the PC as a peripheral. |
| | A **microphone** is used to create data from audio sources and can be combined with **voice recognition technology** to control devices. |
| | Various game controllers, such as joysticks and motion sensors, provide a means to input commands to entertainment systems. |
| | Keypads, ink character recognition systems, AR headsets, and smart textiles that accept input from cuff gestures are other devices used for input. |

*continued...*

| Topic | Key Concepts and Key Terms |
|---|---|
| **3.3 Output Devices** | A **monitor** is a type of screen used to view output. A **video display** connects to a video port on a PC or is integrated into a laptop or mobile device. |
| | Video displays use **liquid crystal display (LCD)** or **light-emitting diode (LED)** technology. |
| | **Organic light-emitting diode (OLED)** displays are thinner, provider better picture quality, and use less power than other video displays. |
| | **Resolution** is the measurement of horizontal and vertical pixels used to render an image. Higher resolutions produce clearer images. A **pixel** is a square with a red, green, and blue color value. |
| | **4K UHD** (ultra-high-definition) is a display technology that uses four times the pixels as high definition. |
| | A **video display projector (VDP)** and an **interactive whiteboard** display computer output on large screens. |
| | Audio output is heard from speakers, headphones, or headsets. A **Bluetooth headset** or Bluetooth speaker is used to listen to audio during voice calls or to listen to music. |
| | The printed copy of computer output is called a **hard copy**. |
| | A **laser printer** uses a laser beam to charge a drum, which causes toner to stick to the paper as it passes through the printer. |
| | An **inkjet printer** forms text and images by spraying drops of ink from ink cartridges onto a page. |
| | Multipurpose printers include printing, scanning, copying, and faxing capabilities in one device. |
| | Blueprints, drawings, and signs are printed on a **plotter** that moves a pen across the page to produce output. Plotters print on large format paper as well as other media such as cardboard or plastic. |
| | A **digital copier** connects to a network and can be used to print output as well as to make photocopies. |
| | Printing high-quality images on photo paper is performed using a **photo printer**. |
| | Most receipts from retail stores are printed on a **thermal printer** that heats coated paper. |
| | A **3-D printer** produces a 3-D object by precisely cutting successive layers of material on top of one another. 3-D printers are used to print everything from replacement parts to synthetic food. |
| **3.4 Network Adapters** | A **network adapter** is any device used to connect to a network and is the interface that allows data to be exchanged between a computer and a modem or router. |
| | To facilitate the exchange of data in a network, a set of rules called a **protocol** is used. |
| | A network connection for a wired connection is called an *Ethernet port*, which refers to the type of cable used for the connection. |
| | A **Wi-Fi adapter** is used to communicate with a wireless network. |
| | The **802.11 protocol** is the name of the standard used for wireless communications in Wi-Fi networks. |
| | An **external wireless adapter** such as a **USB wireless adapter** is used to connect to a wireless network on a PC without an adapter or to connect to a wireless network that uses newer technology. |
| | A Bluetooth adapter is used to provide connectivity within short ranges to other Bluetooth-enabled devices. **Bluetooth 5.0** is a new Bluetooth standard that expands the range up to 800 feet at faster speeds. |
| | Share your smartphone's connectivity with other devices by turning on a feature in the Settings app called *hotspot* or *tethering*. |

*continued...*

| Topic | Key Concepts and Key Terms |
|---|---|
| **3.5 Digital Data** | Computers understand the **binary system**, which uses only 0s and 1s. |
| | The smallest unit in the binary system is a **bit**, which can have one of two values: either 0 or 1. |
| | A group of 8 bits equals a **byte** and represents one character, number, or symbol. |
| | Internet speed is measured in megabits per second, which is 1 million bits transferred per second, or gigabits per second, which is 1 billion bits transferred per second. |
| | A **file** is a permanent copy of a document, photo, or video assigned a **file name** and measured in bytes. |
| | A **kilobyte (KB)** is approximately 1,000 bytes. |
| | A **megabyte (MB)** is approximately 1 million bytes. One MB can store approximately 500 pages of text. Images, music, and videos require many MBs of storage space. |
| | A **gigabyte (GB)** is approximately 1 billion bytes. |
| | A **terabyte (TB)** is approximately 1 trillion bytes. |
| **3.6 Storage Options** | A **storage device** provides the means to read data from and write data to a permanent storage medium and is referred to as a **drive**. |
| | An internal **hard disk** (also referred to as *hard disk drive* or *HDD*) is usually the largest storage device on a standalone PC. Traditional HDDs are magnetic storage media. External HDDs are connected via USB. |
| | A **solid-state drive (SSD)** uses chip-based **flash memory** to store data; it is also more durable and uses less power than an HDD. |
| | Some workplaces require data to be stored on centrally managed network servers, which are considered more efficient and secure than storing business data scattered about the workplace on multiple HDDs. |
| | At home, a wireless media hub lets you store and share data among multiple devices and with other people in the same house. |
| | A **USB flash drive** is a portable flash memory device that is easy to use and comes in all shapes, sizes, and storage capacities. |
| | A **flash memory card** is used in a portable device, such as a digital camera. |
| | Cloud storage providers allow you to store and share files that are accessible from any device with an internet connection. |
| | Some cloud providers allow access to a limited amount of free storage space, with higher capacities available for a subscription fee. |
| | An **optical disc** reads and writes data using a laser beam. Examples include CDs, DVDs, CD-RWs, DVD-RWs, and Blu-ray discs. |

 **Topics Review**

## Chapter 4

# The Operating System and Utility Programs

A computing device needs to have software installed for the device to work. *Software* is the term that describes the set of programs that contain instructions to tell the computer what to do and how to perform each task. **System software** includes the operating system that is designed to work with the hardware that is present. It also includes a set of utility programs that are used to maintain the computer and its devices. The **operating system (OS)** program provides the user interface that allows you to work with the computer, manages all the hardware resources, and provides the platform for managing files and application programs (the programs you use to perform tasks). It is the most important software on your computer, because without an operating system, the hardware and other software programs would not work.

For example, assume you want to sign in to your Facebook account on your PC to create a status post. To do this, you start your computer, launch an internet browser, and connect to the Facebook website. The computer first has to start the OS before you see the interface that allows you to launch the internet browser program, such as Microsoft Edge. Without the OS, the interface would not appear, and the browser would not be able to connect to the internet to load the Facebook page.

In this chapter, you will learn the various OSs available for computers, the typical tasks that OSs perform, and how to use a few utility programs to maintain and troubleshoot your computer.

## Learning Objectives

**4.1** List and describe the major functions of operating system software

**4.2** Identify the operating systems in use for personal computers and describe characteristics that differentiate them from one another

**4.3** Describe the functions of a mobile operating system and list the most common mobile operating systems

**4.4** Explain the purpose of an embedded operating system and the purpose of a cloud operating system, and give an example of each

**4.5** Describe common utility programs used to maintain a computer

**4.6** Recognize and locate tools in the operating system package to solve computer problems

### Watch & Learn
The Operating System and Utility Programs

### Content Online
The online course includes additional training and assessment resources.

## 4.1 Introduction to the OS and OS Functions

The OS manages all the activities within the computer. The OS routes data between the hardware resources and the application programs (Figure 4.1). The OS also starts the user interface and properly shuts down the hardware. The OS manages files that are stored on devices. While you are working, the OS controls the flow of data between memory and the central processing unit (CPU). If you plug a new device into the computer, the OS looks for and installs the software that allows the device to work. Consider the OS the conductor of a symphony; each musician (hardware component and application program) needs the conductor (the OS) to direct when and how to play its piece so the song is performed correctly.

**Read & Learn**

Introduction to the OS and OS Functions

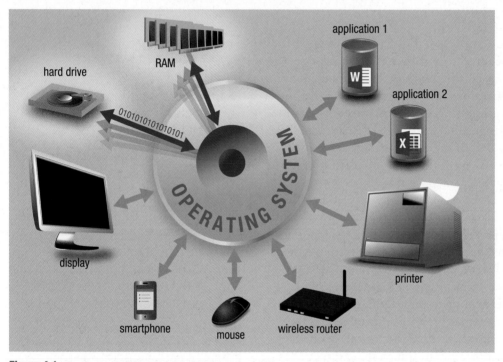

**Figure 4.1**

The OS routes data from RAM to the CPU as well as coordinates the programs and memory that are running.

### Starting and Shutting Down a Computer

Starting a computer is called **booting**. If you turn on a computer that has been shut off, you are doing a **cold boot**. If you restart a computer without turning off the power, you are doing a **warm boot**. The instructions for booting are stored on the basic input/output system (BIOS) or Unified Extensible Firmware Interface (UEFI) chip that resides on the motherboard. At startup, the OS **kernel** is loaded into RAM, where it remains until you turn off the PC. The kernel is the core of the OS that assigns resources and manages memory, devices, and programs. Other parts of the OS are copied into RAM only when needed.

When the boot process is complete, the screen displays the user interface, the lock screen, or a sign-in screen. For example, on a PC running the Microsoft Windows 10 OS, after the lock screen is cleared, a screen prompts you to sign in to use the device. You sign in on a Windows-based PC by using a Microsoft account and password, by using a PIN, or by moving your finger over an image that has been associated with a picture password.

If you decide to turn off the power to your computer, you should always perform a **shutdown** command so that programs and files are properly closed. Depending

**in the know**

The term *booting* is derived from the phrase "to pull oneself up by the bootstraps." The bootstrap program contains the instruction that tells the computer where to find the OS.

on the OS, choose the shutdown function that turns the computer off, or choose a power-saving mode, such as the Sleep command, which saves open documents and programs while turning off unnecessary functions and devices. For example, on a PC running Windows 10, click the Start button, click the Power icon, and then click *Shut down*. On a Mac, click the Apple icon on the menu bar and then click *Shut down*.

Although different OSs are available for computers, most OSs perform similar functions. Most of these functions are performed automatically in the background, often without the user's awareness.

## Providing a User Interface

The **user interface (UI)** is the means to interact with the computer. Most OSs provide the user with a **graphical user interface** (**GUI**, pronounced *gooey*) that presents visual images, such as tiles, icons, or buttons, that you click or double-click with a mouse to tell the OS what you want to do. On a touchscreen, tap or double-tap the tiles, icons, or buttons to perform a command.

The OS also allows for a command to be performed by pressing a key or a combination of keys on the keyboard. For example, on a Windows-compatible PC, pressing the Windows key located left of the spacebar opens the Start menu. On a Mac, the Command key on the keyboard is used in combination with a letter to perform a command.

In Figure 4.2a and Figure 4.2b, the Windows 10 user interface is presented to show the Start menu interface and the use of icons on the desktop/taskbar to start an application or open a window.

Sometimes a **command-line interface** (Figure 4.2c) is used to interact with an OS. In a command-line interface, you type commands on the keyboard at a prompt to tell the OS what to do.

A command-line interface is often used by computer service technicians or network specialists to configure or troubleshoot a PC or network device.

a

b

c

**Figure 4.2**

(a) To open the Windows 10 Start menu interface, click the Start button.
(b) In Windows 10, icons on the desktop or the taskbar can also be used to start applications. (c) In a command-line interface, commands are typed at a prompt, such as C:\Users\Denise> shown here.

## Managing Application Programs and Memory, and Coordinating the Flow of Data

The OS manages the installed application programs and provides you with a utility to install a new program or remove (uninstall) an existing program. When working on a PC, people often have more than one program running at the same time and switch back and forth between open tasks. The OS manages computer resources, allocating data and software to and from the CPU and memory as needed, and determines the order in which tasks are performed.

The OS also manages the flow of data between the CPU, cache memory, RAM, and input and output devices. If the number of programs in operation exceeds the capacity of RAM, space on the hard disk drive stores overflow data. The hard disk drive space allocated to store RAM contents is called **virtual memory**. In Windows, the Task Manager displays information about how the OS manages and coordinates computer resources. Right click the Windows taskbar and then click the *Task Manager* option at the menu that appears to open the Task Manager window (Figure 4.3).

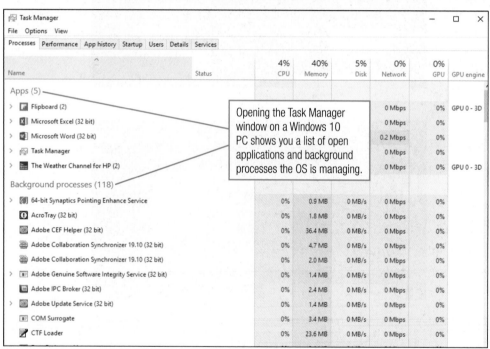

**Figure 4.3**

The Processes tab on the Windows 10 Task Manager shows the many tasks the OS is managing while you are working.

## Configuring Hardware and Peripheral Devices

The OS configures all devices that are installed on or connected to the computer. A small program called a **driver** contains the instructions the OS uses to communicate with and route data to and from the device. When the computer is booted, the OS loads each device driver. If you plug in a new peripheral after the computer is started, the OS searches the device's internal storage for the driver software, loads the software automatically into RAM, and displays a message that the necessary files are being installed. This routine is called **plug and play**. If a device's driver software is not found, the OS displays a message box that prompts you to install the driver from another source, such as a CD, or from a folder on your hard disk drive.

The OS is also used to manage hardware once the PC is operational. You use the OS interface to adjust the screen's resolution, increase or decrease the volume for sound from the speakers, or change the notifications that appear. You can also change the default action that occurs when an external device is plugged in, such as a digital camera. For example, the OS can be instructed to automatically copy the pictures from the camera's internal flash memory card to a folder on the PC, or open a window for you to select and copy the pictures yourself.

You can manage the power plan for the computer, including the time interval before the display is turned off or the computer is put into sleep mode when no activity is detected. These examples of the hardware settings you can change using the OS are just a few of the many hardware options the OS manages.

## Providing a File System

The OS keeps track of the files stored on storage devices and provides tools to find and manage those files by moving, copying, deleting, or renaming them. Utility programs to search, back up, restore, and clean up storage space are also included in the OS package. To open the Windows 10 File Explorer window (Figure 4.4), click the File Explorer button on the taskbar. The window displays buttons to perform file management.

**- - - - in the know**

On a Mac, files are managed in the Finder window, accessed by clicking the Finder icon on the dock (bar along the bottom of a Mac screen).

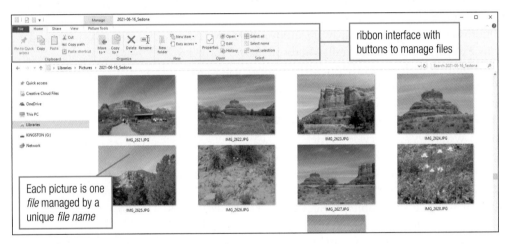

**Figure 4.4**

On the Home tab in a Windows 10 File Explorer Window, click buttons (such as Cut, Copy, Paste, Move to, Copy to, Delete, and Rename) to manage files.

## Managing Software Updates and Security

Software is updated when OS program fixes, security enhancements, and new or modified device drivers become available. By default, the OS is set up to download

and install updates automatically. Click the Start menu button on the Windows taskbar (it's the Windows button at the lower-left corner on the taskbar). Click *Settings* to open the Settings app. Click the Update & Security icon to change the Windows Update options (Figure 4.5).

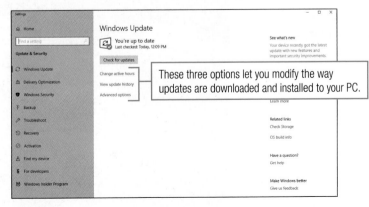

**Figure 4.5**

The Windows Update pane of the Settings window allows you to change the hours you use the device so that a restart does not occur while you're using the PC—the option is called *Change active hours*.

## 4.2 Popular OSs for Computing Systems

When you purchase a new computing device, an OS is preinstalled so the device will work when you turn it on. The OS that is preinstalled may depend on the hardware, since some hardware is designed specifically for a particular OS. The term **computing platform** refers to the combination of hardware architecture and software design that allows applications to run. For example, buying an iMac computer means you will have the macOS installed because Apple computers are not designed to run Windows software. Four OSs typically installed on work, school, and home computers are Windows, macOS, Linux, and UNIX. You may also encounter a network OS on a PC.

### Windows

**Windows**, created by Microsoft Corporation, is the most popular OS because a wide variety of hardware is designed to run Windows and because software applications designed for Windows are plentiful. Windows has evolved from Windows 1.0, released in 1985, to Windows 10, released in 2015.

Windows is updated as Microsoft keeps up with hardware innovations, adds new functionality, and/or redesigns existing functions. Overall, Microsoft strives to make the OS faster and the interface more user-friendly. Before 2015, a new release was assigned a version, such as Windows 7, to differentiate the OS; however, with the release of Windows 10, Microsoft adopted a new strategy. Instead of releasing a new version of Windows every few years, it issues periodic updates to introduce new features and technologies in the Windows as a service model.

Each Windows version generally comes in a variety of editions for various home and business environments. For example, Windows Server is used by businesses that need advanced security and management for networks.

With the popularity of digital assistants on mobile devices that interact with users using voice commands, Windows 10 includes **Cortana**, the Windows personal digital assistant. Cortana can search the web, locate files on your device, and remind you of meetings or events in your calendar. Like other digital assistants, Cortana responds to voice commands, but can also work from keyboard input. Cortana keeps track of your likes and responses to alerts, and uses that information to personalize your interactive experience.

### macOS

The **macOS** is the OS created by Apple, Inc., with the first release in 1984 for the Macintosh computer. The first Macintosh computer used a GUI interface and became the basis for the design of other GUI OSs down the road.

Apple's macOS user interface with the Mojave version on a Macbook is shown in Figure 4.6. The user interface is easily recognizable by its dock (bar along the bottom of the screen) with icons to launch applications and menu bar along the top of the screen. Mojave introduced

**Read & Learn**
Popular OSs for Computing Systems

**Video**
How Has Windows Evolved into the Feature-rich Program of Today?

**Video**
Getting the Most Out of Cortana

**●— — — — in the know**
In July 2018, Statista (a market and consumer data company) reported that 83 percent of world-wide desktop PCs used the Windows OS.

**●— — — — — go online**
https://CC3 .ParadigmEducation .com/MicrosoftHistory
Go here to view a screen-shot and read a brief summary of each version of Windows from version 1.0 to Windows 10.

**Figure 4.6**

Mojave, the macOS released in 2018, introduced a darker color scheme.

Dark Mode, a desktop with a dark color scheme that focuses on content with controls receded into the background.

Apple uses a unique naming convention for its OS, with a version sporting the name of a popular place—El Capitan in 2015 (the name of a rock formation in Yosemite National Park), and Sierra and High Sierra (the name of a mountain range in California) in 2016 and 2017 respectively. Like Microsoft, Apple releases new versions to update functionality, responsiveness, and efficiencies to strive for faster, more user-friendly experiences. Server versions of macOS are also available.

Mac computers have always been popular with graphic artists and those working in multimedia environments because the hardware is designed with high-end graphics processing capability. Consumer demand for mobile products from Apple, such as the iPhone and iPad, have increased brand presence of Macs in the marketplace.

## Linux

**Linux**, which is based on UNIX, was created by Linus Torvalds in 1991. Linux can be used as a standalone OS or as a server OS. What makes Linux preferable to some users is that the OS is an **open-source program**, meaning that the source code is available to the public and can be freely modified or customized. Linux, like UNIX, can run on a variety of devices and is available as a free download on the web, as well as by purchase from Linux vendors. Because of its open-source design, Linux is available in many distributions (versions), such as Red Hat, Ubuntu, and SUSE. The open-source software movement is increasing in large part owing to the lower costs associated with using free software. Linux can be found in use at companies such as IBM and HP, and a popular mobile OS (Android) is based on Linux.

## UNIX

**UNIX**, an OS that has been around since the late 1960s, is designed for servers. Originally designed to be multi-user and multitasking, the OS is often used for web servers because of its robust ability to support many users. UNIX is available in several versions and can run on a variety of devices, including hardware built for Windows and Macs. In fact, the macOS from Apple is based on the UNIX OS. Although UNIX is a highly stable OS, the cost and higher level of skill required to install and maintain a UNIX computer make it less prevalent for many computer users.

## Network OS

A **network operating system (NOS)** incorporates enhanced functions and security features for configuring and managing computers and peripherals connected in a networked environment. Some PC OSs, such as Windows and macOS, have basic networking features built in that let you connect and share files and devices within a home network. An NOS is used in the business world, where more advanced security and management features for connecting, sharing, and providing access to multiple resources are essential for maintaining large networks. An NOS must also interface with other networks that may be using a different NOS. For example, a request received in a Linux network might come from a Windows-based network.

● – – – – in the know

An open-source project gaining traction is an OS based on Android (the OS on many smartphones) called *Android-x86*. Users looking to use their favorite apps and games on a larger screen with mouse and keyboard support are driving this new OS for the PC.

 **Video**

What Types of Tasks Does Network Operating Software Handle?

## 4.3 Mobile Operating Systems and Wearable OSs

A variety of mobile devices, such as smartphones and tablets, use a **mobile operating system**, an OS designed specifically for a mobile device. The OS is stored on a ROM chip. This type of OS is designed to manage hardware and applications with less memory than a personal computer OS yet run fast and efficiently. The mobile OS often manages input from a touchscreen and/or voice commands, and routes data to and from wireless connections including cellular, Bluetooth, and Wi-Fi networks. Mobile devices also have cameras that are used for video calling and taking pictures. Mobile OSs manage small application programs, called *apps*, with support for multitasking among contacts, messages, calendars, music, video, social media updates, web browsing, and numerous other mobile tools. Several hardware manufacturers produce mobile devices, and some have developed their OS for use on their hardware. The two leading mobile OSs are Android and iOS.

The mobile market changes rapidly as wireless technologies and software is developed to take full advantage of faster speeds and capabilities. Competition in the industry is fierce, and companies that lag in new development quickly run into trouble. For example, BlackBerry dominated the mobile market in the early years of smartphones but saw its share of the market decline to less than 1 percent by 2018. Most analysts believe BlackBerry's long delay in releasing a device with touch-enabled multimedia, web browsing, and apps similar to Apple and Android offerings were the reason for BlackBerry's loss of market share.

A wearable device, such as a smartwatch, needs an operating system too. Wearable devices typically use an adapted version of a mobile OS called a **wearable OS**. A wearable OS manages tasks and data similar to those on a smartphone.

### Android

**Android** is the leading worldwide mobile OS, used on many smartphones and tablets (Figure 4.7). Although the global leader, within the United States market, Android is second to Apple's iOS according to market research by StatCounter in May 2018.

Android is based on Linux, making it an open-source OS. In 2007, Android was released by the **Open Handset Alliance**, a consortium of 84 technology and mobile companies. A sampling of alliance members includes Google, Samsung, LG, T-Mobile, and Sprint. The goal of the alliance is to develop open standards for mobile devices. Android is now maintained and

**Figure 4.7**

Android is the OS of choice for mobile devices, including Samsung smartphones.

developed by the **Android Open Source Project (AOSP)**, which is led by Google. Each Android release is named after a dessert. The version in use at the time of writing is Android 8.0 Oreo™.

With the highest worldwide market share for mobile devices, it stands to reason that software developers favor the Android OS when developing new apps. Google Play, the website for Android apps, listed more than 3.6 million apps in March 2018.

**Read & Learn**

Mobile Operating Systems and Wearable OSs

**Video**

How Will 5G Change Our Mobile World?

●- - - - — in the know

The widespread use of smartphones and tablets has resulted in employees bringing their personal devices to the workplace, connecting to the company's network, and even using their own devices to do company work. This trend is called *BYOD—bring your own device*.

## iOS

Apple developed **iOS** as the mobile OS for its iPhone (Figure 4.8), iPod, and iPad. Based on macOS, the mobile OS boasts over 54 percent of the market in the United States.

The first iPhone was released in June 2007. Since then, Apple has released a new iPhone model each September when an updated mobile OS is released.

Users can download apps for their Apple devices only from the App Store on their iPhone or iPad, or from iTunes on a Mac. Development of new apps for Apple devices is tightly controlled by Apple, meaning that new apps are generally virus-free, stable, and reliable.

In the first quarter of 2018, over 2 million apps and games were available for Apple devices from the App Store and iTunes.

**Figure 4.8**

The Apple iPhone uses the iOS interface.

## Wearable OS

Wearable gadgets such as smartwatches, smart rings (Figure 4.9), fitness trackers, smart glasses, and AR headsets vary in size and capability depending on their purpose. Most operate many of the same resources that a smartphone does. A wearable OS manages input, data, processing, connectivity, and notifications within a smaller environment than a smartphone and without sacrificing performance and battery life.

This is an emerging field that changes rapidly. In late fall 2017 and in 2018, technology writers were abuzz about a new wearable OS in development at Apple, called *rOS*, that is based on iOS. The new wearable OS would be used in AR products such as AR glasses or an AR headset. At the time of writing, consumer availability of these new AR devices is estimated to be 2020 or 2021. Table 4.1 lists three common wearable OSs used in devices on the market in 2018.

**Figure 4.9**

Some wearable devices use a proprietary OS—an OS developed by the manufacturer. This smart ring for fitness, heart rate, and sleep tracking made by MOTIV in San Francisco has a proprietary OS.

**Table 4.1**  Wearable OSs in Devices Used in 2018

| Wearable OS | Description |
| --- | --- |
| Wear OS | the Google-supported, open-source OS used in some smartwatches and fitness trackers |
| Tizen | an open-source OS used in some Samsung smartwatches |
| watchOS | the Apple OS used in the Apple smartwatch (based on iOS) |

## 4.4 Embedded OSs and Cloud OSs

**Read & Learn**
Embedded OSs and
Cloud OSs

Computing consoles such as automated tellers, GPS navigation systems, video game controllers, medical devices, point-of-sale cash registers, digital cameras, and a multitude of other consumer and commercial electronics need a specialized OS designed for the device's limited use. In these computing devices, an **embedded operating system** is installed. Embedded OSs are smaller and interact with fewer resources to perform specific tasks fast and reliably. Smart TVs use an embedded OS referred to as a *TV OS* to present a UI and manage the streaming content options.

A cloud OS is a viable option for people with multiple devices who want seamless synchronization of software and documents across all devices wherever they may be (within range of a Wi-Fi network).

### Windows IoT

**Windows IoT** is an OS based on Windows 10 optimized for the smart devices that make up the IoT. One edition in this family of embedded OSs is Windows 10 IoT Core. It is designed for constant connectivity to the cloud, as well as managing, updating, and securing IoT devices and apps. The embedded OS is used for digital signage, wearable devices, smart home appliances, ATMs, interactive kiosks or furniture, and medical devices. Figure 4.10 shows an interactive table running the Windows 10 IoT embedded OS that lets patrons use multi-touch drag-and-drop ordering software and play games while waiting for food.

**Figure 4.10**
This interactive restaurant table developed by Kodisoft uses Windows 10 IoT to facilitate tabletop ordering, playing games, and social media, news, and weather apps.

### Embedded Linux

**Embedded Linux** applications can be found running smart appliances, in-flight entertainment systems, personal navigation systems, and a variety of other consumer and commercial electronics. Similar to the standalone Linux OS, Embedded Linux has several distributions because it is an open-source program.

### Embedded Android

Android is not just for smartphones and tablets. Some home automation systems that control lighting, heating, air conditioning, and interactions with smart appliances, such as washers or dryers, are using an embedded version of Android. Variations of Embedded Android includes Android Things for IoT devices, Android Auto for a console and navigation system in a vehicle (Figure 4.11), and Android TV for a smart TV or game console.

**Figure 4.11**
Android Auto is the embedded OS running this 12-inch touchscreen console in a 2019 Ram truck.

## Windows Embedded

**Windows Embedded** is a family of OSs in various releases still in use in many existing devices. The OS is designed for use in task-specific devices that need to present a UI and manage input, output, and limited resources. Examples are Windows Embedded in an ATM and Windows Embedded Compact in a digital picture frame. Following the release of Windows IoT, Microsoft is expected to eventually phase out Windows Embedded software.

## TV OS Embedded in Smart TV

TVs have evolved into entertainment hubs with a variety of choices for viewing content—traditional cable and satellite signal programming, streamed music, videos, movies, news, and more. Inside smart TVs, an OS is embedded that is a **TV operating system**. The TV OS is the program that runs the set-top box, runs other connected devices, presents the interface for accessing content, and allows the viewer to navigate channels or streamed media. A TV OS uses apps to connect to resources, such as Netflix and YouTube, and accepts input from the remote control device or via wireless keyboard, smartphone, or tablet. A variety of TV OSs are in use, such as the LG webOS, Apple tvOS (Figure 4.12), and Panasonic Firefox OS and My Home Screen 2.0.

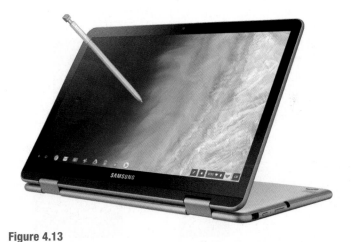

**Figure 4.12**

Version 12 of Apple's tvOS, released in June 2018 for Apple TV 4K, supports Dolby audio for streaming entertainment apps.

## Cloud OS

Google has developed an OS that operates as a virtual desktop (as have other companies). The interface looks and feels like a traditional desktop, but your settings, documents, pictures, contacts, and other files are stored online. A **web-based operating system** is not a true OS like Windows or macOS, because you still need a standalone OS on the PC to start the computer and access the web browser. The appeal of these systems is that extra software for applications does not have to be installed, and synchronizing documents among multiple devices is appealing to people who use more than one device in various locations.

**Figure 4.13**

The Google Pixelbook runs Chrome OS with all settings, applications, and documents stored on cloud servers using Google apps.

The Google **Chrome OS** is a Linux-based OS available on specific hardware called *Chromebooks*, manufactured by Acer, Asus, Toshiba, HP, Dell, and Google (Figure 4.13). Chromebooks boast fast startup with built-in virus protection, automatic updates that promise not to slow down your device over time, and cloud-based applications, such as Google Docs, Gmail, Google Drive, and Google Cloud Print. Chromebooks are popular in schools because of their low cost and simple interface.

Gartner Research forecasted that market share of Chromebooks (small compared to other PCs) is rising with almost 12 million units expected to be sold in 2018.

 **Blog This!**

Is the Future Bright for Web-Based OSs?

 **Job Connect!**

Careers in the Cloud

# 4.5 OS Utilities for Maintaining System Performance

Just as a car needs regular tune-ups to keep it running smoothly and efficiently,
so does a computer. The computer should be regularly maintained by removing
unwanted files and regularly backing up important files to prevent loss of data. File
maintenance utilities often include a file manager, a utility to clean up unwanted files,
a disk defragmenter, and a backup and restore program.

Utilities that protect your computer from malware, such as viruses and spyware, are
discussed in Chapter 7, Computer Security and Privacy.

## File Manager

Open a **File Explorer** window to perform file management tasks, such as moving,
copying, deleting, and renaming files. You also use File Explorer to set up your filing
system by creating a new **folder**, which is a placeholder where you store related
files. A folder on a computer is similar to a paper file folder you use in an office for
paper documents you want to keep together. Open File Explorer in Windows 10
(Figure 4.14) by clicking the File Explorer icon on the Windows taskbar. Click to
select a drive and/or folder name in the left pane to view the list of files stored in
that location in the right pane—called the *content pane*. Click a file in the content
pane and then click the button on the ribbon along the top of the window to
perform the required task, such as Rename or Delete. Options on the View tab are
used to change the ways files are displayed, such as in a list format or by extra large
icons as shown in Figure 4.14.

In macOS, the **Finder** utility is used to perform file management routines. Open a
Finder window by clicking the Finder icon on the dock.

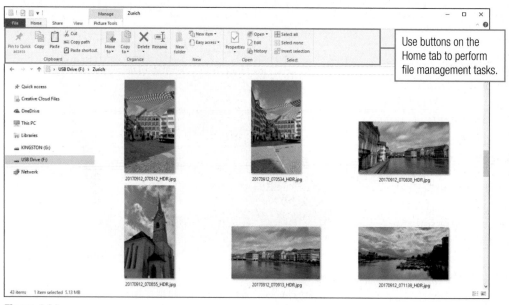

**Figure 4.14**

File maintenance tasks are performed in a File Explorer window on a Windows PC.

## Freeing Up Disk Space

Use the Windows **Storage sense** feature to have Windows free up space on the hard disk drive by deleting files you don't need to keep, such as temporary files and files in the Recycle Bin (deleted files still kept until permanently removed). You can have Storage sense do the maintenance automatically by turning the feature on, or you can choose to run the feature manually. Storage sense only deletes files automatically when the PC becomes low on disk space. Complete these steps to turn on Storage sense:

1. Click the Start button and then click *Settings* to open the Settings app.
2. Click the System icon in the Settings window.
3. Click the *Storage* option in the left pane of the Settings window.
4. Drag the Storage sense toggle switch right to *On* (Figure 4.15).

To free up space manually, click the <u>Change how we free up space</u> hyperlink, change options as needed in the next window, and then click the Clean Now button.

Another way to free up disk space in Windows is with the **Disk Cleanup** utility. Disk Cleanup was used in versions of Windows prior to Windows 10 and is still included in the OS as a desktop app. By default, Disk Cleanup selects downloaded program files and temporary internet files for deletion. To open the Disk Cleanup desktop app, click in the search box on the Windows taskbar, type *disk cleanup*, and then press Enter. Select the drive to be cleaned in the first dialog box, and then choose the types of files to be deleted in the second dialog box by adding and removing check marks next to types of files. Click *OK* to begin the process when you are ready to clean up the selected hard disk drive.

☁ **Let's Explore!**

A Computer Maintenance Checklist

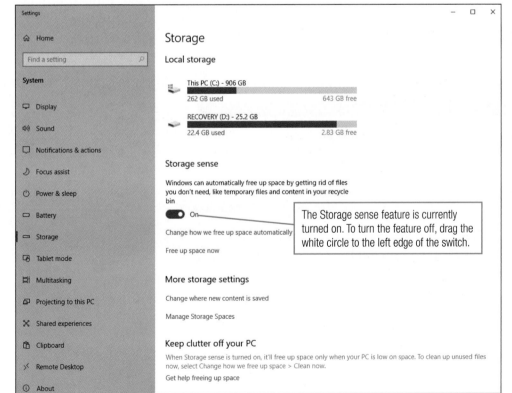

**Figure 4.15**

The Windows Storage sense feature in the Settings app helps remove unnecessary files that clutter a hard disk drive.

## Disk Defragmenter

As time passes, a disk can become fragmented as a result of files being moved or deleted. Files are stored on a disk in groups called *clusters*, and several clusters are sometimes needed to store one document. Sometimes the clusters needed for a document are not stored adjacent to one another. This is called a **fragmentation**. When you reopen a document that has been fragmented, the OS has to gather back together all the fragmented clusters and arrange them in the correct order. As more fragmentation occurs, disk efficiency decreases. A **disk defragmenter** utility is used primarily to rearrange the fragmented files back together to improve file retrieval speed and efficiency. The Windows **Optimize Drives** window (Figure 4.16) performs disk defragmentation manually. Complete these steps to open the Optimize Drives utility:

1. Click in search text box on the taskbar and then type *optimize drives*.

2. Click the *Defragment and Optimize Drives* option in the search results list.

The Disk Defragmenter utility is used primarily by someone with an older PC. Today's hard disk drives are much faster and larger than the HDDs in older computers. If your computer was bought recently and has a large hard disk drive with lots of free space available, or if the PC has a solid-state drive, you do not need to run the Disk Defragmenter utility.

The macOS file system works differently than the file system used on a Windows PC. Mac users do not need to defragment a disk because a Mac automatically defragments files on its own. For those rare instances when a Mac user needs to defragment the hard disk, a third-party utility program is needed.

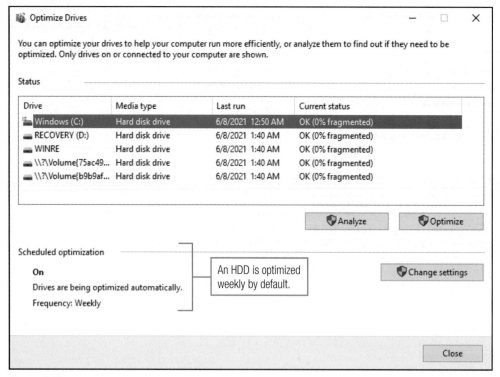

**Figure 4.16**

The Windows Optimize Drives utility is used to analyze a drive for the percentage of fragmentation, optimize the drive, and change optimization settings.

## Backup and Restore Utility

All computer users should back up important files regularly to prevent loss of data from disk corruption, accidental deletion, or other damage that keeps the files from being usable. A **backup utility** is a program that allows you to back up selected files or an entire disk to another storage medium. The **Backup** feature in Windows 10 can be set to perform automatic backups to an external disk drive or a network drive. Complete these steps to set up an automatic backup at regular intervals:

1. Click the Start button and then click *Settings*.
2. Click the Update & Security icon in the Settings window.
3. Click the *Backup* option in the left pane of the Settings window.
4. Click the *Add a drive* option in the Backup pane (Figure 4.17).
5. Click the drive you want to use for the location of backups.

By default, backups are performed hourly. Click the More options hyperlink on the Backup pane to make changes to the time intervals, to add or exclude folders from the backup process, or to perform a manual backup.

Files backed up using a backup utility are not stored in a readable format because they are compressed to save space. In order to copy files back to a disk drive in their original state, the files need to be restored. During a restore, a file or group of files is decompressed. To restore files, display the Backup pane in the Settings window, click the More options hyperlink, scroll down the page if necessary, and then click the Restore files from a current backup hyperlink. Choose the files you want to restore in the File History window and then click the Restore button.

In macOS, the **Time Machine** utility is used to back up and restore files. When you plug in an external hard disk to a Mac, you are prompted in a dialog box to use the disk as the backup storage for Time Machine. Click the Use as Backup Disk button in the dialog box and macOS will start automatic backups every hour for the past 24 hours, daily for the past month, and weekly for all previous months. The oldest backups are deleted as needed when the external drive becomes low on space.

**in the know**

In the IT community, the Backup 3-2-1 rule is considered a best practice. Simply stated the rule is this: Have at least three copies of your data (original plus two backups); keep the backups on two different media (e.g., one on an external hard drive, one on a USB); and keep one copy of a backup off site.

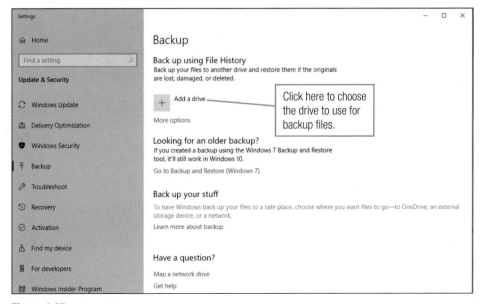

**Figure 4.17**

In Windows 10, the Backup pane in the Settings app is used to choose the drive to save files to or change backup options.

## 4.6 OS Troubleshooting Tools

In a perfect world, the printer would always print, the connection to the internet would always work, and your computer would always operate at blazing fast speed! Causes of computer problems can be complex and not easily resolved. However, knowing how to use tools provided in the OS package that can help you fix some problems will help you get back to work.

### Windows 10 Troubleshooters

To help solve problems with a Windows 10 PC, display the Troubleshoot pane in the Settings app (Figure 4.18): Click the Start button, click *Settings*, click the Update & Security icon, and then click *Troubleshoot* in the left pane. Click the troubleshooter you need, and then click the Run the Troubleshooter button. Depending on the tool you select, a dialog box appears with a progress bar, and messages display as the troubleshooter runs through a series of checks to detect the problem and search for solutions. Solutions will vary depending on the type of problem you are experiencing.

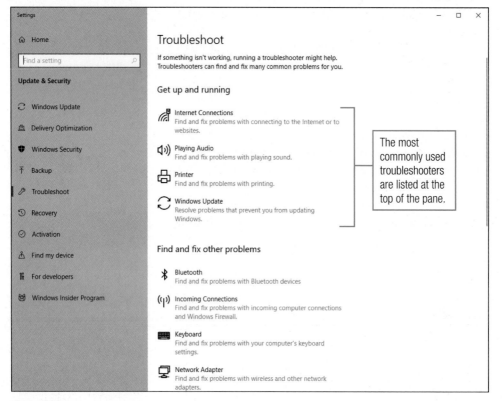

**Figure 4.18**

The Troubleshoot pane in the Windows 10 Settings app provides several troubleshooters that you can use to find and fix problems encountered on your PC.

### Resetting a Windows PC

If your PC is not working well and you were unsuccessful using troubleshooters to find and fix the problem, you may need to use the **Reset this PC** utility. Resetting the PC reinstalls Windows 10. Navigate to the Recovery pane within the Update & Security settings in the Settings app and then click the Get started button in the *Reset this PC* section. On the next screen, choose either the *Keep my files* option or the *Remove everything* option. Click *Keep my files* to remove apps and drivers that you have installed and changes made to settings, but all of your data remains intact.

**Read & Learn**

OS Troubleshooting Tools

●– – – – in the know

According to a help desk software provider, three of the top ten calls to a help desk call center are for a forgotten password, slow internet connectivity, and a file deleted by mistake.

Sometimes you notice a problem occurs immediately after you install something new like a device driver, an application, or an update. Or you may accidentally delete or write over a system file and as a result, cause a program or device to become unusable. In this case, you can use the **System Restore** utility to roll back the PC to the way it was before the problem. Complete these steps to open the System Restore utility:

1. Click in the search text box on the taskbar and then type *system restore*.

2. Click the *Recovery* option in the *Settings* section in the search results list to open the control panel.

3. Click the <u>Open System Restore</u> hyperlink in the Recovery Control Panel window.

4. Click Next in the first System Restore window to start the process of undoing changes made to the PC.

5. In the second System Restore window, select the restore point from the event list and then click *Finish* to begin the restore process (Figure 4.19).

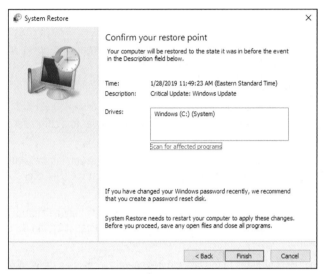

**Figure 4.19**

Use System Restore to undo a change made recently to the PC that has caused the PC to stop working well.

 **Job Connect!**

Careers in Technical Support

**Let's Explore!**

Common Computer Problems

## Help and Support

All OS software includes extensive help and support. In Windows, click in the search text box on the taskbar, type *help*, and then click the **Get Help** app in the search results list. A Get Help window with a Virtual Agent opens. Type a brief question in the bottom of the window and press Enter. The Virtual Agent responds with information about the issue or hyperlinks to related topics. Figure 4.20 shows a Get Help session for an internet connectivity problem.

You can also locate a solution by typing a phrase in the search box on the Windows taskbar and then clicking a hyperlink to a website in the search results list.

On a Mac, click *Help* on the menu bar, type a search phrase in the search text box, and then select a topic in the results list that appears, or choose *Show All Help Topics* to browse help in a window.

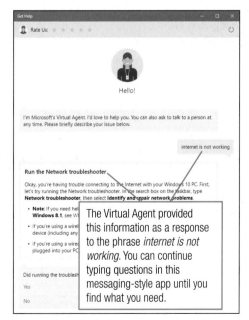

The Virtual Agent provided this information as a response to the phrase *internet is not working*. You can continue typing questions in this messaging-style app until you find what you need.

**Figure 4.20**

Windows 10 provides a virtual agent to locate information for you in Microsoft's support resources.

● ━ ━ ━ ━ ━ go online

**https://CC2.Paradigm College.net/Microsoft Support** and

**https://CC2.Paradigm College.net/Apple Support**

Go to the support website for Microsoft or Apple to find help by searching for articles, fixes, downloads, or other resources. Each company also hosts a community forum at its support website, where users share tips and solutions.

## Topics Review

| Topic | Key Concepts and Key Terms |
|---|---|
| **4.1 Introduction to the OS and OS Functions** | **System software** consists of the **operating system (OS)** plus utility programs used to maintain a computer. |
| | OS software functions include: starting and turning off the PC; providing a user interface (UI); managing the programs, memory, data and hardware devices; keeping track of files; and managing software and security updates. |
| | Starting a computer is called **booting**; starting from a no-power state is a **cold boot**, and restarting a powered-on computer is a **warm boot**. |
| | At startup, the OS core called the **kernel**, which manages memory, devices, and programs, is loaded into RAM. |
| | Use the **shutdown** command to turn off the power to a computer so that programs and files are properly closed. |
| | The **user interface (UI)** is the method with which you interact with the computer. A **graphical user interface (GUI)** uses imagery, such as icons, that the user clicks. With a **command-line interface**, commands are typed at a prompt. |
| | The OS coordinates the flow of data and activities between hardware resources and application programs. For example, when RAM has reached capacity, the OS uses hard disk space, called **virtual memory**, to store the overflow RAM contents. |
| | The OS configures hardware and peripherals by using a driver. A **driver** is a small program with instructions that tell the OS how to communicate with a device. With **plug and play** capability, a driver is automatically installed by the OS when a new peripheral is connected to the PC. |
| | The OS keeps track of all files stored on a system and provides the tools with which to manage those files. |
| | The OS performs automatic updates for program fixes, security enhancements, and new or modified drivers. |
| **4.2 Popular OSs for Computing Systems** | A **computing platform** is a combination of hardware architecture and software design for a specific OS. |
| | **Windows** is the most popular OS used on PCs. A variety of hardware is designed to run Windows and Windows applications. |
| | Windows 10, the most current version of the OS, is updated periodically by Microsoft under the Windows as a service model. |
| | **Cortana**, in Windows 10, is a personal digital assistant that responds to voice commands or keyboard input to help you work and find information. |
| | **macOS** is the OS used on Apple's Mac computers. macOS is easily recognizable by the dock along the bottom of the screen. Mojave, a version released in fall 2018, introduced Dark Mode, a darker color scheme designed to put the focus on content. |
| | **Linux**, created in 1991, is an **open-source program** (meaning that the OS can be freely modified or customized) that runs on a variety of devices. |
| | **UNIX**, developed in the 1960s for servers, is a popular choice for web servers. |
| | A **network operating system (NOS)** has enhanced functions and security features for managing networked environments. An NOS must also be able to interface with different types of networks that may use a different OS. |

*continued...*

| Topic | Key Concepts and Key Terms |
|---|---|
| **4.3 Mobile Operating Systems and Wearable OSs** | A **mobile operating system** is stored on a ROM chip, handles smaller hardware and apps with less memory, and includes support for cameras, touchscreen and/or voice commands for input, and video calling. |
| | A **wearable OS** is installed on a wearable device and is usually an adapted version of a mobile OS. |
| | **Android**, the leading worldwide mobile OS, was released by the **Open Handset Alliance**, a collaboration of 84 technology and mobile companies that develops open standards for mobile devices. |
| | The **Android Open Source Project (AOSP)** is now tasked with maintaining and developing Android. |
| | **iOS** is the mobile OS based on macOS developed by Apple for the iPhone, iPod, and iPad. |
| | Wearable gadgets such as smartwatches, smart rings, and fitness trackers have an OS that operates many of the same resources as a smartphone but within a smaller environment and without sacrificing battery life. |
| **4.4 Embedded OSs and Cloud OSs** | Consumer and commercial electronic devices use an **embedded OS** designed for their limited use. |
| | **Windows IoT** is based on Windows 10 and is optimized for smart IoT devices. |
| | **Embedded Linux** is similar to standalone Linux and is used in a variety of consumer and commercial electronics, such as smart appliances and in-flight entertainment systems. |
| | Android is used as the embedded OS for many electronic devices and includes releases such as Android Things for IoT devices, Android Auto, and Android TV. |
| | **Windows Embedded** is a family of embedded OSs still in use in several task-specific devices, but is expected to eventually be phased out by Microsoft following the release of Windows IoT. |
| | Smart TVs use an embedded OS called a **TV Operating System** to present a UI, manage streaming content, and manage input from a remote control, wireless keyboard, smartphone, or tablet. |
| | A **web-based operating system** is a virtual desktop for accessing files and apps anywhere via the internet. |
| | **Chrome OS** is a web-based OS on Chromebooks that uses Google's apps to store data on cloud servers. |
| **4.5 OS Utilities for Maintaining System Performance** | **File Explorer** for Windows and **Finder** for Mac are utilities for file maintenance, such as copying, moving, or deleting files. Use these utilities to create a **folder**—a place for storing associated files. |
| | **Storage sense** and **Disk Cleanup** are Windows utilities that remove files to free up disk space. |
| | **Fragmentation** occurs when file clusters for the same file are not stored next to each other. A **disk defragmenter** utility, such as Windows **Optimize Drives**, increases disk speed and efficiency by putting file clusters back together. Newer computers with faster and very large hard disks or solid-state drives do not need to be defragmented. |
| | A **backup utility**, such as Windows **Backup** or Mac **Time Machine**, is used to back up files and restore backed-up files to their original state. |
| **4.6 OS Troubleshooting Tools** | Windows includes several troubleshooters that scan the hardware and software to detect and resolve PC problems. Run a troubleshooter by displaying the Troubleshoot pane from the Update & Security options in the Settings app. |
| | Use the **Reset this PC** option to reinstall Windows 10 if a troubleshooter cannot solve an issue. Using this utility lets you keep your data and changes to settings intact, but apps and drivers are removed. |
| | **System Restore** is a Windows utility that reverses a recent system change. |
| | The Windows **Get Help** app and Mac Help are used to find answers to questions or issues. |

**Topics Review**

## Chapter 5

# Application Software

**W**hen you turn on a computing device and start an application for work, school, personal tasks, or entertainment, it is application software that you use to get the task done. **Application software**, also called *applications*, is a category of software used by an individual to carry out tasks or be entertained. For example, to update a budget, open a financial application. To remove the red eye from a picture, open a graphics editing application. An application exists for just about any purpose. Applications make our computers useful to us, providing the ability to create, write, calculate, draw, edit, and otherwise work with text, graphics, audio, and video.

In Chapter 4, you learned about system software to manage data and the hardware installed on the computer. System software runs in the background, providing the platform for applications to work. You turn on a computer and the operating system (OS) is activated. At this time, the OS interface appears, and you can start various applications to complete work, personal, or communication tasks, such as watching movies, playing music, shopping online, or playing video games.

People refer to *software*, *applications*, and *programs* interchangeably. On a mobile device, the term *apps* refers to the various applications for use on a smartphone or tablet. An important distinction applies between system software and application software. You interact with application software to get tasks done or have fun. The application software needs the system software working in the background to be of any use. You do not interact with system software.

In this chapter, you will learn about application software commonly used in the workplace and in personal settings, at home or with family/friends. You will also learn how to manage the applications on a computer and explore types of mobile apps for a smartphone, tablet, or wearable device.

### Watch & Learn

Application Software

### Learning Objectives

**5.1** Describe uses for and features in word processor, spreadsheet, presentation, and database management software applications

**5.2** Identify personal information management, accounting, project management, and document management software applications

**5.3** Distinguish applications used for graphics, publications, multimedia, and web projects

**5.4** List and explain types of applications for personal use

**5.5** Describe the uses and benefits of web-based and open-source applications

**5.6** Explain how to acquire, install, uninstall, and upgrade applications

**5.7** Differentiate functionality in a mobile application and give examples of mobile apps

### Content Online

The online course includes additional training and assessment resources.

# 5.1 Productivity Applications for the Workplace

Basic business tasks are achieved using applications for producing word processing documents (working with text), spreadsheets (calculating and managing numbers), presentations (slide shows), and database management systems (keeping track of data). These applications are called **productivity application software** because they are used in the workplace to perform business-related tasks.

A group of productivity applications is typically bundled together in what is called a **software suite** or **productivity suite**. While many variations of a software suite exist, a minimum bundle includes word processing, spreadsheet, presentation, and email applications. For example, the Microsoft Office Home & Business suite includes Word for word processing documents, Excel for spreadsheets, PowerPoint for presentations, Outlook for email, and OneNote for note-taking. Other bundles of the Microsoft Office suite (called *editions*) have groups of applications to suit a variety of office environments. Microsoft also publishes a version of Office for Mac computers that is designed to run with the Apple macOS.

Alternatives to Microsoft Office and Apple productivity application software include Apache OpenOffice, a free productivity suite made available by the open-source community, and web-based productivity software offered by cloud providers, such as Google and Zoho. These alternatives are introduced in Topic 5.5.

Regardless of the suite you encounter, all productivity applications have similar features. In this topic, you will learn the features for four productivity applications: word processing, spreadsheet, presentation, and database management software.

## Word Processing Applications

A **word processing application** is software used to create documents containing mostly text. Images and other multimedia are often included with text to add visual appeal and promote understanding of content. Figure 5.1 presents a Word document that has text and graphics designed to be a trip planner for the Grand Canyon. The Apple word processor is called *Pages*. With a word processing application, individuals create a variety of documents, such as reports, letters, memos, contracts, brochures,

**Read & Learn**

Productivity Applications for the Workplace

● – – – – – go online

**https://CCA3
.ParadigmEducation
.com/MicrosoftOffice
Online**

Go here to access the free web-based versions of Microsoft Office software: Word Online, Excel Online, and PowerPoint Online. You will learn about web-based applications in Topic 5.5.

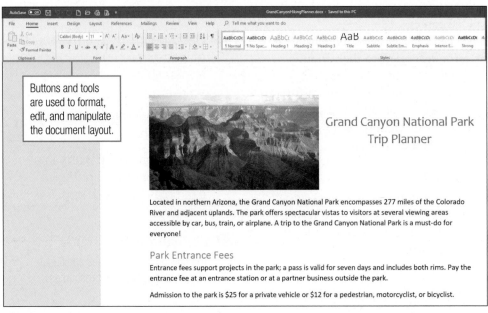

**Figure 5.1**

A document created in a word processor is mostly text. Images add interest and visual appeal to blocks of text and are inserted into the document using the word processor interface comprised of command buttons or menu options.

catalogs, menus, newsletters, mailings, labels, articles, and journals. Word processing applications typically include the following features:

- text formatting options for the type style, color, and size, called *font formatting*
- bold, underline, italic, and other text effects, such as shadow and strikethrough, called *font effects*
- spelling and grammar checking, as well as a thesaurus
- paragraph alignment formatting options, line and paragraph spacing options, tabs, indents, borders, shading, and bulleted and numbered lists
- page formatting options including page numbers, margins, columns, and page breaks
- illustration tools to insert tables, pictures, charts, or other graphics, and draw shapes
- entering and tracking of footnotes, endnotes, captions, citations, and references
- preformatted templates for cover pages, reports, letters, and other documents
- tools to merge standard text with an address list to produce letters
- review tools such as track changes, comments, and sharing functions

## Spreadsheet Applications

Software used to organize, calculate, and chart numbers is a **spreadsheet application**. In a spreadsheet application, data is saved in a file called a *workbook*. Within a workbook, data is organized in separate pages called *worksheets*. A worksheet is a page made up of a grid of columns (vertical) and rows (horizontal), with each column and row intersecting at a placeholder for data called a *cell*. Data is typed into a cell. Calculations performed on values stored in cells can be simple mathematical operations or complex formula statements.

Formulas in spreadsheets automatically recalculate whenever a number in a cell that is part of a formula changes. People perform *what-if analysis* in a spreadsheet by changing a value or formula to see what happens to other values. For example, in a budget worksheet, you could increase the income in one cell and watch how that change affects the amount of money left over after paying bills in another cell. Spreadsheet applications have charting capabilities to graph values in a pie, column, or other type of charts. Figure 5.2 shows a school budget worksheet created in Excel. The Apple spreadsheet application is called Numbers.

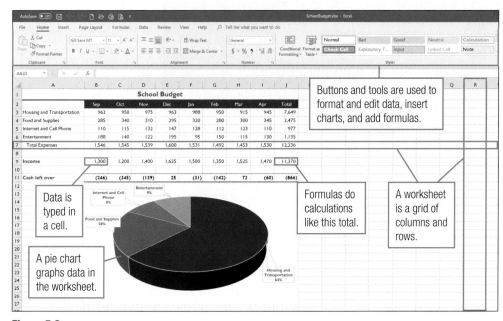

**Figure 5.2**

A spreadsheet application, like Excel shown here, lets you track, analyze, and graph numerical data.

Spreadsheets are used for any work that primarily tracks numbers, such as budgets, revenue, expenses, asset tracking, inventory control, production control, grade books, research data, payroll, billing, costing, estimating, attendance recording, and scheduling. Spreadsheet software includes these features designed specifically for the application:

- preprogrammed formulas used in statistics, finance, math, date, and logic analysis
- data tools for organizing and analyzing large blocks of data
- various chart styles and options to graph data and add other visual elements
- multiple worksheet tabs to group, consolidate, and manage data on separate pages
- tools to perform what-if analysis and decision-making

## Presentation Applications

Slides for an electronic slide show that may be projected on a screen during an oral presentation or on a video display at a self-running kiosk are created in a **presentation application**, such as PowerPoint or Apple Keynote. Presentation software helps you organize, describe, and explain key points, facts, figures, messages, or ideas.

Slides incorporate text, images, audio, and video with animation and transition effects to give an engaging presentation experience. Figure 5.3 shows a group of slides in a presentation created with PowerPoint. Presentation software includes these features designed specifically for creating slide shows:

- several slide layouts for arrangement of titles and content
- presentation designs and themes for backgrounds, colors, fonts, and effects
- tools for inserting graphic objects, and sound and video media
- transition and animation schemes to make and customize dynamic slide shows

Some presentation applications also include an export utility to save a presentation in another file format (like an MP4 video) that can be used for viewing, editing, collaborating, or sharing on a website (like YouTube) or other service.

●－－－－ in the know

Microsoft reports that more than 1.2 billion people use Microsoft Office worldwide—that's close to one in six people on the planet!

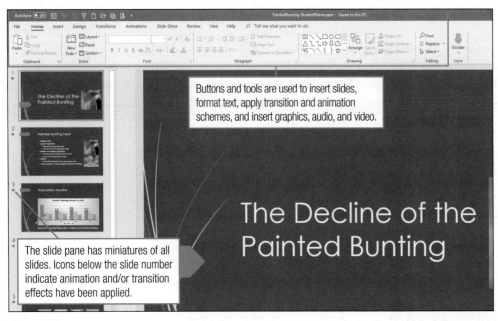

**Figure 5.3**

In a presentation application, a set of slides are presented in a slide show by a speaker or can be viewed in a self-running automated slide show.

## Database Management Applications

A **database management application** is a software application that organizes and keeps track of large amounts of data. The data is stored in such a way that information can be extracted in useful lists and reports. For example, a manager could request a list of all customers residing in a certain zip code, or all employees with no absence reports.

With database management software, you begin by setting up the structure of the database. This setup involves defining the type of data you are going to collect and setting "rules" for how the data will be entered. Creating a rule is a way to make sure that all data is entered accurately and consistently. For example, if you are going to keep track of customers, the information about each customer that you want to enter into the database is itemized, such as the customer's first name, last name, street address, and so on. After the data structure is defined, you can enter and edit data, and then you can produce lists and reports to serve a variety of purposes.

A business uses a database to keep track of customers, patients, clients, vendors, employees, products, inventory, assets, contacts, claims, equipment, service records, and more. Any large amount of data that needs to be organized, analyzed, and stored can be set up in a database. A business chooses a database management application over a spreadsheet application when the quantity of data to be managed is large and when information needs to be extracted using complex analysis tools. Figure 5.4 illustrates a database created in Access to keep track of used textbooks for sale. Database management applications include these features unique to this software:

- tables for defining data structure in columns (called *fields*) and rows (called *records*)
- forms for entering and maintaining data in a user-friendly interface
- queries for extracting and displaying data by criteria
- reports for printing or displaying data with rich text formatting options
- joining related tables for queries and reports that extract data from two or more tables

 **Video**

How Can Databases Make a Difference in Our Lives?

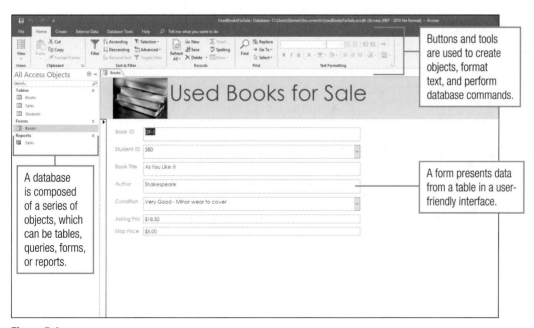

**Figure 5.4**

Database management applications help a business or other organization keep track of large amounts of data. Data in the database can be selected, grouped, and otherwise organized and presented in various views and reports.

## 5.2 Other Workplace Applications

Businesses use applications to do more than enter and edit information through documents, spreadsheets, presentations, and databases. They also use applications to manage communications, calendars, contacts, financial information, projects, and to manage the volume of files that are stored. In some industries, specialized software is designed to complete tasks unique to the industry, and large companies use enterprise-wide applications to manage large-scale operations. In this topic, you will be introduced to some of these other types of applications found in a workplace.

### Personal Information Management Applications

Managing messages, schedules, contacts, and tasks can be onerous with the volume of emails, appointments, contacts, and to-do lists a typical worker encounters. A **personal information management (PIM) application** helps you organize this type of information. Reminders and flags help you remember and follow up on important messages, appointments, events, or tasks.

Figure 5.5 displays the Calendar in Outlook, which is part of the Microsoft Office suite. Outlook can be used by individuals at home to manage personal information. When Outlook is used in a workplace, a server that runs Exchange (Microsoft software to manage email) connects employee accounts. Employees connected to a server running Exchange have additional features not available to other types of users, such as sending and tracking meeting requests and tasks to other employees and designating someone to respond to items on an employee's behalf.

Apple devices include iCal, Contacts, and Mail applications for personal information management. On smartphones and tablets, PIM apps can be set up to synchronize messages, schedules, and contacts with a mobile worker's accounts on a workplace server to let employees stay in touch while away from their desk. PIM applications also include features to share calendar and contact information with other people inside or outside the workplace.

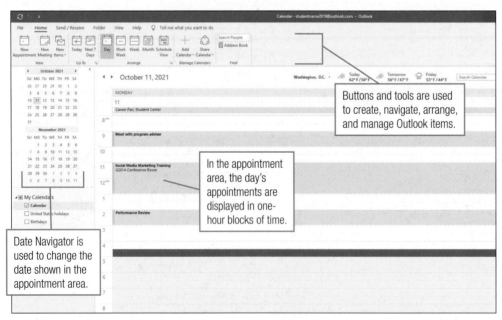

**Figure 5.5**

Outlook is an application you can use to manage messages, schedules, contacts, and tasks. The Calendar module shown here is for scheduling appointments, meetings, and events.

**Read & Learn**

Other Workplace Applications

**Video**

How Do You Identify the Format That a File Is Saved In?

## Accounting Applications

All businesses need an **accounting application** to record and report financial information related to assets, liabilities, equity, revenue, and expenses. These applications provide the basis for financial reporting and tax calculations. Accounting software is generally organized by accounting activity, such as invoicing, purchasing, inventory tracking, payroll accounting and time-tracking, banking, bill paying, payment receiving, and memo processing.

Accounting applications also support costing and estimating features for customer- and vendor-related functions. Small, medium, and large businesses have several accounting applications at their disposal, such as QuickBooks, Quicken, Sage, FreshBooks, and Zero. Figure 5.6 shows the Customer Overview screen in QuickBooks, an accounting applications used by many small businesses.

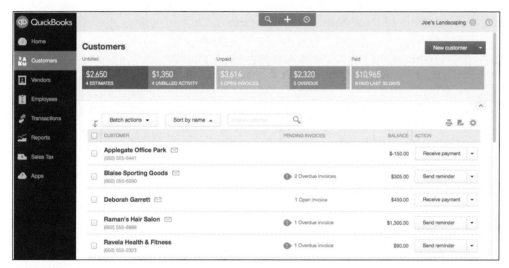

**Figure 5.6**

QuickBooks is an accounting software application for recording and reporting the financial activities for a small business. The Customers screen shown here tells an owner the amounts for unbilled, unpaid, and paid invoices.

Banking institutions and accounting software application developers work together to provide tools in the online banking systems and the accounting applications that allow data to be automatically exchanged. For example, once you have connected your bank account in QuickBooks, transactions can be downloaded automatically from the bank into the accounting application. This feature is called *bank feeds*. For some applications, data is imported into the accounting application by first downloading transaction data from the online banking system to the computer. Direct exchange of bank transactions to the accounting application saves owners and employees time and helps to reduce errors that occur when transactions are manually recorded.

All businesses have physical assets to keep track of and manage throughout their life cycles. These assets range from employer-provided smartphones, laptops, and tablets to vehicles, equipment, and buildings. Each item a company purchases has to be recorded and tracked using a type of accounting software called an **asset management application**. Accounting software companies offer an asset management module either in special editions of their accounting application or as an add-on purchased separately. For example, QuickBooks includes its Fixed Asset Manager only in its desktop editions. A business with complex asset management needs may buy a dedicated asset management application, such as AssetExplorer by the software company ManageEngine or Asset Panda.

●------ green IT

Accounting software features that allow a business to opt for green computing include these tools to omit printed documents:

- emailed invoices and electronic payments to and from customers and suppliers
- emailed point-of-sale receipts
- direct deposit of employee paychecks
- e-filing of tax forms and reports
- online banking and bank statements

# Project Management Applications

Large projects that need several resources and are to be completed on a schedule and within a budget are often managed using a **project management application**. Project management software is used to enter and manage all the individual tasks needed to get a project done. When a task is entered into a project management application, estimated timeframes for completion are identified, resources are assigned, and a budget is allocated. A project manager assigns project team members to each task. A task that has a dependency on another task is identified in the application because this relationship impacts scheduling. For example, when building a new home, the roof has to be completed before the drywall work begins.

Once the project tasks, start date, and completion date are created, the project management application creates the project schedule. The schedule is typically shown in a Gantt chart, which visually shows tasks on a timeline with lines connecting dependent tasks (any task that cannot start until a previous task is completed). Other views in the application are designed to show the project's task list, budgeted and actual costs, allocated resources, status reports, and risk management reports. A project manager can instantly view the impact on the schedule or budget if changes are made to tasks, resources, or timelines.

Another advantage to using project management software is the collaborative tools included with the application. Each team member with access to the project file can update task information and see other statuses or documents related to the project. This means everyone stays up-to-date and can deal with issues that arise immediately.

Numerous project management applications are available, including Project (from Microsoft), Trello, Basecamp, Clarizen, Wrike, Mavenlink, OpenProject, and ProjectLibre. Figure 5.7 shows a task list for a project to create a new website using the project management application OpenProject, an open-source application managed by the OpenProject Foundation in Berlin.

**Job Connect!**
Project Manager

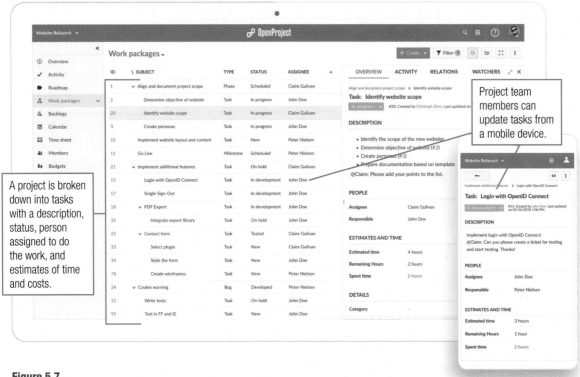

**Figure 5.7**

A project management application like OpenProject is used to keep track of a project's tasks, schedule, resources, and budget. In this view, a task list is shown with the status and person assigned to each task. Team members can update a task from their mobile devices.

## Document Management Applications

A **document management system (DMS)** is an application that manages the creation, storage, retrieval, history, and eventual disposal of documents or other files stored in a central repository. Paper documents are scanned, coded/tagged, and stored electronically. DMS software provides advanced search tools for retrieval as well as sharing of documents. A document's life cycle is managed through the DMS, flagging files that can be disposed of after compliance rules are met. All documents are stored in a common file format, such as **Portable Document Format (PDF)**, a file format that presents a document as it will look if printed, viewable on any computer configuration. A DMS is the cornerstone of a paperless office.

DMS applications and data are stored on secure servers, with access to the software and data controlled to maintain privacy and security rules. A document list view in the LogicalDOC DMS application is shown in Figure 5.8. Other DMS applications include M-Files, FileHold, and Dokmee. DMS applications dedicated to the legal field include Worldox, LegalWorks, and iManage.

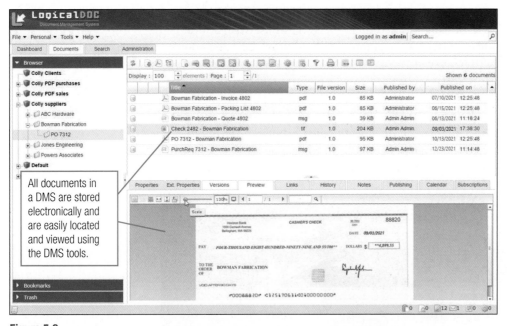

All documents in a DMS are stored electronically and are easily located and viewed using the DMS tools.

**Figure 5.8**

Companies use a DMS application such as LogicalDOC to electronically manage all documents, including paper documents that are scanned, coded, and stored on central servers.

## Industry-Specific Applications

Automotive, construction, manufacturing, creative design, and engineering companies are just a few examples of industries that use computer-aided design (CAD) or computer-aided manufacturing (CAM) applications to create complex designs and drawings for the products they make or buildings or infrastructure projects they build. Medical establishments use medical applications to track patient records, billings, and insurance claims. Hotels and travel agencies have applications to manage bookings, guests, and resources. These are just a few examples of industry-specific applications.

Large companies, called *enterprises*, have higher-level computer needs to manage the activities and processes involved in transactions with customers, vendors, inventories, and human resources. These companies use an **enterprise resource planning (ERP) application**, which is a software solution designed for large-scale organizations. Companies such as SAP and Oracle provide ERP software for these environments, often along with extensive customization, training, and support.

# 5.3 Applications for Creative Content

In this topic, you will be introduced to applications for working with graphics, publications, audio, video, animations, and web pages. **Graphics application software** is used for creating, editing, and manipulating drawings, photos, clip art, and scanned images. **Desktop publishing (DTP) application software** incorporates text, graphics, and the use of colors for creating high-quality documents for marketing, communication, and education. **Multimedia application software** combines text and images with video, audio, and animations for interactive applications that promote products, educate, or entertain people. **Web authoring application software** is for creating content provided over the internet.

## Graphics Applications

Graphic artists, illustrators, photographers, and others use specialized software that has tools for artists to draw pictures, shapes, and images and add special effects to those images. Many drawing applications are available, with Adobe Illustrator and Adobe Photoshop used by many professionals to create, edit, and manipulate illustrations and photographs. CorelDRAW, initially released in 1989, is still in use with a loyal fan base. Inkscape is a free software option for graphics editing from the open-source community.

**Photo editing application software** is used to edit pictures taken with a digital camera. Tools in photo editing software let you crop out unwanted portions of a picture, remove red eye, retouch areas, correct color brightness and contrast, add text, add special effects, and even add in a person or object copied from another photo. Figure 5.9 shows a photo edited in a graphics application to change a model's hair color, adjust her skin tone, and enhance her makeup.

**Figure 5.9**

A professional graphics editor used software to transform the photograph on the left to the photograph on the right by changing the background color and the model's hair color, and retouching her skin tone and makeup.

## Desktop Publishing Applications

Although word processing applications have tools to arrange text and graphics in a document, the software is not used for projects where precise positioning of text with graphic objects is needed. For professional publications, desktop publishing software is used to produce a print-based or digital edition of a document, such as a corporate newsletter, annual report, price list, restaurant menu, catalog, brochure, or book. Adobe InDesign and Publisher by Microsoft are well-known DTP applications.

DTP software has rulers, grids, and guides along with other tools and features that let page layout professionals precisely place text and graphics on a page. The software provides more control over word spacing, space between and around objects and text, and a wider range of fonts, sizes, and colors for text formatting. Figure 5.10 illustrates page 106 and 107 of the textbook (pages you are currently reading) in the InDesign application window on a Mac.

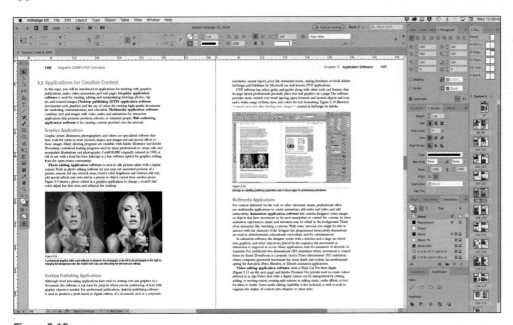

**Figure 5.10**

InDesign is a desktop publishing application for page layout in professional publications.

## Multimedia Applications

For content delivered via the web or other electronic means, professionals often use multimedia applications to create animations, add audio and video, and add interactivity. **Animation application software** lets content designers create images or objects that have movement or lets users manipulate or control the content. In these animation experiences, music and narration may be added in the background. Think of an animation like a cartoon with movement. With some cartoons, you might be able to interact with the character if the designer has programmed interactivity. Animations are used in advertisements, in educational curricula, and for entertainment.

In animation software, the designer works with a timeline and a stage on which text, graphics, and other objects are placed in the sequence the movement or interaction is supposed to occur. Many applications exist for animators of all levels of expertise. For traditional two-dimensional (2-D) animation where movement is created frame by frame, Toon Boom is a popular choice. Three-dimensional (3-D) animation, where computer-generated movement has more depth and realism, has professionals opting for Autodesk Maya, Blender, or ZBrush animation applications.

**Video editing application software**, such as Final Cut Pro from Apple (Figure 5.11 on the next page) and Adobe Premiere Pro, provides tools to create videos referred to as *clips*. Video shot with a digital camera can be manipulated by cutting, adding, or moving scenes; creating split screens; or adding music, audio effects, or text for titles or credits. Some audio editing capability is also included, as well as tools to organize the display of content into chapters or other units.

Closed-captioning text is added here.

Clips and audio are added to the timeline here.

**Figure 5.11**

Final Cut Pro video editing software from Apple is the preferred application by many professional video producers.

Individuals who like to capture video on a smartphone or other digital camera to share the footage may use a video editing application geared toward nonprofessionals. These applications include iMovie (free from Apple) and inexpensive options such as Nero Video, Corel VideoStudio, and Pinnacle Studio.

**Audio recording and editing application software**, such as Audacity and Adobe Audition, offers functions for recording and editing music, sound clips, and podcasts (audio clips posted on the web).

**Screen recording application software**, such as Camtasia or Screencast-O-Matic, captures all or a portion of the computer screen for a determined amount of time, creating a video from the activity. This video can later be edited. Graphical objects can be added over the captured screen images along with text and audio narration to create instructional demonstrations.

## Web Authoring Applications

Web authoring applications like Adobe Dreamweaver and RapidWeaver for Mac computers are used by businesses to create and maintain interactive websites. Alternative open-source applications, such as openElement (Figure 5.12 on the next page), are also available.

Web pages are stored in programming code that is interpreted by a web browser, such as Microsoft Edge or Chrome. Web authoring software provides tools for individuals to create and manage all the pages and external resources for a website by working in either design view or code view. Nonprogrammers can create a web page in design view and let the software automatically generate the programming code needed for browsers. Experienced web developers often work in code view and switch between views as pages are fine-tuned. Figure 5.12 shows a web page in code view, with the software prompting the programmer that an unnecessary semicolon is present.

If you think the screen shown in Figure 5.12 looks daunting, you're in good company! When the web first evolved, a business or individual hired a professional programmer to create a website. The alternative was to learn to create web pages using HTML and other programming languages used on the internet. Web authoring software was developed to help people create and maintain websites, but the software was considered complicated to understand and use by many people.

Today, large organizations still hire inhouse or contract web design teams to create corporate websites, but low-cost or free cloud-based website creation tools have transformed website design and creation into a do-it-yourself activity with little to no programming knowledge needed. Many of these services provide professionally

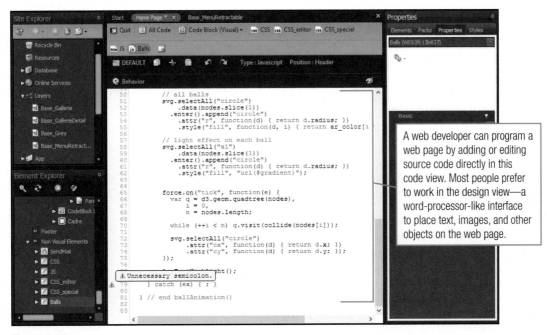

A web developer can program a web page by adding or editing source code directly in this code view. Most people prefer to work in the design view—a word-processor-like interface to place text, images, and other objects on the web page.

**Figure 5.12**

The JavaScript source code for a web page is shown here using the web authoring software application called openElement, an open-source application. Web browsers interpret this code to display the text and multimedia that you see when you view the page.

designed templates that you customize with your content and add-on widgets (like a Buy Now button). Several website builder services are available. These are popular services: WordPress (a free, open-source option), Wix, Weebly, Squarespace, and GoDaddy. Some, like GoDaddy, include domain name registration and web page hosting for a one-stop service. Figure 5.13 shows some of the templates available from Weebly that let someone with no design experience create a modern website.

In summary, the application software featured in this topic is geared toward professionals. Less expensive or free software let nonprofessionals work with creative content at home. Windows 10 includes Paint and Voice Recorder apps and has video editing capability in the Photos app. Apple includes the iPhoto, iMovie, and GarageBand apps on macOS and iOS devices for working with creative content.

**Figure 5.13**

A small business owner or individual with no design or programming experience can create a contemporary website using a cloud-based website builder service like Weebly. Six Weebly website templates are shown here.

## 5.4 Applications for Personal Use

**Read & Learn**
Applications for
Personal Use

**Video**
How Can Software
Help You Learn?

Applications are available to help you manage everything from personal notes, ideas, and thoughts to keeping track of personal finances, preparing tax returns, playing games, designing a garden, and creating a genealogy map. These applications also help you organize recipes, plan trips, and learn new information or master a new language. In this topic, you will survey a variety of applications for personal use.

If you have an interest or a hobby that is not mentioned in this topic, chances are that a related application has already been created. Search the web for an application if you have not already found one.

### Note-Taking Applications

While in meetings, classes, or at home, you might need a place to record notes, thoughts, or ideas. **Note-taking application software** stores, organizes, searches, and shares notes of any type. When the application is open, notes can be entered by clicking anywhere on the screen and typing the details or by using a stylus on a touchscreen. Documents, audio, video, images, emails, appointments, contacts, and web links can be attached or inserted into a note to collect everything related to a topic in one place. Notes can be organized into notebooks and tabs, and search tools let you find information quickly. Notebooks can be shared with others for collaborative work. OneNote, a Windows 10 app (Figure 5.14), and Evernote are two note-taking applications with companion mobile apps that let you synchronize notebooks across multiple devices.

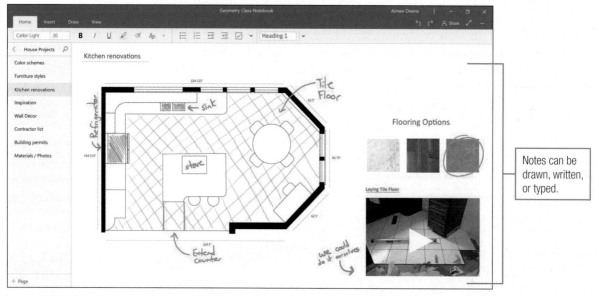

**Figure 5.14**

The OneNote app in Windows 10 is useful to organize notes on any topic related to school, work, or home. The app is also available free for Mac and Android devices.

### Applications to Manage Personal Finances and Taxes

Keeping track of a checking or savings account and personal investments is easier with **personal finance application software**. Most personal finance applications can import activity downloaded from your banking account directly into the personal finance application for reconciling your checking account, analyzing home expenses, and tracking investments. Quicken (Figure 5.15 on the next page) is one of the most commonly used applications for managing personal finances. A personal finance application typically includes a mobile app that automatically synchronizes with the desktop version. Some applications, such as Mint, are free to use.

**Tax preparation application software** guides you through the completion of the complex forms required to file your taxes. These applications generally estimate your tax bill using an interview-style series of questions and input boxes. Applications alert you to missing information and deductions that can help minimize taxes. Tax forms can be printed and can also be filed with the government electronically. The leading tax preparation software is TurboTax, followed by H&R Block. Other alternatives exist with some companies allied with the government offering free filing for low-income taxpayers.

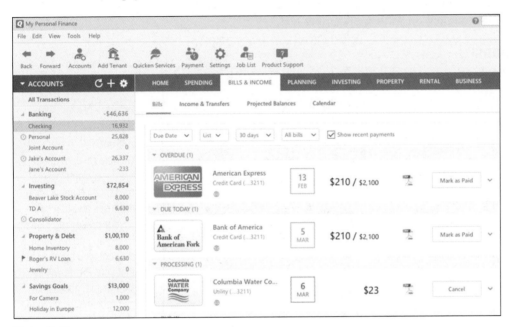

**Figure 5.15**

Quicken is an application program for managing personal finances, such as cash and personal expenses.

## Entertainment Applications

**Entertainment application software** is used for playing interactive games and videos. Some games are designed to be played individually. Others involve subscribing to online services where you can join hundreds of thousands of other game enthusiasts and play against others online. For example, with more than 7 million subscribers, World of Warcraft (WoW) is a highly successful online game involving role-playing and multiple players that boasts a 2017 Guinness World Record for the longest video game marathon at 32 hours and 36 minutes.

## Educational and Reference Software

**Educational application software** is designed to teach children and adults subjects such as math, spelling, grammar, geography, science, and history through an interactive reference or game format. For example, Scholastic education materials include the ScienceFlix application, with over 4,000 objects in an interactive app designed to teach children in grades 4 to 9 about science. You can also find software to help you learn a foreign language and to teach you how to type.

**Reference application software** uses multimedia tools for presenting content in encyclopedias, dictionaries, translation guides, atlases, and other reference resources.

● ━ ━ ━ ━ in the know

According to Newzoo (a games market research company), worldwide spending on digital games is expected to reach $138 billion in 2018, with the mobile game market claiming $70.3 billion, more than half of the total revenue.

## 5.5 Cloud and Open-Source Applications

**Read & Learn**
Cloud and Open-
Source Applications

In Topic 1.6 you were introduced to cloud computing and the SaaS delivery model for software applications. Recall that *Software as a Service (SaaS)* means that the complete software application you are using is not saved on your computing device. SaaS applications are hosted by a cloud service provider and are opened from a web browser.

Open-source applications are often provided for free, meaning that you can download and install the software on a computer without paying a license fee to use the software. In this topic, you will learn about three web-based and two open-source alternatives for productivity applications. Increasingly, businesses and individuals are opting to use these tools to save money on software licensing fees.

### Web-Based Applications

The main advantage to using a web-based application is that you have access to the software and your files from any location at which you have an internet connection. Table 5.1 provides a summary of three web-based productivity suites. The applications in these suites are suitable alternatives to traditional desktop-installed PC applications for word processing, spreadsheets, and presentations.

Some web-based applications are offered

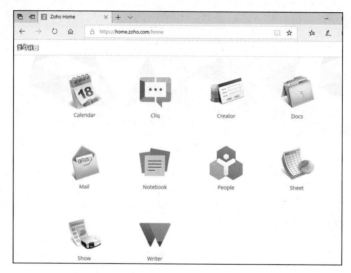

**Figure 5.16**
The Zoho suite has 40 applications. The home screen is shown here after signing in with a Zoho account.

●----- **go online**

**https://CCA3
.ParadigmEducation
.com/Prezi**
Go here to learn about Prezi—a cloud-based presentation application that uses a virtual canvas for presenting ideas. By signing up for a free account, you agree that your Prezis will be searchable, accessible, and reusable by anyone.

entirely free, but may contain ads. Others, like Zoho (Figure 5.16), offer the basic word processor, spreadsheet, and presentation application free, and also offer bundled

**Table 5.1** Three Web-Based Productivity Applications

| Criterion | Google Drive | Office Online | Zoho |
|---|---|---|---|
| Amount of free storage* | 15 GB | 5 GB | 5 GB |
| Applications | Google Docs<br>Google Sheets<br>Google Slides<br>Google Forms<br>Google Drawings | Word Online<br>Excel Online<br>PowerPoint Online<br>OneNote Online | Sheet (Spreadsheet)<br>Show (Presentation)<br>Writer (Word Processor)<br>Plus several apps<br>for collaboration and<br>business processes |
| URL | https://www.google.com | https://www.office.com | http://www.zoho.com |
| Sign-in process | Sign in with Gmail account, click Google apps, Drive, New button, then desired application. | Sign in with Microsoft account, then click the desired application icon. | Sign in with Zoho account, then click the desired application icon. |

*Offered at time of writing.

suites of integrated applications on a subscription basis charged per employee to businesses. Zoho's suite offers enterprise applications for human resources, customer relationship management applications, and IT and help-desk service tracking.

If you decide to use a web-based productivity suite, consider spending time working with a few different options until you find the one with the features and interface that works best for you.

## Open-Source Applications

Apache OpenOffice is a free productivity suite that includes word processor, spreadsheet, presentation, graphics, database, and formula editor applications. Apache OpenOffice is a project of The Apache Software Foundation, providing support for open-source software initiatives. In Figure 5.17, you see a document open in the word processor. The interface for Apache OpenOffice uses a menu bar, standard toolbar, and formatting toolbar familiar to people with experience working in Word before the adoption of the ribbon interface (Word 2003 or earlier). Apache OpenOffice is updated and maintained entirely by volunteers.

Another free open-source productivity suite is LibreOffice, with versions available for Windows, Apple, and Linux computers. The suite has Writer, Calc, Impress, Draw, Base, and Math applications. LibreOffice is a project of the not-for-profit organization The Document Foundation. Similar to Apache OpenOffice, LibreOffice is created and maintained by volunteers, and was born when a team of OpenOffice developers decided to create LibreOffice as a separate suite.

Both web-based and open-source applications provide computer users with free or low-cost access to productivity software with features to create routine documents, spreadsheets, and presentations. Both options have well-established user communities. Advocates in the open-source community maintain that open-source software is the future, noting that Microsoft, Apple, and Adobe participate in open-source projects.

**Video**

How Did the Open Source Movement Evolve?

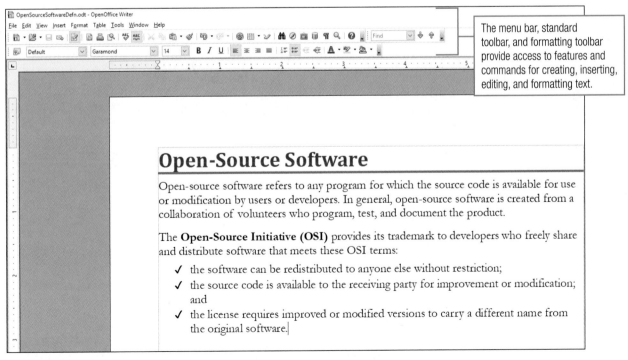

The menu bar, standard toolbar, and formatting toolbar provide access to features and commands for creating, inserting, editing, and formatting text.

**Figure 5.17**

Apache OpenOffice is a free, open-source productivity suite that includes a word processor (shown here) as well as spreadsheet, presentation, drawing, database, and formula editor applications.

## 5.6 Acquiring, Installing, Uninstalling, and Upgrading Applications

Applications can be purchased from a retailer in person or online, or directly from the software publisher at a website. Another option often used to try out software before buying it, or to find free software, is to download an application from a downloads website like Download.com (affiliate of the CBS media corporation). Download.com is one of several websites that offer a repository for applications and games with more than 150,000 free downloads. If you exercise caution at these websites and you need an application for a limited use or timeframe, a download website can be a useful resource. In this topic, you will learn the various models for distributing software and the general steps you take to install, uninstall, or upgrade an application.

### Application Software Distribution Models

Most software is licensed, meaning that installing the software gives you the right to use it, while ownership remains with the software publisher. Historically, software licenses were sold in boxes, with the applications delivered on media such as CDs or DVDs. You inserted the media into your computer to install the application. Today, an installation file to set up software on your computer is typically downloaded from a website once the license fee is paid. A license code (also called a *product ID* or *license key*) is provided after payment of the license fee, and this code allows you to activate the software. Pricing for a software license can vary, so people generally check out a few options before making a purchase. Other ways to acquire software include the following:

- **Shareware** is available to download for a trial period. Once the trial period expires, you need to pay a fee to the software publisher to unlock the application. Other forms of shareware continue to work beyond the trial period but generally have features made inaccessible until the fee is paid.
- **Freeware** can be downloaded and installed at no cost and without restrictions. An example of a freeware application is Apple iTunes.
- Open-source software is usually available for free and downloaded from a website.
- **Subscription software** is purchased from a publisher who charges a monthly or annual fee for access to the software and storage services. Fees vary based on the number of users and the amount of storage space.

### Installing an Application

Software is installed by running an installer application that is downloaded to your computer or copied from some other medium, such as a USB flash drive. If you bought your license in person at a retailer, you would have most likely receive a license card, which has printed instructions to go to a website, download the software, and install the software with the license code printed on the card. If you are downloading software from a website, you have two methods for installing it on a Windows computer:

- You can opt to run the installation application directly from the website. After clicking the download hyperlink, you will have the option to run or save the installation file. Choose the *Run* option to install the software directly from the source website and then follow the prompts that appear. On a Windows PC, allow the installation to proceed by clicking *Yes* when the security warning message box appears.
- Instead, you can opt to save the downloaded file to your hard drive and run the installation later. After clicking the download hyperlink, select the *Save* option to download the file to the Downloads folder, or choose the *Save as* option to specify another location to save the file. When the download is complete, open the File

**Read & Learn**

Acquiring Installing, Uninstalling, and Upgrading Applications

**Video**

How Does a Software as a Service (SaaS) License Differ from a Traditional Software License?

●—————— secure IT

Use caution when installing software from a download website. Some free software is bundled with other applications you do not want. Read installation screens carefully. Deselect options for extra tools or applications. Most download websites advertise that software is virus and spyware free, but these websites have ads that may contain malware if you click on them. You may also find your browser start page is redirected to a different page after installing software from a download website.

Explorer window to view the file, double-click the file name to start the installation routine, and follow the prompts that appear (Figure 5.18). Click the option to allow the installation when the Windows security warning message appears. People who want a copy of the installation file for reuse at a later date use this method.

On a Mac computer, downloaded software is saved automatically to the Downloads folder (next to the Trash icon on the dock). Open the folder and double-click the installer application to start the installation. Some installer applications open a pop-up window instructing you to drag the installer application to the Applications folder, while other installer applications automatically copy the file to the Applications folder and begin the installation. Follow the prompts that appear once the installer application is running.

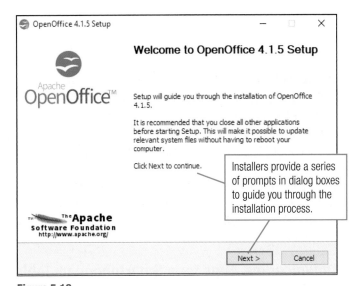

During installation of a new application, you are required to accept a license agreement (unless the software product is open-source). The license agreement, called a **software license agreement (SLA)** or **end-user license agreement (EULA)**, varies by software publisher. Although the terms and conditions are generally long and filled with legal terminology, you should read the agreement and not carelessly click the *I Accept*, *I Agree*, or *Yes* option to move on. SLAs or EULAs specify the number of computers that can legally have the software installed. Some publishers allow installation on five devices; others allow only one. Some agreements also specify that software creators have the right to access files and hardware functions on your device.

**Figure 5.18**

Installing downloaded software typically involves proceeding through an installer wizard that prompts you along the way.

## Uninstalling an Application

Uninstalling an application you no longer use frees up disk space and memory resources. To uninstall an application on a Windows PC, display the Apps & features pane in the Apps category of the Settings app (Figure 5.19). Click to select the application, click the Uninstall button, and confirm Uninstall a second time to start the removal process. Close the Settings app when finished.

On a Mac computer, uninstall an application by dragging the application icon to the Trash. The icon is a shortcut to the application installation bundle, which includes all related application files. If necessary, also drag the application shortcut icon off the dock.

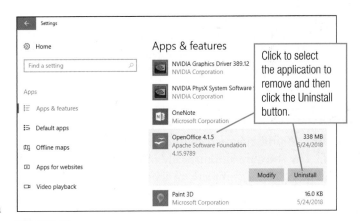

**Figure 5.19**

Display the Apps & features pane in the Settings app to uninstall software.

## Upgrading an Application

Upgrading an application installs a newer version of software already licensed (not to be confused with updating, which installs fixes or patches to the existing version). Some upgrades are free, while others require payment (sometimes at a special price if completed within a limited timeframe). Registered users usually receive notifications when upgrades are available with hyperlinks to download the upgrade software.

## 5.7 Apps for Use on a Mobile Device

Smartphones, tablets, and other mobile and wearable devices have a software edition explicitly designed for mobile use. An application for a smartphone or tablet is called a **mobile application** or *app*. An app is designed for limited purposes, to display on small screens, and to accept touch, voice, or gesture input. Functionality for handwriting recognition and speech recognition capabilities are also part of app design. Other essential features include compatibility and synchronization with PC software and information sharing with other applications.

### Built-in Apps

All smartphones and tablets come equipped with several built-in apps. Typically, PIM apps for keeping track of appointments, contacts, to-do lists, reminders, and notes are standard, as are apps for email, messaging, maps, music, games, and social networks, such as Facebook, Twitter, and Instagram. Other apps installed include a web browser, calculator, clock, and weather app. Some apps are used to operate the hardware, such as the phone and camera apps and a device settings customization app.

People looking to add more functionality to their mobile device install more apps via the app store: Google Play for Android, App Store for Apple, and Store app for Windows.

### Downloading Apps for a Smartphone or Tablet

Apps are available for Android, Apple, and Windows mobile devices to accomplish any type of task you can imagine. In the first half of 2018, Google Play offered more than 3.3 million apps, the Apple App Store had 2 million apps available, and the Windows Store had almost 670,000 apps.

Apps for news, travel, games, banking, currency conversion, shopping, dinner reservations, and reading books are just a few types of apps that are downloaded frequently by mobile device owners.

Smartphones and tablets have an icon or tile that you tap to launch the store where you find and download apps. Figure 5.20 displays the GAMES category in the Google Play store on an Android smartphone. Once you tap the desired application icon, tap the Install button on the next screen to download the app. The software automatically installs. Once the download is complete, an icon or tile automatically appears on the smartphone or tablet screen.

According to Google Play, YouTube Music and Spotify Music were part of the top 10 free Android apps downloaded in June 2018. The top paid apps downloaded in the same period were mostly games.

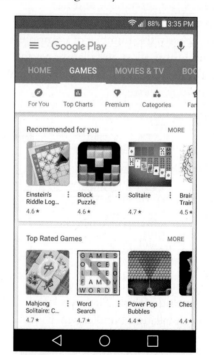

**Figure 5.20**

Android apps are found in the Google Play store. New apps are added frequently, with many offered for free or at a low cost.

## Mobile Productivity Apps

Microsoft has mobile editions of Word, Excel, PowerPoint, and OneNote for use on a smartphone or tablet. You can review, edit, and sync documents to and from OneDrive or other web-based services. Microsoft publishes free Office apps for Android and Apple devices for mobile workers to access their business documents away from the office.

Other mobile productivity suites include mobile editions of Apple Pages, Keynote, and Numbers for iPhones and iPads, and mobile editions of Google Docs, Google Sheets, and Google Slides for both Android and Apple devices. Numerous other productivity apps are available in the app stores for note-taking, time management, project management, accounting, financial management, investing, and more.

Figure 5.21a shows the mobile edition of PowerPoint on an Android smartphone, and Figure 5.21b shows the mobile edition of Excel on the same device.

**Let's Explore!**

Popular Apps for Android or Apple Mobile Devices

a

b

**Figure 5.21**

Mobile editions of the popular Microsoft Office productivity applications include (a) PowerPoint on an Android smartphone and (b) Excel displayed on the same device.

## Wearable Technology Apps

The growth of wearable technology devices is also driving growth in the development of wearable technology apps. The challenges for wearable app developers are the constraints of battery life, a small screen, and less computing power. These are the same challenges encountered by smartphone app developers, but they are magnified for developers of wearable devices. Developers also have to consider the usage of a wearable app. In most cases, the wearer will want a quick interaction, with a display that provides information at a glance and often while withstanding constant movement.

Fitness apps are abundant for bands and smartwatches—everything from apps that measure heart rate to apps that monitor glucose levels and issue fatigue alerts. Alarms, billable-hours trackers, maps, weather forecasts, schedule and note reminders, email alerts, shopping lists, recipe card displays, boarding passes, and conversion tools are just a few of the other wearable tech apps available.

Productivity apps for wearable devices will see growth as the devices become prevalent in the workplace. For example, Microsoft released a smartwatch PowerPoint app that lets the wearer navigate slides in a presentation, view the time elapsed, and see the number of slides in the deck. Watch for more innovative wearable apps as the number of devices and applications for this market continues to grow.

A new category of apps related to the IoT is emerging as products for home automation and smart city resource management increase in demand. Developers for apps installed on IoT devices specialize in writing software that meets the challenges associated with mobile accessibility, security, privacy, connectivity, and compatibility with other technologies. These apps gather data from connected sensors and operate via internet connectivity. While they're designed to be easy to use, the software manages complex processes related to data collection, analysis, and transmission to other services. Some even integrate AR or VR technology.

● — — — — **secure IT**

Be wary when prompted to allow permission for an app to access your camera, microphone, contacts, or other system option or data. Only allow permission requests that make sense. For example, a messaging app needs access to your contacts to facilitate sending text messages. But, a malicious app could steal contacts data to send to a hacker. Go to Settings on your smartphone to check and modify app permissions.

## Topics Review

| Topic | Key Concepts and Key Terms |
|---|---|
| **5.1 Productivity Applications for the Workplace** | **Application software**, or *applications*, is a category of software used by individuals to carry out tasks, manage projects, or otherwise get things done on a computer. |
| | **Productivity application software** is a group of bundled applications that includes word processing, spreadsheet, presentation, and database management applications. |
| | A set of productivity applications bundled together is called a **software suite** or **productivity suite**. An example of a software suite is Microsoft Office, which groups Word, Excel, PowerPoint, Outlook, OneNote, and Access in a variety of bundles. |
| | A **word processing application** is software to create, edit, and format documents that are mostly text but may also include multimedia to improve the appearance or comprehension of content. |
| | A **spreadsheet application** is software to organize, calculate, edit, format, and chart numbers stored in a file called a *workbook*. Within a workbook, data is entered in cells within a grid of columns and rows called a *worksheet*. Spreadsheets are often used to perform what-if analysis. |
| | A **presentation application** is used to create slides for an electronic slide show that communicates key points, facts, figures, messages, or other ideas. |
| | Large amounts of data are organized, stored, and maintained in software called a **database management application**. In a database management application, you set up the structure to store data, input data, maintain data, and then produce lists and reports for various purposes. |
| **5.2 Other Workplace Applications** | A **personal information management (PIM) application** organizes emails, appointments, contacts, and to-do lists. |
| | An **accounting application** is used to record and report the financial activities of a business related to assets, liabilities, equity, revenue, and expenses. Accounting applications also support costing and estimating features, and download data directly from bank feeds. |
| | An **asset management application** is used to record and track physical assets throughout their life cycles. |
| | A business will use a **project management application** to create and maintain a project schedule and project budget. |
| | A **document management system (DMS)** manages the storage, retrieval, history, and eventual disposal of documents stored in a common file format in a central repository. |
| | A DMS uses **portable document format (PDF)**, which stores files as they look when printed. |
| | Industry-specific applications are used to perform tasks unique to a specific business environment, such as computer-aided design or travel bookings. |
| | An **enterprise resource planning (ERP) application** supports large-scale operations with higher level needs for managing transactions and activities with corporation stakeholders. |

*continued...*

| Topic | Key Concepts and Key Terms |
|---|---|
| **5.3 Applications for Creative Content** | **Graphics application software** includes painting and drawing applications with tools to create, edit, and apply special effects to pictures, shapes, or other images. |
| | **Desktop publishing (DTP) application software** is used to place text and graphics on a page with more control over spacing and a wider range of fonts, sizes, and colors for professional, print-based publications. |
| | **Multimedia application software** incorporates text, images, video, audio, and animations to create interactive applications. |
| | A website and its associated web pages are created and maintained using **web authoring application software**. |
| | **Photo editing application software** lets photographers edit pictures and add special effects and text. |
| | **Animation application software** is used by a multimedia professional to create 2-D or 3-D images. The images have movement, and viewers may be able to interact with them. |
| | **Video editing application software** is used to manipulate video recorded with a digital camera and add music, audio effects, narration, and text to create videos called *clips*. |
| | Music clips, sound clips, and podcasts are recorded and edited with **audio recording and editing application software**. |
| | **Screen recording application software** is used to capture all or a portion of a computer screen to create a video from the activity that can be used for instructional demonstrations. |
| **5.4 Applications for Personal Use** | **Note-taking application software** provides a way to store, organize into notebooks and tabs, search, and share notes of any type. |
| | **Personal finance application software** allows you to balance your bank accounts and manage personal investments. |
| | Prepare and file your tax return using **tax preparation application software**. |
| | Playing a game or watching a video uses **entertainment application software**. |
| | Learning about subjects such as math and geography is made easier using **educational application software** that uses multimedia to add interactivity and games to improve comprehension and retention of content. |
| | **Reference application software** uses multimedia tools to provide rich content in an encyclopedia, dictionary, translation guide, atlas, and other references. |
| **5.5 Cloud and Open-Source Applications** | The main advantage to using a web-based application is that you have access to the software and your documents from any location with an internet connection. |
| | Google Drive, Office Online, and Zoho are all cloud providers that provide storage space and a suite of productivity applications for free. Some cloud providers, like Zoho, also offer business-related bundles of productivity applications on a subscription basis. |
| | Apache OpenOffice, a free, open-source productivity suite updated and maintained by volunteers, includes word processor, spreadsheet, presentation, graphics, and database applications. |
| | LibreOffice, a project of The Document Foundation, is another open-source productivity suite with applications called Writer, Calc, Impress, Draw, Base, and Math. |

*continued…*

| Topic | Key Concepts and Key Terms |
|---|---|
| **5.6 Acquiring, Installing, Uninstalling, and Upgrading Applications** | Software is a licensed product, meaning that you pay for the right to use the software but you do not own the software code.<br><br>**Shareware** is a downloadable software application for which you pay a small fee if you decide to keep the application after a trial period.<br><br>**Freeware** is a software application provided for your use without payment (e.g., Apple iTunes).<br><br>**Subscription software** is an application or set of applications that you pay for monthly or annually to access the software and storage services.<br><br>When you purchase software from a retailer or directly from the software publisher, you are given a license code, also called a *product ID* or *license key*, that you use to activate the software license.<br><br>Download and install software directly from a website by running the installer application at the source website or by saving the installer application on your hard disk drive to install at a later time.<br><br>All software except open-source applications require that you agree to or accept a **software license agreement (SLA)** or **end-user license agreement (EULA)**, which contains the terms and conditions for using the application. For example, the agreement specifies the number of computers on which you can legally install the software.<br><br>Uninstall an application from the computer when you are no longer using it to free up disk space and memory resources by displaying the Apps & features pane in the Settings app on a Windows PC, or by dragging the application icon and dock shortcut to the trash on a Mac.<br><br>Installing a newer version of an application is called *upgrading*; upgrades are generally made available for free or at a price lower than purchasing the software new. |
| **5.7 Apps for Use on a Mobile Device** | A **mobile application**, or *app*, is an application designed for use on a mobile device, such as a smartphone, tablet, or smartwatch, that is designed to work on a smaller screen. The app accepts input from touch, voice, gesture, or handwriting, and is compatible and synchronizes with PC software.<br><br>All smartphones and tablets come equipped with several built-in apps that typically include personal information management, communication, messaging, maps, music, and games.<br><br>Microsoft, Google, and Apple all have separate mobile productivity apps that run on smartphones and tablets.<br><br>Wearable technology apps are designed for use on devices, such as fitness bands and smartwatches. These apps accommodate smaller hardware resources that provide for quick interaction, display information at a glance, and withstand constant movement.<br><br>Apps for IoT devices is an emerging category of software that requires developers to build software that is easy to use and handle complex data collection and resources. |

 **Topics Review**

## Chapter 6

# Using Social Media to Connect and Communicate

In Chapter 2, you learned that Web 2.0 refers to second-generation websites that provide for two-way communication, where individuals interact with organizations and with one another by sharing ideas, feedback, content, and multimedia. A large component of Web 2.0 involves social networking sites, such as Facebook, Twitter, Instagram, and Snapchat. They are used daily by millions of people of all ages around the world. Businesses and other organizations also use social media to market their products or services, connect and interact with customers or clients, and source new ideas and content. Media-sharing services, blogs, and wikis are other components of Web 2.0 that focus on user-generated content.

In this chapter, you will learn about social networking, media sharing, blogging, wikis, and how to use social media to promote yourself for work-related purposes. You will also be introduced to ways organizations have incorporated social media technologies into their strategies to increase brand loyalty or attract new business.

## Learning Objectives

**6.1** Describe social networking and differentiate popular social-networking websites

**6.2** List key features of websites used for sharing music, videos, and photos

**6.3** Identify websites used for blogging and explain how to create a blog and a blog post

**6.4** Describe how wikis generate content from users and describe two well-known wiki websites

**6.5** Relate examples of social media strategies used by businesses for marketing and promotion

**6.6** Identify ways to use social networks to post content to enhance your career

### Watch & Learn
Using Social Media to Connect and Communicate

### Content Online
The online course includes additional training and assessment resources.

# 6.1 Social Networking

**Social media** is a term used to describe an online tool for users to generate content and interact with others. A website that provides a platform for people to connect with others and exchange content is called a **social-networking website**. Generally, social networking sites provide a means to build a list of friends, share information, and publish a profile with information about yourself. In this topic, you will be introduced to the social networking sites Facebook, LinkedIn, WhatsApp, Instagram, Snapchat, and Reddit. These sites allow you to connect with friends, family, and business contacts as well as share stories. They can also be used for career networking, job searches, marketing promotions, and other job- or business-related activities.

## Facebook

**Facebook** was ranked first in a 2018 worldwide survey of active social networking users, conducted by Statista (an internet statistical portal). People post updates to their Facebook timeline about what they're doing and share links, photos, videos, music, games, news items, and quotes about anything that is of interest to them. Other people comment on a person's posts and/or show their support by liking the post or reacting with an emoji. Users can send private messages to chat with other people using the separate Facebook Messenger app. Users also create or join Facebook Groups to interact with people who share a similar interest. Facebook Events are also a popular way to send invitations to friends and post event notices and updates.

A new Facebook user starts by creating an account, setting up a profile, and adding friends to their Friend Lists. People are added as friends by searching email contacts or by using the Facebook search feature to send a friend request to someone. Once an individual accepts a friend request, the two people can see each other's Facebook activity.

Businesses and other organizations have Facebook pages to support and build their brand identity. For example, Figure 6.1 shows the Facebook page for GoPro, a

**Read & Learn**
Social Networking

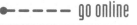
**in the know**

Facebook was founded in February 2004 as a network for college students only. In 2006, Facebook opened up to everyone, and quickly exploded to become the most-used social network worldwide. In June 2017, Facebook reached a milestone with 2 billion users (roughly two-thirds of the world population with internet access).

**go online**

https://CC3.Paradigm
Education.com
/FBSeguinConcepts
Go here to see the Facebook page for this textbook.

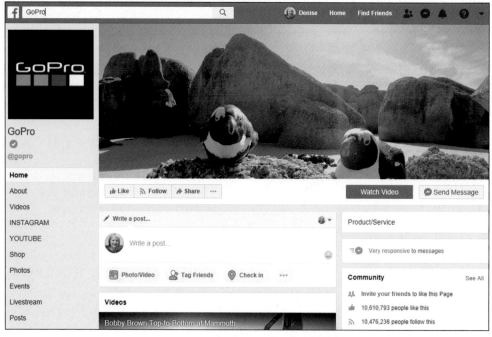

**Figure 6.1**

Businesses create Facebook Pages like the GoPro Page shown here. People interested in the business can follow the page to receive posts in their personal timeline when the business shares an update.

company that makes cameras, drones, and accessories designed for capturing images and video while on the move.

In March 2018, Facebook made headlines when it was reported that a loophole in Facebook's programming interface allowed data in up to 87 million Facebook profiles to be misused during the 2016 US presidential campaign. It was alleged that a company targeted ads to users with the intent to influence the outcome of the election. Facebook was served with several lawsuits when its role was exposed, and the social network found itself fighting off a movement by privacy advocates to encourage people to delete their Facebook accounts. Despite the negative press, Facebook remained the number one social network in November 2018.

## LinkedIn

**LinkedIn** is a business-related social networking site launched in May 2003 and reports that new members join the network at a rate of more than two per second. Microsoft acquired the social network in 2016. LinkedIn reports over 562 million registered members, with students and recent college graduates the fastest growing demographic. It is a professional network where users connect with business contacts.

●----- in the know

More than 70 percent of LinkedIn members reside outside the US.

Creating an account with LinkedIn allows you to create a professional profile with your current and past employment positions, education credentials, and links to websites (such as your current company website). Your profile can be expanded to represent, in résumé-like fashion, a summary of your qualifications and prior work experiences. Once your profile is established, you begin connecting through LinkedIn with people in your workplace and with other professionals in your field and related fields. The search text box in LinkedIn is used to find people by name, job title, company name, or location. You can also join professional association groups to keep up with news in your field.

A Jobs feature includes a search tool to find job listings by job title, keywords, or company name. Job postings can be filtered by date, company, and experience level. Figure 6.2 shows a job search for a social media intern in the United States.

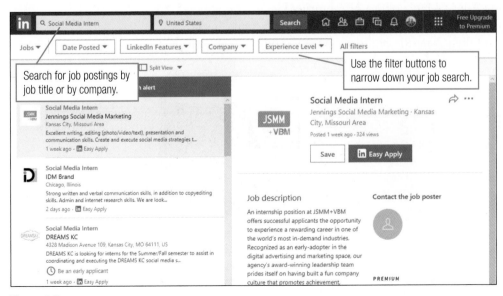

**Figure 6.2**

LinkedIn is often used by job seekers to find current job openings like the postings shown here for a social media intern in the United States.

## WhatsApp

**WhatsApp** was born in 2009 as an instant messaging tool on smartphones to send text messages, photos, video, and documents to other people who use WhatsApp. The app also allows you to have group chats and make voice calls. The name is a pun on the phrase What's up? The app was started as an alternative to SMS text messaging, allowing users to stay in touch with friends and family around the world. At its inception, WhatsApp found a following by people who did not like the charges and limitations incurred when using SMS text messaging to keep in touch. Another appeal of the app is that WhatsApp messages and calls are encrypted, meaning that a third party cannot intercept the transmission to read messages or listen to conversations. The app is easy to use, since anyone in your contacts list can be messaged by using the contact's phone number (as long as the other person also has a WhatsApp account).

Facebook bought WhatsApp in 2014. The app is free to use and does not contain ads (although smartphone users may incur data charges depending on their plan). In July 2018, WhatsApp reported over 1 billion users in over 180 countries. Figure 6.3 shows an example list of chats in WhatsApp for messages sent and received over three days on a user's account.

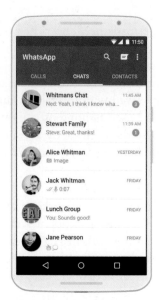

**Figure 6.3**

A list of chats in WhatsApp shows that users can communicate with individuals and groups (the top two messages shown in this image are for groups).

## Instagram

**Instagram** is a social network for sharing photos and videos from an app on a smartphone. In June 2018, Instagram had passed the 1 billion users milestone, with 500 million users active daily. After taking a photo or capturing video with your smartphone or tablet, use the app to add a special effects filter, type a caption, tag other users, tag a location, and then share the photo or video to your Instagram page or other social networking sites. You can follow friends, family members, or other users to see their pictures and videos in your feed and to allow other users who follow you to see your posts.

Instagram started in 2010 and grew to 1 million users in just two months. In 2012, Facebook purchased Instagram. On average, 95 million photos and videos are shared each day.

Figure 6.4a shows a photo with filter effect options, and Figure 6.4b shows editing tools available to adjust a photo in the Instagram app.

a    b

**Figure 6.4**

(a) Filter effects and (b) editing tools are available in the Instagram app for making adjustments before sharing the photo on Instagram.

## Snapchat

In 2011, three Stanford University students developed a new app for mobile devices called **Snapchat**. Today, more than half of the 255 million active Snapchat users (called *Snapchatters*) are between the ages of 12 to 24. Snapchat users capture photos and videos on their device, make edits, add text or drawings, add filters or other special effects (including augmented reality effects using Lens Studio), and then send the media (called a *snap*) to their friends. A snap is viewable for 1 to 10 seconds or until the recipient closes it, at which time the snap is automatically deleted; however, users have the option to save something important to a private storage area or take a screenshot.

In 2017, Wendy's cups with a special Snapchat code gave customers a special filter when the code was scanned. For each shared photo with the filter, Wendy's, along with Coca-Cola and Dr. Pepper, donated $5 (to a maximum of $500,000) to the Dave Thomas Foundation for Adoption®.

After downloading the app from the app store, a user creates an account by providing a name, email address, and birthdate. The user then creates an alias to associate with the snaps and a password to sign in. To add friends, a user can choose to upload contacts and select people who have a Snapchat account, or search for people.

The Stories feature is a collection of pictures and videos that are played back like a slide show. A story can be viewed by friends for a 24-hour period. The Discover feature lets a user browse content from publishers for news, sports, entertainment, and other story sources.

## Reddit

**Reddit** is a social-networking website where registered users submit content with text, links, or photos. Other registered users, called *redditors*, vote to move the content up or down the list. Each page updates stories as new submissions are posted and voting occurs. The position of an entry on the page shifts depending on its submission score, which is the number of upvotes minus the number of downvotes. Content is organized by categories (called *subreddits*)—such as gaming, videos, news, movies, and sports—to which you can subscribe. Each subreddit is independent, with a set of guidelines and rules to follow (moderated by volunteers). A user's posts and comments receive a karma score, which is affected by the number of upvotes—a high karma score indicates the community found the post valuable. In November 2017, Reddit had over 330 million average monthly active users. Figure 6.5 shows part of the Futurology subreddit posts and their votes on July 2, 2018.

**Blog This!**

Is It Fair that Social Media Posts Can Be Used Against You?

> Redditors use these voting buttons to move a story up or down a page.

**Figure 6.5**

The Futurology subreddit has over 13 million subscribers who write posts and comment about the development of humanity, technology, and civilization.

## 6.2 Media Sharing

Some social networks have been developed with a primary purpose of providing a platform to publish and share videos, photos, and music. These websites for **media sharing** allow people to publish content, work with content collaboratively, and distribute content more efficiently than sending files or links as message attachments. For example, YouTube was created in 2005 as a video-sharing website, but in 2015 YouTube Music launched to offer music services as well as video. Spotify emerged in 2008 as a music-sharing website but expanded into video services in 2016. With sharing, commenting, and personalization features at these websites, the difference between a social-networking website and a media-sharing website is no longer distinct.

Some websites have applications for **peer-to-peer (P2P) file sharing**, in which users access each other's hard drives and exchange files directly over the internet. People use P2P applications to share software, games, movies, music, and other content. Some P2P applications have been used to share copyrighted material without permission from the copyright owner, such as exchanging movies or music tracks between users, and are often the target for litigation from copyright holders.

Digital audio and video files are sometimes shared on the internet as podcasts. A **podcast** is an audio or video post that is usually part of a series to which people subscribe. A software application called a **podcatcher** automatically downloads new podcast episodes to your computer or mobile device for you to listen to or watch whenever you like. Other podcasts are streamed from a website. In this topic, you will learn about YouTube, Flickr, Pinterest, and Spotify—websites used primarily for publishing and sharing video, images, and music media.

### YouTube

**YouTube**, owned by Google, is the most popular video-sharing website in the world. A YouTube video becomes a **viral video** when it spreads quickly via internet sharing via social media or email. With a YouTube account, you can search, browse, and watch videos; upload a video you have created; subscribe to someone's YouTube channel; share videos with friends; stream music or watch music videos; and create playlists.

Figure 6.6 shows the official YouTube channel for Windows. A YouTube channel is a set of videos created by one person or organization. A YouTube channel is required if you want to upload a video. To create a channel, sign in to YouTube with a Google account, click your profile icon, click *My channel*, type the name you want associated with videos, and then click the Create Channel button.

**Figure 6.6**

Subscribers to the Windows channel on YouTube, shown here, receive notices when new content is uploaded.

**Read & Learn**

Media Sharing

**Video**

How Can You Safely Share Videos?

● – – – – – go online

**https://CC3.Paradigm Education.com /Mashable**

Go here for up-to-date news on digital culture, social media, and technology.

● – – – – in the know

To stream video or music without ads, you generally have to pay for a subscription. YouTube Premium (formerly called YouTube Red) and Amazon Video (part of Amazon Prime but also available as a standalone service) are two examples of services with large subscriber bases.

According to Hootsuite, over 1.5 billion users watch YouTube videos (second only to Facebook), 96 percent of teens watch YouTube, and more than half of all views come from a mobile device. Approximately 300 hours of content are uploaded to YouTube every minute! Although YouTube is the clear leader for video sharing, Vimeo, another video-sharing website, hosts videos from over 70 million creators.

## Flickr

Most people have digital photos they want to share with family, friends, or other contacts. Uploading pictures to a photo-sharing website lets you easily share a photo album with other people. Although social networking sites such as Facebook and Instagram let you upload and share photos, these sites generally have lower photo storage limits than a photo-sharing website. **Flickr** was among the original photo-sharing services, started in Vancouver, Canada, in 2004. Flickr was owned by Yahoo from 2005 to 2018, when SmugMug acquired Flickr. Figure 6.7 shows the Explore page for photos uploaded to Flickr. See Table 6.1 to review other photo-sharing websites.

Generally, photo-sharing websites let you upload pictures organized into albums, sets, or folders; tag and comment on pictures; and share the photos with others. Most photo-sharing websites offer basic photo editing tools. For example, with a Google account, you can store photos for free at Google Photos, with unlimited storage from Apple or Android devices or via the web, using the *High quality* option. This option compresses photos to save space and applies a maximum resolution of 16 megapixels. Video has a maximum resolution of 1080p (also called Full HD). Users who want to upload photos in the *Original quality* option do not get free, unlimited storage; instead, the 15 GB storage in their Google account is used.

> ●─ ─ ─ ─ ─ SECUre IT
>
> Be careful when sharing photos not to tag people in the pictures without their permission. Some people do not like to have a picture they appear in published on a social-networking website. If in doubt, ask them before sharing or better yet, just choose another photo that won't invade someone's privacy.

**Let's Explore!**
Share a Presentation

**Table 6.1**   Sampling of Photo-Sharing Websites

| Photo-Sharing Website | Restrictions for Free Accounts | URL |
| --- | --- | --- |
| Flickr | Up to 1,000 photos or videos | https://www.flickr.com/ |
| Google Photos | Unlimited photos and videos for images up to 16 megapixels and video up to 1080p | https://photos.google.com/ |
| Photobucket | 2 GB for photos and videos up to 10 minutes each | http://photobucket.com/ |
| Shutterfly | Unlimited photos with no maximum file size | https://www.shutterfly.com/ |
| Snapfish | Unlimited photos for JPG/JPEG and PNG files only | https://www.snapfish.com/ |

**Figure 6.7**
The recent photos on the Explore page at Flickr change constantly as new photos are added.

## Pinterest

**Pinterest** is used to share your pictures with others, to save pictures or ideas that you find on other websites, or to save pictures you see by browsing other people's Pinterest boards. Think of Pinterest as your digital bulletin board. Some items saved to a Pinterest board link to a story, article, recipe, craft, video, or other content.

Pinterest launched in March 2010 and quickly joined the ranks of Facebook and Twitter in the top 10 social-networking websites. By April 2018, Pinterest had over 200 million active users.

Some websites include a Pin It button next to an image or other content that can be copied to Pinterest. You can also add a Pin It button to the toolbar in the web browser that you use. Click the Pin It button, select the image to pin, and then choose the board to pin it to. A board is a collection of related pictures. Users can assign boards unique names. For example, you can create a board for technology and collect pictures of cool gadgets (Figure 6.8). You can browse and follow boards created by other Pinterest users to get ideas for yourself. If you like something you find on someone else's board, you can repin the item to one of your boards.

**in the know**

According to Pinterest, 85 percent of its daily traffic initiates from a smartphone or tablet. Pinterest also says 40 percent of all new Pinterest accounts come from men, despite the widely held belief that Pinterest is a female-dominated social network.

Click here to type what you want to look for on other boards.

Click here to view your own boards and pinned images.

**Figure 6.8**

With Pinterest, users share pictures or other visual content, such as the images pinned to this Technology board. Add content to a board by uploading a picture from your device, by searching within the social network for pins to save to one of your boards, or by pinning content from a website.

## Spotify

**Spotify** is a music- and video-streaming service launched from Sweden in 2008. Spotify is used by over 170 million people to listen to music on internet-connected devices, like smartphones, tablets, gaming consoles, speakers, TVs, cars, wearables, and PCs. Spotify was one of the first websites to offer streaming of music via the web instead of requiring individuals to download tracks to a device. Spotify is also popular because of its vast collection of music, which includes over 30 million tracks and 2 billion playlists. A sampling of other music streaming services is given in Table 6.2 on the next page.

**Table 6.2**  Five Music Streaming Services

| Music Service | URL |
| --- | --- |
| Apple Music | https://www.apple.com/music |
| Deezer | https://www.deezer.com |
| Pandora | https://www.pandora.com |
| Spotify | https://www.spotify.com/ |
| YouTube | https://music.youtube.com |

Spotify offers both a free account and a paid subscription account. A free account has ads between tracks and delivers music at a lower quality. The paid account, called Spotify Premium, is ad-free and streams high-definition music. In addition, Premium users can listen to music offline. A feature called Discover Weekly delivers a personalized playlist of two hours of tracks every Monday to your phone or other devices (Figure 6.9). The weekly playlist is compiled based on the previous tracks you have played and similar tracks listened to by other users.

To use Spotify, download and install the app from the app store for your device or from the Spotify website (see Table 6.2). Create an account or sign in with your Facebook account. Browse for tracks by genre or use the search text box to find a song by artist, title, playlist, or record label. You can easily save favorite tracks to your music library using the save icon next to a track.

Music sharing sites are often the target of litigation by music recording labels for copyright violations. In July 2017, Spotify was served with legal paperwork from two independent music publishers claiming damages totaling $365 million for copyright license infringements. By September that year, Spotify was defending itself in 10 lawsuits for similar infringements with the potential for $1 billion in damages. Spotify was served with yet another lawsuit in January 2018 from Wixen Music Publishing claiming $1.6 billion in damages for failure to secure proper copyright licenses.

YouTube, Flickr, Pinterest, and Spotify featured in this topic are just four options in each category of media-sharing websites. To choose a service for uploading videos, photos, or music, consider using more than one service on a trial basis until you find the website that suits your needs and preferences best.

**go online**

**https://CC3.Paradigm Education.com/Music Resource**
Go here to find a source for music authorized by the Recording Industry Association of America and the Music Business Association—each service listed properly compensates artists and associates involved in making the music offered on the website.

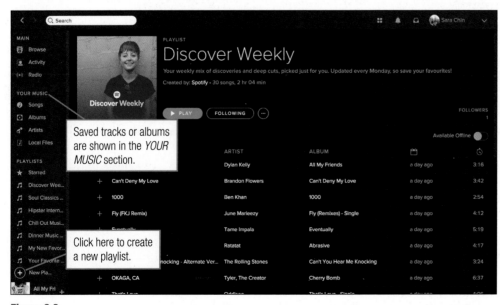

**Figure 6.9**
Spotify plays on most internet-connected devices. The Discover Weekly playlist shown here appears on Mondays and has two hours of music selected for you by Spotify based on your interests.

## 6.3 Blogging

A **blog** is basically a journal posted on a website, with the most recent entry at the top of the page. Older blog entries are usually archived by month and can be displayed by clicking the month to expand the blog entry list. The word *blog* is derived from *web log*. The collection of all of the blogs on the web is referred to as the **blogosphere**. Writing and maintaining a blog is called **blogging**. An entry in a blog is called a **blog post**, *post*, or *entry*. The individual who writes the blog content is referred to as a **blogger**. Blogs exist on just about any topic one can think of and are mostly text and pictures; however, posts also contain links to other web pages and sometimes video and audio. The popularity of YouTube led to a form of blogging that uses videos—called **vlogging**. A **vlog** is a blog that uses video as the primary content. A **vlogger** is a person who creates a video blog.

Blog websites include tools that allow individuals to create, update, and maintain blogs without a programming background. Blogs are often interactive, meaning that people post comments on blog posts, and the authors and their audiences can have conversations about the posts. In this topic, you will learn about two blog hosting websites, WordPress and Blogger, as well as one microblogging website, Twitter.

WordPress and Blogger are two of many blog hosting websites. Table 6.3 provides other blogging services. While most blog hosting websites offer a free account option, expect to pay for extras like customization or extra storage space on the server. Look at more than one option before deciding which blog hosting service to use. Factors to consider when evaluating your options include an intuitive interface, fast server response, and support features that offer assistance if you get stuck. You may decide that paying a fee to host a blog is worth the cost for additonal features or design options.

**Table 6.3** Blog Hosting Services

| Blog Hosting Service | URL |
| --- | --- |
| Blogger | https://www.blogger.com |
| LiveJournal | https://www.livejournal.com |
| Medium | https://medium.com/ |
| Squarespace | https://www.squarespace.com |
| Tumblr | https://www.tumblr.com |
| Twitter | https://twitter.com |
| Weebly | https://www.weebly.com |
| WordPress | https://wordpress.com |

### Microblogs

A few blog services such as Twitter and Tumblr are microblogging hosting services. A **microblog** is a blog that restricts posts to a smaller amount of text than a traditional blog. A **microblogger** is someone who typically posts about what he or she is doing, what he or she is thinking about at the moment, or links to interesting content (stories, pictures, or videos). Tumblr is a microblog that includes social networking features. A post on Tumblr is called a *tumblelog*. Many Tumblr users have blogs to bookmark videos or pictures they have found on the web and share them with their followers. Tumblr users also frequently reblog items seen on someone else's page at their own pages.

**Twitter** is by far the most popular microblog, with posts limited to 280 characters or less. A post on Twitter is called a **tweet** and contains information about a breaking news story, opinion, status update, or other topic the author

**Read & Learn**

Blogging

**Video**

When Do Blogs Go Beyond Words?

**in the know**

In July 2018, the YouTube channel with the most subscribers was created by PewDiePie, a vlogger from Sweden whose real name is Felix Kjellberg. The PewDiePie channel launched in 2010 amassed 64 million subscribers by mid-2018 with 3,500 videos.

**in the know**

Twitter was launched in July 2006 and by 2018 had 330 million monthly active users that collectively sent 500 million tweets per day—nearing 6,000 tweets per second!

finds interesting. Because of the short nature of a tweet, many users post to Twitter from their mobile devices. Clicking the Tweet button opens a small window in which the user types a new post. Twitter users type the **hashtag symbol (#)** preceding a keyword or topic (with no space) in a tweet to categorize posts or replies. Clicking a hashtagged word in any post will show you all the other tweets categorized with the same hashtag.

Many people like to follow other Twitter users to see the tweets of friends, family, business associates, celebrities, public figures, or news agencies. Twitter users also follow business accounts to find out about new products, sales, or other specials (Figure 6.10a). Many businesses set up a support account for their customers to tweet questions or issues to the customer service team or to post information to customers about common customer support questions (Figure 6.10b).

According to Hootsuite, 80 percent of users access Twitter from a smartphone or tablet, with 90 percent of video views from Twitter happening on a mobile device.

Some Twitter accounts reach millions of followers. In July 2018, Katy Perry recorded the most Twitter followers at 109.5 million. Twitter is also a frequent resource to learn about breaking news or other trending topics.

In August 2017, residents in flood-ravaged Texas used Twitter and Facebook to post updates and requests for help when 911 and other emergency call centers were overwhelmed. Despite pleas from authorities to not use social media and continue to try calling 911, Texans experienced success posting pleas for assistance on social media platforms. The hashtags *#SOSHarvey* and *#HelpHouston* in tweets directed rescuers and others wanting to help.

For example, one woman tweeted a plea for help along with her address when she could not get through to a 911 operator. Just over an hour later, the same woman tweeted a thank you after she and her two children were picked up by a fire rescue truck.

Figure 6.10c shows a tweet by Reuters alerting residents to the imminent danger from the flood waters. George Washington University media professor, Nikki Usher, remarked that Hurricane Harvey was the first major US natural disaster of the social media age, "where you see rescuing via social media."

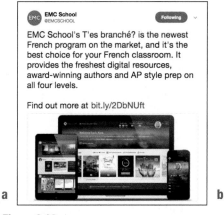

**Figure 6.10**
A business will tweet about (a) new products or (b) to offer tips and helpful links for common issues identified by customer support. Twitter became a (c) news resource for aid by victims of flood waters in Houston in August 2017.

## WordPress

**WordPress** (https://wordpress.com) is a blog hosting website that uses the WordPress open-source blogging software. To create a website or blog, click the Get Started button on the main page and then answer the questions posed. For example, provide the name for your site, the topic, the primary goal for the site, and your knowledge level for creating a website. Enter a name or keyphrase to use as your website address and click the option for a free account. Next, provide your email address and a password.

By default, a WordPress blog address begins with the domain you want to name your blog and ends with *.wordpress.com*. For example, if you type your name as the blog address, the URL you would give to people to find your blog would be *yourname.wordpress.com*. After registering for a new blog, go to your email to activate the account from the confirmation email message. At the screen that appears after you activate the account, click the <u>Write</u> hyperlink to publish your first blog post. Figure 6.11 shows the screen for creating a new post. WordPress prompts pop up to guide you through each step in a word processor–style window.

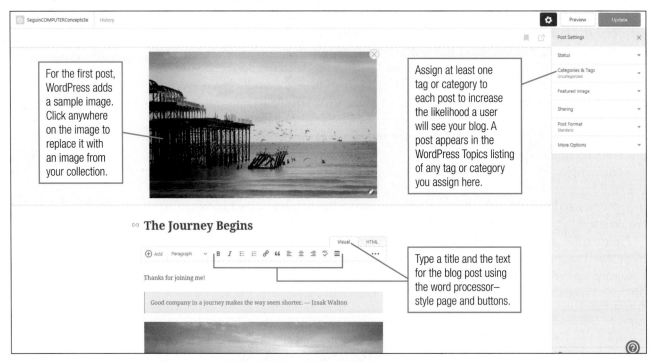

**Figure 6.11**

A blog post is created by typing the title and text on the new post page using an interface similar to that of a word processor like Word.

## Blogger

**Blogger** (https://www.blogger.com) is another popular choice for creating a blog. Sign up for a Gmail account first if you don't already have one. When ready, navigate to the Blogger web page, scroll down and then click the Create Your Blog button. Sign in with your Gmail account and set up your profile. At the Welcome page, click the Create New Blog button, type a title for the blog, enter a blog address, choose the theme with the colors and style you want to use, and then click the Create blog button. Blog addresses in Blogger end with *.blogspot.com* (*.blogspot.ca* if you reside in Canada), so the URL for a new blog would be *yourblogaddress.blogspot.com* (*yourblogaddress.blogspot.ca* in Canada). When the blog is created, click the <u>*Create a new post*</u> hyperlink to start your first post. Begin by typing the post title and text, as shown in Figure 6.12. Format the text, add links, pictures, video, or quotes as needed, and then when finished, click the Publish button.

Once the first post has been added to the blog, additional posts are created by clicking the New post button. You can change the theme (background, layout, and set of colors and fonts) at any time and preview how the blog will look in both PC and mobile environments by clicking the *Theme* option in the left pane (Figure 6.13).

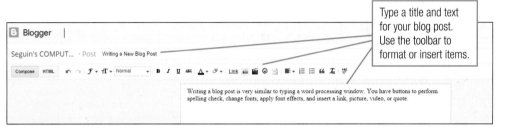

**Figure 6.12**

Blogger is a popular blog hosting service because the tools are easy to use and navigate for beginner bloggers.

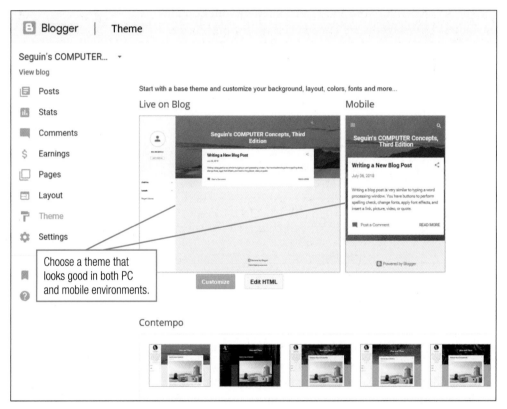

**Figure 6.13**

The blog design scheme can be changed at any time on the Theme page, and you can preview how the blog will look on both a PC and a mobile device.

In 2018, Statista reported the number of bloggers in the US who update a blog at least once each month to be 31.7 million people in 2020. Surveys of bloggers reveal that most start their blog as a hobby for reasons that include to share ideas with like-minded people, to connect with new people locally and globally, to help people going through a similar phase or circumstance, or to create awareness about an issue. Successful bloggers advise those that are considering starting a new blog to write original, compelling content that will appeal to the ideal reader within the target market for the blog topic and to post consistently to grow readership.

---

**Let's Explore!**

Tweet a Social Media Fact

●━━━━━ secure IT

If you use content in a blog post that is not your original writing, make sure you credit the author and provide a link back to the source. You don't want to violate copyright. Also, be sure you don't include copyrighted photographs or other media without permission from the owner of the work.

# 6.4 Wikis for User-Generated Content

A **wiki** is a website with **wiki software** that allows anyone (except those under a ban restriction) to create and edit content at the website. Clicking the edit tab, edit button, or edit link in a wiki opens the text in a word processor–style window, where you can add, delete, or edit content. This type of open content is used for collaboration and knowledge sharing. The postings on wikis that provide encyclopedic or other reference content are monitored by people to make sure entries are well written and accurate. In this topic, you will explore two well-known wiki websites: Wikipedia and WikiHow.

## Wikipedia

The most used online encyclopedia is **Wikipedia**, supported by the not-for-profit Wikimedia Foundation. Individuals looking for new information about a subject often start at Wikipedia. Anyone can create content at Wikipedia; however, thousands of editors regularly review and update pages.

People who want to edit at Wikipedia are encouraged to create an account so that their postings will be identified with their name or alias. If you edit a page without being logged in to an account, the IP address for your PC is attributed to your content in the page history. New contributors to Wikipedia are encouraged to first post in the **Wikipedia Sandbox** (Figure 6.14), where they can learn how to edit content.

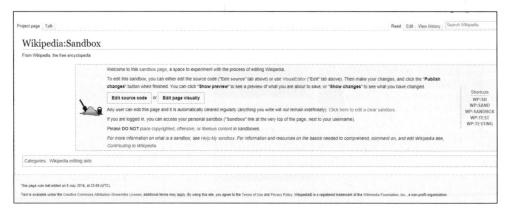

**Figure 6.14**

The Wikipedia Sandbox contains instructions on how to edit content, and new contributors are encouraged to post a test edit here before modifying a real article.

Wikipedia is a frequent starting point for students and teachers alike to gather quick information and to get a grasp of a new topic. Most schools have a policy or general practice that prevent students from using Wikipedia as a source in a research paper, essay, or other academic work. The reasoning for disallowing citations of Wikipedia content is that content in articles on Wikipedia can be written by anyone. Although articles are reviewed and errors are corrected when found, an error may go unnoticed and remain until discovered and corrected. Thousands of edits are made each day, and it would be impossible for Wikipedia to monitor each one promptly. However, some teachers permit Wikipedia sources, depending on the topic or purpose of the assessment; check with your instructor, if necessary, before citing Wikipedia in a graded assignment.

The Featured content portal at Wikipedia contains articles that have undergone a thorough review and meet the highest standards. An article that is part of the Featured content portal displays a bronze star at the top right corner of a page, as shown in Figure 6.15 on the next page. These articles are reviewed extensively and edited many times before being assigned the bronze star.

**Read & Learn**

Wikis for User Generated Content

**Video**

What Are the Differences between Wikis, Blogs, and Microblogs?

●  ━ ━ ━ ━ in the know

At the time of writing, over 5.67 million articles were available on Wikipedia.

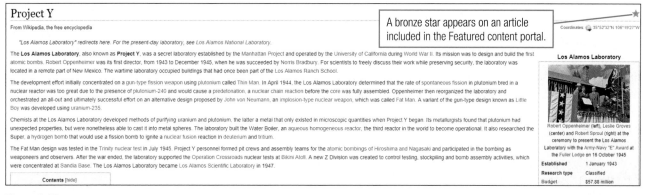

A bronze star appears on an article included in the Featured content portal.

**Figure 6.15**

Featured content articles display a bronze star at the top right of the page indicating the content has been thoroughly reviewed.

## WikiHow

When people need a how-to instruction manual, **WikiHow** is a free resource. The website contains more than 187,000 how-to articles. According to wikiHow, the mission of the website is that "by creating the world's most helpful instructions, we will empower every person on the planet to learn how to do anything." You can search for the instructions you need using keywords or browse articles by category at the home page.

An article with a green check mark signifies an expert has reviewed the article. For example, a medical article with a green check mark has been reviewed by a doctor, or a legal article has been reviewed by a lawyer. According to WikiHow, the average article "has been edited by 23 people and reviewed by 16 people." Volunteers also read feedback on articles and will rewrite instructions or redo illustrations to make the article more easily understood.

Figure 6.16 shows the mobile app. WikiHow is a for-profit wiki that sells ads to finance the operation of the website and shows ads to users who are not registered.

Wikis to investigate are listed in Table 6.4.

A green check mark indicates article has been reviewed by an expert.

**Figure 6.16**

WikiHow is available as a free mobile app. This article, shown on an Android device, shows the start of instructions on how to perform the Heimlich maneuver to assist a choking victim.

**Table 6.4**  Popular Wiki Websites

| Wiki | URL | Purpose |
| --- | --- | --- |
| Wikibooks | https://www.wikibooks.org | Free textbooks and manuals |
| Wikimedia Commons | https://commons.wikimedia.org/wiki/ | Freely usable images, photographs, or videos |
| WikiHow | https://www.wikihow.com | How-to instructions |
| Wikipedia | https://www.wikipedia.org | Encyclopedia |
| Wiktionary | https://www.wiktionary.org | Dictionary for words in all languages |
| Wikitravel | https://wikitravel.org | Worldwide travel guide |

# 6.5 Social Media Strategies in Business

Business strategies for using social media incorporate a wide variety of techniques, such as using Facebook pages to provide customer support, invite feedback, offer a forum for new ideas, and allow users to exchange tips, post reviews, or otherwise interact with interested people. Businesses also offer coupons or information on specials using social media. Social media content designed to encourage customer engagement and brand loyalty employ hashtag campaigns with video or photo-sharing contests/campaigns to engage customers.

A category of e-commerce called **social commerce (s-commerce)** refers to the use of social networks to support the buying and selling of products and donations or services. Social commerce employs techniques that encourage user-generated assistance in online selling, such as user ratings, referrals, reviews, and participation in online communities. Social networks offer assistance to businesses and non-profits with social commerce activities such as these three examples:

- The Facebook Business page gives over 11.9 million followers marketing tips, strategies, and success stories, as well as hosts live video streams with experts that speak on a variety of topics to increase business engagement (Figure 6.17a).
- Twitter hosts a business blog and tweets daily to 1.01 million followers on their @TwitterBusiness page. They offer tips for using Twitter in s-commerce activities (Figure 6.17b).
- Instagram's website https://business.instagram.com/ is dedicated to helping businesses use Instagram for promotion with information, tips, and help for creating ads.

This topic describes s-commerce strategies employed by National Geographic, iHeartDogs, and Lowe's.

**Read & Learn**
HSocial Media Strategie in Business

**Video**
How Are Businesses Using Social Networking?

a

b

**Figure 6.17**

Social networks provide advice to businesses to help them effectively use the services offered for marketing, publicity, branding, s-commerce, customer engagement, and brand loyalty. (a) Facebook's Business page hosts live stream interviews and (b) Twitter's business account tweets hyperlinks to insights daily.

**Let's Explore!**
Find a Business in Social Media

## National Geographic

Statista gave top billing to National Geographic in 2017 for the highest user engagement in social media platforms, with over 1.6 billion actions. The American Marketing Association reported that its most "liked" image in 2017 (just under 2 million) was a photograph of an indigenous child balancing a pet monkey on his head.

National Geographic also earned 1 million new subscribers a month for three consecutive months after launching its Snapchat Discover page. *Safari Live* is an interactive live video stream of wild animals in Africa that broadcasts twice daily on YouTube, Twitter, and Facebook, with viewers able to comment and pose questions to the guides in real time using the hashtag *#safarilive*.

## iHeartDogs

In November 2017, iHeartDogs ran a Facebook ad campaign that advertised their Black Friday sale was 0% (Figure 6.18). The novel ad campaign that went against all consumer expectations for a Black Friday sale resulted in the company quadrupling its sales. By clicking a link from the ad, readers were told that instead of giving a discount, the company would double their donations from purchases made over the Black Friday buying weekend with a goal to provide 100,000 meals to dogs in need. The campaign was so successful that the company was able to double their donation to 200,000 meals.

**Job Connect!**
Social Media Careers

**Figure 6.18**

The unconventional iHeartDogs Facebook ad for Black Friday 2017 led to the company quadrupling sales and exceeding their charitable giving goal by telling customers that instead of a sales discount they would double their donations of meals for dogs.

## Lowe's

Using Instagram and Twitter, Lowe's publishes DIY home project videos using the hashtag *#hypermade* (Figure 6.19). The Lowe's 2017 "Made in a Minute" Facebook campaign used 360-degree video to show viewers how to complete a DIY project. The company actively engages with commenters on its Facebook page, which boasts over 4 million followers, as well as through Twitter. Most comments receive a reply from a Lowe's employee the same day or the next day. The company also runs a DIY-focused YouTube channel, as well as a channel highlighting the company's community involvement called "Lowe's Open House."

**Figure 6.19**

Lowe's uses multiple social media channels to publish DIY project videos with the hashtag *#hypermade*, such as this planter project posted on Instagram.

Lowe's social media campaigns win industry awards. The company was recognized by Facebook with a global award for the "Made in a Minute" campaign.

# 6.6 Leveraging Social Networks for Your Career

Employers are frequently turning to social media to look at prospective employees and decide whether the person is a good fit. If you have applied for a job, chances are someone in human resources has searched your name on social networks to see what your public persona looks like and what you have accomplished. The employer may also use social networks to fact-check your résumé.

In today's world, *not* having a presence on social media can hinder your career advancement. If you do not have a few basic social network accounts, you should consider starting to build a personal brand suitable for the type of career you want. Employers want to see how you present yourself to the world, since it will reflect how you will represent them if they hire you. If you are currently active on social media, take some time to review your profile, privacy settings, and activity to make sure a potential employer will find only what you want them to see.

Social media can also help you as you develop your career by providing venues for you to build new contacts, explore new ideas and topics related to your field, and let you establish yourself as someone with current knowledge and skills who is a reliable, trustworthy, top performer.

Be aware of your **digital footprint**, which is the accumulation of your online activities. Your digital footprint includes your social media activity, but also extends to an umbrella of your online shopping habits, online subscriptions, photos, and videos that you have uploaded or shared. In this topic, you will learn how to use activity on social networks to enhance your career.

Regularly using social media can help you stay on top of current trends in your industry, like the LinkedIn feed shown here from Microsoft to its LinkedIn followers.

## Building Your Professional Network on LinkedIn

As you learned in Topic 6.1, LinkedIn is the social network where you connect with business contacts. Start with your profile and add text in each section, if possible. At a minimum, keep up-to-date information in the *Experience*, *Education*, *Accomplishments*, and *Interests* sections. Use words in the *Headline* box that are keywords an employer would use to search by. It is worth the investment to pay for a headshot photo by a professional photographer if you can afford it. If not, have a friend or family member take a good-quality photo of you that will show well in your profile. Check the public profile settings to see how your page looks to someone else (Figure 6.20 on the next page).

Search for people that you can connect with at your current or former workplace, look for friends, and seek out family members to expand your personal network. Next, look for someone in your field who has posted content. Reply to a few posts, making sure your comments extend the conversation and/or are related to an article or link they shared. The comments should not be short replies that say, "Great article! Thanks." The idea is to establish yourself as someone with similar knowledge in the same field. Then, reach out and send the person a connection request. Always include a brief note in the request letting the person know who you are and why you are reaching out. Chances are they will have read your replies and are more likely to accept the request. They will likely check your profile before accepting, so polish up your profile, if necessary, before sending a connection request to someone you don't know. In order to ensure successful results, use a targeted approach instead of blanketing a large group.

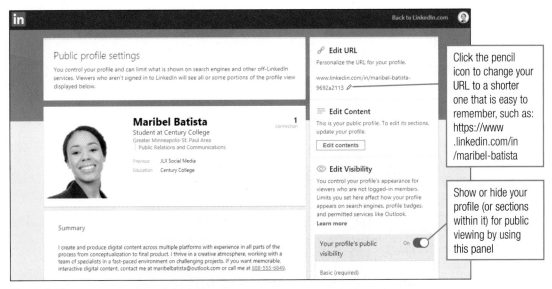

**Figure 6.20**

Click the *Edit public profile & URL* hyperlink at the top right of your profile page to get to the page shown here and change what the general public sees in your LinkedIn profile. Public profile settings affect results that appear in search engines for people not already connected to you.

Post your own content on a regular basis once you are familiar with the writing style and content posted by others in your field. LinkedIn content is geared toward the workplace, so post a link to a new and relevant article or video, or share a recent experience or tip. When posting links, tell readers why you are sharing the content, or pose questions to invite replies.

Ask people who know you and have worked with you to use the Recommendation feature to endorse you. You can do the same for them, but only give an endorsement for someone with skills that you can attest to and are willing to stand behind. Employers look at endorsements and treat them as bona fide references.

LinkedIn is a resource for job searches. Use the Jobs feature to look for openings. Filters let you drill down the results list by date, company name, experience level, and features, for example: Under 10 Applicants, Easy Apply, and In Your Network.

## Facebook Beyond Friends and Family

Most people think of Facebook as a social network for friends and family, but it can be used to build a professional brand. Facebook content can be a bit more conversational in nature. Join a Facebook group suited to your career to develop new connections. Follow businesses in your field and comment on their posts about products, events, or other information, or ask questions to extend your knowledge.

Update your privacy settings so personal photos and content are not publicly visible. The Privacy Checkup feature helps you review settings. Complete these steps to use Privacy Checkup:

1. Click the Quick Help icon in the blue banner at the top of the page (displays as a question mark inside a circle).

2. Click the *Privacy Checkup* option.

3. Change the setting for who can see your posts and your profile, if necessary, and then click the Next button. (You can change this option for an individual post you want visible to anyone on the internet.)

4. Change the options for who can see your email, birthday, year of birth, and relationship status from your profile, if necessary, and then click the Next button.

● – – – – in the know

According to Express Writers (a content creation agency), a Facebook post best practice is to keep the text within 80 to 120 words and include a visual. Positive statements do better in Facebook when they end with an exclamation mark!

5. Edit the options for the apps and websites you have used with Facebook, if necessary, and then click Finish.

## Using Instagram and Snapchat for Visuals

People go to Instagram and Snapchat for pictures, videos, or other visual content. If your career gives you an opportunity to take a compelling photograph or create a new infographic with interesting content, share those visuals if you have permission to post the content on one or both of these networks. Instagram has the larger following, so choose that network first. Tag the content with hashtags and use appropriate emojis (Figure 6.21). These two social networks lean toward light-hearted content. Ask a friend who knows how to create graphics to help you if you are not familiar with photo editing software so that visuals you post are of good quality.

Look for hashtags used by others that might apply to your images and incorporate them into your posts. Hashtags help expand your audience. Follow businesses and people that are related to your career. Comment strategically with positive statements to extend the conversation in progress.

This photo caption has a relevant quote, hashtag, appropriate emojis, and uses a question that sparked 91 comments.

**Figure 6.21**

Post visuals with hashtags and emojis. This photo related to Earth Day also ends with a question to invite comments.

## Tweeting with a Purpose

Posting thoughts or ideas along with hashtags, your Twitter handle, and links in under 280 characters can be a challenge. The character restriction means you have to think and plan a tweet carefully. Tweets with visuals do better to attract interest. Tweet something that is topical and of current interest to followers. Or, consider asking a question about an issue to get people to respond with suggestions or advice. As you browse your feed, retweet anything that has interesting content. When you retweet, add a brief note about why you're retweeting (Figure 6.22). Comment on other tweets if you can add to the conversation. Avoid contentious or political topics that can alienate some of your audience.

**Figure 6.22**

When retweeting, include a statement to personalize it and let readers know why you think the tweet is important.

● – – – – – go online

**https://CC3. ParadigmEducation .com/TinyURL**

Go here to use a free service that converts a long URL to a short one that won't take up your entire post!

Follow businesses and people in your field. By reading their tweets, you will get ideas for content and stay current on industry innovations. Job seekers learn about openings, and a memorable tweet mentioned in a job interview helps you stand out.

## Managing Social Networks with Hootsuite

**Hootsuite** (https://hootsuite.com) is a company that helps you manage all your social network content in one place. An individual plan to manage up to three social profiles is free. Managing four to ten social profiles requires a subscription payment.

With a single sign-in at Hootsuite, you can create all your posts for all social networks and push the content to each one from the Hootsuite dashboard. Using the content scheduling feature, you can specify when each item is to be published, and Hootsuite takes care of automatically posting the status update, tweet, or post on the days and times you specify (Figure 6.23). With Hootsuite, you can set aside a block of time to create bulk content posts in one sitting so that you are not trying to fit social media content creation into each work day or each work week.

**Video**

How Can Hootsuite Help You Organize Your Communications?

AutoSchedule lets you set up your posts in advance and then let Hootsuite take care of publishing them.

**Figure 6.23**

In the Hootsuite dashboard, you can set the day and time you want posts automatically published to your social networks.

## Guidelines for Content Posts

Creating content for social media designed to establish yourself as a professional in your field will take some practice and research. Aim to put your stamp on content. Be patient, since it will take time to develop your style and gain followers.

The following guidelines will help you enhance your personal brand:

- Post consistently. Articles by social marketers give guidelines such as post to Facebook and LinkedIn once per day and tweet three times per day, but it is more important that you keep a consistent schedule and avoid posting three times one week and then being absent for the next two weeks.

- Quality rules. The content you post must be relevant and timely. Outdated or low-value material will cause people to unfollow you.

- Spelling and grammar count. Have someone proofread your posts before publishing if you are not good at editing your work.

- Content is king. People are looking for unique content and tend to respond more positively to personal stories. No matter how frequently you post, if your content is dull or too repetitive, you'll lose followers.

- Mix up the way you present content in your posts. Write about an experience, observation, or idea; share links; use an infographic or photograph; create or share a video; quote someone noteworthy; or ask questions to engage comments.

- Monitor the times when you get the most engagement from your audience to define which days and times are best for publishing content.

- Reply promptly to people who comment on your posts.

- Start with just one or two social networks and expand to others as you build confidence in your social media strategy.

Adjust your approach as changes in social networks occur. Stay alert to a new social network that pops up or one that falls out of favor so that your time is well spent.

## Topics Review

| Topic | Key Concepts and Key Terms |
|---|---|
| **6.1 Social Networking** | An online tool used by individuals to generate content and interact with others is referred to as **social media**. A **social-networking website** provides a platform for people to connect and communicate with one another, individually or within a group. |
| | **Facebook** is the most popular social networking site and is typically used by people who want to connect with friends and family. |
| | **LinkedIn** is a social network used to connect with business contacts. |
| | **WhatsApp** connects friends and family who want to exchange messages, photos, videos, or documents, individually or in group chats. The app can also be used to make voice calls. |
| | **Instagram** is a social network used to share photos and videos, some of which are edited with special effects filters and text. |
| | **Snapchat** is an app for a mobile device to capture photos or videos; add text, drawings, and filters; and send the content, called a *snap*, that is viewable for 1 to 10 seconds to other users. |
| | **Reddit** is a website where users submit content to pages organized into categories, called *subreddits*, and other registered users vote a story up or down the list. |
| **6.2 Media Sharing** | Websites for **media sharing** let registered users upload and share photos, or music. |
| | A **peer-to-peer (P2P) file sharing** application lets users access each other's hard drives to exchange files online. |
| | A **podcast** is an audio or video series to which people subscribe. Software that automatically downloads new podcast episodes for a subscriber is called a **podcatcher**. |
| | **YouTube** is owned by Google and is the most popular video-sharing website in the world. |
| | A **viral video** is a video that spreads quickly via forwarded links, shared links, or word-of-mouth. |
| | **Flickr** is a photo-sharing website that provides tools for users to upload, organize, and share photos. |
| | **Pinterest** is used to pin images that you want to save and share to a board in your Pinterest account. |
| | **Spotify** is a streaming music and video service. The service can be used for free or with a paid subscription. Playlists and tracks can be saved and shared with friends. |
| **6.3 Blogging** | A **blog** is an online journal created at a blog-hosting website, with the most recent post at the top of the page. |
| | All the blogs on the web collectively are referred to as the **blogosphere**. |
| | Writing and maintaining a blog is called **blogging**; each entry on a blog is called a **blog post**, *post*, or *entry*, and the person who wrote the blog is called a **blogger**. |
| | A form of blogging that uses videos is called **vlogging**. A **vlog** is a blog with video as the primary content created by a **vlogger**. |
| | A **microblog** is a blog in which posts are restricted to minimal text created by a **microblogger** who typically writes posts about what is happening at the moment, with links to interesting content, pictures, or videos. |
| | **Twitter** is the most popular microblog. A post on Twitter, called a **tweet**, is limited to 280 characters or less, with the **hashtag symbol (#)** preceding a keyword the author uses to categorize the tweet. |
| | **WordPress** is a blog website that uses the open-source WordPress blogging software, in which a blog is typed inside a word processor–style window. |
| | **Blogger**, owned by Google, is a blogging tool that uses templates and a user-friendly interface. |

*continued...*

| Topic | Key Concepts and Key Terms |
|---|---|
| **6.4 Wikis for User-Generated Content** | A **wiki** website uses **wiki software** that allows any user not under a ban restriction to add or edit content on websites that are used for collaboration and knowledge sharing. |
| | Wikis that provide encyclopedic information have thousands of editors who review and update content to ensure postings are well written and accurate. |
| | **Wikipedia** is the most used online encyclopedia. |
| | New contributors are encouraged to do a test edit in the **Wikipedia Sandbox** before editing a real Wikipedia article. |
| | **WikiHow** is a wiki website with thousands of how-to articles on a wide range of topics. |
| **6.5 Social Media Strategies in Business** | A category of e-commerce called **social commerce (s-commerce)**, refers to the use of social networks to support the buying and selling of products and services. |
| | National Geographic experienced the highest user engagement in 2017 on social networks, with posts including unique photographs, their Snapchapt Discover page, and their interactive *Safari Live* broadcasts twice daily on YouTube, Twitter, and Facebook. |
| | A novel 2017 Black Friday ad campaign that advertised a 0 percent discount on Facebook by iHeartDogs resulted in the company quadrupling sales and doubling a charitable gift of meals for dogs in need. |
| | Lowe's is known for DIY videos and pictures using the hashtag *#hypermade* on Instagram, Twitter, and Snapchat. The company has a "Made in a Minute" campaign on Facebook that uses 360-degree video to showcase a project, and also publishes videos on multiple YouTube channels. |
| **6.6 Leveraging Social Networks for Your Career** | Employers conduct searches of job candidates on social networks to view their public profiles and fact check résumés. |
| | Your **digital footprint** is the accumulation of all your online activities. |
| | To build a professional network on LinkedIn, start with a complete profile, add connections, post relevant content on a regular basis geared towards the workplace, and ask connections to endorse you. |
| | LinkedIn is a resource for job hunting with filters that let you drill down to openings that interest you. |
| | In Facebook, follow businesses in your field and comment on their posts, join appropriate Facebook groups to build new connections, and update your privacy settings to make sure personal photos and content is not publicly visible. |
| | The Privacy Checkup feature in Facebook assists you with reviewing and changing privacy settings. |
| | Post good-quality visuals that are lighthearted in nature to Instagram and Snapchat that make effective use of hashtags and emojis. |
| | Plan tweets that are topical and with current information, or pose questions to engage followers. Tweets with visuals attract more interest. |
| | When retweeting someone's post, add a brief note about why you are retweeting. |
| | **Hootsuite** is a company that provides software to manage multiple social networks in one dashboard. You can create bulk posts and use the scheduling feature to specify the days and times to post the content of each social network. |
| | Guidelines for content posts include: post consistently, quality rules, spelling and grammar count, vary the type of content, monitor times to determine best days and times for publishing, reply promptly to commenters, and begin with one or two social networks until you are confident with the strategy before expanding to other social networks. |

**Topics Review**

## Chapter 7

# Computer Security and Privacy

Chances are you spend most of your online time browsing the web, social networking, messaging, shopping, or researching. A network intrusion can occur anytime you are online, leaving you vulnerable to security and privacy breaches. According to Interpol, **cybercrime** (illegal acts involving a computing device that occur on the internet) is one of the fastest-growing crime areas, with organized crime groups increasingly controlling these illegal activities. Given the prevalence of cybercrime and other malicious computer threats, every person needs to understand and protect personal computing equipment, data, and information from theft or misuse. Some people believe each person has a responsibility to make sure his or her computer or other connected devices are protected because the security practices each person follows affect others. For example, if one of your computing devices becomes infected with malware, it could spread to infect others. Another possibility is that your infected device may be used to facilitate a network attack on another system without your knowledge.

**Computer security**, also known as *information security*, includes all activities related to protecting hardware, software, and data from loss due to unauthorized access or use, theft, fraud, natural disaster, and human error. A 2018 survey of 620 IT professionals in North America and Europe by ESG Cybersecurity revealed computer security is the area where organizations have continual problematic shortages of computer security professionals. Steady employment at a good salary is available for those who pursue a career in cybersecurity.

In this chapter, you will learn about the various security and privacy concerns associated with using a computer connected to a network or the internet, and strategies you can employ to secure your devices and data. You will also examine computer security and privacy issues related to mobile and IoT devices and then learn how to secure a smartphone, tablet, wearable device, or smart device in your home.

## Watch & Learn

Computer Security and Privacy

## Learning Objectives

**7.1** Explain various types of network risks that occur when computers are connected to a network or the internet, and list preventive security strategies

**7.2** Describe a botnet and a denial of service attack and the techniques used to protect against these types of intrusions

**7.3** Distinguish various types of malware and discuss methods to prevent malware infections

**7.4** Differentiate phishing, pharming, and clickjacking threats, and identify best practices for thwarting these attacks

**7.5** Recognize privacy concerns when using the internet and strategies for safeguarding personal information

**7.6** Identify mobile device security risks and techniques for minimizing risk

## Content Online

The online course includes additional training and assessment resources.

# 7.1 Unauthorized Access and Unauthorized Use of Computer Resources

Connecting to a network at home, at work, or at school has many advantages, such as sharing access to the internet and the network resources, storage, and application software. However, these advantages come with risks. Network attacks occur often in business and government organizations by hackers looking to steal large amounts of data (about customers, employees, and citizens) that they can use to make a profit from illegal activities. In this topic, you will learn about the types of risks associated with using a network, which include unauthorized access and unauthorized use, and ways in which a network can be protected from these threats. While the focus of hackers is on networks that are gateways to big data, a smaller network in your home should not be ignored when it comes to computer security.

## Unauthorized Access

Using a computing or network device without permission is referred to as **unauthorized access**. Unauthorized access occurs when someone using software programming skills or other technical means gains entry without permission into the hardware and data stored on a network resource. The term **hacker** refers to an individual who accesses a network without permission, and **hacking** describes activities involved in gaining unauthorized entry into network resources. Table 7.1 describes five network intrusions that have made recent news headlines. The personal information stolen in these events can have devastating consequences if bank or credit card data is used to steal cash or identities from victims. Theft of intellectual property (inventions or designs for new products) puts future jobs and investments at risk from canceled product developments.

Some hackers have good intentions; they attempt to pinpoint weaknesses in network security and may even be hired by an organization to hack into the organization's own network. This type of hacker is called a **white hat**. A hacker who gains unauthorized access with malicious intent to steal data, or for other personal gain, is called a **black hat**.

Wireless networks become a target for hackers because they can be easy to infiltrate through methods such as wardriving and piggybacking. Driving around with a portable computing device and trying to connect to someone else's unsecured wireless network is **wardriving**. Connecting to someone else's wireless network without the network owner's knowledge or consent to provide access is called **piggybacking** or *Wi-Fi piggybacking*. This can occur when a neighbor is within range of the unsecured wireless network of another neighbor or business.

**Read & Learn**

Unauthorized Access and Unauthorized Use of Computer Resources

**Job Connect!**

Computer Security Experts

**Table 7.1** Network Hacks That Have Made the Headlines

| When the Hack Was Reported | Organization That Was Hacked | Data That Was Affected |
|---|---|---|
| June 2018 | Adidas | Information for millions of customers stolen from a US website |
| March 2018 | 144 US universities and 5 US government agencies | Intellectual property estimated at $3.4 billion |
| March 2018 | Under Armour | Personal information of 150 million users of MyFitnessPal app |
| November 2017 | Uber | Personal information for 57 million Uber customers; Uber paid a hacker $100,000 ransom to cover up the breach. Uber is the defendant in lawsuits after the breach was revealed. |
| September 2017 | Equifax | Personal information for 145.5 million American, Canadian, and British citizens |

## Unauthorized Use

Using a computer, network, or other resource for purposes other than those intended is referred to as **unauthorized use**. Unauthorized use can occur when an employee uses the employer's computer for activities such as personal email, personal printing, or personal online shopping without the employer's permission. An example of unauthorized use at a school is when a student uses a computer on a school campus to send inappropriate email or instant messages to another student or teacher.

*Wardriving* is driving around and then attempting to connect to an unsecured wireless network.

## Strategies to Prevent Unauthorized Access

Multiple strategies are available to prevent unauthorized access, such as: sign-on credentials for each authorized user; secured and monitored facilities; and firewalls, passwords, and encryption standards on networking equipment.

**Securing User Accounts with Passwords**  Family members and employees are advised to choose a password that is not easy to guess and to never reveal it to anyone else or write it down in a visible location. Some workplaces use technology that requires employees to change their passwords every month, and the reuse of an old password may be prohibited. When a device is not in use, the screen should be locked so that the password is required to sign into the computer. Every computer user should choose a **strong password**, which is a password that is difficult to hack by humans or password-detection software. Technology tools called *password generators* help create strong passwords. Password generators are available via websites (Figure 7.1) or apps. A strong password meets the following criteria:

- is a minimum of eight characters
- uses a combination of uppercase letters, lowercase letters, numbers, and symbols
- does not contain any dictionary words or words spelled backward
- does not contain consecutive or repeated numbers or letters
- has no personal information, such as a birthdate

● – – – – in the know

In the spring of 2018, Microsoft announced that it is moving forward with plans to remove password authentication in Windows. The company will replace passwords with other user sign-ons that cannot be "cracked, breached, or phished." Timelines and details of the new credentials are yet to be finalized.

**Let's Explore!**

How to Create a Strong Password That Is Easy to Remember

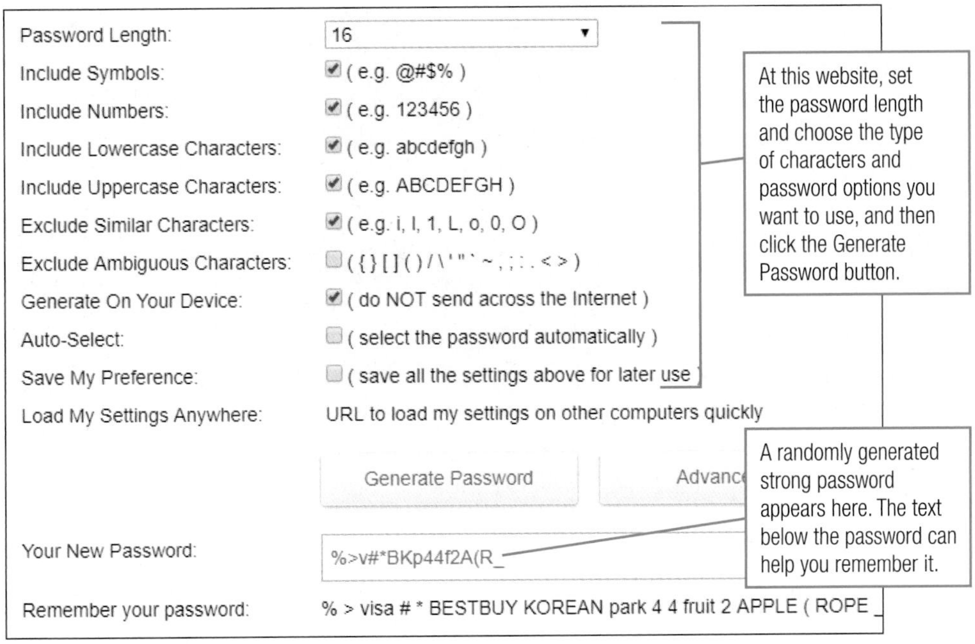

| | |
|---|---|
| Password Length: | 16 ▼ |
| Include Symbols: | ☑ ( e.g. @#$% ) |
| Include Numbers: | ☑ ( e.g. 123456 ) |
| Include Lowercase Characters: | ☑ ( e.g. abcdefgh ) |
| Include Uppercase Characters: | ☑ ( e.g. ABCDEFGH ) |
| Exclude Similar Characters: | ☑ ( e.g. i, l, 1, L, o, 0, O ) |
| Exclude Ambiguous Characters: | ☐ ( { } [ ] ( ) / \ ' " ` ~ , ; : . < > ) |
| Generate On Your Device: | ☑ ( do NOT send across the Internet ) |
| Auto-Select: | ☐ ( select the password automatically ) |
| Save My Preference: | ☐ ( save all the settings above for later use ) |
| Load My Settings Anywhere: | URL to load my settings on other computers quickly |
| | Generate Password        Advanc |
| Your New Password: | %>v#*BKp44f2A(R_ |
| Remember your password: | % > visa # * BESTBUY KOREAN park 4 4 fruit 2 APPLE ( ROPE |

At this website, set the password length and choose the type of characters and password options you want to use, and then click the Generate Password button.

A randomly generated strong password appears here. The text below the password can help you remember it.

**Figure 7.1**

Websites that generate random secure passwords help you create a strong password. The one shown here is from https://passwordgenerator.net.

Many companies and financial institutions offer **two-factor authentication (2FA)**, also called *two-step verification*, for signing into an account. 2FA is a more secure authentication process. The most common 2FA is typing your password and then receiving a randomly generated code—via email message, text message, or an app on your smartphone—that you enter on a second screen. Other types of 2FAs require you to enter information you know, such as an identification number or the answer to a security question like, "What was the name of your first school?" A hacker is not likely to know this personal information in addition to your phone number and password.

**Securing a Device Using a PIN**  A difference between using a **personal identification number (PIN)** and a password to secure a computing device is that the PIN is associated with the device; if you want to use the same PIN on another device, you must set it up on each desktop, laptop, smartphone, or tablet.

A password is associated with an account, such as a Microsoft account, and can be used on any PC. A password is transmitted to a server for authentication, while most PIN authentication occurs on the local device. If someone steals or guesses a PIN, the person still needs the physical device to access data. But if someone gets hold of a password for an account, he or she can sign on to the account from anywhere on any device.

In Windows 10, a PIN is part of a security design specific to a device that can be up to six digits long. After four failed PIN attempts, the device must be restarted. If a few more failed attempts occur, the PIN on the locked device is disabled and the device can only be used by signing in with another authentication method, such as a password.

Apple uses similar passcode technology on any iOS device. The iPhone and iPad allow six attempts to sign on with the passcode before the phone is disabled. Android devices also use PINs to lock devices; ten attempts to enter the correct PIN are allowed before the data is deleted and the phone reset.

Most people like PINs more than passwords, so security experts offer this advice: aim for six digits, don't choose something obvious such as a word spelled on the keypad, and don't use any of these PINs that account for nearly 20 percent of four-digit PINs: 1234, 1111, and 0000!

**Securing Access with Biometrics**  As you learned in Chapter 3, a biometric device authenticates a person's identity using physical characteristics, such as a fingerprint, iris, facial recognition, or voice pattern. The use of biometric authentication for accessing computing devices or workplace facilities is becoming more common. A 2018 survey by Spiceworks (an IT industry organization) of IT professionals in North America and Europe revealed that 62 percent currently have biometric security in place for various purposes. By 2020, a further 24 percent plan to have biometric authentication. A fingerprint scan (Figure 7.2), followed by facial recognition, were the top two biometrics used.

**Figure 7.2**

A fingerprint scanner like the one shown here is the most commonly used biometric authentication.

**Securing Computing Equipment**  In the workplace, entry to rooms with network servers and network equipment is often secured with physical locks. Entry to these rooms requires authentication before the door is unlocked. An employee gains access by using a card that must be swiped or scanned by a reader to unlock the door. Biometric scans of a fingerprint, iris, or face are also commonplace. Some employers in the Spiceworks survey reported using hand-geometry recognition and palm-vein recognition biometrics.

**●----- secure IT**

A four-digit PIN is easy to shoulder surf—someone nearby can easily figure out the digits you enter. Many people also use the same PIN on their smartphone, their ATM card, credit card, and home security alarm. Be aware of shoulder surfers when you type a PIN, and use a different PIN on your home security system than you use for a banking or credit card.

**●---- in the know**

One security firm offers tools for detecting unauthorized use by learning an employee's digital behavior. BioCatch software runs in the background, building a profile of the employee's habits. For example, the software learns the employee's typing style (with one or two hands), scrolling method, and screen-viewing preferences (e.g., how the employee switches screens).

All network computing devices should have a **firewall**, which is hardware, software, or a combination of both that blocks unwanted access to the network (Figure 7.3). All operating systems by default turn on the firewall feature. Incoming and outgoing network traffic is routed through firewalls that examine each message and block communication that does not meet security criteria. Some network routers have a built-in firewall (called a *hardware firewall*), adding an extra layer of protection. A large network uses, a piece of networking equipment that is a dedicated firewall.

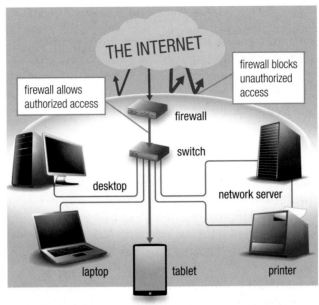

**Figure 7.3**

In the network shown here, all of the devices are protected behind a firewall to block unauthorized access.

**Securing a Wireless Network** Wireless routers and access points are secured from wardrivers and piggybackers by using a network security key to gain access. The key is given only to authorized users. A device sold with a factory-set password should have the password changed to a strong password as soon as it is plugged in. Wireless equipment should also use the highest security and encryption standard available for the device. **Encryption** scrambles communications between devices so that the data is not readable by anyone other than the sender and intended recipient. **Wi-Fi Protected Access (WPA)** and **Wi-Fi Protected Access 2 (WPA2)** are security standards found in most wireless devices currently in use. Each standard uses encryption, but WPA2 employs a more sophisticated technique than WPA. A router or access point should have WPA2 enabled; however, WPA provides good encryption if WPA2 is not available. An older encryption technology that may still be found in some equipment is Wired Equivalent Privacy (WEP). A router that only offers WEP encryption should be replaced with a new device. In 2018, the **Wi-Fi Protected Access 3 (WPA3)** standard was finalized. WPA3 includes features to make it harder for hackers to gain access by using password-cracking tools and improves the security of public Wi-Fi networks. Expect to see WPA3 devices appear on the market in late 2019. A device with WPA3 encryption will provide the best security.

## Strategies to Prevent Unauthorized Use

An **acceptable use policy (AUP)** describes for employees, students, or other network users the permitted uses and inappropriate uses for computing equipment and networks. Some organizations allow employees to use a workplace computer for personal use while the employee is on break; other organizations prohibit personal use of any kind. Some companies also use monitoring email or software that provides reports on activities (like social networks visited).

Video surveillance, although usually part of a broader safety and security strategy, is another method organizations can use to monitor authorized access and authorized use of computers and networks.

 **Video**

What Is the Difference between a Hardware and Software Firewall?

●---- in the know

Large organizations also use security software called *intrusion detection software* that analyzes network traffic for suspicious data while it passes through a firewall and routers. An alert is sent to the network administrator when a possible threat is detected.

 **Video**

How Does Encryption Work?

●----- go online

**https://CC3 .ParadigmEducation .com/ProtectNetwork**

Go here to watch a video on steps to take to protect a home wireless network.

 **Blog This!**

Should Employers Monitor Personal Use of a Business Smartphone?

# 7.2 Botnets and Denial of Service Attacks

Two types of attacks on a computer network that disrupt network service are botnets and denial of service attacks. These malicious activities cause widespread network slowdowns and sometimes block authorized activity from taking place. In this topic, you will learn how a botnet spreads and how to prevent your computer from becoming compromised. You will also learn how a denial of service attack disrupts a network.

## Botnets

A computer that is connected to the internet and is controlled by a hacker or other cybercriminal without the owner's knowledge is called a **zombie** computer. A collection of zombie computers that work together to conduct an attack on another network is a **botnet**. Botnets are used by hackers or cybercriminals to send spam emails, spread viruses, or steal personal data. In some cases, the botnet may be used to conduct a denial of service attack.

A computer becomes a zombie when it is infected with software (usually a virus) that is hidden from the user. The computer can then be controlled remotely by the hacker or cybercriminal. Generally, the infection happens from one of the following events:

- The user clicks a link in an email or opens an email attachment that contains the malicious code, which is then installed on the computer.
- The user downloads a video, image, or software from a website without realizing the file is a cover for the malicious code; this method is often used at media-sharing and peer-to-peer–sharing websites.
- The user visits a website without realizing that the malicious code is being downloaded in the background.

Figure 7.4 illustrates a botnet spreading malicious code to a target computer. Each zombie device in the botnet is attempting to deliver malware to the target laptop on the right. One zombie (the desktop PC) successfully delivers the malware to the target laptop when the laptop user clicks a link, downloads an infected object, or visits a website co-opted by the hacker. Once the target computer is infected, it joins the botnet. Table 7.2 on the next page lists signs that may indicate your computer has become a zombie. Keep in mind that some of the symptoms listed in Table 7.2 could indicate other issues, such as lack of memory. Have the suspected PC checked by a professional to determine the cause of any symptoms.

 **Read & Learn**

Botnets and Denial of Service Attacks

 **Video**

How Are Botnets Used on the Internet?

●----- in the know

An IoT botnet is made up of malware-infected IoT devices, such as security cameras, DVRs, appliances, and other smart devices. The increasing volume of IoT devices makes this type of botnet a much larger threat—as much as 10 times larger than a traditional PC-based botnet.

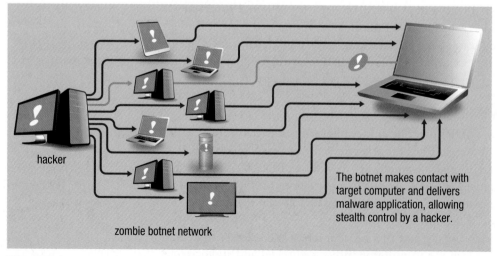

hacker

zombie botnet network

The botnet makes contact with target computer and delivers malware application, allowing stealth control by a hacker.

**Figure 7.4**

Your computer can be taken over by bots and used to send spam or malware to other PCs.

**Table 7.2** You've Been Hacked!

| Activities that Indicate a Device Could Be a Zombie |
|---|
| Your internet connectivity slows down consistently and dramatically and/or the browser closes frequently for no reason. |
| Hard disk activity occurs when you are not running any applications (including automatic antivirus or software updates). |
| Your computer or device becomes unresponsive, crashes frequently, or gives you unexplained error messages. |
| Hard disk space is filling up unexpectedly. |
| A different website (not the one you have set as your default) appears when you open the browser window. |
| New desktop icons or toolbars appear that you did not install. |
| Your email inbox contains undeliverable messages for people you did not email. |

To prevent your computer from becoming a zombie, make sure you have antivirus software automatically scheduled to scan and update, make sure your computer is set to install updates automatically, and have a firewall active at all times. If you think your computer is a zombie, immediately update your antivirus software and perform a full scan. In some cases, you may need to hire professional help to erase your hard disk drive and start from scratch by reinstalling only legitimate software and data.

## Denial of Service Attacks

A **denial of service (DoS) attack** occurs when a network or web server is overwhelmed with network traffic in the form of a constant stream of bogus emails or other messages to the point that the server response time becomes slow or shuts down completely. Legitimate users are denied access, usually receiving a message that the server is busy. Well-known companies and government websites are often the targets of DoS attacks conducted by botnets. Figure 7.5 displays how a DoS attack is orchestrated.

Hackers or other cybercriminals perform DoS attacks for a variety of reasons, such as to draw attention to a social or political cause, to embarrass a company or government, to gain notoriety, or to make demands on a company. Organizations attempt to prevent DoS attacks by employing firewalls, intrusion detection software, and antivirus software tools.

●– – – – in the know

Imperva (a cybersecurity firm) reported that DoS attacks in the last quarter of 2017 mostly hit internet service providers (58.4 percent) and gambling services (23.6 percent), with organizations in Hong Kong, the United States, Taiwan, the Philippines, and Malaysia the frequent targets.

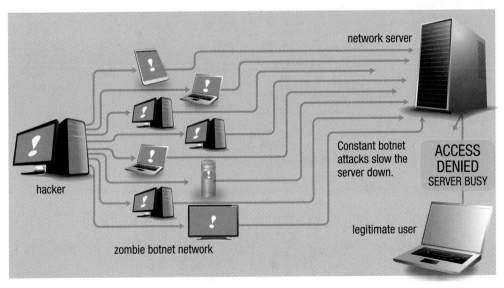

**Figure 7.5**

Denial of service attacks prevent legitimate users from accessing a server when the server becomes overwhelmed with a constant stream of network requests initiated by a hacker for the purpose of causing a shutdown in service.

# 7.3 Malware Infections

☁ **Read & Learn**
Malware Infections

Any malicious software that is designed to damage, disable, or steal data is called **malware**. Malware installs on your computer without your knowledge, usually from email or online activities. Malware can delete files, damage files, steal personal data, track your activities, display pop-up windows or messages, or turn your computer into a zombie. In this topic, you will learn about common malware that exists in the form of viruses, worms, Trojan horses, rootkits, and ransomware, as well as strategies to prevent malware from infecting your computer.

## Viruses

A **virus** is a form of malware that, once installed on a host computer without the owner's knowledge, can replicate itself and spread to other media on the infected computer and to other computers on the network. A virus can infect a computer through a variety of means:

- downloading and installing infected software
- opening an infected file attachment in an email message (Figure 7.6)
- visiting an infected website
- plugging in an infected USB or other peripheral device
- clicking links in messages (including instant messages) or at untrustworthy websites

A **macro virus** is embedded in a document and infects the computer when the user opens the document and enables the infected macro. Microsoft Office documents were once a favorite target for macro viruses because of the popularity of the software suite. Microsoft improved security in its Office applications so that macros are automatically disabled when a document is opened. Always exercise caution when enabling a macro in a document that was received via email.

In a **drive-by download**, the user's device is infected by visiting a compromised web page. The malware installs in the background when it encounters a browser, app, or OS that is out of date and has a security flaw. Always keep your OS and browser up to date, and use browser add-ons that block or disable scripts.

●– – – – – **go online**

**https://CC3
.ParadigmEducation
.com/VirusInformation**
Go here for up-to-date virus information, including recent threats, a virus threat meter, and a global virus map.

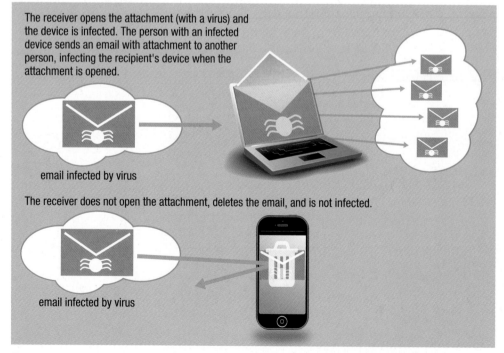

The receiver opens the attachment (with a virus) and the device is infected. The person with an infected device sends an email with attachment to another person, infecting the recipient's device when the attachment is opened.

email infected by virus

The receiver does not open the attachment, deletes the email, and is not infected.

email infected by virus

**Figure 7.6**

Opening an email attachment with a virus infects your computer and starts spreading the virus to others.

## Worms

Self-replicating malware that requires no action on the part of the user to copy itself to another computer on a network is called a **worm**. Your computer can become infected with a worm simply by being connected to an infected network (Figure 7.7). Worms typically exist to backlog network traffic or performance, or they may even shut down a network. In some cases, a worm is spread via a social network. For example, a worm was identified that propagated via Facebook when a user posted an infected video that spread the worm to all the user's friends and groups—effectively becoming a botnet.

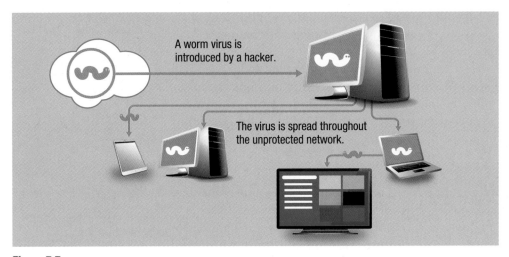

A worm virus is introduced by a hacker.

The virus is spread throughout the unprotected network.

**Figure 7.7**

A worm automatically sends copies of itself to other computers on the network.

## Trojan Horses

Malware that disguises itself as a useful application but then infects your computer when you run the application is called a **Trojan horse** (Figure 7.8) This type of malware is named after the legend in which Greeks concealed themselves inside a hollow wooden horse to invade the city of Troy. Residents of Troy thought the horse was a victory trophy and brought it inside the gates of the city. Troy was overtaken when the invaders came out of the hollow horse inside the city. Similarly, Trojan horse malware presents as a useful application in an ad, message window, or on a web page. Often, these applications resemble antivirus software or an application to fix a computer that is running slowly. This is meant to entice you to run the software that then installs the malware.

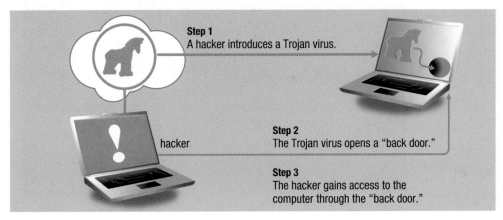

**Step 1**
A hacker introduces a Trojan virus.

hacker

**Step 2**
The Trojan virus opens a "back door."

**Step 3**
The hacker gains access to the computer through the "back door."

**Figure 7.8**

A Trojan virus is embedded inside an application that you think is useful. When you install the software, the virus opens a back door for the hacker, who then gains control of your PC to join a botnet or to steal your data.

●－－－－ in the know

A program that is not classified as *malware* but is considered nuisance software, or within a "gray zone" of malware, is a potentially unwanted program (PUP). A PUP is usually not malicious but uses system resources, slowing down PC performance. PUPs can also covertly track your activities and modify system settings. A PUP is installed when you download some other software. Avoid PUPs by paying attention during installation of software. Deselect check boxes for additional software or add-ons, or say no when asked if you want to install these items.

Trojans do not replicate themselves, but the damage done by them can be disastrous. Some Trojans send personal information back to a cybercriminal who uses the personal information or sells it to a third party that uses the data for fraud or theft activities.

## Rootkits

A **rootkit** hides on the infected computer and provides a **back door** (way to bypass computer security) for a hacker or other cybercriminal to remotely monitor or take over control of the PC. Using the remote access, the hacker or cybercriminal can run damaging software or steal personal information. Unlike viruses and worms, the objective of a rootkit is intended not to spread to other computers but to control the target PC. Rootkits may end up on your computer by piggybacking on the installation of some other software or via a virus.

A rootkit is very hard to detect because the software is designed to be invisible by masquerading as an essential OS file. Antivirus software may overlook the file, and even the OS may think the file is a system file and prevent its removal. Companies that sell antivirus and security software usually offer a standalone utility designed to detect and remove rootkits.

## Ransomware

A type of malware that locks or restricts access to the computer, demanding that the user pay the malware creator to unlock the device is called **ransomware**. In some cases, the contents of the hard drive are encrypted by the malware, meaning the files are unreadable without the encryption key. Ransomware infects a device when the user opens an email attachment containing the malware. It also infects a computer when the user clicks on a compromised website. Users are lured to these websites via an email message or pop-up ad.

In May 2017, hundreds of thousands of computers were attacked by the WannaCry ransomware. WannaCry exploited a vulnerability in older Windows PCs that had not been updated to the latest OS, rendering files on those PCs unusable unless the owner paid a ransom. The attack spread to 150 countries with damages estimated to be hundreds of millions before it was stopped.

Security companies report that ransomware is rising, with cybercriminals shifting focus from individual users to enterprises. In March 2018, the city of Atlanta computer systems were attacked. City computing systems were down for more than a week. Although costs were not publicly revealed, one estimate by *Wired* magazine is that the city spent upwards of "$2.6 million on emergency efforts to respond to a ransomware attack that destabilized municipal operations." Figure 7.9 shows an example of a ransomware payment demand screen (provided by ESET, an internet security company). The ransomware shown in Figure 7.9 circulated in Ukraine in the fall of 2017.

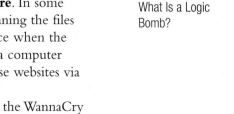

in the know

Not all rootkits are malware. Parents or employers may use a legitimate rootkit to remotely command, control, or monitor activity on a computer system used by a child or employee.

**Let's Explore!**
What Is a Logic Bomb?

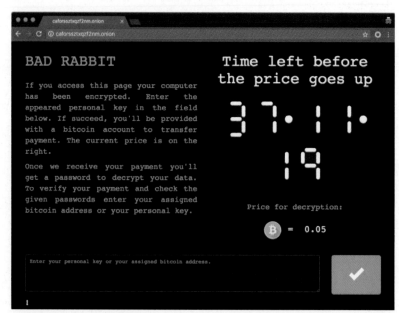

**Figure 7.9**
ESET's screen shot of the payment demand page for the Bad Rabbit ransomware attack.

Although cybercriminals are increasingly attacking corporations and government websites, individuals are still subject to attack and need to remain vigilant.

## Malware Protection

All computer users need to have up-to-date **antivirus software** running on their computers at all times. This software is set up to run automatic scans on a regular basis during a time when the computer is not likely to be in use. Most antivirus software applications also scan all incoming emails automatically. Check the settings in the application to ensure "real-time protection" is turned on, which means the software is continuously active in the background while you work.

New computers are sold with antivirus software already installed for a trial period. OSs also include malware protection. For example, Windows 10 installs with Windows Defender (malware protection from Microsoft) turned on by default. You may decide to use different software or upgrade to a comprehensive suite of threat protection. Check reviews and look for information from trusted sources about the antivirus software that detects and removes malware the best. Most antivirus software companies offer software packaged in a suite of internet security utilities or separate, standalone security applications. If you have one or more mobile devices, look for a suite that includes protection for smartphones and tablets.

You will also find free antivirus software from several computer security software companies, such as AVG, Avast, and Kaspersky. Search for information about free software before installing it to make sure the software is recommended by a reputable publication, such as *Consumer Reports* or *PC* magazine. Be wary of downloading free antivirus software from an ad or pop-up window that appears when you are on a website. These ads are often disguising Trojan malware. Instead, go directly to the website of a company that you know is trusted, and download free software from there. Regardless of the malware protection application you choose, make sure it runs automatically and scans regularly, and that a firewall is always active.

●---- in the know

One of your malware defense strategies should be to make regular backups of important files to another device or a cloud provider so that your data is safe in the event of a malware infection.

## What to Do If You Suspect Malware Is on Your PC

In Topic 7.2, you learned the way in which malware is delivered to a PC or other device and the symptoms that might indicate your device is a zombie computer. Remember that a computer that suddenly runs slower than before does not necessarily mean malware is the reason. But, ruling out malware is a first step to diagnosing a computer issue.

Open the antivirus application on your computer and run an update to make sure the software has the latest virus definitions and removal tools. Next, run a full computer scan. If malware is detected, choose the option to clean the device. When finished, you may want to run a second full scan using an online scanning application. Figure 7.10 shows the ESET Online Scanner with a threat found. These scanners do not require you to download and install new software—they run directly from a website.

Most malware is successfully removed with antivirus tools. If you still notice a computer issue afterward, take the PC or device to a computer repair technician. Paying a fee for a technician to check out the PC and clean viruses may be the only way to ensure your device is malware-free.

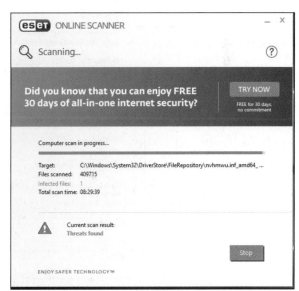

**Figure 7.10**

Some computer security companies, like ESET, offer a free online scanner to check your device for malware. The scanner offers removal tools when an infection like the one shown here is detected. ESET also offers free cybersecurity training at https://www.eset.com /us/cybertraining.

## 7.4 Phishing, Pharming, and Clickjacking Threats

A growing area of cybercrime involves online fraud or scams designed to trick unsuspecting users into revealing personal information so the criminal can steal money from the individual or engage in identity theft. **Identity theft** occurs when an individual's personal information is obtained by a criminal who then uses the information to buy products or services under the victim's name or to pose as the victim for financial gain. In this topic, you will learn about three scams used to obtain personal information: phishing, pharming, and clickjacking.

### Phishing

**Phishing** (pronounced *fishing*) is the term that describes activities that appear to be initiated by a legitimate organization (such as a bank) in an attempt to obtain personal information that can be used in fraud, theft, or identity theft. A popular email phishing technique involves sending a message that appears to be from a bank or credit card company that directs the user to click a link (Figure 7.11). The link goes to a bogus website that appears to be valid, and may even include a real bank logo and color scheme. At the phishing website, the user is prompted to enter personal information, such as a username, password, bank account number, or credit card number. If the individual complies, a criminal now has critical personal information and will use it to transfer money from the individual's bank accounts or to charge purchases to a credit card.

Phishing is growing on social networking sites, owing to the high volume of activity on these sites. For instance, coinciding with the excitement of a launch of the Apple Watch, Facebook and Twitter were used to lure people into entering personal information in a fake giveaway. Those who succumbed to the scam entered their full names and Facebook handle and were then asked to invite 100 friends to do the same—with 100 invites earning the user an Apple Watch. Twitter users who mentioned the smartwatch in a tweet were also directed to the scam at fake accounts named Apple Giveaways. If you see an offer like this on a social networking website, search the company web page associated with the offer to make sure the offer is legit.

> Note: This is a service message with information related to your Chase account(s). It may include specific details about transactions, products or online services. If you recently closed your account, please disregard this message.
>
> **CHASE**
>
> Dear Chase Online℠ Customer: **Chase Online℠ Payment Pending.**
> **You have an unconfirmed payment Pending on your account**
> **Please verify your account information for payment approval**
> **We implore you to follow the link below to verify your account details.**
>
> Account Verification
>
> **NOTE: You are strictly advised to match your information correctly to avoid service suspension.**
>
> **Thank you for your co-operation.**
> To start the Re-activate process click on Chase Online℠.
>
> **You May Visit Our Web** www.chase.com
> Once you have completed the process, we will send you an email notifying that your account is available again. After that you can access your account online at any time.
>
> The information provided will be treated in confidence and stored in our secure database. **If you fail to provide required information your account will be automatically deleted from Our online database**
>
> Sincerely,
>
> Cathy J. Marinelli
> Senior Vice President
> Online Banking Team
>
> JPMorgan Chase Bank, N.A. Member FDIC
> ni$2012 JPMorgan Chase & Co.
> Your personal information is protected by advanced online technology. For more detailed information, view our Online Privacy Policy. To request in writing: Chase Privacy Operations, 451 Florida Street, Fourth Floor, LA2-9376, Baton Rouge, LA 70801 .

**Figure 7.11**

This message received by Chase customers was a phishing attempt to trick users into clicking the links and revealing account information.

**Read & Learn**

Phishing, Pharming, and Clickjacking Threats

**Video**

What Are Current Trends in Phishing Attacks?

●‒ ‒ ‒ ‒ in the know

The February 2018 *Phishing Activity Trends Report* from the Anti-Phishing Working Group revealed that phishers are using the https protocol in their websites to trick people into thinking that the site is secure. Consumers who check for *https* at e-commerce pages to make sure their payment information is encrypted may fall for this latest ploy.

●‒ ‒ ‒ ‒ go online

**https://CC3 .ParadigmEducation .com/ReportPhishing**

Go here to submit a report about a phishing message you have received.

## Pharming

**Pharming** is similar to phishing except that the individual is tricked into typing in personal information at a phony website that appears to be the real website. One type of pharming attack generates an email that installs code on the user's computer without the user's knowledge, and the code then modifies files on the computer. In other pharming cases, the attack compromises the IP address for a Domain Name System (DNS) server. When an individual enters the correct web address for a website, such as his or her bank, the malicious code redirects the URL, sending the victim to a fake web page. The web page at the pharming website looks legitimate, so the individual proceeds to enter personal information.

Phishing and pharming scams employ **spoofing** techniques, in which a sender's email address is altered to a phony address that appears legitimate to the email recipient, or an IP address is altered to appear to be a trusted source.

## Clickjacking

**Clickjacking** occurs when a button, graphic, or link on a web page appears to be real but, when clicked, causes malicious software to run. The malicious software code is hidden so that users have no idea that they are not clicking a real link. The bogus button, graphic, or link directs users to another website, where personal information is requested or malware is installed.

In some Facebook clickjacking schemes, a bogus Like button redirects users to advertising pages where the scammers receive money for each wayward clicker. Techniques used in clickjacking scams often involve a breaking news story, exclusive content or contest lure, or a trending news item. Be cautious of clicking links or buttons for something that appears too good to be true or seems unlikely, such as a celebrity death hoax.

## Tips to Avoiding Identity Theft

Identity theft scams often initiate in an email. Delete messages you receive that ask you to click a link to update information. Never click links in unsolicited email messages. Before typing personal information on any web page, examine the URL and other text on the page. Phony websites usually have spelling or grammar errors, may contain extra words not necessary to identify the domain, or the URL will have a slightly different character substitution within it, such as the number *1* instead of the letter *i*. A long string of meaningless characters in the URL should also make you wary. Hover the mouse pointer over hyperlink text like "Click here to update your account information" in the message or on the website. Examine the text that appears, looking for the red flags previously mentioned.

If you've clicked a link and are being asked to enter personal information, take a moment to recheck the URL to see whether the URL changed. If the URL has changed, then check whether the new URL is suspicious. Many legitimate e-commerce websites have a **secure site seal** (also called a *trust seal*). The seals are issued by a third-party company that has verified the website is a safe site. Use a search engine to verify a secure site seal if you see one you do not recognize. Figure 7.12 shows a secure site seal issued by Symantec, the company that bought VeriSign (the most recognized trust seal used on websites) and that produces Norton AntiVirus software.

To be absolutely safe, always close an email message or browser before typing any personal information into a website form field. Alternatively (or additionally), call the company that is requesting your information and speak to someone who can tell you whether the request is legitimate.

 **Let's Explore!**

What Is Social Engineering?

**Figure 7.12**

A badge like the Norton Secured seal shown here, issued by Symantec, is a sign that the website is a safe site.

## 7.5 Information Privacy

🌥 **Read & Learn**
Information Privacy

With so much information collected and stored online, individuals and companies are increasingly concerned with information privacy. **Information privacy** is the right of individuals or organizations to control the information collected about them. Consider all the websites at which you have set up accounts and the personal information you provided at each site. Consider also that some websites track the pages you visit and store information about you. Software may be installed on your computer and tracking everything you do. In this topic, you will learn about privacy concerns related to online activities.

### Cookies

A web server may send a small text file to be stored on your device that contains data about you, such as your username and the pages you visited. This text file is called a **cookie** and is used to identify you when you return to the website. In many cases, the use of cookies is welcomed; they are helpful when used to prefill a login screen or customize your viewing preferences at an e-commerce site. However, cookies might also be used for unwanted purposes, such as tracking your activities or gathering information about you without your permission.

All browsers provide the ability for you to control how cookies are handled. Figure 7.13 shows the Cookies section in the Advanced settings panel for the Microsoft Edge internet browser, where you can choose how to manage cookies. This setting can be changed to *Block all cookies* or *Block only third-party cookies*. Blocking all cookies is not recommended because signing in to websites that you use frequently becomes difficult without the website cookie installed on your hard drive. A third-party cookie is a cookie placed on your hard drive by a domain that is different from the domain of the web page you are visiting. For example, if you are viewing a web page from the domain shoppingsite.com and a cookie is placed on your system from adtracker.com while viewing that page, the adtracker.com cookie is considered a third-party cookie. Third-party cookies are often designed to track your surfing habits. Many people choose to block third-party cookies as a precaution against behind-the-scenes tracking by websites.

● – – – – – secure IT

On a device using the mobile version of Google Chrome, you cannot block third-party cookies. But, you can periodically clear browser data on the smartphone or tablet. To do this, open the mobile browser, tap the menu (three vertical dots), tap *Settings*, tap *Privacy*, tap *Clear browsing data*, and then tap *Clear Data*. Confirm the action by tapping *Clear*.

Settings and more button

Follow these steps to open the Privacy and security pane in the settings panel:
1. Click the Settings and more button.
2. Click *Settings*.
3. Click the Privacy and security option in the left pane of the settings panel.

These options appear after clicking the *Cookies* option box arrow. Click in the browser window away from the panel to close the panel when finished.

**Figure 7.13**

All browsers have a *Settings* function to change privacy options, usually in an Advanced category on the page. Shown here is the Advanced settings panel for Microsoft Edge, with the *Block all cookies*, *Block only third-party cookies*, and *Don't block cookies* options visible.

## Spyware

Software that exist on your computer without your knowledge and track your activities are referred to as **spyware**. Some spyware may be on your workplace computer so that your employer can monitor your online activities. Other spyware may be used by websites to target ads to your preferences. Antivirus applications typically include spyware detection and removal.

A version of spyware called a **keystroke logger**, or *keylogger*, may be activated as part of a rootkit or Trojan horse and records every keystroke you type and sends it back to a cybercriminal. Keystroke loggers may be used to obtain your bank account number and password or other personal information that leads to identity theft. Some keystroke logger software is promoted on the internet to businesses as a computer security tool and to parents who want to monitor their children's online activity.

**Figure 7.14**

MyCouponize is an example of adware. The enticing ad promises to save you money when online shopping. However, the adware was found to continuously monitor browsing activity, record personal information from that activity, and share the data with third parties. The user is at risk of identity theft if a cybercriminal acquires personal information.

Software responsible for pop-up ads that appear on your desktop or while viewing web pages are the result of software referred to as **adware**. These unwelcome ads often piggyback spyware onto your computer. Figure 7.14 shows an ad for MyCouponize, targeted at Mac users, that was risky adware for anyone who installed it.

Technology company Lenovo made headlines when adware called Superfish, preinstalled on new Lenovo PCs, was found to have a flawed design that left users vulnerable to a **man-in-the-middle (MITM) attack**. An MITM attack occurs when a hacker intercepts communications between a user and a website for purposes of capturing personal information. In the Lenovo case, Superfish adware left a security hole on the user's PC that meant a hacker could install malware on the device. Lenovo was served with a class action lawsuit over the adware debacle. On July 18, 2018, Lenovo agreed to pay $8.3 million in damages to settle the lawsuit.

## Spam

**Spam** is electronic junk mail—unsolicited email sent to a large group of people at the same time. Spam messages entice readers to buy something, are phishing messages, or are vehicles to distribute Trojan viruses in attachments or via a link the user clicks in the message. Spam messaging has infiltrated social networks and text messages. According to SpamLaws.com (an internet security information company), 14.5 billion spam messages are sent globally each day, about 45 percent of all email, with the United States the number one country that initiates spam.

A **Twitter bot** is software programmed to follow people on Twitter based on popular keywords. When someone clicks a link in a Twitter bot comment that appears as a response to a tweet they post, they might receive spam or malware. Beware of followers with strange handles.

Obtaining a store loyalty card or registering for free accounts often requires you to provide your email address. These databases of names and email addresses are sold for marketing purposes, which is how you end up with spam. Many people use free web-based email accounts, such as Outlook or Gmail, for online activities only. This helps to reduce the amount of spam invading your personal or work email inbox. Sign up for a new email account when spam at the free account becomes onerous.

## Ways to Protect Personal Information

When shopping online or conducting other business that requires a financial transaction, make sure the URL at the website begins with *https* and that you see a small closed padlock next to the Address bar or on the Status bar of the browser window. These signs indicate a secure website using a protocol that protects data sent to the site with **Transport Layer Security (TLS)**, which encrypts transmitted data so that the data is unreadable if intercepted.

Table 7.3 provides strategies you can use to protect personal information while pursuing online activities. Ultimately, you are responsible for keeping sensitive information private and being diligent at websites where you provide personal data. Check privacy policies and do not reveal information that you do not feel is necessary for the purpose of the visit.

**go online**

**https://CC3 .ParadigmEducation .com/ElectronicPrivacy**
Go here for up-to-date information on privacy issues from the Electronic Privacy Information Center.

**Table 7.3** Ways to Minimize Tracking of Your Personal Information

| Internet Activity | Ways to Minimize Threats |
| --- | --- |
| Web browsing | • Change the privacy option in the browser that you use to block third-party cookies.<br>• Regularly clear the history in your browser. |
| Emailing | • Use a free web-based email account for online shopping and social networks and keep your regular email address for other purposes.<br>• Delete spam messages right away without opening them. |
| Shopping/social networking | • Fill in only the fields marked with asterisks at websites that ask you for personal information when registering for an account; these are the minimum fields required to obtain an account.<br>• Unsubscribe to mailing lists that you think are selling your email address for marketing purposes.<br>• Hide sensitive data in your personal profiles (such as your birth date) at websites and social media sites such as Facebook.<br>• Use different usernames and strong passwords for online shopping, social networks, and banking. Make sure the answers to security questions you have set up to recover your password from websites are not easily gleaned from other information about you that is online.<br>• Be strategic when allowing apps or websites to tag your location. Geolocation apps on your smartphone may sound cool, but only allow those that are necessary for a device you are using (like a fitness tracker). Consider that a third party could be tracking your every movement.<br>• Do not reveal personal details in status updates or other posts that may allow someone to determine where you live or collect enough personal information about you for identity theft. |
| All online activities | • Make sure the antivirus application you use includes antispyware and is set to automatically update and scan your computer regularly.<br>• Make sure a firewall is active at all times you are online.<br>• Shred documents with personal information before discarding them. |

**green IT**

Before recycling or donating old computing devices, be sure to securely remove all personal data from the internal storage media. The internal HDD has lots of personal information stored on it, and you won't want the data falling into the wrong hands. To do this properly requires that you download an application that wipes a drive clean. Free applications are available. Don't just delete files using the OS interface—a wipe-clean application is more secure.

## Data Privacy in the Cloud

When storing pictures, documents, or other data at a cloud provider website, such as Flickr or Dropbox, find out the privacy and security policies that affect your data. Typically, you control the data privacy by restricting who can see a file. However, the popularity of cloud storage is making cloud providers a target for cybercriminals looking to scoop large quantities of data that can be mined for personal information. Employee error at these websites may also lead to accidentally leaking personal information. In 2017, World Wrestling Entertainment exposed the data of three million fans on an Amazon Web Services server. Data was set to *Public* by mistake by

 **Video**
How Can You Protect Your Personal Information When Online?

an employee. WWE customer information was available for anyone to download until the error was discovered. To be safe, never post at a cloud provider website files containing sensitive information, such as your birthdate and social security number, that could make you vulnerable to identity theft.

At social networks, such as Facebook and Twitter, make sure you review and change privacy settings so that only the information you want public is viewable by anyone. In Facebook, check the information about you that is public. An easy way to do this is to open a web browser, go to your favorite search engine page, such as Bing or Google, and do a search for *YourName Facebook* (substituting your first and last name for *YourName*). Click the link in the search results to facebook profiles with your name. Click your account in the Facebook profiles list to view what everyone sees about you. Change your profile if you see personal information that should not be shared with the general public. When posting information on a social network page, also consider that friends who see the information could share what you post—people you don't know will see it too.

Cybercriminals are trolling social networks for personal data they can use for financial gain. Three popular methods used on social networks are to place offers for gifts (Figure 7.15), post scandalous news or a death hoax about a celebrity, or post news about a tragic world event. Generally, once you click one of these items in your news feed, you are asked to fill out a survey, invited to download software that is really malware, or redirected to a website with a drive-by download. Often you are encouraged to share the post on your timeline to entice your friends to do the same activity. Don't respond to surveys in social networks and always practice the advice to think before you click!

**Figure 7.15**

A frequent lure on social networks is the promise of a gift card after filling out a survey. Cybercriminals gain your personal information and entice you to share the post with friends— you expose your friends to the same scam. This Starbucks gift card scam often appears on social networks.

## Data Privacy and Wearables

Wearable devices that do more than count steps or track your heart rate are entering the mainstream consumer market. Consider these products announced recently:

- a video camera that resembles a button on a shirt or blouse
- a medication with an ingestible digital sensor (called a *digital pill*) to monitor patient medication compliance and physical data
- a device worn under a woman's garments that detects the earliest signs of invasive cancer in breast tissue
- a smart swimsuit that records your pool laps for you

These devices collect information about the wearer and the physical health and wellbeing of the individual. Symantec, a security software company, identified unique risks for data from a wearable device to be hacked. A hacker could use malware to access personal data stored on the device. The hacker could also use malware to intercept data as the data is transmitted from the device to the owner's smartphone or from the device to a cloud storage website. Owners of these devices need to stay alert to new information about these devices and read the privacy policy from the manufacturer before creating accounts on their websites. Wearable devices should be turned off when not in use.

## 7.6 Mobile Devices and the Internet of Things Security

Mobile devices such as laptops, tablets, and smartphones have untethered people from their homes and offices; however, the portable nature of these devices makes them vulnerable to security risks. The rising tide of smart devices connected to the internet in homes and cars is raising alarms for some people concerned with the security and privacy of data. In this topic, you will examine the various issues for mobile users and IoT-device owners, and methods used to protect hardware and data from loss or theft.

**Read & Learn**

Mobile Devices and the Internet of Things Security

### Mobile Malware

McAfee Labs (an internet security company) detected over 25 million **mobile malware** (viruses designed for mobile devices) incidents in the first quarter of 2018—more than double the number the company saw the previous year. Android devices are a popular target for malware because of the open-source nature of the OS; however, iOS devices are not without risk. In the *10 Years of (Hacking) iOS* report by Skycure (a mobile security company), iOS malicious apps tripled since 2016.

A smartphone or tablet can get a virus via an infected app downloaded from an app store or the web, an infected computer connected to the device, an email attachment opened on a smartphone, a text message, or a Bluetooth connection in close proximity.

Smartphones and tablets should have a mobile security app installed to shield the devices from malware. App stores offer free mobile security apps like the Avast app shown in Figure 7.16. If you subscribe to antivirus software on your PC, look for the companion mobile app for your smartphone or tablet.

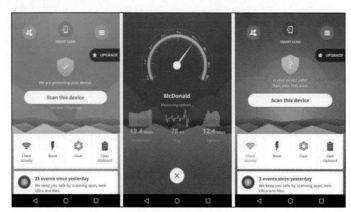

**Figure 7.16**

TechRadar, a UK-based technology news company, rated this free Avast mobile security app as the top choice for protecting Android smartphones and tablets.

### Lost or Stolen Mobile Devices

Securing a mobile device with locks, biometrics, and/or remote tools is necessary to protect against the loss of a device due to theft or absentmindedness. While the hardware may be easily replaced, the personal information and data stored on the device usually has the higher value for individuals and employers. The following tools assist with securing mobile devices and data in the events of loss or theft:

- Many devices are equipped with fingerprint readers that restrict access to the authenticated user only (Figure 7.17a). Some phones are using facial recognition.
- A PIN or passcode for access to the device (Figure 7.17b) should be enabled; should the device be stolen or lost, the PIN/passcode may provide enough time to employ remote wiping utilities.
- Physical locks with cables that attach a laptop to a table or desk in a public place are a deterrent to thieves looking for an easy target (Figure 7.17c).
- Technology for remote wiping, locking, and tracking of a lost or stolen mobile device lets the owner wipe the device clean of data and track its location.
- Regular backups of data stored on mobile devices are recommended.

a           b           c

**Figure 7.17**

Mobile devices can be secured with (a) fingerprint readers, (b) passcodes or passwords, and (c) physical cable locks to restrict access.

## Bluetooth Risks

Bluetooth technology, which wirelessly connects and exchanges data between two devices in close proximity, is subject to risk from intrusion by others within range. Bluetooth range can be from 3 to 800 feet, depending on the power class of the device. When using Bluetooth in a public space with many people nearby, a risk exists that someone else can connect to your device and send you a virus or access personal data. Turn Bluetooth on only when needed and turn it off as soon as you are finished with it. Consider installing a Bluetooth firewall app that will secure your device from unwanted intrusions.

Mobile devices carry risks unique to their portable nature. Employers may require employees with mobile devices to abide by a mobile computing policy. A mobile computing policy usually requires employees to take responsible measures to secure laptops, tablets, or smartphones while away from the office, make frequent backups to a secure site, and password protect all sensitive company documents stored on the device.

**Video**

What Risks Are Involved in Using Bluetooth?

## Security in the Internet of Things

Amazon's digital assistant, Alexa, made news headlines in the spring of 2018 when the smart speaker recorded a private conversation between a husband and wife in their home and then emailed the recording to a person in the husband's contacts list without the owner's knowledge. The highly publicized incident caused people to question whether the smart devices designed to make life easier are presenting security and privacy risks for homeowners. Consider these other events that also occurred:

- A smart fridge was hijacked to send spam.
- A baby monitor was hacked to spy on a toddler.
- Thousands of security cameras were hacked and then used to overwhelm a web server to the point where the website crashed.

Security concerns for IoT devices include cybercriminals stealing personal data or adding IoT devices to their botnet network. Most IoT devices are set up to transmit data to a cloud storage website and/or a smartphone. Personal data could be exposed or intercepted during or after transmission. Smart devices may not be set up to regularly update software that closes loopholes against new vulnerabilities. If you have a smart device connected to your home network, learn about the device's security settings, how the software is updated, and steps you can take to make sure your home network is secure.

Smart devices, such as this video baby monitor, pose security and privacy risks to homeowners from cybercriminals or data breaches on cloud-provider websites.

## Topics Review

| Topic | Key Concepts and Key Terms |
|---|---|
| **7.1 Unauthorized Access and Unauthorized Use of Computer Resources** | Acts involving a computing device on the internet is **cybercrime**.<br><br>**Computer security** activities are related to protecting hardware, software, and data from loss due to unauthorized access, unauthorized use, theft, fraud, natural disaster, or human error.<br><br>**Unauthorized access** occurs when someone gains entry without permission to a computer through a network.<br><br>A **hacker** is an individual who accesses a network without permission, while **hacking** describes activities used to gain unauthorized access to a network.<br><br>A **white hat** is a hacker who gains entry into a network to pinpoint weaknesses in the network security.<br><br>A **black hat** is a hacker who gains entry into a network with malicious intent.<br><br>**Wardriving** occurs when someone drives around with a portable device trying to connect to someone else's unsecured wireless network.<br><br>**Piggybacking** refers to connecting to a wireless network without the network owner's permission.<br><br>**Unauthorized use** is performing activities on a computer other than the intended uses.<br><br>A **strong password** is difficult to hack by humans or password-detection software.<br><br>**Two-factor authentication (2FA)** requires the user to enter a randomly-generated code or PIN or answer a security question after signing in with a username and password.<br><br>A **personal identification number (PIN)** is associated with the device and must be set up on each computer. Using a PIN is considered more secure because the PIN does not pass to a server—a hacker needs access to the device if he steals or guesses a PIN.<br><br>Access to a computing device or entry to rooms with computing equipment at workplaces can be secured using biometrics; computing facilities may also use a card reader.<br><br>A **firewall** is hardware, software, or a combination thereof that blocks unauthorized access.<br><br>**Encryption** scrambles communications between devices so that the data is unreadable to anyone except the originator and intended recipient of a transmission.<br><br>**Wi-Fi Protected Access (WPA)** and **Wi-Fi Protected Access 2 (WPA2)** are encryption standards used in wireless routers and access points. **Wi-Fi Protected Access 3 (WPA3)** provides the best security by making it harder for hackers to use password-cracking tools.<br><br>An **acceptable use policy (AUP)** spells out appropriate and inappropriate use of computing equipment in the workplace. |
| **7.2 Botnets and Denial of Service Attacks** | A **zombie** is a computer controlled by a hacker or cybercriminal without the owner's knowledge.<br><br>A collection of zombie computers that attack a network is called a **botnet**.<br><br>A computer becomes a zombie by being infected with software, usually from a virus, that is hidden from the user.<br><br>To prevent a computer from becoming a zombie, regularly scan with up-to-date antivirus software, update the OS regularly, and make sure a firewall is always active.<br><br>A **denial of service (DoS) attack** occurs when a botnet attacks a network by sending a constant stream of messages that slow down or completely shut down a server.<br><br>Organizations use firewalls, intrusion detection software, and antivirus tools to prevent DoS attacks. |

*continued...*

| Topic | Key Concepts and Key Terms |
|---|---|
| **7.3 Malware Infections** | **Malware** refers to any type of malicious software.<br><br>A **virus** is a form of malware that replicates itself, spreading to other media on the infected computer and to other computers.<br><br>A **macro virus** is a virus embedded in a document macro; when the user enables the macro, the virus infects the computer.<br><br>A **drive-by download** occurs when the user's computer is infected simply by visiting a compromised website using an out-of-date browser, app, or OS that has a security flaw.<br><br>A **worm** is self-replicating software that infects other computers on a network without user action.<br><br>A **Trojan horse** is malware that appears to the user to be a useful application but which in fact is malware.<br><br>**Rootkit** is software that provides a **back door** (a way to bypass security) for a hacker or cybercriminal to control a computer remotely.<br><br>**Ransomware** is a form of malware where the target device is locked or encrypted and a demand for payment is made by the malware creator to provide the owner with an unlock code.<br><br>**Antivirus software** detects and removes malware. If your PC or other device exhibits symptoms of malware, make sure the antivirus application on the device is up-to-date with the latest virus definitions and removal tools and run a full computer scan. |
| **7.4 Phishing, Pharming, and Clickjacking Threats** | **Identity theft** occurs when personal information is obtained by a criminal who uses the information for financial gain.<br><br>**Phishing** activities usually occur in messages that appear to be legitimate but are intended to steal personal information by convincing the user to click a link to a phishing website where the user enters personal information.<br><br>**Pharming** is where malicious code redirects a URL to a fake web page after a person types a web address and at which the unsuspecting user enters personal information.<br><br>**Spoofing** techniques alter a sender's email address or a web address so that it appears legitimate.<br><br>**Clickjacking** occurs when a user clicks a phony button, image, or link and is redirected to another website or a virus downloads to the device.<br><br>To avoid identity theft, never click links in unsolicited email messages, and carefully examine a URL before entering personal information at a website arrived at by clicking a link.<br><br>A **secure site seal** provides an indicator that the website is a safe site because the badge means that a third-party company has verified the website. |

*continued...*

| Topic | Key Concepts and Key Terms |
|---|---|
| **7.5 Information Privacy** | **Information privacy** refers to the rights of individuals to control information that is collected about them. |
| | A **cookie** is a small text file placed on your hard drive by a web server with information about your username and the pages you visited. |
| | **Spyware** is software that tracks your activities without your knowledge. |
| | A **keystroke logger**, or *keylogger*, records every keystroke you type and sends the data to a cybercriminal. |
| | **Adware** displays pop-up ads on your device or at web pages and will sometimes piggyback spyware onto your computer. |
| | A **man-in-the-middle (MITM) attack** occurs when a hacker intercepts communications between a user and a website for purposes of capturing personal information. |
| | **Spam** is unsolicited email sent to large groups of people. |
| | A **Twitter bot** is software programmed to follow people based on popular keywords and can result in spam or malware when a user clicks a link in a tweet posted by a bot. |
| | A closed padlock next to the Address bar and *https* at the beginning of a website address are signs indicating that the website is a secure website using **Transport Layer Security (TLS)**, which encrypts transmitted data. |
| | Find out the privacy and security policies in effect at cloud-providing websites where you store photos and documents. If necessary, review and change privacy settings at social networks. |
| | Cybercriminals troll social networks for personal data they can use for financial gain. Be cautious about what you post and avoid common scams, such as offers for gifts. |
| | Wearable devices collect personal and health history data that can be intercepted by hackers or be subject to another type of breach. Read privacy policies from manufacturers and turn off wearable devices when they are not in use. |
| **7.6 Mobile Devices and the Internet of Things Security** | **Mobile malware** are viruses designed for mobile devices. |
| | Android devices are attacked more frequently due to their open-source mobile operating system. |
| | A smartphone or tablet gets a virus by downloading an infected app, by being connected to an infected computer, from a message or message attachment, or via a Bluetooth connection. |
| | All tablets and smartphones should have mobile security software installed. |
| | Mobile devices should be secured with locks, biometrics, and/or remote wiping and tracking tools in case of loss or theft. Perform regular backups to ensure a copy of important data is safe. |
| | Using Bluetooth in a public space leaves you vulnerable; someone else with a Bluetooth device can access your smartphone, tablet, or laptop and send you a virus or steal personal data. |
| | Employers may require mobile workers to abide by a mobile computing policy that requires them to take responsible measures to secure their devices. |
| | Smart devices connected in our homes and cars have security and privacy risks that include cybercriminals stealing personal data, adding IoT devices to their botnet network, or intercepting data as the device transmits to a cloud storage website or smartphone. |

**Topics Review**

# Buying a New Computing Device

**B**uying a new laptop, tablet, or smartphone requires an understanding of the technical jargon that often appears in computer ads. An experienced electronics sales associate will help you choose the right device; however, some people like to shop online or find out in advance the features that are important to consider. In this appendix, you will learn to shop for a device that meets your needs using a step-by-step buying process model.

## Buying a New PC or Tablet

Are you in the market for a new computer or are you looking to add a second device to supplement a computer you already own? Many people own more than one computing device. Perhaps you have a desktop PC and want a laptop or tablet to use for school. The following pages will guide you through a decision-making process that will help you make an informed purchase. Spending a little time planning your requirements will save time and money and help you avoid making an impulse purchase you may regret.

A section with tips for making a smartphone purchase is included at the end of the appendix. The following steps are geared toward deciding how to buy a personal computer or tablet.

● – – – – **go online**

**https://CC3
.ParadigmEducation
.com/Ximix**

Go here for an online buying guide that has suggestions for devices by user profile, such as Families & Everyday Consumers or Students.

## Step 1: Determine Your Maximum Budget

Before you start browsing computer ads, know how much money you can afford to spend. Computers are sold in a wide range of prices, from a few hundred dollars to a few thousand dollars, to suit a variety of budgets. Setting your maximum price will help narrow the search when you are ready to shop.

## Step 2: Decide on a Desktop, Laptop, or Tablet

Once you know your budget, the next step is to focus on the type of device that best meets your needs. Figure A.1 illustrates four types of computing devices that are a good place to start. Desktop PCs are declining in usage, but nevertheless, some people prefer the stationary system unit that sits on or under a desk with an attached monitor, keyboard, and mouse. If you normally work in one place and have the physical space to accommodate the larger PC, a desktop is a good value. The desktop PCs of today last longer than earlier desktops and are easier to upgrade than most mobile devices. All-in-one PCs pack all the system unit components behind the monitor, which requires less space. Desktop PCs also include a larger screen for easier viewing of content.

If you are looking for a portable computer that you can move around the house, that has a full-size keyboard with all the power and capability of a traditional desktop, and that has a medium-size screen, then a laptop is what you need. A laptop computer provides everything a traditional desktop PC provides, in a portable case that you can carry around. Laptops are available in a wide variety of configurations and screen sizes. The weight of a laptop varies: some smaller units weigh in around 2 pounds, but those with a larger screen can top the scales at over 7 pounds. The lowest weight laptops are sometimes called *ultrabooks*. An ultrabook is lighter and slimmer than other

a

b

c

d

**Figure A.1**

You need to decide between (a) a desktop PC, which is a good value for working in one place; (b) a laptop, which offers portability; (c) a 2-in-1 laptop, which combines the computing power and features of a laptop with the convenience of a detachable touchscreen; or (d) a tablet for lightweight, touchscreen portability (a good choice for when a full-powered PC is not needed).

laptops, with most models weighing in under 3 pounds. An ultrabook will not have an optical disc drive (because the optical drive adds weight), and the battery lasts longer than a battery in a traditional laptop.

Tablets are smaller and lighter than laptops. Typically, people buy a tablet as a secondary device, with a desktop or laptop serving as their primary device. If you want to use a touchscreen, or a stylus to handwrite notes, a tablet is usually the preferred device. Other options to consider include 2-in-1 laptops (with screens that rotate 180 degrees to fold over a keyboard) and tablets supplied with a docking station and a full-size keyboard.

## Step 3: Choose Applications and Operating System

Computers come with an operating system and some applications preinstalled, so you may be tempted to skip this step. However, this step is important because it can help you keep within your budget. Watch for systems that preinstall limited trial editions of applications. These applications usually expire within a short timeframe, such as 90 days, or at some point within the first year after purchase. Following expiration, you have to pay a license fee to continue using the software. You need to find out in advance whether you will have to set aside money to pay for application software separately.

Check your school's computer store to see whether you can buy applications at academic pricing or even get software for free. For example, Microsoft provides free software, such as Microsoft Office, to eligible students at qualifying schools (Figure A.2). Some academic licenses have to be purchased through your school to qualify for student discounts; however, increasingly, retailers are able to offer the same pricing upon proof of student registration.

It pays to shop around for the best price on the application you want to install. Some software publishers offer coupons for web-based purchases, and some retailers can offer attractive pricing when you package applications with hardware. Look for a software suite that will bundle together multiple applications in one product. These are always a better value than buying individual applications.

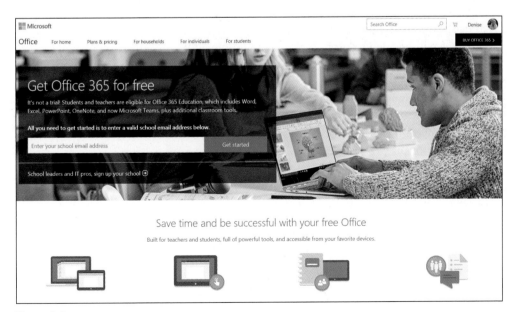

**Figure A.2**

The Microsoft Education program provides some Microsoft software free to teachers and students at qualifying schools—enter your school email address at **https://CC3.ParadigmEducation.com/MicrosoftEducation** (shown here) to get Office 365 for free.

Deciding on the applications you want to use will help you choose the appropriate hardware and OS. All applications specify the system requirements for running the software. You will notice that software companies specify minimum and recommended system requirements. Go for the recommended requirements in a new purchase, because performance is usually slower at the minimum requirements. Pay particular attention to the specified operating system that is needed to run the application, the amount of RAM, the processor speed, and the disk storage used by the application(s).

Applications are typically available in a Mac- or a Windows-compatible edition. With the popularity of cloud-based software that functions on either platform, deciding between Mac and PC is becoming less dependent on software and more of a personal choice and budget issue. Apple PCs are more expensive than Windows-compatible PCs. The Apple iPad and iPhone are also more expensive than Android tablets and smartphones and Windows tablets.

Apple fans are fiercely loyal to the brand and will argue that the extra money for Apple is well worth the investment. Ultimately, if you are not partial to either platform or OS family of products, you should try both environments at your school or at a store to determine whether you become a Mac or a PC owner. You may also want to investigate the operating system that is standard for your desired career. For example, many design jobs require you to work with a Mac, yet many engineering jobs expect you to work with a PC.

If your budget is tight, consider buying a Chromebook—the laptop that uses Chrome OS with all web-based applications such as Google's free productivity suite. If you do not qualify for free software or academic discounts, free cloud software is a good alternative to mainstream applications, and you'll save money.

## Step 4: Compare Hardware Components to Match Software Needs

Table A.1 on the next page provides an overview of specifications you should check in a typical ad for a Windows-compatible laptop computer. If necessary, review Chapter 3 for more information on any of the hardware components that are listed.

If you can afford to spend more, buy more than you need to meet the recommended requirements. This will ensure you have a computer that will meet your needs into the future. In a nutshell, go for the computer with the fastest processor, most RAM, and largest internal hard disk drive. These systems will typically be packaged with other components that will be adequate for an average computer user working with desktop applications and the internet. If you cannot afford to spend more now, look for a computer ad that says the RAM is expandable. This will be a better choice than one that cannot be upgraded or that limits the amount of RAM that can be added later.

Look for the Energy Star logo or something in the ad that says the device is Energy Star qualified. This means the computer will use less energy than one that is not rated.

Look for this logo, which tells you the device is energy efficient.

Examine the width, height, and weight specifications to get an idea of the physical characteristics. Pay particular attention to weight and screen size if you will be carrying the computer around at school in a backpack.

**Table A.1** Hardware Components Typically Advertised for a Laptop

| Features | Description |
|---|---|
| Processor | Intel Core, Atom, AMD, or ARM processors are commonly found in laptops and tablets. Laptops and tablets are sold with multiple-core CPUs, usually dual- or quad-core. Look for the amount and type of cache memory (L1, L2, or L3) with the CPU and the highest processor speed (1.5 GHz and higher). |
| RAM | RAM works with the CPU to give you high-speed performance. The more RAM you have, the better your system will perform. Look for a minimum of 4 GB of RAM on a laptop and 2 GB on a tablet. Check whether RAM is expandable in modules that can be added to the device at a later time. A system with a higher level of expandable RAM will last longer. |
| Screen | A desktop PC will have a larger screen than a laptop or tablet, typically 19 inches or more. A laptop or ultrabook will have a screen size in the range of 15 inches or more. Tablet screens are smaller, ranging from 9 to 12 inches. Screen size is measured diagonally from corner to corner. |
| Video card | In most cases, the video card supplied with a laptop is adequate. If you are going to use your computer for gaming or high-end graphics design, pay attention to the graphics card, video memory, and GPU. Look for higher specifications for these items to achieve faster graphics processing capabilities. |
| Hard disk space | The larger the amount of hard disk storage space you can get, the better. However, keep in mind that you can always buy external hard disk drives to store digital photos, videos, and music files. Many laptops are now supplied with a 1 TB HDD, but capacity varies. Look for a minimum of 500 GB. |
| Optical disc drives | Some laptops include a DVD or Blu-ray disc drive. Ultrabooks do not have optical drives (although you can purchase one as a peripheral). Optical drives are mostly considered obsolete because people opt to stream movies and music; however, some people who have a large library of music and movies still use discs. |
| Digital media reader | Transferring data from a digital camera or smartphone is easier if the laptop has a digital media card reader. Most laptops are equipped with a reader or include a slot for one. |
| Networking capabilities | Laptops are equipped with integrated Wi-Fi networking capabilities, and most also have integrated Bluetooth connectivity. An Ethernet port provides the added capability of plugging in to a wired network using a twisted-pair cable (sometimes called an *RJ-45 connector*). At school, some classrooms, lounges, or libraries may be equipped with network plug-ins that will give you faster internet access than the wireless network. |
| Battery life | Check the battery life to see how long the computer will operate before requiring power from an outlet. Battery life is important if you will be using the laptop away from power outlets for long periods of time. |
| Ports | Check the number and type of ports that are available. The more USB ports the better, though you can buy an inexpensive USB hub if you need to expand the port availability. An HDMI port will allow you to plug in a second screen for watching movies or extending the display capability for dual-screen work. If you want to plug in a second monitor that you already own, check the type of display port included—you may need an adapter cable for older monitors. |
| Integrated webcam | Laptops are usually equipped with a webcam. If you plan to use video calling, this feature will be important to you. |

**go online**

https://CC3
.ParadigmEducation
.com/ComputerBuying

Go here for the Consumer Reports *Computer Buying Guide*. Scroll down the page to find the interactive video buying guide.

## Step 5: Choose Customization Options

If you are purchasing a laptop to use in place of a desktop PC, you may want to add an external monitor, keyboard, and mouse. These peripherals will allow you to design an ergonomically correct workspace at home to avoid the repetitive strain and eye strain that occur from prolonged use of a computer. Wireless keyboards and mice allow you to adjust your workspace to more comfortable positions.

If you like listening to music or watching movies, you may want to invest in external speakers or high-quality, noise-canceling headphones. A headset with a microphone is desirable if you plan to use voice technology in many applications.

If you need to print at home, consider whether you want a laser or inkjet printer. Compare the cost of toner versus ink and the number of pages you can print before replacing supplies to help you decide on the type of printer best for you. Some computers are bundled with a printer. Make sure

When considering the customization options you need, consider how you will carry the device and allocate money for a well-designed backpack or roller case.

the printer included in such a package is one you want; if you would prefer a better quality printer, try to negotiate a discount on it rather than accepting the packaged device. Consider an all-in-one printer that includes scanning and copying capabilities. All-in-one printers are a good value and will use less desk space than multiple devices.

If you are purchasing a laptop or tablet, think about how you will carry it, and buy a well-padded case or backpack. Backpacks designed to carry laptops have padded dividers and wide, cushioned straps to distribute the weight on your shoulders. Consider buying a case with wheels if you will be carrying a laptop along with several textbooks at school. For tablets, consider buying a sleeve or case that you can use to protect the screen surface.

Many mobile devices are subject to theft or loss. Consider biometric options, such as a fingerprint reader, or tracking software and hardware that will allow you to remotely monitor or wipe the device should it be stolen or lost.

## Step 6: Decide Whether to Buy an Extended Warranty

Computers come with a limited parts and labor warranty that is typically good for one year. Consider whether you want to buy the extended warranty plans that are offered when you buy a new computer. These warranties are priced depending on the value of the device, and some retailers' warranties will fix mishaps not covered by the manufacturer. Decide whether you want to spend the money on these plans. Some people believe extended warranties are a waste of money, while others prefer the peace of mind that comes with knowing they are covered for anything that might happen.

If you have a credit card that includes a buyer protection plan, find out whether you must make the purchase using the card, and check the policy to see whether the credit card insurance will cover you for any accidental damage or loss. If you have coverage, you may not need the extra protection from a retailer's extended warranty.

Most computer equipment carries a one-year parts and labor warranty.

## Step 7: Shop Around, Ask About Service, and Check Reviews

Spend time online before deciding what to purchase, comparing two or three devices in the same price range that you can afford. Also, compare the pricing at in-store electronics retailers and online etailers. Be mindful of shipping costs that you may incur if you buy from a web-based store. Consider creating a spreadsheet to compare a few options side-by-side.

Make sure to visit your school's computer store. Schools often offer attractive deals from hardware manufacturers that have discounts negotiated for the large school audience. School staff will also be able to help you choose a computer for programs with specialized hardware or software requirements.

If you are comfortable negotiating at an electronics store, you may be able to convince the retailer to upgrade some options for you in a package price. For example, it would not hurt to ask the sales associate to throw in a wireless keyboard or mouse, or a set of speakers, with a new laptop.

Make sure to inquire about service options. Ask where warranty work and other repairs are performed. Some stores will do warranty repairs or out-of-warranty repairs onsite, while others have to ship the PC back to the manufacturer. If the retailer ships the computer elsewhere, ask how long a typical repair will take. In some cases, you could be without your PC for more than a week if the store ships the repair offsite to be fixed. Ask who pays for the shipping to and from the repair center. Finally, some stores do not offer any assistance with repairs—avoid these locations if you are not technically savvy or are averse to risk.

You don't want to wait until you need your computer serviced to find out where to take it. Some sellers service what they sell onsite, some will look after it for you but send it away for service, and others may refer you to a third-party service company. If you cannot be without a computer for long, the availability of onsite repairs will be important to you.

Some people prefer to pay a little more for a computer that they can have serviced at the store of purchase in case of malfunction. You will have to decide how important the repair facility is to you before making a purchase. Be sure to ask all these questions to help you make a choice you are most comfortable with.

Search the web for reviews of the product you are considering and the retailer location where you want to purchase it. While reviews are not always helpful (many are negative, and the positive ones may not be authentic), they do provide insight into the questions you should ask and the issues for which you want to be on the alert.

## Considerations for Buying a Smartphone

Before buying a smartphone, choose the provider with which you want to subscribe, the data plan that will meet your needs, and the features you want in a smartphone. Consider whether the provider you have chosen includes the cost of the smartphone in the contract. Some wireless providers will provide a base model smartphone free of charge or at a discounted price, or an upgraded model at a low price, if you sign up for a specified contract period. Watch for limits, caps, and time-of-use charges that may apply. For example, a plan may advertise unlimited use, but the unlimited use may apply in certain hours only. Note that some tablets may also include 4G wireless access and will need a data plan from a wireless carrier to take advantage of a cellular network when you are not in range of Wi-Fi conectivity, so you may wish to consider a data plan that allows for multiple devices.

Once you have chosen a data plan, decide whether you are partial to an Android or an Apple device. For example, if you already own a MacBook and an iPad, then an iPhone makes the most sense for you because you will already be familiar with the interface. Apple devices by design share data seamlessly via iCloud. If you are not partial to Apple, more smartphone choices are available for the Android operating system and at different price points. Visit a store or kiosk and try out different models to get a feel for how the device operates and whether it feels natural as you use it.

●— — — — in the know

Some people prefer to buy an unlocked smartphone. An unlocked smartphone is not tied to a specific wireless carrier. This means you can buy a month-to-month voice and data plan without being locked into a long-term contract. To activate the phone, you will need to purchase a SIM card (a card you insert into the phone with your phone number, contacts, and call history) from the wireless carrier you choose.

The features of a smartphone that matter to you are the next area to focus on. Is it the screen size, the camera, the storage space, or the battery life? All of these factors will influence the models you want to compare.

For example, if the screen is important to you, try out models with a few different screen sizes. Larger screens might be nice, but consider how you will carry the smartphone and whether it will fit in your hand, pocket, or purse conveniently. You will want to try out messaging and browsing on different size screens to find the one with which you are comfortable. Some people like to operate a smartphone one-handed and will opt for a smaller size, while others want the largest screen size they can get.

The camera offered on the phone may be a deciding factor for you. Many people look for the best camera they can get because they don't want to carry two devices when traveling. For others, the storage space is key for having their music library at their fingertips.

Compare the battery life of various smartphone models. Battery technology is improving and newer batteries last longer, but often battery life does not live up to the advertised specifications. Talk time is the standard by which to base comparisons. Look for reviews that have tested models and provide actual talk time statistics. You may want to inquire about purchasing a second battery if you will often be away from charging stations and think you will use the phone more than the advertised specifications. Another option is to purchase an external battery pack, which can be used to charge your smartphone and other devices. New phones are compatible with wireless charging pads. Look for this feature in a device and check into buying a charging pad for home use.

People use their smartphones to take pictures, record video, play music, communicate in messaging apps, play games, and more. The apps for performing these tasks need storage space on a memory card. Check the size of the memory card that comes with the smartphone and consider whether you will need to buy a larger memory card. Check the compatibility of the memory card with your laptop or other computing devices so that you can easily exchange media files among them.

A variety of other accessories you may want to purchase with the smartphone include headphones, keyboards, speakers, screen protectors, and cases. Figure A.3 shows two accessories, a keyboard and a speaker, you can connect to a smartphone using Bluetooth (a wireless connectivity standard).

a

b

**Figure A.3**

Accessories to consider for a smartphone include (a) a keyboard, for those occasions when you have a lot to type (look for a foldable version if you want to carry it with you), and (b) a waterproof speaker for listening to music.

## Replacing an Older Device?

When you buy a new device, try to sell your older device, donate it to charity, or trade it in if the seller accepts trade-ins. If you cannot sell, give away, or trade in older electronics, remember to look into disposal options for ewaste that will properly recycle the device. Go to **https://CC3.ParadigmEducation.com/EWaste** for more information on recycling electronics.

# Wireless Networking

A wireless network is standard in many homes. People prefer a wireless network so that a network cable does not have to be installed from room to room through floors or ceilings in their home. With a wireless network, you can move freely around the house and outside to the yard with a computing device and not lose a network connection. You can connect PCs, tablets, smartphones, wearable devices, gaming consoles, smart appliances, and smart TVs. Family members browse the internet, watch YouTube, and stream movies and music from any room in the house.

In this appendix, you will learn the basic steps for setting up a wireless network and connecting devices to the network in a home. You will also learn how to set up a wireless network on Apple equipment, and how to connect to a public wireless network.

## Setting Up a Wireless Network at Home

Wi-Fi uses a combination wireless router/wireless access point and radio signals to transmit data within a limited range. The speed you will experience in a wireless network will vary depending on the equipment, the ISP service, the number of people sharing the connection, and the type of bandwidth being used by each connected user. Most wireless routers in use are either Wi-Fi 5 (also called *wireless ac* or *802.11ac*) or Wi-Fi 4 (also called *wireless n* or *802.11n*) routers. In 2019, Wi-Fi 6 devices became available (also called *wireless ax* or *802.11ax*). Many wireless routers available today can operate at speeds near 1 Gbps. While this speed sounds lightning fast, bear in mind that the equipment can only deliver the internet speed for which you have subscribed (see Step 1). You also need to have devices that are compatible with a wireless router capable of Gbps speed to take advantage of newer technology.

Even with a wireless network, some homes may have a desktop PC that is connected by plugging in a network cable to the wireless router. The following steps describe the general process for setting up a wireless network. Different routers provide slightly different software interfaces; however, the general steps will be the same.

**Video**

How Do I Set Up a Wireless Home Network?

175

## Step 1: Subscribe to High-Speed Internet Service with an ISP

As you learned in Topic 2.2 in Chapter 2, to connect to the internet, you need an account with an internet service provider (ISP). Contact a telephone company, cable company, or dedicated internet service company for high-speed internet access. Each provider will offer different pricing levels that vary according to the speed, with faster speeds costing more per month.

A telephone company will provide you with a DSL modem, and a cable company will provide you with a cable modem. Some companies call the device a *gateway*. Newer modems provided by ISPs are a combination wireless router and modem in one piece of equipment called a *modem router*. Some installations may require two pieces of equipment: a DSL or cable modem, plus an additional wireless router. To proceed to Step 2, you need the ISP service activated, as well as a DSL or cable modem router, or a modem plus wireless router. Depending on the service provider, a modem will be supplied at no cost. Other ISPs provide the equipment but add a monthly rental cost or separate equipment charge to your bill. Some ISPs let you buy your equipment and give you a list of compatible

A cable modem (left) or DSL modem (right) may be provided by the ISP, or you may have to purchase your own. If an ISP-provided modem malfunctions, the ISP will be required to fix or replace it for you. If you buy your own modem and/or router, it will be up to you to fix or replace it in the event of an equipment failure.

devices. Brand names for commonly used wireless routers you can purchase at an electronics store are ARRIS, Actiontec, Linksys, D-Link, and Cisco.

Your ISP may also provide instruction documentation when you subscribe for service if the company's technician does not go to your residence to do an onsite installation. New subscribers generally have a technician visit their home to set up the service and install and configure the equipment. Subscribers with existing service who are upgrading a router or modem will generally do the installation themselves, possibly with telephone assistance from the ISP. If the documentation from your ISP contains instructions that vary from the steps below, complete the steps according to the instructions from the ISP.

## Step 2: Connect the Networking Hardware Equipment

Connect one end of a coaxial cable (cable modem) or twisted-pair cable (DSL modem) to the modem and the other end to the cable or telephone line coming into your house. Plug the power adapter into the modem and into a power outlet (Figure B.1a). You should see lights illuminate on the front of the modem, indicating power and connectivity are live.

If you have a standalone router (ISP did not provide a single modem router device), connect the modem to the wireless router by plugging one end of a network cable into the back of the modem and the other end into the back of the wireless router. Plug the power adapter into the wireless router and into a power outlet (Figure B.1b). You should also see lights illuminate on the front of the standalone router.

If you have a desktop PC in the same room as your networking equipment, you can connect the PC to the integrated modem router or standalone router by plugging one end of a network cable into the PC network port and the other end into the back of the modem router or standalone router. This allows the PC to use a wired connection to the router, while the remaining devices in your home will use a wireless connection.

**———— in the know**

Try to pick a spot inside your home that is centrally located and will not have anything that causes interference with the Wi-Fi signals. Placing the equipment too close to other electronic devices or along a shared wall with a neighbor who has wireless devices, may create some interference with the signal. If you experience poor performance after setting up the equipment, you may need to change the wireless channel the modem is using in the router's administration application.

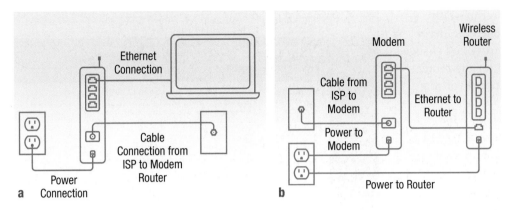

**Figure B.1**

(a) Connection diagram for plugging in a single modem/router device. (b) Connection diagram for plugging in a cable or DSL modem and a wireless router as two devices.

Once the networking equipment is connected and has power, wait for the indicator lights on the device(s) to remain solid and then proceed to Step 3.

## Step 3: Configure the Wireless Router

Integrated modem routers or standalone routers have an application that you access through a web browser that lets you change the device configuration. For some devices, one computer may need to be plugged into the wireless router while you configure it. After this step, you can unplug the PC from the router and connect wirelessly afterward.

Some ISPs require that the device is activated the first time it is powered on in your home. To activate the modem, you may need to call the ISP service department. The ISP technician will ask you for the serial number or other identification code from the modem and then he will authorize the device (also called *provision* the device). Other ISPs provide for automatic activation the first time the device is powered on.

Start with a computer connected to the router through the Ethernet port and a web browser open, or connect a PC wirelessly using the instructions provided by the ISP. Some devices are programmed to step you through a setup process as soon as you connect a device for the first time. In that case, follow the steps as prompted through the setup application.

If a setup application does not automatically start, look in the instructions that came with the device for a URL that launches the administration application. Open a web browser and type the URL into the Address bar. If you cannot find the URL in the instructions, look on the device itself for a label with printed information like the one shown in Figure B.2.

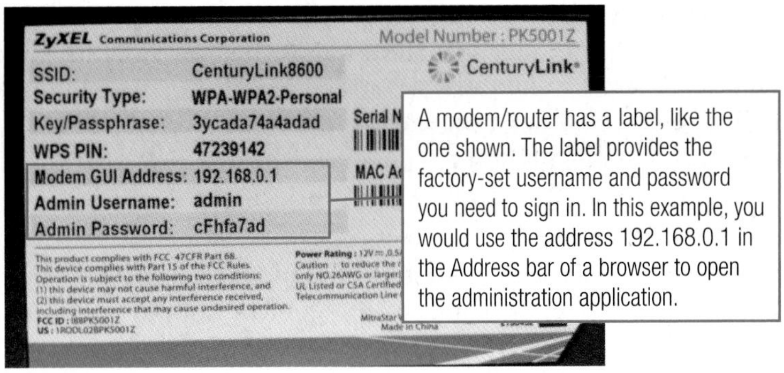

Proceed by following the instructions that appear in the setup program. The software interface is designed to be user-friendly, and prompts assist you at each step. If the application begins with a login screen (Figure B.3), the factory-set account is usually *admin* (username) and *password* (password), or a password is printed on the device label.

**Figure B.2**

When no instructions are provided, look on the modem/router for a label like the one shown here for a URL to type into the Address bar of a browser. The URL launches the modem/router administration application.

Generally, you need to do three tasks: assign the router a name (called the *service set identifier [SSID]*) and security key, change the administrative password for the router, and choose the security standard you want to use.

Note that the software interface for each router will vary. If in doubt about how to proceed, call the technical support assistance telephone number given to you.

When giving the router a new name, try not to use an obvious name that will identify you with the router. For example, do not use your name or your address. If you use an obscure name, then when someone is within range of your router, they will not associate the network name with you and try to break into your internet

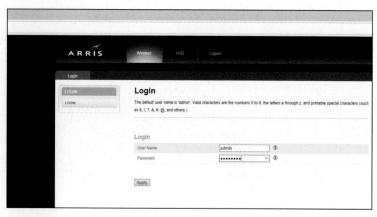

**Figure B.3**

A sign-in screen like the one shown here is required to start the modem/router administration application. Once signed in, proceed to change the SSID, administration password, Wi-Fi password, and security standard. Each interface is slightly different but is easy to follow.

access (this can help you avoid situations such as someone nearby trying to piggyback your connection). A good idea is to prevent the router from broadcasting the router SSID. Although free public Wi-Fi networks, such as those at airports, coffee shops, and libraries, use a recognizable name to provide unsecured network access to everyone who is within range, home users should choose a name that is not easily associated with the location. Also, assign a security key that your neighbors or anyone who knows you would not be able to guess.

Change the default administrator password to a strong password known only to you. The administrative password will be needed if you ever want to change the router settings in the future. Write down the username and password for administering the router and store it in a secure location.

Activate the security standard for the router, choosing the most secure standard available. Wi-Fi Protected Access 3 (WPA3) provides the strongest security for your wireless network. If you did not get a new router recently, WPA3 will not be available to you. In that case, choose Wi-Fi Protected Access WPA2. Older equipment before 2006 will only have Wi-Fi Protected Access (WPA), which is a less secure standard. If the router you are using only shows WPA, consider replacing it. WPA3, WPA2, and WPA secure your wireless network with encryption. The differences in the standards refer to internal features that make it more difficult for a hacker to break into the wireless network using password-cracking tools. People will not be able to connect to the wireless router unless they know the security key, and data transmitted across the wireless network will be encrypted.

If you used a laptop to configure the wireless router, you can unplug it when finished and proceed to connect each wireless device to the router as described in Step 4. You can leave a desktop PC plugged into the router if the desktop PC does not have a wireless network adapter or you do not want to use it wirelessly.

## Step 4: Connect Individual Devices to the Wireless Router

Laptops, tablets, wearable devices, and smartphones come equipped with a built-in wireless interface card, called a *network adapter*. Generally, the wireless network adapter turns on by default when the device is powered on. If necessary, turn on Wi-Fi connectivity, if it is not currently active. For each PC you want to connect to the wireless router, follow steps similar to those illustrated in Figure B.4.

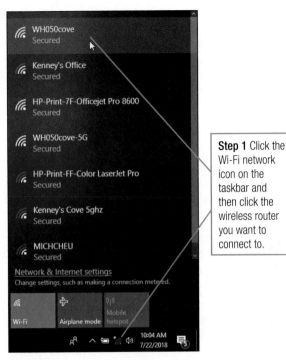

Step 1 Click the Wi-Fi network icon on the taskbar and then click the wireless router you want to connect to.

Step 2 Click to insert a check mark in the *Connect automatically* check box so that the computer will connect automatically in the future whenever you are within range of the router, and then click the Connect button. Type the security key or passphrase when prompted and then click Next.

**Figure B.4**

To connect a PC to a wireless router in Windows 10, click the Network icon on the taskbar, click the SSID you created for the modem/router, click the option to connect automatically, and then click the Connect button. At the next screen, type the password you created for the device.

Connect a tablet or smartphone by opening the Settings app on the device. On the Networks page, tap the Wi-Fi option to open the Wi-Fi Networks page—here you will see a list of available wireless networks within range. Tap the SSID for your network and then tap the *Connect* option. Type the password and then tap the Connect button.

Connect a television, gaming console, or other smart device by following instructions provided by the device manufacturer. In most cases, you have two options. One is to use a push button to instruct the device to automatically detect the router. The other option is to use a remote control to enter the device name and security key in an on-screen menu on the appliance.

## Setting Up a Wireless Network on Apple Equipment

Prior to April 2018, to set up a wireless network with Apple computers, you needed a Wi-Fi router made by Apple, called an *AirPort*. AirPort Express was the name of Apple's Wi-Fi 4 (802.11n) router, and AirPort Extreme was the name of Apple's Wi-Fi 5 (802.11ac) router. Another device, called AirPort Time Capsule, combined a Wi-Fi 5 router with up to 3 TB of storage for backing up data in one device. To set up the wireless network, you plugged in the AirPort Express, AirPort Extreme, or Time Capsule device into the high-speed cable or DSL modem from your ISP and then followed the instructions that appeared in the AirPort Utility program. In 2018, Apple discontinued making the AirPort devices, but the devices are still available until existing inventory is sold. If you do not have an Apple AirPort router, you will need to buy a third-party router and follow the steps described in the previous section.

# Connecting to a Public Wireless Network

To connect a laptop to a public wireless network, follow the steps in Figure B.4 on the previous page. Figure B.5 illustrates the steps for connecting a smartphone to a public wireless network.

A public network name should be obvious in the list of available networks; however, if necessary, you can ask a staff person (such as a store clerk, hotel clerk, or airport worker) for the network name and password. Exercise caution when using a public Wi-Fi hotspot, and do not transmit personal or confidential data, because public networks are not secured with encryption and are at risk of interception by hackers. Never perform financial transactions using a public Wi-Fi hotspot. If you must check email or sign in to a social network in a public Wi-Fi hotspot, do so with awareness of your surroundings and avoid entering confidential information in an email or post. If you frequently sign in to email or social networks using public Wi-Fi networks, consider having a different username and password for these services than the ones you use for other online activities—especially banking or online shopping accounts.

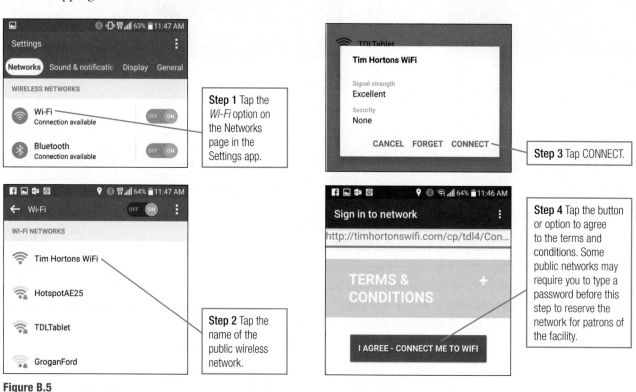

**Figure B.5**

Follow the four steps shown here to connect an Android device to a public wireless network. Apple devices follow a similar connection process.

# Glossary/Index

data caps, 30, 68–69
Deezer, 129
Dell, 87
**denial of service (DoS) attack** an attack that occurs when a server is overwhelmed with traffic from a constant stream of messages from a botnet that slow down the server's response time or shut down the server entirely, 150
**desktop computer** a PC that includes system unit and a separate monitor, keyboard, and mouse. An all-in-one case houses the system unit and monitor in one device, 4–5
  buying new computer choices, 167–169
  memory module used in, 56
**desktop publishing (DTP) application software** an application that incorporates text, graphics, and colors to create high-quality documents for marketing, communication, education, or for other commercial purposes, 106–107
**dial-up** internet access in which you connect your PC to your telephone system via a modem built into your PC, 33
digital cameras, 60
**digital copier** an output device connected to a network that can be used as a traditional paper photocopier and also accept output from computers for printing, 64
digital data, 68–69
**digital footprint** the accumulation of all your online activities, 138
**digital literacy** the ability to use various computer devices, communications technologies, and the internet to find, assess, use, create, and share content with others, 3
digital media reader
  specification when buying computer, 171
digital pill, 161
**digital subscriber line (DSL)** internet access provided by your telephone company that connects using telephone lines, 32
**Disk Cleanup** a Windows utility that allows you to scan a particular storage device or drive to select various types of files to be deleted, 89
**disk defragmenter** a utility that rearranges fragmented files back together to improve file retrieval speed and efficiency, 90
The Document Foundation, 113
**document management system (DMS)** a program that manages the creation, storage, retrieval, history, and eventual disposal of documents and other files stored in a central repository, 105
Dogpile, 38, 40
Dokmee, 105
**Domain Name System (DNS) server** a server that holds the directory that associates an IP address with a web address, 36
  pharming and, 157
**domain name** the part of the web address that is the organization name, an abbreviation of the organization name, or an alias associated with the owner of the computer. Also referred to as a *second-level domain name,* 37
  evaluating web content by, 42
**drive** a storage device that is identified by a unique letter and label or assigned a name, 70
**drive-by download** a type of malware that installs automatically on a user's device when the user simply visits a compromised web page and is using a browser, app, or OS that is out of date with a security flaw, 152

**driver** a small program that contains the instructions the OS uses to communicate and route data to and from a device, 80
Dropbox, 15, 73
DSL modem, 32

## E

**ebook reader** a device larger than a smartphone but smaller than a tablet, used to read electronic versions of books, newspapers, magazines, or other digital media, 6
ebooks, 6
**e-commerce (electronic commerce)** buying or selling over the internet, 44
  social commerce, 136–137
  types of, 45
**educational application software** programs designed for children and adults to learn about subjects such as math, spelling, grammar, geography, science, and history, 111
educational search engines, 40–41
**802.11 protocol** the standard developed to facilitate wireless communication among several hardware providers. Letters after 802.11 indicate the Wi-Fi adapter version. As data transfer rates improve, a new version letter is added, 66–67, 175
El Capitan, 83
elearning, 47
**email (electronic mail)** the sending and receiving of digital messages, sometimes with documents or photos attached, 45–46
  web-based, 46
**email address** unique identifier used to connect to your email account at the ISP server in which your messages are collected and stored. An email address is in the format username@mailserver.com, 45
**email client** a program, such as Microsoft Outlook or Apple Mail, that is used to create, send, store, and receive messages, 45
**embedded computer** a processor programmed to perform a particular task or set of tasks that is incorporated into a household appliance or other consumer electronic device. See also *smart device,* 10
**Embedded Linux** open source Linux operating system program that runs smart appliances, in-flight entertainment systems, personal navigation systems, and a variety of other consumer and commercial electronics, 86
**embedded OS** a specialized operating system that is smaller than a desktop OS and with fewer resources to manage that is designed for a specific device, 86–87
**encryption** the process of scrambling data transmitted between devices so that the data is not readable to anyone but the sender and the authorized recipient, 149
**end-user license agreement (EULA)** a license agreement you are required to accept when you install software. Also known as *software license agreement (SLA),* 115
Energy Star-certified devices, 19, 170
energy vampires, 18
**enterprise resource planning (ERP) application** a software solution designed for large-scale organizations to manage activities and processes involved with transactions for customers, vendors, inventories, and human resources, 105

**entertainment application software** programs for playing interactive games and videos, 111
entertainment controllers, 61
EPEAT program, 18, 19
Equifax, 146
ereader, 6
**ergonomics** the design of equipment and a person's work space to promote safe and comfortable operation of the equipment by the individual, 21
ESET online scanner, 155
**e-tailer (electronic retailer)** a business that sell online to consumers, 45
**Ethernet port** a circuit board or card in a computer that facilitates connection to the internet. See also *network interface card (NIC),* 30, 57, 66
  data speed for, 66
etrash, 18
**ewaste (electronic waste)** discarded or unwanted electrical items. Also known as *e-trash,* 18
  recycling, 19
**expansion card** a circuit board that plugs into an expansion slot on the motherboard to provide or improve functionality, 55
external hard drive, 14
**external wireless adapter** a device that has a built-in radio transmitter and receiver that allow you to connect to a wireless modem or wireless router, 67

## F

**Facebook** one of the most popular social networking sites used by people who want to connect with friends and family, 122–123
  building professional network on, 139
  Messenger, 46
  Privacy Checkup tool, 139
  social commerce and, 136
facial recognition, 60
fiber-optic service (FiOS), 32
fiber to the home (FTTH), 32
**fiber to the premises (FTTP)** internet access that involves the ISP running a fiber-optic cable directly to your home or business. Also known as *fiber to the home (FTTH)* or *fiber-optic service (FiOS),* 32
**file** the collection of bytes that comprises the saved document, 69
  operating system managing, 81
**File Explorer** a file management system in Windows in which you move, copy, delete, and rename files, 88
FileHold, 105
file management system
  back up and restore utility, 91
  disk cleanup utility, 89
  disk defragmenter, 90
  file manager, 88
**file name** the unique name assigned to a document when you save, 69
**Finder** the window you open on a Mac device to perform file management routines, 88
**Firefox** a web browser from the Mozilla Foundation for PCs and mobile devices that uses open source software (program code is free to share and modify), 35
**firewall** hardware, software, or a combination thereof that blocks unwanted access to your network, 149

## Image Credits

com/OJO_Images, 60 © Pressureua/Dreamstime.com, 60 Shutterstock/AlexLMX, 60 Used with permission from GoPro.; 61 Courtesy of Meta, 61 Used with permission from Google; 62 Courtesy of Dell., 62 Shutterstock/ Scanrail1, 62 Used with permission from Apple., 62 Used with permission from Asus.; 63 © bikeriderlondon/ Shutterstock.com, 63 © SMART Technologies, Inc.; 64 © Sylvie Bouchard/Shutterstock.com, 64 123rf/Sergey Ryzhov, 64 Courtesy of knight news.com; 65 Courtesy of ICON build website and theverge.com, 65 If you are considering writing a story on ICON, please feel free to get in touch. We also have high-res assets for download in our electronic press kit. (https://www.iconbuild.com/press/); 66 © iStockphoto.com/Visualfield, 66 123rf/Piotr Adamowicz; 67 © iStockphoto.com/scanrail, 67 Google and the Google logo are registered trademarks of Google Inc., used with permission.; 70 © iStockphoto.com/sqback; 71 © ekler/ Shutterstock.com, 71 © iStockphoto.com/scanrail, 71 © nikolamaster/Shutterstock.com, 71 ©Yevgen Romanenko/ Shutterstock.com; 72 ©Kovalchuk Oleksandr/Shutterstock. com; 72 Shutterstock/Jozsef Bagota;

## Chapter 4

Page 81 Used with permission from Apple, Inc.; 82 Used with permission from Apple, Inc.; 84 shutterstock/Framesira; 85 shutterstock/ DR-images, 85 Used with permission from Motiv.; 86 Used with permission from Interarctive Restaurant Technology., 86 Used with permission from Ram Truck Brand.; 87 Image property of Apple., 87 Image provided by Samsung

## Chapter 5

Page 103 Reprinted with permission © Intuit Inc. All rights reserved.; 104 Used with permission from OpenSource.; 105 Courtesy of Logical doc.; 106 (C)Shutterstock/irbis pictures; 108 Courtesy of Apple; 109 Courtesy of OpenElement., 109 Courtesy of Weebly.; 111 Used with permission from Quicken.; 112 Used with permission from Zoho.; 113 Apache, OpenOffice, OpenOffice.org and the seagull logo are trademarks or registered trademarks of The Apache Software Foundation.; 116 Google and the Google logo are registered trademarks of Google Inc., used with permission.

## Chapter 6

Page 122 Images courtesy of GoPro; 123 Courtesy of LinkedIN; 124 Courtesy of Instagram, 124 Courtesy of Instagram, 124 Courtesy of WhatsApp.; 125 Courtesy of Reddit., 125 Shutterstock; Leonel Calara; 127 Courtesy of Yahoo.com and Flickr.com.; 128 Courtesy of Pinterest.;

129 Image is the property of Spotify.; 131 Courtesy of Twitter., 131 Shutterstock/ Sattalat Phukkum, 131 Tweet property of Reuters., 131 Used with permission from LG.; 132 Courtesy of Wordpress.org, 133 Google and the Google logo are registered trademarks of Google Inc., used with permission., 133 Google and the Google logo are registered trademarks of Google Inc., used with permission.; 134 Courtesy of Wikipedia.org.; 135 Article provided by wikiHow, a wiki building the world's largest, highest quality how-to manual. Please edit this article and find author credits at wikiHow.com. Content on wikiHow can be shared under a Creative Commons License., 135 Courtesy of Wikipedia.org.; 136 Courtesy of Twitter., 136 Courtesy of Twitter.; 137 Used with permission from Fig and First., 137 Used with permission from Lowe's.; 138 Courtesy of LinkedIn and Microsoft., 139 Courtesy of LinkedIn.; 140 Used with permission from CityTurtles., 140 Used with permission from Sprout Social.; 141 Used with permission from Hootsuite.com.

## Chapter 7

Page 147 © iStockphoto.com/RonBailey, 147 Courtesy of passwordgenerator.net.; 148 © Andrea Danti/Shutterstock. com; 154 Photo courtesy of cybersecurity company ESET; 155 Photo courtesy of cybersecurity company ESET; 157 Courtesy of Symantec; 159 Courtesy of the MacSecurity blog, property of Apple, Inc.; 161 Courtesy of OnlineThreatAlerts.com; 162 Courtesy of Avast.com; 163 (C) Shutterstock/Audrey Popov, 163 © iStockphoto.com/ Khosrownia, 163 © ptnphoto/Shutterstock.com, 163 © ymgerman/Shutterstock.com

## Appendix A

Page 167 © Oleksiy Mark/Shutterstock.com; 168 © iStockphoto.com/alexsl, 168 © iStockphoto.com/CostinT, 168 © Kathy Burns-Millyard/Shutterstock.com, 168 © iStockphoto.com/LL28; 169 Used with permission from Microsoft.; 170 Courtesy of Energy Star; 172 Courtesy of DigitalTrends.com, 172 © GalaStudio/Shutterstock.com; 173 © iStockphoto.com/Andrew_Howe; 174 Used with permission from Jelly Comb., 174 Used with permission from Ultimate Ears., 174 (c)iStock/adventtr

## Appendix B

Page 175 image by Krish Dulal, distributed under a CC-By 2.0 license, 175 © iStock.com/pictafolio; 176 Used with permission from Arris., 176 Used with permission from Arris.; 179 Used with permission from Microsoft., 179 Used with permission from Microsoft.; 181 © istock/flladendron

# Preface

## Course Overview

Today's students arrive in the classroom with more confidence in using technology than any generation before them. You have grown up with technology as a part of your life and the internet as a source of information, entertainment, and communication. Chances are you have used a word processor and a presentation program for several years to prepare materials for school projects. You may have learned your way around a computer application by trial and error. However, to be efficient and successful, you need to learn how to use software applications in a way that saves you time and makes the best use of the available feature set. To that end, *Seguin's COMPUTER Applications with Microsoft® Office 365,* 2019 Edition, provides the tools you need to succeed immediately in your academic and personal lives as well as prepare yourself for success in your future career. In this course, you will learn skills you can apply immediately to accomplishing projects and assignments for school and to organizing, scheduling, recording, planning, and budgeting for your personal needs. You will find the work done in this course to be relevant and useful, with the content presented in a straightforward approach.

Along with well-designed pedagogy, practice and problem solving will help you learn and apply computer topics and skills. Technology provides opportunities for interactive learning as well as excellent ways to quickly and accurately assess your performance. To this end, this course is supported by Cirrus, Paradigm Education Solution's web-based training and assessment learning management system.

## Course Goals

*Seguin's COMPUTER Applications* offers instruction that will guide you to achieve entry-level competence with the latest editions of Microsoft Windows, web browsers, and the Microsoft Office productivity suite, including OneNote, Outlook, Word, Excel, PowerPoint, and Access. You will also be introduced to cloud computing alternatives to the traditional desktop suite. No prior experience with these software programs is required. Even those with some technological savvy can benefit from completing the course by learning new ways to perform tasks or reinforce skills. After completing this course, you will be able to:

- navigate the Windows operating system and manage files and folders.
- use web browsers such as Microsoft Edge, Google Chrome, and Mozilla Firefox to navigate and search the web, as well as download content to a PC or mobile device.
- use navigation, file management, commands, and features within the Microsoft Office suite that are standard across all applications.
- organize and manage class notes in OneNote.
- communicate and manage personal information in Outlook.
- create, edit, format, and enhance documents in Word.
- create, edit, analyze, format, and enhance workbooks in Excel.
- create, edit, format, and enhance slides and set up a slideshow in PowerPoint.
- create and edit tables, forms, queries, and reports in Access.
- integrate information among the applications within the Microsoft Office suite.
- use cloud computing technologies to create, edit, store, and share documents.

Reading time is minimized; you will learn just what you need to know to succeed within these programs. You will practice features with step-by-step instruction interspersed with text that explains why a feature is used or how the feature can be beneficial to you. You should work through each chapter at a PC or with a tablet, so that you can complete the steps as you learn.

## Course Organization and Methodology

This course is divided into 15 chapters that are best completed in sequence; however, after assigning the essential skills learned in Chapters 1 through 5, your instructor may opt to assign Word, Excel, PowerPoint, Access, integration, and cloud computing technologies in the order of his or her choice.

Each chapter begins with a brief introduction to the chapter content and a list of learning objectives. Following the chapter opener, each chapter topic begins with a list of skills to be mastered. Each topic is presented in two or four pages. The topics were developed using features available in Microsoft Office 365. You may find that with your computer and version of Office, the appearance of the software and the steps needed to complete an activity vary slightly from what is presented in the courseware. A variety of marginal notes and other features expand or clarify the content. You will gain experience with topic features by working through hands-on exercises, which consist of step-by-step instructions and illustrative screen shots. At the end of each chapter, you will have a chance to review a summary of the chapter topics.

## What Makes This Course Different from Others?

Many courses that teach computer applications were designed and organized for software that was in effect one or two decades ago. As software evolves and becomes more flexible and streamlined, so too should software courses. With this mandate, this course has been designed and organized with a fresh look at the skills a student should know to be successful in today's world. The author has chosen and placed in a logical sequence those skills that are considered essential for today's student. Consider this course a "software survival kit for school and life." Nothing more, nothing less!

Many of the student data files in this course are based on files created by students for projects or assignments in courses similar to those students may be enrolled in now. Students will open and manipulate real work completed by someone just like them. Other files include practical examples of documents that students can readily relate to their school and personal experiences.

## This Course Is Green!

Instructions to print results have been intentionally omitted for all exercises and assessments. This approach is consistent with a green computing initiative to minimize wasteful printing for nongraded topics or assessment work and also provides instructors with maximum flexibility in designing their course structure.

## Course Features

The following guide shows how this course uses a visual approach combined with hands-on activities to help you learn and master key skills and topics.

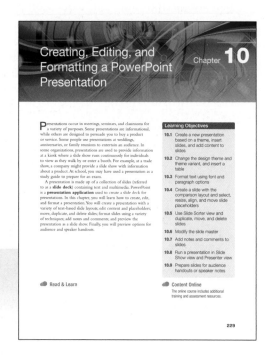

**Chapter openers** provide an overview of the software, give students a preview of the software features, and explain how the software is used in the workplace. The openers also provide a list of **learning objectives**, which align to each topic.

# Chapter Elements

**Topics** are presented in two or four pages.

**Skills** learned in each topic are listed for review and reinforcement.

**Step-by-step instructional text** accompanied with frequent illustrative screen shots presents instructions for completing the tasks in simple, easy-to-follow steps.

**Screen captures** with step numbers provide visual confirmation and guidance.

**App Tips** extend or add to your knowledge of a feature.

**Quick Steps** briefly summarize the steps to complete major tasks, for quick reference and review.

Text to be typed is set in **magenta font** to stand out from the instructional text.

**Oops!** hints anticipate common challenges and provide solutions to help you succeed with the topic.

---

## Sample page 230

230  Seguin's COMPUTER Applications with Microsoft® Office 365, 2019 Edition

### 10.1 Creating a New Presentation and Inserting Slides

Begin creating a new presentation on the PowerPoint Start screen by choosing a template, theme, and variant on a theme, or by starting with a blank presentation. The first slide in a presentation is a **title slide** with a text **placeholder** for a title and a subtitle. A placeholder is a rectangular container on a slide that can hold text or other content. Each placeholder on a slide can be manipulated independently.

PowerPoint starts a new presentation with a title slide displayed in Normal view. In Normal view, the current slide displays in widescreen format in the **slide pane**. Numbered slide thumbnails display in the **slide thumbnails pane**. A notes pane at the bottom of the slide pane and a Comments pane at the right of the slide pane can be opened as needed.

1. Start PowerPoint and then click the More themes hyperlink near the top right of the PowerPoint Start screen.
2. Click the *Ion* theme in the New backstage area.

The New backstage area has a gallery of design themes. Click to preview a theme along with the theme variants. A **variant** is a different style and color scheme included in the theme family.

*Watch & Learn*
*Hands On Activity*

**skills**
Create new presentation
Choose theme and theme variant
Insert slide
Edit text on slide

**app tip**
Double-click a theme to start a new presentation using the theme default style and color scheme.

**Tutorial**
Opening a Presentation Based on a Template

Microsoft automatically updates templates and themes. Your screen may vary.

3. Click the last variant (orange color scheme), and then click the right-pointing arrow below the preview slide next to *More Images*.

Browse through the *More Images* slides to see a variety of content in the color scheme and view the theme or variant style and colors before making a selection.

Title Layout

Click here if you want to close the preview to select another theme.

4. Click the second variant (blue color scheme), and then click the right-pointing arrow below the preview slide to view the blue color scheme with a Title and Content layout depicting a chart.

---

## Sample page 235

Chapter 10  Creating, Editing, and Formatting a PowerPoint Presentation  235

10. Select *5* in the *Number of columns* text box in the Insert Table dialog box and then type 2.
11. Select *2* in the *Number of rows* text box, type 6, and then click OK.

PowerPoint inserts a table on the slide with the colors in the theme variant.

12. With the insertion point positioned in the first cell in the table, type Type of Car and then press Tab or click in the second cell.
13. Type Cost.
14. Type the remaining entries in the table by pressing Tab to move to the next cell or by clicking in the next cell and then typing the text as follows:

| | |
|---|---|
| Small size | $600 |
| Medium size | $675 |
| Large family sedan | $750 |
| Minivan | $775 |
| SUV | $825 |

### Modifying a Table

The Table Tools Design and Table Tools Layout tabs provide options for modifying a table. These tools are the same table tools you learned in Word. Use the handles to enlarge or shrink the table size, or drag the border of a table to move it on the slide.

15. Drag the right-middle sizing handle to the left until the right border of the table ends approximately below the *c* in *Maintenance* in the title text, as shown in the image at right.
16. Click in any cell in the second column of the table.
17. Click the Table Tools Layout tab, click the Select button in the Table group, and then click *Select Column*.
18. Click the Center button in the Alignment group.
19. Click in any cell in the first column of the table, select the current entry in the *Width* text box in the Cell Size group, type 3.5, and then press Enter.
20. Drag the top border of the table to move it to the approximate location shown in the image at right.
21. Save the revised presentation using the same name (**CarMaintenance-YourName**). Leave the presentation open for the next topic.

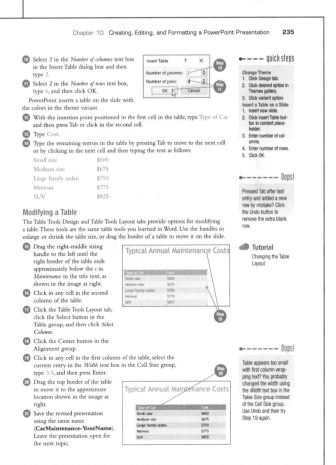

**quick steps**
Change Theme
1. Click Design tab.
2. Click desired option in Themes gallery.
3. Click variant option.
Insert a Table on a Slide
1. Insert new slide.
2. Click Insert Table button in content placeholder.
3. Enter number of columns.
4. Enter number of rows.
5. Click OK.

**Oops!**
Pressed Tab after last entry and added a new row by mistake? Click the Undo button to remove the extra blank row.

**Tutorial**
Changing the Table Layout

**Oops!**
Table appears too small with first column wrapping text? You probably changed the width using the Width text box in the Table Size group instead of the Cell Size group. Use Undo and then try Step 19 again.

Insert Table dialog box: Number of columns: 2 — Number of rows: 6 — OK — Cancel

Typical Annual Maintenance Costs

| Type of Car | Cost |
|---|---|
| Small size | $600 |
| Medium size | $675 |
| Large family sedan | $750 |
| Minivan | $775 |
| SUV | $825 |

A **Topics Review** chart at the end of each chapter summarizes the main chapter content and key words learned.

**Topics Review**

| Topic | Key Concepts |
|---|---|
| 7.1 Inserting, Editing, and Labeling Images in a Document | Graphic elements assist with comprehension and/or add visual appeal to documents. |
| | Insert an image from a web resource or OneDrive using the **Online Pictures button** in the Illustrations group on the Insert tab. |
| | Click the **Layout Options button** to control how text wraps around the image and to change the position of the image on the page. |
| | Use the **Pictures button** to insert an image from a file stored on your computer or other storage medium. |
| | Edit the appearance of an image and/or add special effects using buttons on the Picture Tools Format tab. |
| | A caption is explanatory text above or below an image that is added using the **Insert Caption** feature on the References tab. |
| | Word automatically numbers images as Figures and can generate a Table of Figures automatically. |
| 7.2 Adding Borders and Shading, and Inserting a Text Box | Add a border or **shading** to paragraphs to make text stand out on a page. Shading is color applied to the page behind the text. |
| | A page border surrounds the entire page and is added using the **Page Borders button** on the Design tab. |
| | Apply a border to selected text using the **Borders gallery** from the Borders button arrow in the Paragraph group on the Home tab. |
| | Shading is added using the Shading button arrow in the Paragraph group. |
| | A **pull quote** is a quote typed inside a text box. |
| | Insert text inside a box using the **Text Box button** in the Text group on the Insert tab. |
| 7.3 Inserting a Table | A table is a grid of columns and rows in which you type text and is used when you want to arrange text side by side or in rows. |
| | A **table cell** is a rectangular-shaped box in the table grid that is the intersection of a column and a row. Text in the table is typed inside table cells. |
| | A **Quick Table** is a predesigned table with sample data, such as a calendar or a tabular list. |
| | Create a table by clicking a square in the table grid accessed from the Insert tab that represents the number of columns and rows or by entering the number of columns and rows in the Insert Table dialog box. |
| | Pressing Tab in the last table cell automatically adds a new row to the table. |

*continued...*

5 Click the right-pointing arrow below the preview slide two more times to view other types of content with the blue color scheme.

6 With the Photo Layout preview displayed, click the Create button.

7 Compare your screen with the one shown in Figure 10.1.

**Figure 10.1**
A new PowerPoint presentation with Ion theme and blue color variant in the default Normal view is shown above. See Table 10.1 for a description of screen elements.

**Table 10.1** PowerPoint Features

| Feature | Description |
|---|---|
| Notes button | Button to turn on or turn off the notes pane at the bottom of the slide pane |
| Placeholders | Containers in which you type or edit text, or insert other content such as an image or audio clip |
| Slide pane | Pane that displays the active slide; add or edit content on a slide in this area |
| Slide thumbnails pane | Pane that displays numbered thumbnails of the slides in the presentation; navigate to, insert, delete, or duplicate a slide in this pane |
| Status bar | Bar that displays active slide number with total number of slides in the presentation and displays a message about an action in progress |
| View and Zoom buttons | These buttons change the display of the PowerPoint window. View buttons in order are: Normal, Slide Sorter, Reading View, and Slide Show. Zoom buttons enlarge or shrink the display of the active slide. |

**quick steps**

Start a New Presentation
1. Start PowerPoint.
2. Click More themes hyperlink.
3. Click theme.
4. Click variant.
5. Click Create button.
Insert a Slide
Click New Slide button in Slides group.
OR
1. Click New Slide button arrow.
2. Click required slide layout.
Edit Text
1. Activate slide.
2. Select text or click in placeholder and move insertion point as needed.
3. Type new text or change text as required.

**Tutorial**
Exploring the PowerPoint Screen

**Tables** and **figures** organize information in a streamlined format to minimize your reading load.

# Review & Assessment

Chapter Assessments allow you to put your new knowledge to work and demonstrate the kind of thinking and problem-solving needed in today's workplace. For each chapter you will find the following assessments.

An **Audio assessment** asks you to listen to instructions provided in an **Audio File** and then to compose the document, workbook, or presentation as instructed.

A **Mobile assessment** allows you to explore your new skills using mobile devices.

A **Jobs assessment** specifically challenges you to apply your skills to real-world job experiences.

---

**Audio** ▶ Assessment 6.6  Audio—Internet Research and Composing a New Document

**Type:** Individual or Pairs
**Deliverable:** Word Document

You are asked to help with a project on social media by creating a document that describes what you read online after researching two to three recent events where social media was used to promote social good. The project manager has left you a voicemail with information about the project.

1. Listen to the audio file **SocialMediaForSocialGood_instructions**.
2. Complete the research and compose the document as instructed.
3. Save the document in the Ch6 folder within the Assessments folder on your storage medium as **SocialMediaResearch-YourName**.
4. Submit the assessment to your instructor in the manner requested.
5. Close the document.

**Mobile** ▶ Assessment 6.7  Go Mobile—Creating a Document with Conference Locations

**Type:** Individual
**Deliverable:** Word Document

1. Open the Word mobile app on your smartphone and start a new blank document.
2. Type today's date and then press Enter to move down to the next line.
3. Type Conference Venue Ideas and then press Enter to move down to the next line.
4. In a bulleted list, type four locations near you that are suitable for hosting a conference attended by approximately 150 people.
5. Press Enter twice after the text typed at Step 4 and then type your name.
6. Use Save As to change the document name to **ConferenceLocations -YourName**, saving it in your OneDrive personal storage account.
7. Submit the assessment to your instructor in the manner requested.
8. Exit the Word mobile app.

**Jobs** ▶ Assessment 6.8  Job Ready—Composing a Letter

**Type:** Individual, Pairs, or Teams
**Deliverable:** Document

You are a member of a conference planning team organizing a conference to be held next year during the week after the winter semester ends. The conference will be attended by student leaders from all the community colleges in your state or province. The focus of the conference is to learn about current trends in online learning. Your task is to help the committee responsible for choosing the keynote speaker for the conference opening night dinner.

1. The chair of the committee wants you to draft the invitation letter to be sent to the person the committee chooses to invite to be the keynote speaker. Begin by searching online for sample text for a keynote speaker invitation.
2. In a new blank document, type a sample invitation letter with details about your conference. Insert a fictitious name and address for the letter recipient. Include your name in the letter as the writer.
3. Apply font and paragraph format options appropriate for a business letter.
4. Save the document as **SpeakerInvitationLtr-YourName** in the Ch6 folder in Assessments on your storage medium.
5. Submit the assessment to your instructor in the manner requested.

---

**Visual** ▶ Assessment 6.5  Visual—Campus Flyer from Template

**Type:** Individual
**Deliverable:** Campus Flyer

1. Create a flyer for your school campus similar to the one shown in the Assessment 6.5 Campus Flyer below. Use a current date and a location suitable for concerts on or near your campus. Add current popular band names to the *FEATURING* section. Enter a fictitious web address and sponsor information. Make any other changes you think are necessary.
*Note: Search for the template shown using the search phrase simple flyer in the New backstage area.*
2. Save the flyer in the Ch6 folder within the Assessments folder on your storage medium as **CampusBandFlyer-YourName**.
3. Submit the assessment to your instructor in the manner requested.
4. Close the document.

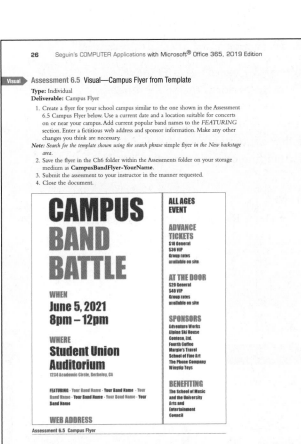

Assessment 6.5  Campus Flyer

A **Visual assessment** challenges you to create a document similar to a model document. This type of assessment provides less step-by-step instruction than other assessments and thus requires attention to detail and creative problem solving. Some visual assessments will require internet research or creative writing.

# The Cirrus Solution

*Elevating student success and instructor efficiency*

Powered by Paradigm, Cirrus is the next–generation learning solution for developing skills in Microsoft Office and computer concepts. Cirrus seamlessly delivers complete course content in a cloud-based learning environment that puts students on the fast-track to success. Students can access their content from any device anywhere, through a live internet connection. Cirrus is platform independent, ensuring that students get the same learning experience whether they are using PCs, Macs, or Chromebook computers. Cirrus provides access to all the *Seguin's COMPUTER Applications* content, delivered in a series of scheduled assignments that report to a grade book to track student progress and achievement.

## Dynamic Training

Cirrus online versions of *Seguin's COMPUTER Application* course include interactive assignments to guide student learning.

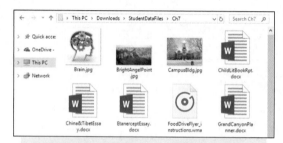

**Student Data Files** for each section provide the files needed to complete each task. **Audio Files** needed for some assessments are also provided.

**Watch and Learn Lessons** show you how to perform the application topic tasks in a video, allow you to read the topic content, and include a short quiz that allows you to check your understanding of the content.

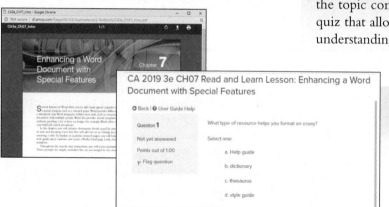

**Read and Learn Lessons** provide the chapter opening content to read and a short quiz allows you to check your understanding of the content.

**Hands On Activities** enable you to complete the Topic tasks in provided data files, compare your solutions against a model answer image, and submit your work for instructor review.

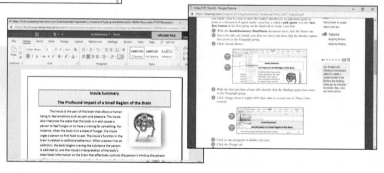

**Tutorial Lessons** provide interactive, guided training and measured practice.

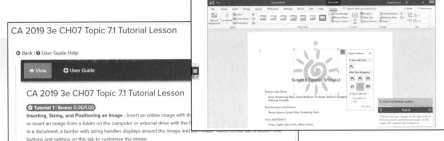

Interactive study tools, including a **Glossary** (including Key Terms, their definitions, and their location in the course), **Crossword Puzzle**, **Infographic** with quiz, and **Review Quizzes** (Multiple Choice, Completion, and Matching) give you an opportunity to help reinforce your understanding of the skills and topics in a fun, visual, game-like way.

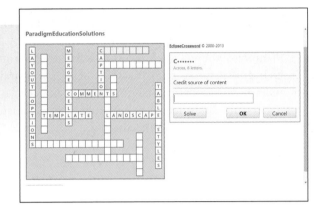

**Exercises** and **Project Exams** provide opportunities for students to further develop and demonstrate topic skills. Exercises and Project Exams are completed live in the Office application and are automatically scored by Cirrus. Detailed feedback and how-to videos help students evaluate and improve their performance.

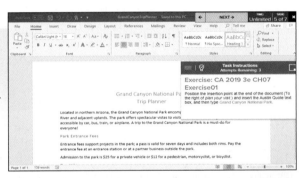

**Skills Check Exams** evaluate your ability to complete specific tasks. Skills Check Exams are completed live in the Office application and are automatically scored.

Multiple-choice **Chapter Exams** test your overall comprehension of the information learned in the chapters.

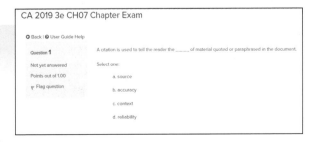

## Student eBook

The Student eBook makes *Seguin's COMPUTER Applications* content available from any device (desktop, laptop, tablet, or smartphone) anywhere. The ebook is accessed through the Cirrus online course.

## Instructor eResources

Cirrus tracks students' step-by-step interactions, giving instructors visibility into students' progress and missteps. With Exam Watch, instructors can remotely join a virtual, live, skills-based exam with students—a helpful option for struggling students who need one-to-one coaching, or for distance learners. In addition to these Cirrus-specific tools, other instructor materials are available. Accessed through Cirrus and visible only to instructors, the *Instructor eResources* for *Seguin's COMPUTER Applications* include the following materials:

- Planning resources, such as lesson plans, teaching hints, and sample course syllabi
- Delivery resources, such as activity answer keys and online images and templates
- Assessment resources, including live and annotated PDF model answers for topic work and review and assessment activities, rubrics for evaluating student work, a skills exam for each topic, and chapter-based exam files

# Acknowledgments

The author and editors would like to thank the following individuals for their contributions to and feedback on the course: Julia Basham and Becky Jones; and Patti Ann Reynolds, Toni McBride, and Nicole Oke, Fanshawe College. A special thank you is extended to Michael Seguin for his contributions.

# About the Author

Denise Seguin has served on the Faculty of Business at Fanshawe College of Applied Arts and Technology in London, Ontario, from 1986 until her retirement from full-time teaching in December 2012. She developed curriculum and taught a variety of office technology, software applications, and accounting courses to students in postsecondary Information Technology diploma programs and Continuing Education courses. Seguin served as Program Coordinator for Computer Systems Technician, Computer Systems Technology, Office Administration, and Law Clerk programs and was acting Chair of the School of Information Technology in 2001. Along with authoring *Seguin's COMPUTER Concepts*, First and Second Editions, and *Seguin's COMPUTER Applications with Microsoft® Office*, First and Second Editions, she has also authored Paradigm Education Solution's *Microsoft Outlook* and co-authored *Our Digital World*, *Benchmark Series Microsoft® Excel®*, *Benchmark Series Microsoft® Access®*, *Marquee Series Microsoft® Office*, and *Using Computers in the Medical Office*.

In 2007, Seguin earned her Masters in Business Administration specializing in Technology Management, choosing to take her degree at an online university. She has an appreciation for those who are juggling work and life responsibilities while furthering their education, and she has taken her online student experiences into account when designing instruction and assessment activities for this course.

# Seguin's

# COMPUTER
## Applications

## for Microsoft® Office 365

### 2019 Edition

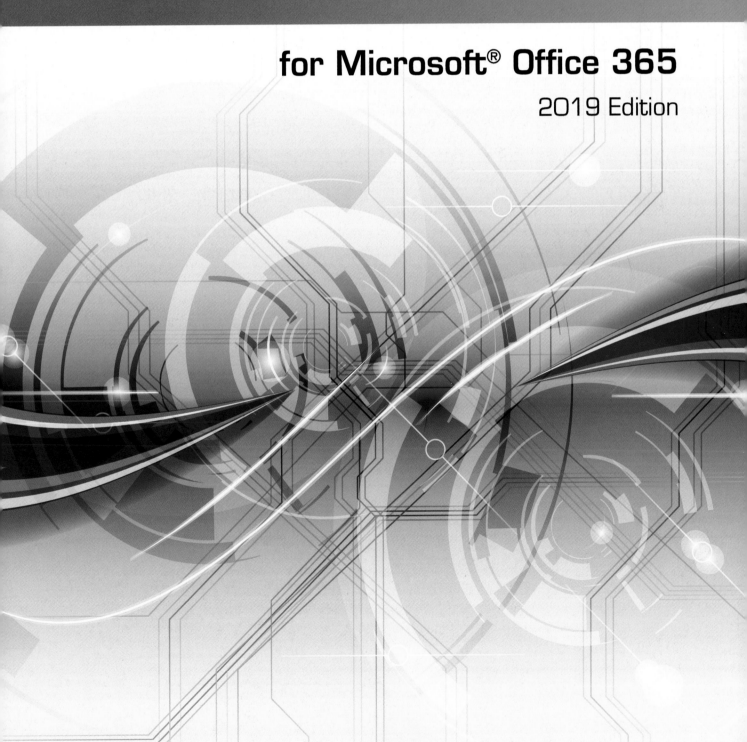

## Chapter 1

# Using Windows 10 and Managing Files

Windows 10 is the operating system (OS) software published by Microsoft Corporation. An OS provides the user interface that allows you to work with a computer or mobile device. The OS also manages all the hardware resources, routes data between devices and applications, and provides the tools for managing files and application software. Every computing device requires an OS; without one, your computer would not function. Think of the OS as the data and device manager that ensures data flows to and from each device and application. When you touch the screen, click the mouse, type words on a keyboard, or activate Cortana to say voice commands, the OS recognizes the input and sends the data to the application or device that needs it. If a piece of hardware is not working, the OS senses the problem and displays a message to you. For example, if a printer is not turned on, the OS communicates that the printer is offline. When you power on a computer or mobile device, the OS displays the user interface when the computer is ready.

Computers and mobile devices have an OS preloaded and ready to use. Some tasks require you to interact with the OS directly, such as when you launch an application, switch to another open window, and manage your files and folders. In this chapter, you will learn to navigate in Windows 10 and manage files on a storage medium.

*Note: You will need a removable storage medium with enough storage space to download a copy of the student data files for this textbook.*

## Learning Objectives

**1.1** Navigate Windows 10 using touch, a mouse, or a keyboard

**1.2** Sign in to Windows 10, launch an app, and switch between apps

**1.3** Use the search text box and minimize, maximize, close, and snap windows

**1.4** Use icons in the notification area to view the status of a setting and view customization options for the notification area

**1.5** Lock the screen, sign out, and shut down Windows 10

**1.6** Use and customize the Start menu

**1.7** Personalize the Lock screen, desktop background, and color scheme

**1.8** Describe *file, folder, download,* and extract student data files

**1.9** Browse files with File Explorer

**1.10** Create new folders and copy files and folders

**1.11** Move, rename, and delete files and folders and safely eject a USB flash drive

Read & Learn

Content Online

The online course includes additional training and assessment resources.

## 1.1 Using Touch, Mouse, and Keyboard Input to Navigate Windows 10

**Watch & Learn**

●—————— skills

Describe Windows 10 touch gestures

Describe basic mouse actions

List common keyboard commands for Windows 10

**Windows 10** is the latest edition of the Microsoft OS for personal computers (PCs) and mobile devices. The Windows 10 **user interface (UI)** uses icons and a Start menu to interact with users. The **Start menu** is a pop-up menu that displays when you click the **Start button** located at the bottom left corner. In the left pane of the Start menu are two columns. The first column has icons to change your account settings, display the Documents or Pictures folder, open the Settings app, or access Power options. In the top section of the second column are shortcuts to the most-used applications, followed by all of the applications installed on your PC in alphabetical order. The right pane contains tiles. A **tile** is a square or rectangle that, when clicked, starts an app or an application. An **app** is an application used on a smartphone, tablet, or PC that is designed usually for one purpose, to work on a touchscreen, to use less power than a desktop application, and to use fewer network resources. An **application**, which is more powerful than an app, is software used to get a task done with a wide range of functions and features. Applications make use of a full-size keyboard and a mouse, and they are designed for a larger screen and more powerful computing resources.

Windows 10 works with touchscreen input, mouse input, keyboard input, and voice recognition. In this topic, you will learn how to navigate the UI using touch gestures, the mouse, and the keyboard.

### Using Touch to Navigate Windows 10

On a touchscreen, you perform actions using gestures. A **gesture** is an action or motion you perform with one or two fingers, a thumb, or a stylus. Table 1.1 on the next page explains the gestures used with the Windows interface.

On a touchscreen device with no attached keyboard, when a task requires typed characters, such as an email address, message text, or web address, the **touch keyboard** (Figure 1.1) is used. Tapping the area that requires typed input generally displays the touch keyboard. If the keyboard does not display when you need to type, tap the keyboard icon at the right end of the taskbar next to the date and time.

The touch keyboard is available in five layouts and two sizes. Tap the keyboard settings icon at the top left of the touch keyboard to choose an alternate keyboard layout or to make the keyboard narrower. Alternate keyboard layouts include a thumb keyboard, a miniature keyboard, a handwriting touchpad for writing with your finger or by using a stylus, and a full layout keyboard that provides access to the full range of keys on a standard keyboard (Figure 1.2 on the next page).

**Figure 1.1**

The touch keyboard in lowercase mode is the default layout that appears when typed characters are expected.

**Table 1.1**  Windows Touch Gestures

| Gesture | Description and Mouse Equivalent | What It Does | What It Looks Like |
|---|---|---|---|
| Tap or Double-tap | One finger touches the screen and immediately lifts off the screen once or twice in succession. **Mouse:** Click or double-click left mouse button. | • Launches an app or application<br>• Follows a link<br>• Performs a command or selection from a button or icon | |
| Press and hold | One finger touches the screen and stays in place for a few seconds. **Mouse:** Point or right-click. | • Shows a context menu<br>• Shows pop-up information or details | |
| Slide | Move one or two fingers in the same direction. **Mouse:** Drag (may need to drag or scroll using scroll bars). | • Drags, pans, or scrolls through lists or pages | |
| Swipe | Move one finger in from the left or right a short distance.<br><br>**Mouse:** Click the Notifications icon on the taskbar.<br><br>**Mouse:** Click the Task View button on the taskbar. | • In from the right reveals the Action Center panel with Notifications and quick action tiles at the bottom<br><br>• In from the left shows thumbnails of open windows so you can switch to another window | |
| Zoom Out | Two fingers touch the screen apart from each other and move closer together. **Mouse:** Ctrl + scroll mouse wheel toward you. | • Shrinks the size of text, an item, or tiles on the screen | |
| Zoom In | Two fingers touch the screen together and move farther apart. **Mouse:** Ctrl + scroll mouse wheel away from you. | • Expands the size of text, an item, or tiles on the screen | |

**Figure 1.2**

The full Windows layout touch keyboard with access to letters, numbers, and special keys.

For some purposes, you still use a mouse and traditional keyboard with a touch-enabled device. For example, typing an essay for a school project is easier using a traditional keyboard, and doing precise graphics editing is easier with a mouse. Connect a universal serial bus (USB) or wireless mouse and/or keyboard for intensive work like this.

## Using a Pointing Device to Navigate Windows 10

For traditional desktops or laptops, navigating the Windows user interface requires the use of a **mouse**, **trackball**, **touchpad**, or other pointing device. These devices are used to move and manipulate a **pointer** (displayed as ⌖ ) on the screen; however, the white arrow pointer can change appearance depending on the action being performed.

To operate the mouse, move the device up, down, right, or left on the desk surface to move the pointer on the screen in the corresponding direction. A scroll wheel on the top of the mouse can be used to move the screen up or down by rolling the finger upward or downward on the wheel. Left and right buttons on the top of the mouse are used to perform actions when the pointer is resting on an item. Table 1.2 provides a list and description of mouse actions.

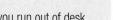 **app tip**

If you run out of desk surface, lift the mouse up off the desk, place it back down, and then continue moving in the same direction to extend the mouse movement farther.

**Table 1.2**  Mouse Movements and Actions

| Term or Action | Description |
| --- | --- |
| Point | Move the mouse in the direction required to rest the white arrow pointer on a button, icon, option, tab, link, or other screen item. |
| Click | Quickly tap the left mouse button once while the pointer is resting on a button, icon, option, tab, link, or other screen item. |
| Double-click | Quickly tap the left mouse button twice. On the desktop, an application is launched by double-clicking the application's icon. |
| Right-click | Quickly tap the right mouse button. Within a software application such as Word or Excel, a right-click causes a shortcut menu to appear. Shortcut menus in software applications are context-sensitive, meaning that the menu that appears varies depending on the item the pointer is resting upon when the right-click occurs. |
| Drag | Hold down the left mouse button, move the mouse up, down, left, or right, and then release the mouse button. Dragging is an action often used to move or resize an object. |
| Scroll | Use the scroll wheel on the mouse to scroll in a window. If the pointing device you are using does not include a scroll wheel, click the scroll arrows on a horizontal or vertical scroll bar at the right or bottom of a window, or drag the scroll box in the middle of the scroll bar up, down, left, or right. |

With a trackball pointing device (Figure 1.3a), you roll your thumb or fingers over the ball to move the pointer on the screen. Buttons on the trackball operate like a mouse for performing actions. Some people prefer a trackball because it responds to smaller movements and the device stays stationary on the desk. A worker with a repetitive strain injury may switch to a trackball because the device is not moved on the desk to move the pointer.

To use a touchpad (Figure 1.3b), move your finger up, down, right, or left across the surface of the touchpad to move the pointer on the screen in the same direction. Tap the touchpad or a button below it to perform an action. Some touchpads support touch gestures.

a                                        b

**Figure 1.3**

(a) A trackball (the bright blue ball on the mouse shown) responds to smaller movements of the ball to position the pointer. (b) A touchpad lets you move the pointer using your finger.

## Using a Keyboard to Navigate Windows 10

Some people like to use a **keyboard command** (also called a *keyboard shortcut*) to perform an action because it is fast and easy to use. The Start key (Figure 1.4) is positioned at the bottom left of a keyboard between the Ctrl or Function key (FN) and the Alt key. Many keyboard shortcuts use the Start key. For example, press the Start key at any time to bring up the Start menu. Useful keyboard commands are described in Table 1.3. If you prefer using keyboard commands, search online for articles that give other Windows 10 keyboard navigational commands.

**Figure 1.4**

Some keyboard commands listed in Table 1.3 use the Start key, which is found in the bottom row of the keyboard, left of the spacebar.

*Note: Instructions in this textbook are written with mouse actions. If necessary, check with your instructor for the equivalent touchpad or other pointing device action. If you prefer to use touch gestures, refer to the gestures with mouse equivalent actions provided in Table 1.1 on page 5.*

**Table 1.3** Keyboard Commands or Shortcuts

| Keyboard Shortcut | What It Does |
| --- | --- |
| Start key | Displays Start menu |
| Start key + a | Opens the Action Center panel at the right side of the screen |
| Start key + d | Returns to the desktop from an app or application window |
| Start key + e | Opens a File Explorer window |
| Start key + i | Opens the Settings app |
| Start key + l | Locks the screen |
| Start key + m | Minimizes all windows |
| Start key + q | Starts a search for apps, files, settings, and web links |
| Start key + Tab | Displays Task View to switch to another window |
| Alt + F4 | Closes app, application, or other active window |
| Alt + Tab | Switches between open apps |
| Ctrl + D | Deletes selected folder(s) or file(s) |
| Up, Down, Left, or Right Arrow keys | Moves selection on Start menu to an app or application name, setting, folder name, or tile; pressing the Enter key launches the selection |

●‑ ‑ ‑ ‑ ‑ ‑ app tip

To use a keyboard command, hold down the Start, Alt, or Ctrl key, press and release the letter or function key, and then release the Start, Alt, or Ctrl key.

## 1.2 Starting Windows 10 and Exploring Apps

If you are turning on your PC from a no-power state, the **Lock screen** shown in Figure 1.5 appears. The Lock screen also appears if you resume computer use after the system has gone into sleep mode. Depending on your PC or mobile device, turning on or resuming system use from sleep mode involves pressing the Power button or moving a mouse.

Each person using a computer with Windows will have a **user account**. A user account includes a username and a password or personal identification number (PIN). Windows stores settings information for each user account so that each person can customize options without conflicting with the settings for other people who use the same device. Microsoft offers two types of user accounts at sign-in: a Microsoft account or a local account.

### Signing In with a Microsoft Account

To sign in to Windows using a phone number or an email address, set up the account as a **Microsoft account**. A Microsoft account allows you to download new apps from the Windows store, see live updates from messaging and social media services in tiles on the Start menu, and sync your Windows and browser settings online so that they are the same across all devices. To create a Microsoft account, open a browser, navigate to https://login.live.com, and then click <u>Create one!</u>

### Signing In with a Local Account

If you prefer not to share Windows and browser settings on a PC or mobile device with other devices, set up an account as a **local account**. Automatic connections to messaging and social media services also do not work with a local account.

*Note: The screen images in this textbook were made with a user signed in to Windows with a Microsoft account.*

1. Turn on the computer or mobile device, or resume system use from sleep mode.
2. At the Lock screen (Figure 1.5), click anywhere on the screen or press any key on the keyboard to reveal the sign-in screen.

The Lock screen image can be changed. Your Lock screen may show a different image.

10:29
Friday, July 10

Icons show power and network connectivity status. Other notifications may also appear here.

**Figure 1.5**
The Windows 10 Lock screen displays landscape photos by default.

**Watch & Learn**

●━━━━━━ Skills

Start and sign in to Windows 10

Launch an app

Switch apps

**Tutorial**
Signing In to and Out of Windows 10

●━━━━━━ app tip

Touch user? When the Lock screen displays, swipe up from the bottom edge of the screen to reveal the sign-in screen.

**3** Depending on the system, the next step will vary. Complete the sign in by following the instructions in 3a, 3b, or 3c that match the configuration of your device:

a. Type your password in the *Password* text box below your account name and press Enter or click the Submit button (right-pointing arrow).

b. Type your PIN in the text box below your account name, perform the touch gestures over your account's picture password, or perform another biometric sign-in, such as scanning your fingerprint.

A picture password uses an image from your Pictures library. You must perform three touch gestures over the image to sign in. A PIN code lets you sign in faster using a numeric code.

● – – – – – **app tip**

A PIN, picture password, or biometric authentication is set up in the Accounts category of the Settings app. You have to enroll in Windows Hello, which uses biometric authentication to unlock a device using facial recognition, an iris scan, or a fingerprint scan. Note that not all systems support these sign-in options.

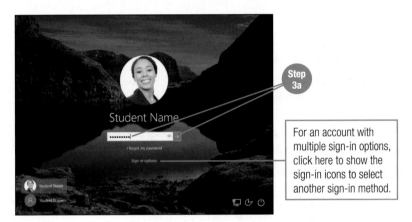

Student Name

**Step 3a**

For an account with multiple sign-in options, click here to show the sign-in icons to select another sign-in method.

c. If the active account is not your account, click the correct account name, picture, or silhouette icon at the bottom left of the sign-in screen, type your password in the *Password* text box, and then press Enter or click the Submit button (right-pointing arrow).

## The Desktop

Once you are signed in, the desktop, similar to the one shown in Figure 1.6, appears. The **desktop** displays icons and buttons that are used to launch applications or open

● – – – – – **app tip**

Signing in to Windows 10 on a tablet or other touchscreen device may convert the display to tablet mode, in which the Start screen displays tiles for common apps, a menu button, a power button, and a back button to navigate the UI.

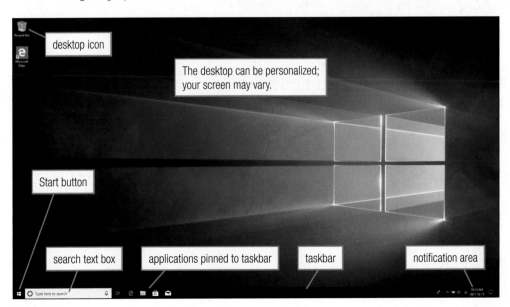

desktop icon

The desktop can be personalized; your screen may vary.

Start button

search text box        applications pinned to taskbar        taskbar        notification area

**Figure 1.6**

The Windows 10 desktop displays a Start button, search text box, pinned icons, and a notification area on the taskbar along the bottom of the window. Icons on the desktop above the taskbar launch applications or open a window.

other windows and displays a taskbar at the bottom of the screen. The **taskbar** has the Start button, the search text box, icons to start applications, and the notification area. The **search text box** is used to locate applications, settings, files on your PC, or information online. The **notification area** at the right of the taskbar shows the current date and time and displays icons to view the status of system settings. When you sign in, Windows may open the Start menu on the desktop for you.

## Launching an App

Built-in apps for Windows 10 include Photos, Maps, Mail, Calendar, Groove Music, and Movies & TV, to name a few. The Windows 10 family of apps is designed to look and operate consistently on all devices. To launch an app, click the Start button and then click the tile for the app in the right pane.

*Note: If you are using a tablet with tablet mode active, swipe in from the right edge of the screen and tap the Tablet mode quick action tile to turn tablet mode off.*

④ Click the Start button if the Start menu is not open. Click the Photos tile in the right pane on the Start menu to launch the Photos app. Messages or prompts may appear when you launch an app with information features or requesting permission to access your location. Close or navigate the prompts as desired.

The **Photos app** shows thumbnails of the photos stored in the Pictures library on a PC, tablet, or smartphone, and in OneDrive, grouped by date.

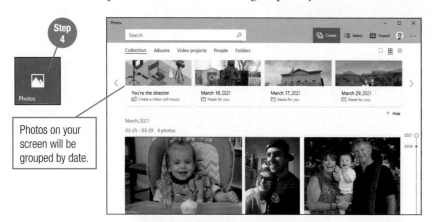

Step 4

Photos on your screen will be grouped by date.

⑤ Click the Start button and then click the Calendar tile in the right pane.

The **Calendar app** shows appointments and reminders for events stored in your Microsoft account. Birthday reminders for accounts connected to Facebook are shown in the calendar.

Step 5

Monday 19

The current month displays.

☁ **Tutorials**

Exploring the Windows 10 Desktop, Taskbar, and Start Menu

Opening and Closing Applications from the Start Menu

●------- **Oops!**

Can't find the Photos tile? Your Photos tile may be a different size or be live, which means the tile shows thumbnails of images instead of the tile shown here. If you still can't find it, type Photos in the search text box and then press Enter. If the Photos app or any other app in this topic or in topics that follow is not installed on your device, open any available app to practice the remaining steps.

6  Click the Store icon on the taskbar.

In the **Store app**, you can search for and download new apps for your PC or mobile device.

## Switching Windows

You can switch between open apps by clicking the button on the taskbar for the app, by viewing open windows in Task View, or by using the keyboard shortcut Start key + Tab. Each application or other open window has a button on the taskbar. Pointing to a button on the taskbar displays a thumbnail above the button with a preview of the window and its active contents. The preview is helpful to select the correct window when you have multiple documents open in the same application.

7  Click the Photos button on the taskbar.

The Photos app moves to the foreground.

8  Click the Task View button on the taskbar.

9  Click the Calendar window.

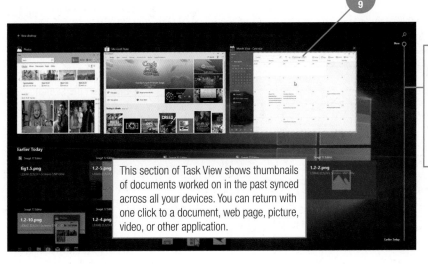

This section of Task View shows thumbnails of documents worked on in the past synced across all your devices. You can return with one click to a document, web page, picture, video, or other application.

The Timeline feature allows you to scroll down to view items opened in the past 30 days.

**Task View** shows all open apps with large thumbnails so you can easily see the active contents in each window. Click the thumbnail for the app you want to display full-size. The **Timeline** feature in Task View displays along the right side of Task View. To view everything you have had open today, yesterday, and over the past 30 days, scroll down the Timeline scroll bar.

Click the magnifying glass at the top right of the scroll bar to open a search text box and type a name of a document or the name of an application. Windows filters the view to show results that match the search criterion.

**quick steps**

**Start Windows 10 and Sign In**
1. Turn on power or resume system from sleep mode.
2. Click anywhere on screen or press any key to remove Lock screen.
3. Type password and press Enter or click Submit button, type PIN, or perform other sign-in authentication.

**Launch an App**
Click Start button and then click desired tile in right pane at Start menu.

**Switch Windows**
Click desired button on taskbar.
OR
1. Click Task View on taskbar.
2. Click desired window.

**Oops!**

No Task View icon on taskbar? Right-click over the search text box or on an empty area of the taskbar and then click *Show Task View button*.

**app tip**

On a touchscreen, swipe in from the left edge of the screen to display Task View.

**Tutorial**
Moving between Open Windows

## 1.3 Using the Search Text Box and Working with Windows

The search text box is a quick way to start an app, launch an application, locate a file or setting, or find information online. Click in the search text box and start typing the first few letters of what you are looking for—app name, file name, or other topic. The search text box is the fastest way to locate help on anything in Windows 10. The search text box contains the dimmed text *Type here to search*.

**1** Click in the search text box and then type al.

Watch the entries appear in the results list in two panes above the search text box. Each character you add in the search text box refines the list. Options appear grouped by categories such as *Best match* in the left pane, and Bing search results appear in the right pane. The options along the top filter the search results by Apps, Documents, Email, Web, or More.

**2** Click *Alarms & Clock* in the *Apps* section in the left pane of the results list.

You should now have four windows overlapping each other on your desktop.

Each app, application, setting, or other feature opens in a window, which allows you to view and work on multiple tasks at the same time. Each window contains standard features for moving, resizing, and closing a window so you can arrange items on your desktop to suit your work preferences. Figure 1.7 identifies the standard window features.

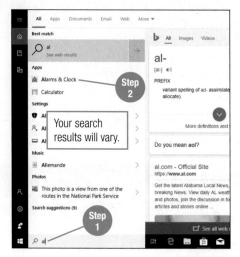

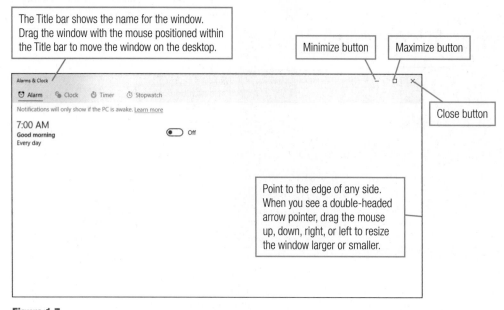

The Title bar shows the name for the window. Drag the window with the mouse positioned within the Title bar to move the window on the desktop.

Minimize button    Maximize button

Close button

Point to the edge of any side. When you see a double-headed arrow pointer, drag the mouse up, down, right, or left to resize the window larger or smaller.

**Figure 1.7**
Shown above is the Alarms & Clock window displaying standard features.

**3** Click the Minimize button located on the Alarms & Clock Title bar.

The **Minimize button** is used to reduce the window to a button on the taskbar.

**4** Click the Close button on the Calendar Title bar.

The **Close button** is used to close the window. In a window that contains an open document, the document is also closed, and you are prompted to save changes before closing if changes have been made since the last Save command.

**5** Click the Maximize button on the Photos Title bar.

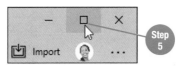

Clicking the **Maximize button** expands the window size to fill the entire desktop. At the same time, the Maximize button changes to the Restore Down button. Click the **Restore Down button** to return the window to its previous size.

**6** Click the Restore Down button on the Photos Title bar.

**7** Position the mouse pointer on the Photos Title bar and then drag the window up, down, left, or right a few inches to practice moving the window on the desktop.

**8** Click the Alarms & Clock button on the taskbar to restore the window to the desktop.

The **Snap** feature in Windows lets you dock a window to one side of the screen without moving and then resizing the window manually. To "snap" a window to the left or right side of the screen, drag the Title bar of the current window to the left or right edge, releasing the mouse when you see the window outline fill the right or left half. Snap a window to one of four quadrants by dragging it to a corner. The **Snap Assist** feature pops up when a window is snapped, leaving part of the screen empty and other windows open. In the empty portion of the screen, Snap Assist shows thumbnails of the remaining windows. Click the thumbnail for the window you want to fill the remainder of the screen.

**9** Drag the Alarms & Clock Title bar to the left edge of the screen until the outline of a window in the left half of the screen appears (drag off the left edge if necessary) and then release the mouse.

The Alarms & Clock window now fills the left half of the desktop, and Snap Assist shows you thumbnails for the Photos and Store windows in the empty space on the right.

**10** Click the Photos window thumbnail.

The screen is now split in half, with the Alarms & Clock window in the left half and the Photos window in the right half.

**11** Close the Photos, Alarms & Clock, and Store windows.

Alarms & Clock window is snapped to the left half of screen at Step 9.

## 1.4 Using the Notification Area

The notification area at the right end of the taskbar shows the current date and time, the Notifications icon that opens the **Action Center** panel, the Speaker icon used to adjust the volume, and the Network icon, which shows network connectivity. If additional icons are available but not visible, the Show hidden icons arrow is used to reveal other settings. Next to the Show hidden icons arrow is a People icon, which is used to pin contacts to the taskbar. A Power icon is included on devices that have a battery. Additional icons may appear depending on the installed applications. Use the notification area to view or manage the status of a setting or other feature.

**1** Click the current date and time in the notification area at the right end of the taskbar.

A calendar appears above the date and time displaying the current month. Change the month using the up- and down-pointing arrows. Appointments or reminders appear in the area below the calendar.

**2** Click in an unused area on the desktop to hide the calendar.

**3** Click the Speaker icon in the notification area.

The volume control slider opens. Take note of the current value for the volume level.

**4** Drag the slider right to adjust the volume higher.

A chime sounds when you release the mouse at the new volume level.

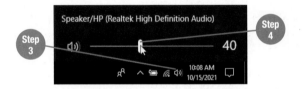

**5** Drag the slider left to restore the volume control to the original level.

**6** Click in an unused area on the desktop to hide the volume control slider.

**7** Click the Network icon in the notification area.

A list of Wi-Fi networks within range are shown, with the active network shaded blue. At the bottom of the panel are quick action tiles. A **quick action tile** is a shortcut to a setting. Clicking the quick action tile for a wireless network you are connected to turns the Wi-Fi connection off. Click the Airplane mode quick action tile to turn on the airplane mode setting, which turns off all wireless communications for your device. You use the Network panel to manage your network connectivity by disconnecting from your current Wi-Fi network or by connecting to another one. To connect to another Wi-Fi network, click the network name and then follow the prompts that appear.

**8** Click in an unused area on the desktop to hide the Network panel.

9  Click the Power icon in the notification area. Skip to Step 11 if you do not see a Power icon because you are using a PC that does not use a battery.

In the Power panel, the Battery settings link opens the Settings app with the Battery pane active, where you can adjust options that affect battery life.

10  Click in an unused area on the desktop to remove the Power panel.

11  Click the Notifications icon to open the Action Center panel.

In the top section of the Action Center panel, notifications appear from settings, apps, applications, and your connected accounts. For example, you will see messages regarding installed updates, new mail, or a birthday reminder from the Calendar app. At the bottom of the panel are rows of quick action tiles. The tiles shown vary depending on the type of device in use.

12  Click the All settings quick action tile in the Action Center panel to display the Settings window.

13  Click the System icon.

14  Click *Notifications & actions* in the left pane.

15  Scroll down and review the options for customizing Notifications.

16  Drag the switch slider left to turn a feature off or right to turn a feature on for an option that you would like to change, and then scroll back to the top of the page.

17  Click <u>Add or remove quick actions</u>.

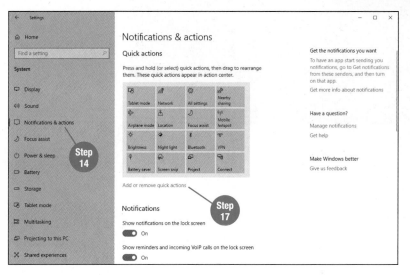

18  Review the options for quick action tiles and drag to slide a switch from On to Off or vice versa for any quick action that you would like to change.

19  Click the Back button (displays as a left-pointing arrow) to return to the previous screen.

20  Close the Settings app window.

## 1.5 Locking the Screen, Signing Out, and Shutting Down Windows 10

If you need to leave your PC or mobile device for a short period of time and do not want to close all the apps and documents, you can lock the device. Locking the system causes the Lock screen image to appear full screen so someone else cannot see your work. All your apps, applications, and documents are left open in the background, and you can resume work right away once you unlock the device with your password, PIN, picture password, or biometric authentication.

1. Click the Start button, type photos in the search text box and press Enter, or click the Photos tile to launch the Photos app.
2. Click the Start button.
3. Click the icon for your account at the top of the first column in the left pane.
4. Click *Lock*.

The Windows Lock screen appears with the current date and time and notifications, if any are available.

Step 4

 Watch & Learn

●────── skills

Lock the Screen
Sign out of Windows 10
Shut down Windows 10

🌥 **Tutorials**
Locking the Screen
Signing Out of a
Windows Session

The Lock screen may vary.

10:29
Friday, July 10

5. Click anywhere on the Lock screen or press any key on the keyboard to display the sign-in screen.
6. Sign in with your account authentication credential.

Notice when you are back at the desktop that the Photos app remained open while the device was locked.

### Signing Out of a Windows Session

When you are finished with a Windows session, you should **sign out** of the PC or mobile device. Signing out is also referred to as *logging off*. Signing out closes all apps, applications, and files. If a computer or mobile device is shared with other people, signing out is expected so that other users can sign in to their accounts. Signing out and locking also provide security for your device because someone would need to know your password, PIN, or picture password touch gestures to access applications or files.

●────── app tip

If you have documents open with unsaved changes, Windows prompts you to save the documents before proceeding with the sign-out process.

7  If necessary, click the Start button.

8  Click the icon for your account at the top of the first column in the left pane, and then click *Sign out*.

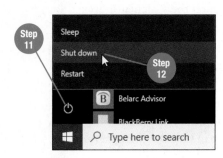

The Windows Lock screen appears; however, this time Windows closes the Photos app automatically.

9  Display the sign-in screen and sign in with your account authentication credential.

Notice the Photos app is not on the desktop because it was closed when you signed out of the previous Windows session.

## Shutting Down the PC or Mobile Device

If you want to turn off the power to your computer or mobile device, perform a **shut down** command. Shutting down the system ensures that all Windows files are properly closed. The power turns off automatically when shut down is complete.

10  If necessary, click the Start button.

11  Click the Power icon at the bottom of the first column in the left pane.

*Note: Check with your instructor before proceeding to Step 12, because some schools do not allow students to shut down computers. If necessary, click in an unused area on the desktop to remove the Start menu and proceed to the next topic.*

12  Click *Shut down*. If someone else is signed in on the device you are shutting down, a message displays informing you the other user could lose unsaved work if you proceed (Figure 1.8). Click the Shut down anyway button to proceed with turning off the device, or click in an unused area on the desktop to remove the message and cancel the shutdown operation.

In a few seconds, the power will turn off. In some instances, system updates will be installed during a shutdown operation.

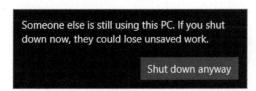

**Figure 1.8**

This warning message appears if you are about to power off a PC and someone else using the same computer has not signed out. Proceeding to shut down may cause her or him to lose changes that were not saved.

13  Wait a moment or two and then press the power button to turn the PC back on.

14  When the Lock screen appears, display the sign-in screen and then sign back in to Windows.

# 1.6 Using and Customizing the Start Menu

The Start menu can be customized to suit your preferences by adding or removing tiles, rearranging tiles, resizing a tile, and turning a live tile on or off to enable or stop notifications or other content from appearing on the tile.

*Note: In some school computer labs, the ability to change Windows settings is disabled. If necessary, complete this topic and the next topic on a home computer.*

1. Double-click the Recycle Bin icon on the desktop to open the Recycle Bin window.

2. Click the Start button, and then click the Settings icon (first icon above the Power icon) to open the Settings app.

3. Click the Start button, scroll down the list of applications, and then click *Money* to launch the Money app. Choose another app if you do not have the Money app installed.

4. Click in the search text box, type calc, and then press Enter to launch the Calculator app.

5. Close the Calculator, Money, Settings, and Recycle Bin windows.

The methods you just practiced to start apps and open a window included the following: double-clicking an icon on the desktop, clicking an option in the left pane on the Start menu, clicking a tile in the right pane on the Start menu, and typing the first few characters in the search text box. The method you decide to use most often is a personal preference. You will develop your own style for using the Start menu and desktop interface as you gain experience.

## Resizing the Start Menu

You can make the Start menu wider or taller by dragging the top of the Start menu up or the right border right. Drag the top border down or right border left to make the Start menu shorter or narrower.

6. Click the Start button, point at the right border of the Start menu, and then drag right when the pointer changes to a right-and-left-pointing arrow to make the menu wider. If your menu is already at the maximum width, drag the border left to make the menu narrower.

7. With the Start menu still open, point at the top border and then drag up when the pointer changes to an up-and-down-pointing arrow to make the menu taller.

## Pinning and Unpinning Tiles to and from the Start Menu

Right-click a tile to reveal the shortcut menu with the options *Unpin from Start*, *Resize*, and *More*. The More menu is used to access the options *Turn Live Tile on* or *Turn Live Tile off* (depending on its current state), *Pin to taskbar*, *App settings*, *Rate and review*, and *Share*. Click **Unpin from Start** to remove the tile from the Start menu. A tile is added to the Start menu by right-clicking the app or application name in the application list on the Start menu, and then clicking **Pin to Start**. The shortcut menu offers more or fewer choices depending on the app or application name that you right-click on the Start menu.

8  With the Start menu open, right-click the Photos tile and then click *Unpin from Start* on the shortcut menu.

   The tile is removed, and the Start menu remains open.

9  Scroll down the list of applications in the left pane, and then right-click *Photos*.

10  Click *Pin to Start* on the shortcut menu.

   The Photos tile is added back to the Start menu, and the Start menu remains open. Notice that the tile is not added back to the same location from which it originated.

## Rearranging Tiles

Tiles can be moved to a new location on the Start menu by dragging the tile to the desired location.

11  Drag the Photos tile back to its original location in the right pane on the Start menu.

   As you drag a tile around on the Start menu, the tiles are dimmed, and you will notice other tiles shifting around to make room for the tile.

## Resizing a Tile and Turning Live Updates Off or On

Tiles are either square or rectangular in shape and can be sized Small, Medium, Wide, or Large using the *Resize* option on the shortcut menu. Some tiles do not offer all size options. Some tiles display live updates, with notifications, headlines, or pictures displayed on the tile. Change the live update status using the **Turn Live Tile off** or **Turn Live Tile on** option on the shortcut menu.

12  With the Start menu open, right-click the Calendar tile, point to *Resize*, and then click *Large*.

13  Right-click the Calendar tile, point to *Resize*, and then click *Medium*. If necessary, drag the Mail or another tile back upward to fill in the space next to the Calendar tile.

14  Right-click the Mail tile, point to *More*, and then click *Turn Live Tile off*.

15  Right-click the Mail tile, point to *More*, and then click *Turn Live Tile on*.

16  Resize the Start menu back to its original height and width, if desired.

17  Rearrange tiles or make other changes as needed to restore the Start menu to its original settings. Click on an unused area of the desktop to close the Start menu.

**quick steps**

**Remove a Tile from the Start Menu**
1. Right-click tile.
2. Click *Unpin from Start*.

**Add a Tile to the Start Menu**
1. Click Start button.
2. Scroll down left pane list to locate desired app or application name.
3. Right-click app or application name.
4. Click *Pin to Start*.

**Rearrange Tiles**
Drag tile to desired location.

**Resize a Tile**
1. Right-click tile.
2. Point to *Resize*.
3. Click desired size.

**Turn Off/On Live Updates**
1. Right-click tile, then point to *More*.
2. Click *Turn Live Tile off* or *Turn Live Tile on*.

**Tutorials**

Pinning and Unpinning Tiles to and from the Start Menu

Rearranging Tiles

Resizing a Tile and Turning Live Updates Off or On

# 1.7 Personalizing the Lock Screen and the Desktop

 **Watch & Learn**

Most people like to put a personal stamp on their device. In Windows 10, you can personalize the Lock screen by changing the picture that displays and the apps that provide notifications when the screen locks. The desktop background can be changed to another image, and a different color can be chosen for tiles and window accents. A theme changes the background image, color, and sound scheme in one step. The Start menu by default shows the most used and recently added apps and applications in the left pane. You can turn one or both of these categories off. Open the Settings app and click *Personalization* to make the changes described.

● - - - - - - - Skills

Change the Lock screen background image

Add an app to the Lock screen notifications

Change the desktop background image

Change the color scheme

1  Click the Start button and then click the Settings icon.

2  Click *Personalization* in the Settings window.

3  Click *Lock screen* in the left pane.

   The current Lock screen image is shown above five thumbnails for other background images on the Lock screen pane. Click a thumbnail to preview the image or click the Browse button to choose an image on your computer to use as the Lock screen image. You can also opt to show a slide show for the Lock screen by changing the *Background* option from *Picture* to *Slideshow*.

☁ **Tutorial**

   Personalizing the Desktop

4  Click one of the five background thumbnails below the current Lock screen image. You determine the background image you want to view.

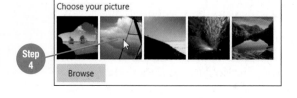

● - - - - - - - Oops!

No background thumbnails? The *Background* option below the preview is at a different setting (such as *Windows spotlight*). Windows spotlight downloads pictures and advertisements automatically from Bing. Click the *Background* option box arrow and then click *Picture*.

The app buttons below the background thumbnails control which apps give notifications on the Lock screen image.

5 If necessary, scroll down to view the *Choose apps to show quick status* section, click the first button with a plus symbol and then click *Weather*.

The Weather app will now provide notifications when the screen is locked.

6 Click the *Background* option in the left pane.

7 Click one of the five background thumbnails in the *Choose your picture* section. You determine which picture you want for your background. If the *Choose your picture* thumbnails are not visible, click the *Background* option box arrow and then click *Picture*.

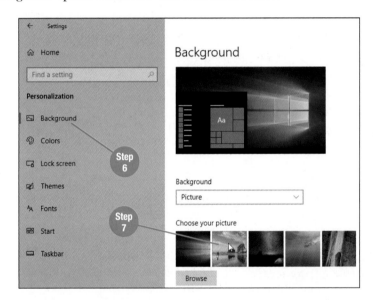

8 Click *Colors* in the left pane.

9 Click a color square in the *Windows colors* section and then close the app window. You determine the color you want to use for tiles and window accents.

The new background image is shown on the desktop.

10 Click the Start button and look at the new accent color used on tiles.

11 Lock the screen.

The picture you chose at Step 4 appears as the new Lock screen image.

12 Sign back in to unlock the system.

13 Open the Settings app and repeat Step 2 through Step 9 to restore the Lock screen, background, and color scheme to their original settings. At Step 5, scroll up the list and click *None* on the Weather button to remove the app from the Lock screen.

14 Close the Settings window.

**quick steps**

**Change the Lock Screen Options**
1. Click Start button.
2. Click Settings.
3. Click *Personalization*.
4. Click *Lock screen* in left pane.
5. Change desired Lock screen options.

**Change Background**
1. Click Start button.
2. Click Settings.
3. Click *Personalization*.
4. Click *Background* in left pane.
5. Click desired thumbnail image.

**Change Color Scheme**
1. Click Start button.
2. Click Settings.
3. Click *Personalization*.
4. Click *Colors* in left pane.
5. Click desired color square.

**app tip**

When you are signed in with a Microsoft account, your personalization settings are saved online. Sign in from another PC or mobile device and your new Lock screen, background, and color scheme will appear on the other device.

# 1.8 Understanding Files and Folders and Downloading and Extracting Data Files

A **file** is a document, spreadsheet, presentation, photo, or any other data that you create and save in digital format. Files are also videos and music that you play on a device. Each file has a unique **file name**, which is a series of characters you assign to the file when you save it that allows you to identify and retrieve the file later.

A system of organizing files so that you can find and retrieve a specific document or photo when you need it is necessary. Folders are created on a storage device to organize electronic files. A **folder** is a name assigned to a placeholder or container in which you store a group of related files. Think of a folder on the computer in the same way you consider a paper file folder in a desk drawer in your home. You might have one file folder for your household bills and another file folder for your school documents. Separating documents into file folders makes it easy to put a document away when you are done with it and to locate the document when you need it again. In Figure 1.9, files and folders are shown to help you understand how digital data is organized on a storage device.

**Watch & Learn**

●- - - - - - - skills

Define *file*, *file name*, and *folder*

Download and extract student data files to a USB flash drive

**Tutorials**

Creating a Compressed Archive (ZIP) File

Downloading and Extracting Student Data Files

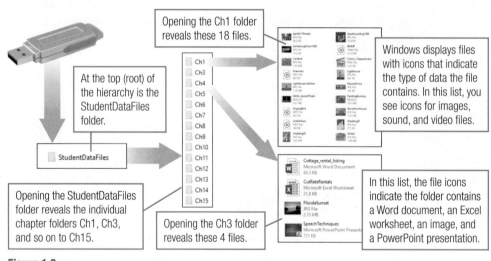

**Figure 1.9**
Digital data is stored in files, which are organized into folders.

To complete the tasks in the rest of this textbook, a set of student data files must be copied to a USB flash drive or other storage medium. Check with your instructor for the storage medium to use for your course. For example, your instructor may prefer that you save files to a network folder at your school.

The student data files are copied from the Cirrus course. The files are packaged so that you need to download only one file, called an *archive file* or a ZIP file. A **ZIP file** assembles and compresses a group of files under one name that looks like a folder name with a .zip file extension. The files are compressed so that the ZIP file size is as small as possible, meaning the file will copy faster over a network. A ZIP file should be uncompressed before the files stored within it are used so that each file is restored to its original size and put back in the folder from which the file originated. To restore the files within a ZIP file to their original size and folder structure, you perform a process called *extracting*.

*Note: You will need a USB flash drive to complete the steps in this topic and the next three topics. Check with your instructor for alternative instructions if he or she prefers that you save files somewhere else, such as in a network folder on a school server.*

●- - - - - - app tip

You can recognize a ZIP file in a file list because Windows displays the icon as a file folder with a zipper.

1. Insert a USB flash drive into an available USB port on your PC. Skip this step if you are saving files to another storage medium.

2. Go to your Cirrus course online and navigate to the course resources.

3. Click Student Data Files, and on the next page, click https://CA3.ParadigmEducation .com/StudentDataFiles.

Student Data Files
Click https://CA3.ParadigmEducation.com/StudentDataFiles link to open resource.

4. If the files do not start downloading automatically, copy the hyperlink and paste it in the address bar.

The ZIP archive file will copy to the Downloads folder. The download will take a few moments depending on the speed of your network connection. Watch the progress of the download at the bottom of the browser window.

5. Click the button at the bottom of the browser for the StudentDataFiles ZIP folder when the message displays that StudentDataFiles.zip finished downloading. A File Explorer window opens. The Extract Compressed Folder Tools tab appears because the file opened is a ZIP archive file.

**File Explorer** is the utility used to browse files and folders on storage devices and perform file management routines. You will learn more about File Explorer in the next three topics.

6. If necessary, click the Extract Compressed Folder Tools tab.

7. Click the Extract all button.

8. Click the Browse button in the Extract Compressed (Zipped) Folders dialog box.

9. Click your USB flash drive in the left pane of the Select a destination dialog box, and then click the Select Folder button.

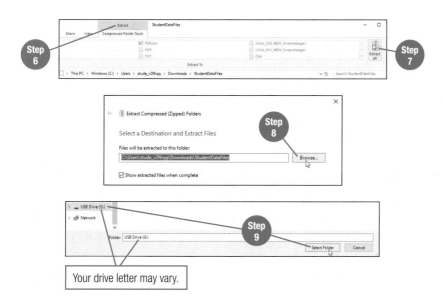

Your drive letter may vary.

10. Click the Extract button in the Extract Compressed (Zipped) Folders dialog box.

A progress message displays the status of the extraction. When finished, a new File Explorer window opens with the StudentDataFiles folder shown.

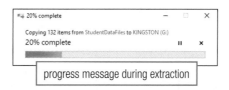

progress message during extraction

11. Close all open windows, including the OneDrive browser window.

**quick steps**

**Download and Extract Student Data Files**
1. Navigate to the Cirrus course.
2. Click Student Data Files and then click https://CA3. ParadigmEducation. com/StudentDataFiles.
3. If necessary, copy and paste the hyperlink in the address bar.
4. Click file button to open File Explorer.
5. If necessary, click Compressed Folder Tools Extract tab.
6. Click Extract all button.
7. Click Browse button.
8. Click USB flash drive, then Select Folder.
9. Click Extract button.

**Oops!**

A window other than File Explorer opens at Step 5? Another ZIP archive utility installed on your device (such as 7-Zip), may open the downloaded item in the archive application. In that case, close the window and click the File Explorer icon on the taskbar. Click *Downloads* in the left pane. Right-click StudentDataFiles.zip in the right pane, point to *Open with*, click *Windows Explorer*, and then proceed to Step 7.

# 1.9 Browsing Files with File Explorer

To view or manage the files on a storage device, open a File Explorer window by clicking the File Explorer button on the taskbar. In File Explorer, you perform file maintenance tasks, such as creating new folders to organize files, renaming files, copying files, moving files, and deleting files. To begin, you will learn how to navigate the File Explorer window shown in Figure 1.10.

1  Click the File Explorer button on the taskbar.

File Explorer is divided into two panes: the Navigation pane (left pane) displays the list of places associated with the device, and Content pane (right pane) displays files and folders stored on the device or folder selected in the Navigation pane. File Explorer opens by default in a view called **Quick access**, which by default shows a list of frequently used and recently used files, as well as the contents stored in three places: Downloads, Documents, and Pictures.

**Watch & Learn**

━━━━━━━ **skills**

Identify features in a File Explorer window

Browse files on a device using File Explorer

**Tutorial**

Navigating between Local Volumes and Folders in File Explorer

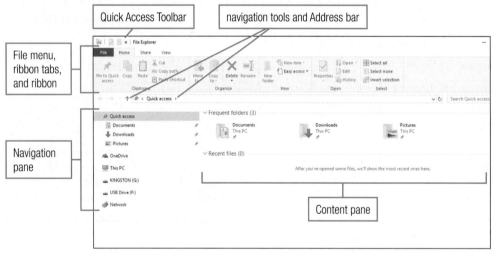

**Figure 1.10**

Shown here is the File Explorer window.

2  Click This PC in the Navigation pane.

The This PC window displays all the available storage options on the computer.

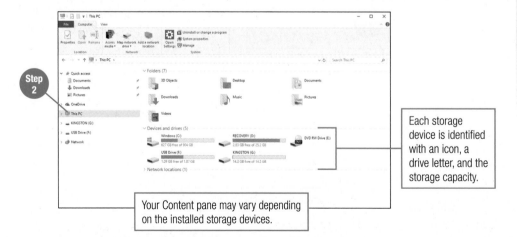

━━━━━━━ **app tip**

You can set File Explorer to always open with This PC active instead of the Quick access view. To do this, right-click Quick access in the Navigation pane, click *Options*, change the *Open File Explorer* option to *This PC*, and then click OK.

③ If necessary, insert your USB flash drive into an available USB port. If a new window opens when you insert the USB flash drive, close the window.

④ Double-click the icon representing your USB flash drive in the *Devices and drives* section to view the contents in the Content pane.

The Content pane shows the name of the folder you extracted from the ZIP file in the previous topic. Your Content pane may show additional files if you did not use an empty USB flash drive.

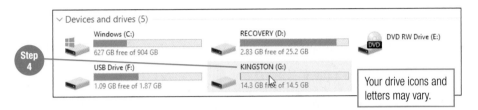

⑤ Double-click the folder *StudentDataFiles*.

A list of folders stored within the StudentDataFiles folder appears. A folder within another folder is sometimes referred to as a **subfolder**.

⑥ Double-click the folder *Ch1*.

A list of files stored in the Ch1 subfolder appears in the Content pane.

⑦ Click the View tab.

⑧ Click the *Tiles* option in the Layout group.

Use the Layout options to view files in the Content pane as small, medium, large, or extra large icons; as a list format with or without file details, such as the date and time the file was created or last updated and the file size; as a list with the file type and file properties such as author (Content); or as tiles (shown below).

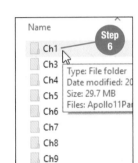

⑨ Click Desktop in the Navigation pane.

Browse content in File Explorer by clicking names in the Navigation pane, or by double-clicking device or folder names in the Content pane.

⑩ Close the File Explorer window.

*Note: Leave the USB flash drive in the device for the remaining topics.*

---

**quick steps**

**Browse Files in File Explorer**
1. Click File Explorer icon on taskbar.
2. If necessary, insert a removable storage medium such as a USB flash drive.
3. Click desired location or device in Navigation pane, *or* double-click a folder or device name in Content pane.

---

**app tip**

Double-clicking a file name opens the application associated with the file and automatically opens/plays the document, image, video, or sound.

# 1.10 Creating Folders and Copying Files and Folders

As you work with software applications, such as Word or Excel, you will create many files that are documents or spreadsheets. You may also download files from a digital camera or a website and receive files from email or text messages. Storing the files in an organized manner with recognizable names will allow you to easily locate the files later. If you want to have folders created in advance of creating files, you need to have an organizational structure already in place. From time to time, you also need to rename, copy, move, or delete files and folders to maintain a storage medium in good order.

## Creating a Folder

Creating a folder on a computer is like labeling a file folder in your filing cabinet. The title provides a brief description of the type of documents that will be stored inside the file folder. On the computer, in File Explorer, click the **New folder button** in the New group on the Home tab, and then type a name for the folder to set up the electronic equivalent of a paper filing system.

1 Click the File Explorer icon on the taskbar.

2 Click This PC in the Navigation pane.

3 Double-click the icon representing your USB flash drive.

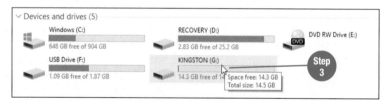

4 If necessary, click the Home tab to display the ribbon interface.

The **ribbon** provides the buttons you need to perform file management tasks. Buttons are organized into the tabs Home, Share, View, and Manage. Within each tab, buttons are further organized into groups, such as Clipboard, Organize, New, Open, and Select (on the Home tab).

5 Click the New folder button in the New group.

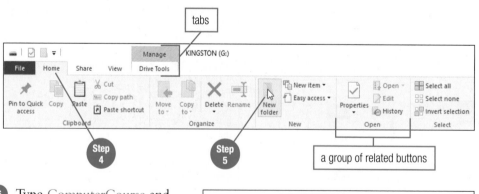

6 Type ComputerCourse and then press Enter.

------- app tip

Tabs and/or buttons on the ribbon change and become available or unavailable depending on what is selected in the window.

------- app tip

In the folder structure you are creating, no spaces between words are used. Windows allows the use of spaces; however, an industry best practice is to avoid spaces in folder names.

**7** Click the View tab and then click the *List* option in the Layout group.

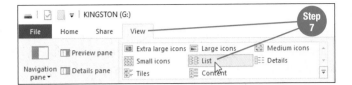

List view displays names with small icons representing each file or folder and without details such as the date or time the file or folder was created or modified, the file type, or the file size. Each folder can have a different view.

**8** Click the Home tab. If the ribbon is not currently expanded, a down-pointing arrow (Expand the Ribbon icon) displays below the Close button near the top right of the window. Click the down-pointing arrow to expand the ribbon; otherwise, proceed to Step 9.

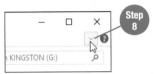

The down-pointing arrow expands the ribbon permanently in the window. Once it has been clicked, it displays as an up-pointing arrow (like a caret ^ symbol), which is the Minimize the Ribbon icon.

**9** Double-click the *ComputerCourse* folder in the Content pane.

You have now opened the ComputerCourse folder. File management tasks performed next will occur inside this folder.

**10** Click the New folder button, type ChapterTopicsWork, and then press Enter.

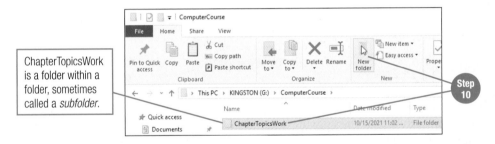

ChapterTopicsWork is a folder within a folder, sometimes called a *subfolder*.

**11** Click the New folder button, type Assessments, and then press Enter.

You now have two folders (or subfolders) within the ComputerCourse folder.

**12** Click the Up arrow button on the Address bar.

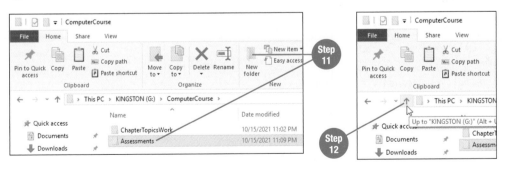

After clicking the Up arrow, the Address bar shows that you moved the display up one level in the folder hierarchy.

Notice you now see only the ComputerCourse folder and the StudentDataFiles folder. The two folders created in Steps 10 and 11 are no longer visible because they are *inside* the ComputerCourse folder.

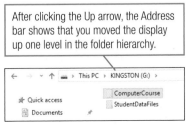

**quick steps**

**Create a Folder**
1. Click File Explorer icon on taskbar.
2. Navigate to desired storage medium and/or folder.
3. If necessary, click Home tab.
4. Click New folder button.
5. Type folder name.
6. Press Enter.

**Oops!**

No down-pointing arrow below the Close button? Then your ribbon is already expanded in the window. Skip Step 8.

**app tip**

The terms *folder* and *subfolder* are used interchangeably when a subfolder exists. Windows does not distinguish a subfolder from a folder; however, some people prefer to use *subfolder* to differentiate a folder within the hierarchy on a disk.

## Copying Files and Folders

Copying a file or folder from one storage medium to another makes an exact copy of a document, spreadsheet, presentation, image, video, music file, or other object. Copying is one way to make a backup of an important file on another storage medium. Use the **Copy button** and the **Paste button** in the Clipboard group on the Home tab to copy a selected file or folder.

**Tutorial**

Moving and Copying Using Copy and Paste

**13** Click to select the StudentDataFiles folder.

Before you can copy, you must first select a file or folder—a single-click selects a file. A selected file or folder displays with a blue background in the Content pane. On touchscreen devices, a check box also displays next to folder and file names; a check mark in the check box indicates the file is selected.

**14** Click the Copy button in the Clipboard group on the Home tab.

**15** Double-click the folder *ComputerCourse* in the Content pane.

**16** Click the Paste button in the Clipboard group.

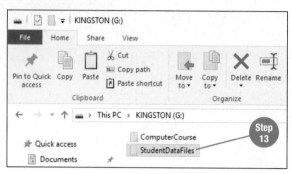

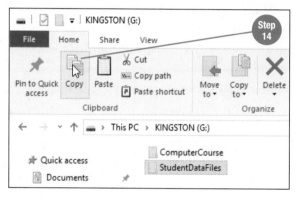

**app tip**

Ctrl + C is the universal keyboard shortcut to Copy.

**app tip**

Ctrl + V is the universal keyboard shortcut to Paste.

This step is copying all the student data files and placing the copy of the folder and all its contents in the ComputerCourse folder. As the copying takes place, Windows displays a progress message. When the message disappears, the copy is complete.

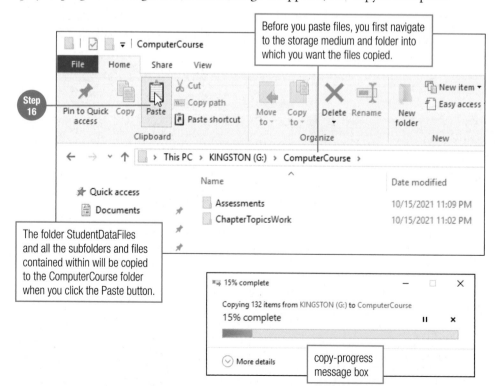

**17** Double-click the StudentDataFiles folder name.

**18** Double-click the *Ch1* folder name.

In the next steps, you will copy individual files to another folder.

**19** Click the View tab and then click the *List* option in the Layout group. Skip this step if List view is already active.

**20** Click *Lighthouse*, hold down the Shift key, and then click *Winter*.

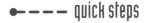
**quick steps**

**Copy Files or Folders**
1. Open File Explorer.
2. Navigate to source data storage medium and folder.
3. Select file(s) or folder(s) to be copied.
4. If necessary, click Home tab.
5. Click Copy button.
6. Navigate to destination storage medium and folder.
7. Click Paste button.

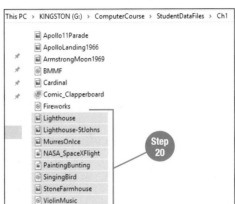

This action selects all the files, starting with **Lighthouse** and ending with **Winter**. Holding down the Shift key while clicking a file name instructs Windows to select all files from the file name selected first to the file name selected second.

Another way to select multiple files is to hold down the Ctrl key while clicking a file name. Use the Ctrl key if you do not want all the files in between two selected files to be selected. In other words, Ctrl + click is used to select multiple files that are not next to each other.

**21** Click the Home tab.

**22** Click the Copy button.

**23** Click *ComputerCourse* in the Address bar.

Clicking a device or folder name in the Address bar is another way to navigate to devices or folders.

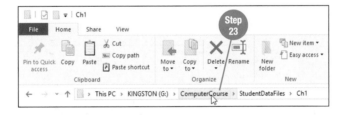

**24** Double-click the folder *ChapterTopicsWork* in the Content pane.

**25** Click the Paste button.

The selected files are copied into the ChapterTopicsWork folder.

**26** Click the View tab and then click *List*.

**27** Click the Close button to close the File Explorer window.

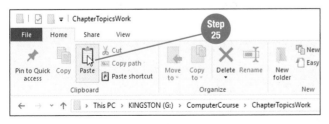

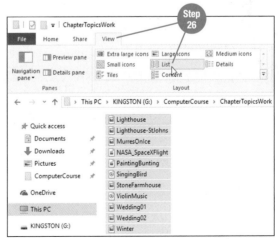

# 1.11 Moving, Renaming, and Deleting Files and Folders and Ejecting a USB Flash Drive

**Watch & Learn**

━ ━ ━ ━ ━ ━ Skills

Move a file or folder
Rename a file or folder
Delete a file or folder
Empty the Recycle Bin
Eject a USB flash drive

**Tutorial**
Moving Files

File Explorer is also used to move, rename, and delete files and folders. Sometimes you will copy a file or save a file in a folder and later decide you want to move it elsewhere. You may also assign a file name to a file or folder and later decide you want to change the name. Files or folders no longer needed can be deleted to clean up the disk. When you are finished using a USB flash drive, you should properly eject the drive to avoid problems that can occur when files are not properly closed.

## Moving Files

Files are moved using a process similar to copying. Begin by selecting files and clicking the **Cut button** in the Clipboard group on the Home tab in File Explorer. Navigate to the destination location and click the Paste button. When files are cut they are removed from the source location.

**1** Right-click the Start button and then click the *File Explorer* option.

**2** Click This PC in the Navigation pane and then double-click the icon for your USB flash drive in the *Devices and drives* section.

**3** Double-click *ComputerCourse* in the Content pane.

**4** Double-click *ChapterTopicsWork* in the Content pane.

Assume that you decide that the files copied from the StudentDataFiles Ch1 folder in the last topic should be stored inside a folder within ChapterTopicsWork.

**5** Click the New folder button in the New group on the Home tab, type Ch1, and then press Enter.

**6** Click *Lighthouse*, hold down the Shift key, and then click *Winter*.

**7** Click the Cut button in the Clipboard group.

**8** Double-click *Ch1*.

**9** Click the Paste button in the Clipboard group.

The files are removed from the ChapterTopicsWork folder and placed within the Ch1 folder.

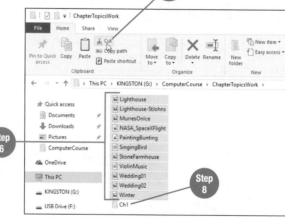

━ ━ ━ ━ ━ ━ app tip

Ctrl + X is the universal keyboard shortcut to Cut.

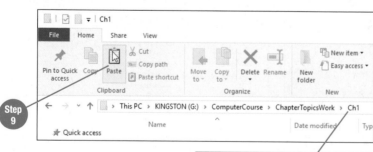

The Address bar shows the destination folder is Ch1 within ChapterTopicsWork.

**10** Click the Back button to return to the previous list.

Notice the files are no longer in the ChapterTopicsWork folder.

**11** Click the Forward button (right-pointing arrow next to Back button) to return to the Ch1 folder.

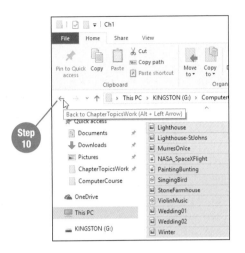

## Renaming Files and Folders

At times you will receive a file from someone else and decide you want to rename the file to something more meaningful to you, or you may decide upon a new name for a file or folder after the file or folder was created. Change the name of a file or folder with the **Rename button** in the Organize group on the Home tab.

**12** Click the View tab and then click the *Extra large icons* option in the Layout group.

**13** If necessary, scroll down the Content pane to view the contents shown for each file in the folder.

**14** Click to select the file *Winter*.

Assume a friend sent you this image of a weeping birch tree laden with snow in a winter scene. You decide to rename the image.

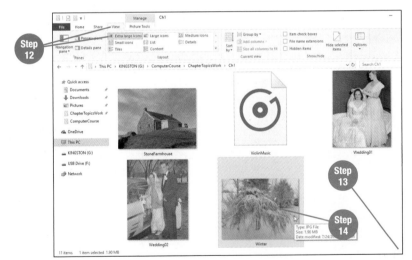

**15** Click the Home tab and then click the Rename button in the Organize group.

**16** Type BirchTreeInWinter and then press Enter.

**17** Click to select the file *ViolinMusic*.

Assume this file is the song "Danny Boy" played on a violin and recorded by you as you heard the song at an outdoor event.

**quick steps**

**Move Files or Folders**
1. Open File Explorer.
2. Navigate to source data storage medium and folder.
3. Select files or folder to be moved.
4. If necessary, click Home tab.
5. Click Cut button.
6. Navigate to destination storage medium and folder.
7. Click Paste button.

**Rename Files or Folders**
1. Open File Explorer.
2. Navigate to data storage medium and folder.
3. Select file to be renamed.
4. If necessary, click Home tab.
5. Click Rename button.
6. Type new name and press Enter.

**Tutorial**
Renaming Files and Folders

18. Click the Rename button, type DannyBoyViolin, and then press Enter.

19. Click *ComputerCourse* in the Address bar.

You can rename a folder as well as a file.

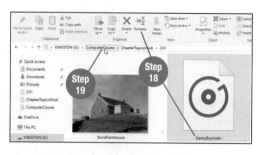

20. If necessary, click to select the folder named *ChapterTopicsWork*.

21. Click the Rename button, type CompletedTopics, and then press Enter.

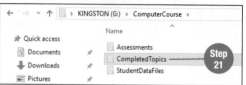

## Deleting Files and Folders

Delete files or folders when you no longer need to keep them. You should also delete files or folders if you have copied them to a removable storage medium for archive purposes and want to free up space on the local disk. In File Explorer, select the files or folders to be deleted and then click the **Delete button** in the Organize group on the Home tab. Files and folders deleted from the local hard disk are moved to the Recycle Bin. While it is in the Recycle Bin, you can restore the file to its original location if you deleted the file in error. Files deleted from a USB flash drive are not sent to the Recycle Bin; therefore, exercise caution when deleting a file from a USB flash drive.

22. Double-click *CompletedTopics* and then double-click *Ch1*.

23. Click **Lighthouse**, hold down the left mouse button, and then drag the file to Documents in the Navigation pane.

Dragging a file from a folder on one storage medium to a location on another storage medium copies the file.

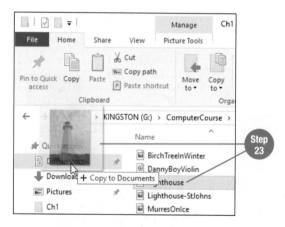

24. Click Documents in the Navigation pane.

25. Click **Lighthouse** and then click the Delete button in the Organize group (click the top of the button—not the down-pointing arrow).

26. Click This PC in the Navigation pane and then double-click the icon for your USB flash drive in the *Devices and drives* section.

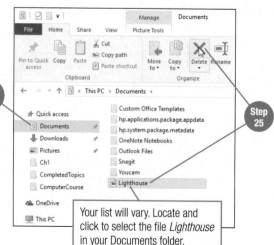

Your list will vary. Locate and click to select the file *Lighthouse* in your Documents folder.

app tip

Dragging a file or folder on the same storage medium moves the file. To be sure of the operation you want to perform, drag a file or folder to another location using the right mouse button. When you release the mouse, a shortcut menu appears. Choose the Copy or Move option on the menu.

**Tutorial**

Deleting Files and Using the Recycle Bin

27  Double-click *ComputerCourse*, double-click *CompletedTopics*, and then double-click *Ch1*.

28  Click **SingingBird** and then click the Delete button.

29  Click the Yes button in the Delete File message box that appears.

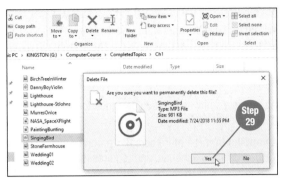

The confirmation message appears when you delete a file from a USB flash drive because the files are permanently deleted, not sent to the Recycle Bin from which the file could be restored.

30  Close the File Explorer window.

31  Double-click the Recycle Bin icon on the desktop.

32  Click the Empty Recycle Bin button in the Manage group on the Manage Recycle Bin Tools tab.

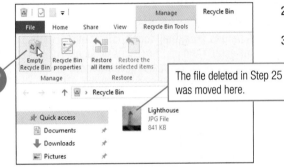

The file deleted in Step 25 was moved here.

33  Click the Yes button to permanently delete the file in the Delete File message box.

34  Close the Recycle Bin window.

**quick steps**

**Delete Files or Folders**
1. Open File Explorer.
2. Navigate to data storage medium and folder.
3. Select file(s) or folder(s) to be deleted.
4. If necessary, click Home tab.
5. Click Delete button.
6. If prompted, click Yes button.

**Empty the Recycle Bin**
1. Double-click Recycle Bin icon on desktop.
2. Click Empty Recycle Bin button.
3. Click Yes button.

When you are finished with a USB flash drive, you should always use the **Safely Remove Hardware and Eject Media** feature before pulling the drive out of the USB port. This ensures all files are properly closed to avoid data corruption.

35  Click the Show hidden icons arrow in the notification area (up-pointing arrow like caret ^ symbol) on the taskbar and then click the Safely Remove Hardware and Eject Media icon.

36  Click the *Eject* option in the pop-up menu. Your menu may show *Eject USB Flash Drive*, *Eject Storage Media*, or some other label after *Eject*, depending on the device. If more than one USB flash drive appears in the list, click to select the correct drive before ejecting the drive.

37  Remove your USB flash drive from the computer when the message appears that it is safe to do so.

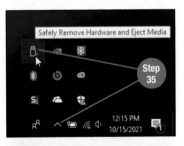

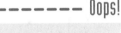

**Oops!**

The Safely Remove Hardware and Eject Media icon is not always hidden; you may not have to reveal it with the Show hidden icons up-pointing arrow.

A message appears when the USB device can be safely removed.

## Topics Review

| Topic | Key Concepts and Key Terms |
|---|---|
| **1.1 Using Touch, Mouse, and Keyboard Input to Navigate Windows 10** | **Windows 10** is the operating system from Microsoft. The **user interface (UI)** is a desktop that includes icons, a **Start menu** accessed from the **Start button**, and a taskbar along the bottom of the screen. |
| | A **tile** is a square or rectangle on the Start menu used to launch an app, application, or other feature. |
| | An **app** is a small application designed for a smartphone, tablet, or PC with touchscreen input and is mainly for one task. |
| | An **application** is a more complex program that does more tasks than an app, uses full-size keyboards, uses a mouse, and requires more extensive hardware resources. |
| | A **gesture** is an action, such as a tap or swipe motion, that you perform with your finger or stylus on a touch-enabled device. |
| | The **touch keyboard** appears on-screen when typed characters are expected, such as in text messages. |
| | A **mouse**, **trackball**, **touchpad**, or other pointing device is used to move the white arrow **pointer** on the screen and perform mouse actions, such as click or double-click. A trackball has a ball that you roll your thumb or fingers over to move the pointer. A touchpad on a laptop lets you move your finger over the surface to move the pointer. |
| | Pressing the Start key brings up the Start menu. |
| | A **keyboard command** involves pressing the Start key, Ctrl, Alt, or a function key with a letter to perform an action. |
| **1.2 Starting Windows 10 and Exploring Apps** | The **Lock screen** appears when you start Windows or resume system use from sleep mode. |
| | A **user account** is a username with a password, PIN, or other authentication method used to sign in to Windows. |
| | Signing in to Windows using a **Microsoft account** (a phone number or email address) means that your settings are stored online and synced with other devices. |
| | Signing in with a **local account** means that your settings are not shared with other devices, and you do not see live updates on some of the tiles on the Start menu. |
| | Once signed in, the **desktop** appears and the Start menu opens. The desktop has icons to launch applications and a **taskbar** along the bottom of the screen. |
| | On the taskbar, the Start button, **search text box**, and icons are pinned to quickly start apps or applications. The **notification area** at the right end has icons to show system status. |
| | The **Photos app** shows images stored on the local PC and from OneDrive. |
| | The **Calendar app** is used to enter appointments or events. |
| | Use the **Store app** to search for and download new apps. |
| | Switch between apps using the app button on the taskbar or by displaying **Task View** using the Task View button on the taskbar to click the desired thumbnail. |
| | The **Timeline** feature in Task View shows what you have opened today, yesterday, and further back up to 30 days. |

*continued…*

| Topic | Key Concepts and Key Terms |
|---|---|
| **1.3 Using the Search Text Box and Working with Windows** | Click in the search text box, type the first few letters of what you are looking for, and then click the item in the search results. |
| | The **Minimize button** reduces a window to a button on the taskbar. |
| | Click the **Close button** to close a window and any content within the window. |
| | The **Maximize button** expands a window size to fill the entire desktop. |
| | The Maximize button changes to the **Restore Down button** when a window is maximized. Use Restore Down to return the window to its previous size. |
| | The **Snap** feature lets you dock a window to the left or right half of the window, or to one of four quadrants by dragging the window left, right, or to a corner. |
| | **Snap Assist** shows thumbnails of open windows when you snap a window and leave part of the screen empty. Click one of the thumbnails to fill the empty part of the screen with another window. |
| **1.4 Using the Notification Area** | Click the current date and time at the right end of the taskbar to view a pop-up calendar. |
| | Click the Notifications icon to open the **Action Center** panel where you can view system messages or access several quick action tiles for managing settings. |
| | Click the Speaker icon to adjust the volume. |
| | Click the Network icon to manage network connectivity. |
| | A **quick action tile** displays in Notification panels and is used to turn on or turn off a setting or open a setting window. |
| | The Power icon displays the battery status and a link to manage battery settings. |
| **1.5 Locking the Screen, Signing Out, and Shutting Down Windows 10** | Lock the device if you need to leave your PC or mobile device for a short period of time by clicking Start, your account name, and then Lock. |
| | Locking leaves all documents and apps open but unavailable to anyone but yourself. |
| | Click Start, your account name, and then **sign out** to close all apps and documents (also referred to as *logging off*). |
| | Perform a **shut down** command if you want to turn off the power to the computer. |
| | The *Shut down* option is accessed from the Power option on the Start menu. |
| **1.6 Using and Customizing the Start Menu** | Apps or applications can be started by double-clicking an icon on the desktop, by clicking a tile on the Start menu, by scrolling the list in the left pane on the Start menu, or by typing the first few letters of the name in the search text box. |
| | Resize the Start menu by dragging the right border or the top border left, right, up, or down as desired. |
| | Right-click a tile and click **Unpin from Start** to remove the tile from the Start menu; locate an app or application in the list on the Start menu, right-click the name, and click **Pin to Start** to add a tile for the software to the Start menu. |
| | Move tiles to a new location on the Start menu by dragging the tile. |
| | Tiles are either square or rectangular in shape and can be resized larger or smaller. |
| | The **Turn Live Tile off** command stops a tile from displaying notifications or status updates; use **Turn Live Tile on** to restore the tile's notifications. |

*continued…*

| Topic | Key Concepts and Key Terms |
|---|---|
| **1.7 Personalizing the Lock Screen and the Desktop** | Click the Start button, click Settings, then click *Personalization* in the Settings app to personalize the Lock screen and desktop options. |
| | You can choose from five other pictures for the Lock screen or browse to a picture on your PC or mobile device. You can also change the apps that display notifications on the Lock screen. |
| | Change the desktop background image and/or the color scheme in the Settings app using the Background and Colors categories. |
| **1.8 Understanding Files and Folders and Downloading and Extracting Data Files** | A **file** is any document, spreadsheet, picture, or other text or image saved as digital data. |
| | When you create a file, you assign a unique **file name** that allows you to identify and retrieve the file. |
| | A **folder** is a name assigned to a placeholder where a group of related files are stored. |
| | A group of files and folders can be compressed and saved as a **ZIP file**, which bundles everything together in one file with a .zip file extension. |
| | When you open a ZIP file, a window opens with the Extract all button used to restore the files and folders from the zipped package, allowing you to use them in their original state and folder structure. |
| | **File Explorer** is the Windows utility used to browse files and perform file management tasks. |
| **1.9 Browsing Files with File Explorer** | File Explorer opens in **Quick access** view, which shows frequently used and recently used files. |
| | The left pane in File Explorer is the Navigation pane, and the right pane is the Content pane. |
| | Click This PC in the Navigation pane to view all the available storage devices. |
| | To browse content, click names in the Navigation pane or double click names in the Content pane. |
| | Use options in the Layout group on the View tab to change the Content pane display of files. |
| | A folder created within another folder is sometimes called a **subfolder**. |
| **1.10 Creating Folders and Copying Files and Folders** | Click the **New folder button** in the New group on the Home tab in File Explorer to create a new folder. |
| | The **ribbon** in File Explorer provides buttons organized into tabs and groups. Buttons are used to carry out file management tasks. |
| | Copying a file or folder makes an exact duplicate of a document, spreadsheet, presentation, image, video, music file, or other object in another folder and/or storage medium. |
| | Begin a copy task by first selecting the file or folder to be copied and then use the **Copy button** and the **Paste button** in the Clipboard group on the Home tab. |
| **1.11 Moving, Renaming, and Deleting Files and Folders and Ejecting a USB Flash Drive** | Files are moved by selecting the files or folders, clicking the **Cut button** in the Clipboard group on the Home tab, navigating to the new destination drive and/or folder, and clicking the Paste button. |
| | Click the **Rename button** in the Organize group on the Home tab, type a new name, and then press Enter to change the name of the selected file or folder. |
| | Select files or folders no longer needed; then click the **Delete button** in the Organize group on the Home tab. |
| | Files deleted from a hard disk drive are sent to the Recycle Bin and remain there until the Recycle Bin is emptied. |
| | Files deleted from a USB flash drive are not sent to the Recycle Bin. |
| | Open the Recycle Bin from the desktop to view files deleted from the hard disk drive. |
| | Emptying the Recycle Bin permanently deletes the files or folders. |
| | Eject a USB flash drive using the **Safely Remove Hardware and Eject Media** icon in the notification area on the taskbar. Click the Show hidden icons arrow to locate invisible icons. |

**Topics Review**

# Navigating and Searching the Web

## Chapter 2

For many people reading this textbook, the internet is part of daily life, used to search for information, connect with friends and relatives, watch videos, listen to music, play games, or shop. Mobile devices, such as tablets and smartphones, are used to browse the web anywhere they happen to be via cellular or Wi-Fi connections. According to Statista (a market and consumer data company), 52 percent of website visits originated on a mobile device in 2018. In the workplace, being able to effectively navigate and search the web is a requirement for all workers.

In this chapter, you will learn definitions for internet terminology and learn how to navigate the web using the two most popular web browsers. You will also learn how to use search tools to find and print information, view multiple websites in a browsing session, bookmark web pages you visit often, and copy an object from a website to your computer.

*Note: No student data files are required to complete this chapter.*

### Read & Learn

### Learning Objectives

**2.1** Define *internet*, *World Wide Web*, *web browser*, *web page*, *hyperlink*, and *web address*

**2.2** Navigate the web using Microsoft Edge, use tabs to view multiple websites, and add a page to the Favorites list

**2.3** Navigate the web using Chrome, use tabs to view multiple websites, bookmark a page, and use the Find bar

**2.4** Use a search engine website to find information, refine a search using advanced search tools, and print a web page

**2.5** Download a picture from a web page

### Content Online

The online course includes additional training and assessment resources.

## 2.1 Introduction to the Internet and the World Wide Web

**Watch & Learn**

●────── Skills

Define *internet*

Define *World Wide Web*

Explain *web browser*, *web page*, and *hyperlink*

Describe *web address*

The **internet**, or **net**, is a global network that links other networks such as individuals, businesses, schools, government departments, nonprofit organizations, and research institutions. The worldwide system of interconnected high-speed communications and network equipment, computing devices, and communication protocols to send and receive data is known as the *internet*. When you want to look up flight times on the internet, the request sent from your device must travel via communications media, communications equipment, and computers owned by a telephone, cable, or satellite company to reach the airline web server.

### The World Wide Web

The collection of electronic documents circulated on the internet in the form of web pages make up the **World Wide Web**, or **web**. A **web page** is a document that contains text and multimedia content, such as images, video, sound, and animation. Web pages also contain hyperlinks. A **hyperlink**, also called a **link**, is text, an image, or an icon on a web page that moves you to another related page when clicked. Web pages are stored in a format that is read and interpreted for display within a **web browser**, which is application software used to view web pages. A website is a collection of related web pages for one organization or individual. For example, the collection of web pages linked to the main page for your school make up your school's website. All the web pages and resources such as photos, videos, sounds, and animations that make the website work are stored on a computer called a *web server*. Web servers are connected to the internet continuously.

You connect your PC or mobile device to the internet by subscribing to internet service through an **internet service provider (ISP)**, a company that provides access to internet infrastructure for a fee. The ISP will provide you with the equipment needed to connect to its network, as well as instructions for installing and setting up the equipment to work with your computer and other mobile devices. Once your account and equipment are set up, you can start browsing the web using a browser, such as Microsoft Edge or Google Chrome.

### Web Addresses

Similar to your postal address, each web page has a unique text-based **web address** that is used to navigate to the page. A web address is also called a **uniform resource locator (URL)**. One way to navigate the web is to type a web address into the **Address bar** of a web browser. For example, to view the main web page for the publisher of this textbook, you would use the web address https://ParadigmEducation.com. If *http* or *https* (*s* indicates a secure connection that encrypts data) is left out when typing a web address, the browser still finds the web page by assuming *http* or *https*. For this reason, many people often go to a page by typing only the portion of the address after *http://*. To illustrate all of this, the parts of the web address (URL) https://www.nasa.gov/topics/moon-to-mars/index.html shown in Figure 2.1 on the next page are explained in Table 2.1.

Although you navigate to a web page using a friendly text-based address, the real address for the computing device that stores the web page is called an **internet protocol (IP) address**. An IP address is either four groups of numbers 0 to 255 separated by periods (called *IPv4*), or eight groups of characters (0 to 9 or *a* to *f*) separated by colons (called *IPv6*). A server called a Domain Name System (DNS) server, owned by a business or ISP, maintains a directory that references the IP address to the web address. When you type a web address into the browser, a DNS server locates the IP address to transmit the data to the correct computer.

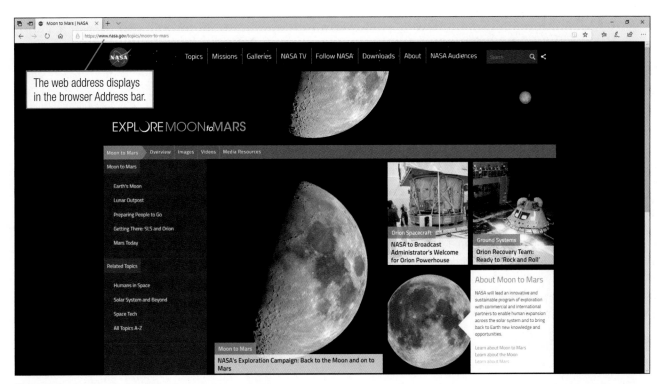

The web address displays in the browser Address bar.

**Figure 2.1**

A web page from NASA is shown in the Microsoft Edge browser window. Your page may look different if NASA has updated the page since publication.

**Table 2.1** The Parts of the Web Address (URL) https://www.nasa.gov/topics/moon-to-mars/index.html

| Part of URL | What It Means | Examples |
|---|---|---|
| https:// | Hypertext Transfer Protocol Secure. A set of rules for transmitting data in encrypted format over the internet. *Https* is used for exchanging data in online transactions. Other websites (like nasa.gov) are starting to use the protocol to protect data on their websites and to authenticate their websites for the general public. Other protocols used are http (to transfer data without encryption) and ftp (to transfer a file). | http<br>ftp<br>https |
| www.nasa.gov | Domain name. A text-based address for the web server that stores the web page. The last three letters are the extension and generally tell you the type of organization; e.g., nasa.gov is a government website. Some domain names end with a two-character country code, such as eBay.ca for the Canadian auction site eBay. Using *www* in a domain name is becoming less common.<br>   In the *Examples* column are five domain names to show a sampling of a variety of extensions. Other extensions, such as *.net, .biz, .legal,* and *.realtor* are also used. Extensions are either restricted or urestricted. A restricted extension means that registration of a domain that uses the extension has to meet specific criteria. For example, only accredited educational institutions can register for a domain with the extension *.edu*. An unrestricted extension, like *.com*, can be registered by anyone. | www.google.com (Google)<br>www.harvard.edu (Harvard University)<br>www.navy.mil (US Navy)<br>www.linux.org (Linux open-source community)<br>driftaway.coffee (Driftaway coffee bean subscription service) |
| /topics/moon-to-mars/ | Folder path to the web page. The forward slash (/) and text after the slash are the folder names on the web server where the page is stored. | Folder path names will vary. |
| index.html | Web page file name. The file name extension indicates the language used to create the web page. *Html* stands for *Hypertext Markup Language*, which uses tags to describe content. | File names will vary. |

Many times when you go to the web to look for information, you do not know the web addresses for the pages you need. In this case, search tools help you find web pages. You will learn to find web pages using search tools in Topic 2.4.

## 2.2 Navigating the Web Using Microsoft Edge

**Microsoft Edge** is the web browser included with Windows 10. The browser replaced the Internet Explorer web browser included with earlier Windows versions. Similar to the Windows apps you learned about in Chapter 1, where an app looks and feels the same on all devices, Microsoft Edge is the browser for all Windows 10 devices, including tablets. A mobile app for Microsoft Edge is also available for Android and Apple smartphones.

When you launch the web browser, the Start page displays with news and information feeds unless the browser has been customized to display another page. For example, many schools customize browsers to first display the student portal. Microsoft Edge is designed to be a fast and secure browser, with a clean, distraction-free interface. Reading view strips away the sidebars and navigation on a web page to show only the main content on a page. The Reading view icon on the Address bar is active for a page that can be viewed without the navigation items. Microsoft Edge also has tools to annotate a web page with markup, highlights, handwriting, or typed notes, saving the page as a web note for reading in the Reading list, a OneNote notebook, or shared via email.

**Watch & Learn**

------- Skills

Navigate the web using Microsoft Edge

Use Reading view

View multiple websites within the same window

Add a page to Favorites

**Tutorial**

Getting Started with the Microsoft Edge Browser

### Starting Microsoft Edge and Displaying a Web Page

**1** Click the Microsoft Edge icon on the taskbar.

The Microsoft Edge browser window displays the Start page shown in Figure 2.2 unless the browser has been customized to display another page.

**2** Click the Address bar (contains the dimmed text *Search or enter web address*), type nps.gov/grca, and then press Enter. If a message box appears asking you to sign up for national park news, close the message using the × at the top right corner. Review the layout and tools in the Microsoft Edge window shown in Figure 2.3 on the next page.

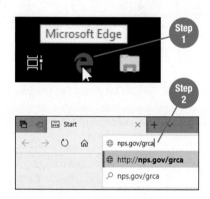

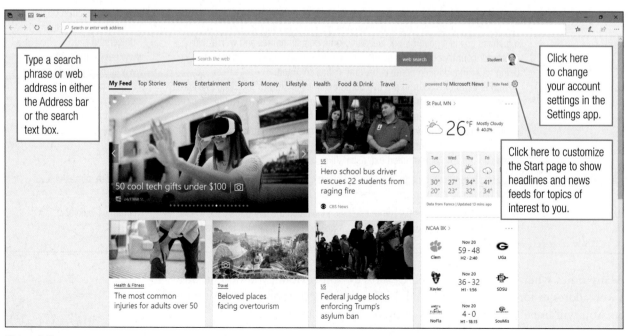

**Figure 2.2**

The Microsoft Edge Start page is displayed. The Start page provides hyperlinks to business, entertainment, health, and finance news and current weather forecast and sports scores.

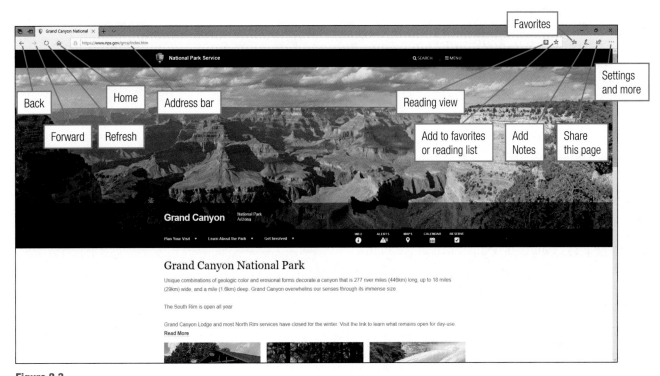

**Figure 2.3**

The National Park Service Grand Canyon web page is shown in the Microsoft Edge browser window.

③ Click <u>MENU</u> near the top right of the web page.

④ Click <u>About Us</u>.

⑤ Click <u>News</u>.

Navigating web pages involves clicking links to move from one page to another on a website or to a web page stored on another website. Click **Back** to return to the previous page viewed, or click **Forward** to move forward one web page viewed. To display a different web page, click the Address bar to select the current entry (or drag to select the current text in the Address bar), type the web address, and then press Enter.

⑥ Click in the Address bar or drag to select the existing web address, type loc.gov, and then press Enter. If a message box appears asking you to support the Library of Congress, close the message.

The Library of Congress web page appears. The starting page for a website is called the **home page**.

⑦ Click the icon with three bars at the top right of the page.

A button or icon with three bars is called a *hamburger button* because it resembles a hamburger patty between a top and bottom bun. The hamburger button indicates a menu.

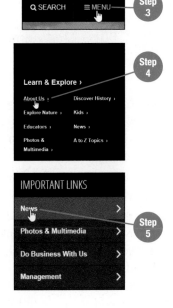

**8** Click <u>Visit</u>.

**9** Click Back to return to the Library of Congress home page (move back one web page).

**10** Click Forward to return to the Visiting the Library page (move forward one web page).

**11** Click <u>Guidelines & Tips</u>.

**12** Click Reading view on the toolbar.

The page reformats within Microsoft Edge to remove all the sidebars and navigational items and show the text in a distraction-free reading layout. Some web pages cannot be reformatted in Reading view. A web page that is not available in Reading view displays a dimmed Reading view icon on the toolbar.

**13** Scroll right to view the second screen in Reading view.

**14** Click Reading view on the toolbar to turn off the view.

**15** Click in the Address bar or drag to select the existing web address, type nasa.gov/topics/history, and then press Enter.

**app tip**

Click Refresh (next to Forward) to update a web page. Sometimes you need to do this if the page stops loading properly or otherwise hangs, or you want to reload the page because the page you are viewing is not the current version.

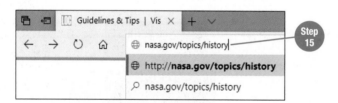

**16** Scroll down and click a link to a story that interests you. Read the story using Reading view, if the view is available.

**17** If necessary, turn off Reading view and then click Back to return to the NASA History page.

## Displaying Multiple Web Pages and Adding a Page to Favorites List

Open multiple web pages within the same Microsoft Edge window by displaying each page in its own tab. Click **New tab** (displays as a plus symbol) to open a new page in the window. Switch between web pages by clicking the tab for the page you want to view. This is called **tabbed browsing**. The web address for a page you visit frequently can be saved to the **Favorites** list by displaying the page and then using the star on the toolbar to add the page to Favorites.

**Tutorial**

Using Tabbed Browsing

**18** Click New tab (plus symbol) next to the tab for the NASA History page at the top of the window.

**19** Type flickr.com/commons in the Address bar on the new tab and then press Enter.

The Commons contains public photography archives. Images in The Commons have no known copyright restrictions. The next time you need a photo for a project, consider sourcing an image from this website.

**20** Click the NASA History page tab to change the web page displayed in the window.

When the pointer is resting on a tab, a thumbnail of the web page appears.

**21** Click the Close tab icon (displays × on tab) on the NASA History tab.

**22** With The Commons page displayed in the window, click the star at the right end of the Address bar.

**23** Click the Add button to accept the default option to save the page in Favorites.

The page will now display in the Favorites list with the name Flickr: The Commons. The star turns gold for a page that is added to the Favorites list.

**24** Click Favorites to open the Favorites panel at the right side of the window.

Display the Favorites panel to view the Favorites list, Reading list, History list, or Downloads list.

**25** Click *Favorites* in the left pane of the panel to view the Favorites list, if another list is currently active.

**26** Click Favorites to close the panel.

**27** Close the Microsoft Edge window.

The Commons page added to the Favorites list in Step 23

## 2.3 Navigating the Web Using Chrome

**Chrome** is a web browser for PCs and mobile devices from Google. In 2018, StatCounter (a company that measures web traffic) reported that Chrome was the browser of choice for almost 50 percent of people within the United States and Canada. Since the release of Chrome in 2008, the browser has gained a significant following due to its fast page loading and direct searching capability from the Address bar using the Google search engine.

### Starting Chrome and Displaying a Web Page

1. Click the Chrome icon on the taskbar or double-click the icon on the desktop, and then review the layout and tools in the Chrome window shown in Figure 2.4.

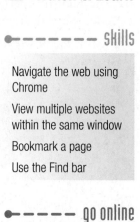

If Chrome is not currently set as the default browser, you may be prompted with a message bar below the Address bar to set it as the default browser.

2. With the insertion point positioned in the Address bar, type techmeme.com and then press Enter.

Techmeme is a technology news site that posts each day the top news headlines from the technology industry.

**Watch & Learn**

------- Skills

Navigate the web using Chrome

View multiple websites within the same window

Bookmark a page

Use the Find bar

------ go online

**https://CA3
.ParadigmEducation
.com/Chrome**

Go here to download and install Chrome on your PC or mobile device.

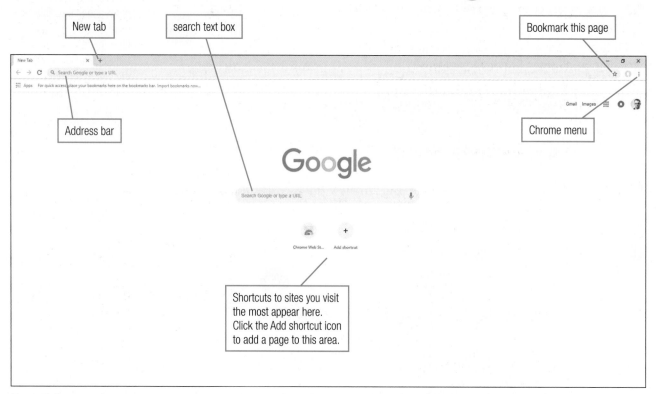

**Figure 2.4**

The Google search engine is the default search engine used for new search requests in the Chrome browser. Chrome is regularly updated—your image may vary from the one shown here.

**3** Click a link to a technology story that interests you on the Techmeme page.

Techmeme is a content aggregator—a company that searches, collects, and organizes content in one place. Clicking a link to a story on the Techmeme page takes you to another website.

**4** Click Back to return to the Techmeme page.

Your web address will vary depending on the story you clicked in Step 3.

**Tutorials**

Starting Chrome and Displaying a Web Page

Displaying Multiple Web Pages and Bookmarking Pages

## Displaying Multiple Web Pages and Bookmarking Pages

Similar to Microsoft Edge, Chrome uses tabbed browsing to display more than one web page within the same browser window. Open a new tab and navigate to a web address to browse a new site without closing the existing web page. Chrome displays a page name at the top of the tab along with a close control. Web pages you visit frequently can be bookmarked by clicking the white star at the end of the Address bar (displays the ScreenTip *Bookmark this page*).

**5** With techmeme.com the active web page, click the white star at the end of the Address bar.

The white star changes to blue, and the Bookmark dialog box appears.

**6** Click Done to close the Bookmark dialog box.

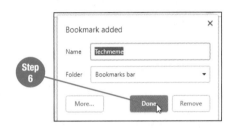

The **Bookmarks bar** displays below the Address bar. The Booksmarks bar is turned off or on using the **Chrome menu** (three vertical dots) near the top right of the window. You will practice showing the Booksmarks bar in the next step.

**7** Click the Chrome menu (three vertical dots with the ScreenTip *Customize and control Google Chrome*) and then point to *Bookmarks*. A check mark next to *Show bookmarks* bar means that the bar is turned on. In that case, click in a blank area away from the menu to close the menu; otherwise, click *Show bookmarks bar* to turn the bar on.

**app tip**

By default, bookmarks are added to the Bookmarks bar, which is docked below the Address bar when the bar is turned on.

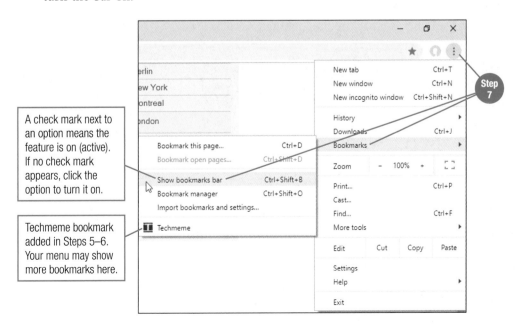

A check mark next to an option means the feature is on (active). If no check mark appears, click the option to turn it on.

Techmeme bookmark added in Steps 5–6. Your menu may show more bookmarks here.

8  Click New tab (plus symbol) next to the Techmeme tab to open a new page.

9  Type youtube.com and then press Enter.

10 Click the white star and then click Done to add youtube.com to the
   Bookmarks bar.

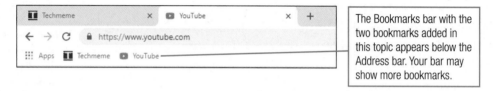

The Bookmarks bar with the
two bookmarks added in
this topic appears below the
Address bar. Your bar may
show more bookmarks.

## Searching the Web from the Address Bar

 **Tutorial**

Searching the Web
from the Address Bar

Web browsers like Microsoft Edge and Chrome allow you to type a search phrase in
the Address bar. In Chrome, search results from Google appear in the window. The
default search engine can be changed if you prefer to use another search engine
website, such as bing.com. To change the default search engine, click the Chrome
menu, click *Settings*, and then change the *Search engine used in the address bar* option
by clicking the down-pointing arrow next to the current search engine (such as
Google) in the *Search engine* section. Close the Settings tab when finished.

11 Click New tab next to the YouTube tab, type seven wonders of the world, and
   then press Enter.

   Notice that as you type, searches that match your entry appear in a drop-down list
below the Address bar. If one of the searches is what you are looking for, click the
entry in the list.

12 Click the Images hyperlink below the Address bar to filter search results to
   display images only.

13 Scroll down a few screens to view images, and then scroll back to the top of the
   page and click the All hyperlink below the Address bar.

14 Click in the Address bar or drag to select the current web address, type great wall
   of china, and then press Enter.

**15** Click <u>Great Wall of China - Wikipedia</u>, then scroll down to review the *Wikipedia* page about the Great Wall of China.

**16** Scroll back up to the top of the *Wikipedia* page.

## Finding Text on a Page

Use the **Find bar** to locate a specific word or phrase on a web page. The Find bar is turned on from the Chrome menu. Using the Find bar is helpful when viewing a long story or article and you need to locate a reference that you know exists in the text.

**17** Click the Chrome menu and then click *Find*.

The Find bar opens at the right side of the window, below the Address bar, with the insertion point positioned in the Find bar search text box.

**18** Type *first emperor* in the Find bar search text box.

The currently selected match is shaded with an orange background on the page, and the number of matches found on the web page displays in the Find bar next to the search text box. Additional matches below the current selection are shaded yellow (scroll to see the yellow highlights).

**19** Click Next (down-pointing arrow) on the Find bar to scroll the page to the next occurrence of *first emperor*.

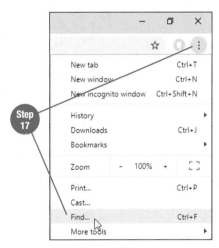

**quick steps**

**Start Chrome**
Click Chrome icon on the taskbar or double-click the icon on the desktop.

**Display a Web Page**
Click in Address bar, type web address, and then press Enter.

**Display Multiple Web Pages**
1. Click New tab.
2. Type web address for new web page.
3. Press Enter.

**Bookmark a Web Page**
1. Display desired web page.
2. Click white star.
3. Click Done.

**Find Text on a Web Page**
1. Click Chrome menu.
2. Click *Find*.
3. Type text to find.
4. Click Next for each occurrence.

**Tutorial**

Finding Text on a Page

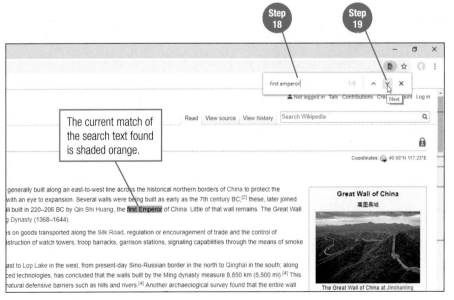

The current match of the search text found is shaded orange.

**20** Continue clicking Next until you have seen all occurrences.

**21** Click Close (displays ✕) on the Find bar to remove the bar from the screen.

**22** Close the Chrome window.

**app tip**

In Microsoft Edge, the Find bar is accessed by clicking Settings and more, then *Find on page*.

**app tip**

Ctrl + F is the universal keyboard shortcut for the Find command in any browser or application.

## 2.4 Searching for Information and Printing Web Pages

A **search engine** is managed by a company that indexes web pages by keywords and provides a search tool with which you can search their indexes to find web pages. To create indexes, search engines use software called a **spider**, or **crawler**, that reads web pages and other information supplied by the website owner to generate index entries. Search engines also provide advanced tools, which you can use to narrow search results.

### Using a Search Engine to Locate Information on the Web

Several search engines are available, with Google and Bing the two leading companies; however, consider searching the web using other search engines, such as Dogpile, Ask, and DuckDuckGo, because you get different results from each company. Spider and crawler application capabilities and timing for indexing create differences in search results among companies. Depending on the information you are seeking, performing a search in more than one search engine is a good idea.

**1** Click the Microsoft Edge icon on the taskbar.

By default, Microsoft Edge performs searches using Bing, the search engine from Microsoft.

**2** With the insertion point positioned in the search text box (which displays the dimmed text *Search the web*) on the Start page, type cover letter examples and then press Enter or click the web search button.

Notice that as you type, Bing provides search suggestions in a drop-down list as soon as you begin typing. Click a search suggestion in the list if you see a close match.

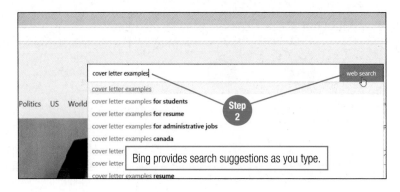

Bing provides search suggestions as you type.

**3** Scroll down the search results page and click a link to a page that interests you.

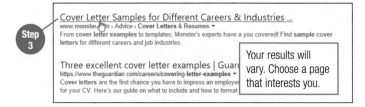

Your results will vary. Choose a page that interests you.

**4** Read a few paragraphs about cover letters at the page you selected.

**Watch & Learn**

 skills

Use search tools to find information on the web

Narrow search results by applying advanced search options

Print a web page, including selected pages only

**Tutorial**

Using Search Engines to Find Content

go online

**https://CA3 .ParadigmEducation .com/Dogpile**
Go here to search using a metasearch search engine. A metasearch search engine sends the search phrase to other search engines and displays one list of search results from the wider group.

5  Click Back to return to the search results list.

6  Click in the Address bar or drag to select the current web address in the Address bar, type duckduckgo.com, and then press Enter.

Many people who want to browse the web in private like to use DuckDuckGo. While most search engines collect your search data to show you personalized results and ads, DuckDuckGo does not track your activity and does not personalize search results.

7  With the insertion point positioned in the search text box, type cover letter examples and then press Enter or click the search button (magnifying glass).

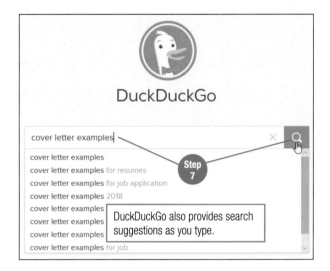

● – – – – – – app tip

Click a category, such as Images, Videos, Maps, or News, on a search engine website to restrict search results to a specific type of content.

8  Scroll down the search results page. Notice that some links are to the same web pages that you saw in the Bing search results; however, the same page may be in a different location in the search results list, or you may notice new links that were not shown by Bing.

## Using Advanced Search Options

In both the Bing and DuckDuckGo "cover letter examples" search results, thousands of links were located and shown on the results pages. To help you narrow a list by specifying additional criteria, search engines provide advanced search tools or filters. Each search engine provides different tools. Explore the options at the search engine you prefer, or look for a help link that provides information on how to use advanced search tools.

☁ **Tutorial**

Using Advanced Search Options

9  Click in the Address bar or drag to select the current web address, type google.com, and then press Enter.

10  With the insertion point positioned in the search text box, type cover letter examples and then press Enter.

11  Click Settings (below search text box), and then click Advanced search in the drop-down list.

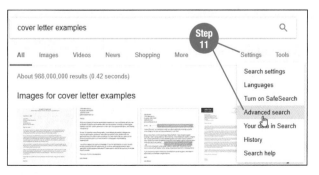

⑫ Click in the *site or domain* text box and then type edu. (You may need to scroll down the screen to locate the *site or domain* option.)

A .edu domain is restricted to US-accredited postsecondary institutions.

⑬ Click the Advanced Search button near the bottom of the Advanced Search page.

⑭ Click <u>Tools</u> below the search text box, click <u>Any time</u>, and then click <u>Past month</u>.

The search results list is now significantly reduced from the millions in the prior results list.

⑮ Scroll down the search results page and then click a link to a page that interests you. Notice that all URLs are .edu domains and all dates are within the past month.

## Printing a Web Page

Printing a web page can sometimes be frustrating, because web pages are designed for optimal screen viewing (not printing). Many times you may print a web page only to discard a second or third page that you did not need, or the content printed did not fit the width of the paper and the printout was unusable.

⑯ Click in the Address bar or select the current web address, type https://www.rollins.edu/career-life -planning/documents/samples-cover-letter.pdf and then press Enter.

The web address provides four sample cover letters in the portable document format (PDF), suitable for printing. Assume you decide to print two of the example letters for later use when you are looking for a job.

⑰ Click Settings and more (three dots) and then click *Print*.

☁ **Tutorial**

Printing a Web Page

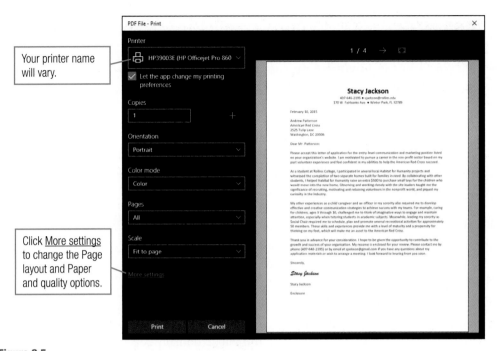

Your printer name will vary.

Click <u>More settings</u> to change the Page layout and Paper and quality options.

**Figure 2.5**
The Print dialog box in Microsoft Edge.

Page 1 in the printout (first cover letter example) is previewed in the Print dialog box shown in Figure 2.5. The Next page arrow (right-pointing arrow above the preview area) is used to navigate through all the pages that will print. Option boxes are available to change the destination printer, number of copies, page orientation, color mode, pages to print, and scale. The options available vary depending on the page selected for printing. For example, since this is a PDF page, options to change the margins and header/footer are omitted.

18  Click the Next page arrow (right-pointing arrow) above the preview page to view the second page that will be printed.

19  Click the Next page arrow two more times to view the remaining letters.

Assume you decide to print only the first and third letters.

20  Click the *Pages* option box (by default it displays *All* or *All pages*) and then click *Page range*.

21  Click in the *Range* text box and then type 1,3.

In the dialog box, the red text displayed above the *Range* text box illustrates acceptable ways to enter a print range.

22  Click the Print button, or if you are not connected to a printer, click the Cancel button to close the Print dialog box.

Only pages 1 and 3 of the web page print. Avoid wasting paper by previewing a web page before printing to make sure you only print the pages that you intend to use.

23  Close the Microsoft Edge window. Click *Close all* if prompted to close all tabs.

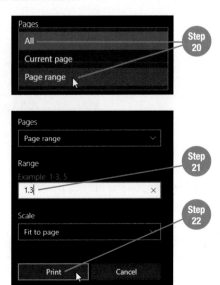

## 2.5 Downloading Content from a Web Page

Copying an image, audio clip, video clip, or music file (such as MP3) from a web page to your PC or mobile device is referred to as **downloading** content. Most content is protected by copyright law from being used by someone without permission. Before you download content, check the website for restrictions against copying information. Look for a contact link and request permission from the website owner to use the content if no restrictions are shown. When you include work created by someone else in an academic aasignment or in any other publication created by you, always provide a citation for the source of that work. Some authors will request that you provide a link to the page where the original work resides.

### Saving an Image from a Web Page

Saving content from a web page generally involves right-clicking an object to display a context menu with a save option. In most cases, a file for an image, video clip, or audio clip is saved to the Downloads folder on the local device.

1. Click the Microsoft Edge icon on the taskbar.

   Assume you want to find an image from the Grand Canyon for a project. You decide to use the Flickr: The Commons page to find an image in the public domain that can be used without copyright restrictions.

2. Click Favorites.

3. Click Favorites in the left pane if the Favorites list is not currently shown.

4. Click *Flickr: The Commons* in the Favorites list (you added this link to the Favorites list in Topic 2.2).

5. If necessary, scroll down to the section titled *A Commons Sampler*.

6. Click in the *Search The Commons* text box, type grand canyon, and then press Enter or click the Search button.

7. Scroll down and view the images in the search results page.

8. Click an image that you like and want to save.

9. On the next page, scroll down and read the information below the photograph. Look for entries about copyright restrictions. The entry *Unrestricted* means you are free to copy and use the image.

10. If necessary, scroll back up the page to the image preview.

11. Click the Download this photo icon (the down-pointing arrow above a bar) at the bottom right of the image preview area and then click Medium in the pop-up list. Note that the dimensions for each size option will vary depending on the image that is currently selected.

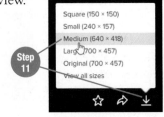

**Watch & Learn**

**SKIIIS**

Download an image from a web page

**Tutorial**

Downloading a File from a Website

**Oops!**

Don't have Flickr: The Commons in your Favorites list? If necessary, type flickr.com/commons in the Address bar and then press Enter.

**in the Know**

The Commons at flickr.com was started in 2008 as a joint project between Flickr and the Library of Congress with a goal to increase access to publicly held photographs. The project expanded to over 100 library, museum, and other public collections worldwide.

**12** Click the Save button in the message bar that appears at the bottom of the window, then click Close (×) when the message displays that the image finished downloading.

**13** Click in the Address bar to select the current web address, type facebook.com/GrandCanyonNationalPark, and then press Enter. You may be prompted to sign in to Facebook to view the page. If necessary, skip to Step 18 if you cannot view the Facebook page.

**14** Scroll down the timeline to a picture that you would like to save, and then click on the picture to view it in a larger window.

**15** Right-click the picture and then click *Save picture as*. (You may need to right-click in the center of the image to get the menu shown at the right.)

**16** In the Save As dialog box, click Downloads in the Navigation pane, select the current text in the *File name* text box, type GrandCanyon, and then click Save.

The timeline pictures will vary. Choose any picture you want to save.

**quick steps**

**Save a Picture from a Web Page**
1. Display desired web page.
2. Right-click image.
3. Click *Save picture as* or *Save image as*.
4. Navigate to desired storage medium and folder, type file name if necessary, then click Save.

**Oops!**

Different context menu? In Chrome, choose the option *Save image as* on the shortcut menu in Step 15.

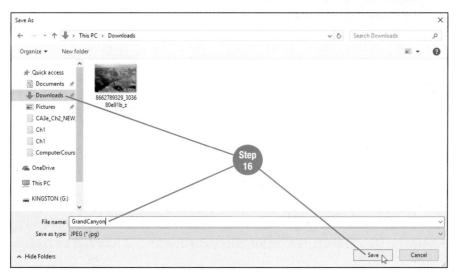

**17** Close the Microsoft Edge window. Click Close all if prompted to close all tabs.

**18** Click the File Explorer icon on the taskbar.

**19** Click Downloads in the Navigation pane.

You will see the two downloaded pictures in the Content pane.

**20** Double-click each image you downloaded to view the image in the Photos app or other image viewing window, and then close the window.

**21** Close File Explorer.

## Topics Review

| Topic | Key Concepts |
| --- | --- |
| **2.1 Introduction to the Internet and the World Wide Web** | The **internet**, or **net**, is the infrastructure of interconnected communications equipment, network equipment, and computing devices that represents the global network linking individuals and organizations. |
| | All the web pages circulated on the internet form the **World Wide Web**, or **web**. |
| | A document stored on a web server that contains text and multimedia is called a **web page**. |
| | Text, an object, or icon on a web page can be a **hyperlink**, or **link**, that moves you from one page to another when clicked. |
| | A **web browser** is software, such as Microsoft Edge or Google Chrome, that is used to view web pages. |
| | An **internet service provider (ISP)** is a company that provides internet service to individuals and businesses for a fee. |
| | A **web address** is a unique text-based address, also called a **uniform resource locator (URL)**, that identifies each web page on the internet. |
| | One way to display a web page in a browser is to type the web address in the browser's **Address bar**, located at the top of the window. |
| | Every device connected to the internet is given a unique **internet protocol (IP) address**, which is either four groups of numbers 0 to 255, or eight groups of characters 0 to 9 or *a* to *f*. A Domain Name System (DNS) server looks up in a directory the associated IP address for a requested web address. |
| | A web address is made up of parts that define the protocol, domain, folder(s), and file name for the web page. |
| **2.2 Navigating the Web Using Edge** | **Microsoft Edge** is the web browser included with Windows 10. |
| | Start Microsoft Edge by clicking the Microsoft Edge icon on the taskbar. |
| | To display a web page, type the web address (URL) in the Address bar and then press Enter. |
| | Click **Back** to move back one page; click **Forward** to move forward one page. |
| | The first page that displays for a website is called the **home page**. |
| | Display a web page in Reading view to read the page content without the sidebars and navigational items. |
| | Click **New tab** (plus symbol) to open a new page in the window. |
| | **Tabbed browsing** means you have more than one web page open in a window. Each web page has a tab that you click to switch from viewing one page to another page. |
| | Display a web page, click the star at the right end of the Address bar, and then click the Add button to save a page to the **Favorites** list. |
| | Click **Favorites** to show Favorites, Reading list, History, and Downloads in a panel at the right side of the window. |
| **2.3 Navigating the Web Using Chrome** | **Chrome** is a web browser from Google for PCs and mobile devices. |
| | Start Chrome by clicking the Chrome icon on the taskbar or by double-clicking the Chrome icon on the desktop. |
| | Display a web page by typing a web address or search phrase in the Address bar. |
| | Click the white star to bookmark the current web page on the **Bookmarks bar**. |
| | Click the **Chrome menu** to turn on or turn off the display of the Bookmarks bar that docks below the Address bar. |
| | Open the **Find bar** from the Chrome menu to locate all occurrences of a word or phrase on the current web page. |

*continued…*

| Topic | Key Concepts |
|---|---|
| **2.4 Searching for Information and Printing Web Pages** | A search tool to find indexed pages is called a **search engine**. A company indexes web pages by keywords using software called a **spider**, or a **crawler**.<br><br>Consider searching for information by using more than one search engine because you will get different results from each due to differences in spider and crawler software and timing differentials.<br><br>To find information using a search engine, launch a web browser, type the web address for the desired search engine, type a search word or phrase in the search text box, and then press Enter.<br><br>Each search engine provides tools for advanced searching, which narrows the search results list. For example, at Google, the Advanced Search page provides options to include or exclude words in the search and restrict search results to types of domains, file formats, a country, or a language.<br><br>Print a web page by accessing the Print option from a menu in the browser (in Microsoft Edge, click Settings and more, and then click *Print*).<br><br>In the preview area of the Print dialog box, preview each page to avoid unnecessary printing of pages that are wasteful. |
| **2.5 Downloading Content from a Web Page** | Saving a copy of an image, audio clip, video clip, or music file from a web page to your PC or mobile device is called **downloading**.<br><br>Some websites, such as flickr.com, provide a download tool that facilitates downloading an image on their website to the Downloads folder.<br><br>If no download tool is on the web page, saving content generally involves right-clicking the object and selecting a save option on the shortcut menu. |

 **Topics Review**

# Exploring Office 2019 Essentials

Office 2019 is a suite of software applications that includes the applications Word, Excel, PowerPoint, Access, and Outlook. The suite is available in various editions that package the applications in collections geared toward a home, business, or student customer using the software under a one-time purchase for a PC installation, or a subscription installation called Office 365. The Office Professional 2019 edition for a PC installation and the Office 365 Home subscription include all the applications used in this textbook.

One reason the Office suite is popular is because several features or elements are common to all the applications. Once you learn your way around one of the applications in the suite, another application looks and operates similarly, making the learning process faster and easier.

In this chapter, you will navigate the Office 2019 interface and perform file-related tasks or routines common to all the applications. You will customize the Quick Access Toolbar and choose options using multiple methods. You will also save and open files to and from OneDrive, an online file storage option.

## Learning Objectives

**3.1** Start an Office application, identify common features, and explore the ribbon

**3.2** Open, save, print, export, close, and start new documents in the backstage area

**3.3** Customize and use the Quick Access Toolbar

**3.4** Select text and objects, perform commands using the ribbon and Mini toolbar, and select options in a dialog box

**3.5** Copy and paste using buttons in the Clipboard and format an object using a task pane

**3.6** Use the Tell Me feature and the Help task pane to find options and help resources

**3.7** Save files to and open files from OneDrive, navigate longer documents, and use Undo

**3.8** Change the Zoom setting and screen resolution

**Read & Learn**

**Content Online**
The online course includes additional training and assessment resources.

## 3.1 Starting and Switching Applications, Starting a New Presentation, and Exploring the Ribbon

**Watch & Learn**

● ------- Skills

Start an Office application

Switch between applications

Start a new presentation

Explore the ribbon

Word, Excel, PowerPoint, and Access begin on a Start screen and share some common features and elements. An application can be started using an icon on the desktop, from a tile or application name on the Start menu, or by using the search text box on the taskbar.

### Office 2019 Editions

The Office 2019 suite is packaged in a variety of application bundles (called *editions*) as one-time purchased software. Table 3.1 lists three of these editions. The suite is also available through a monthly or annual subscription as Office 365. **Office 365** includes Word, Excel, PowerPoint, Outlook, Publisher, and Access, along with Skype calling minutes, additional OneDrive storage, and technical support from Microsoft. Other advantages to an Office 365 subscription are the suite is accessible for up to six users and updates to the software are automatically pushed to each installation.

Microsoft also offers a special edition of Office 365 for students called Office 365 University, available only to verified students and faculty of postsecondary institutions. Other editions for installation on a Mac computer and collections of software and services geared toward business environments are available.

**Table 3.1**  Office 2019 One-Time Purchase Editions

| Edition | What It Includes |
|---|---|
| Office Home and Student 2019 | Word, Excel, and PowerPoint |
| Office Home and Business 2019 | Word, Excel, PowerPoint, and Outlook |
| Office Professional 2019 | Word, Excel, PowerPoint, Access, Publisher, and Outlook |

### Office 2019 System Requirements

Table 3.2 provides the standard system requirements for installing Office 2019. Generally, if the PC can successfully run Windows 10, then Office 2019 will also work on the same hardware, provided there is enough free disk space for the application software files.

**Table 3.2**  Office 2019 System Requirements

| Hardware Component | Requirement |
|---|---|
| Processor | 1 gigahertz (GHz) or faster processor |
| Memory | 2 gigabytes (GB) of RAM<br>Mac computer requires 4 GB RAM |
| Disk space | 3 GB available free space<br>Mac computer requires 6.0 GB available disk space |
| Operating system | Windows 10<br>Mac computer requires Mac OS X 10.10 or newer |
| Display setting | 1024 x 768 screen resolution or higher (PC)<br>1280 x 800 screen resolution or higher (Mac) |
| Browser | Current versions of Edge, Chrome, Firefox, or Safari |

### Starting an Application in the Office 2019 Suite

To start an application in the Office 2019 suite, double-click the application icon on the desktop or click the Start button and then click the application tile in the right pane on the Start menu. If a tile for the application is not on the Start menu, scroll down the list of application names in the left pane on the Start menu and then click the application name.

**Tutorial**

Opening a Blank Document

A quick way to find and start an application is to type the application name in the search text box on the taskbar and then click the name in the search results list. In many instances, you will see the desired application name appear in the search results list after typing only the first few letters in the name.

**1** Click in the search text box on the taskbar, type word, and then click *Word* in the right pane. Close any message or task pane that may appear about new features.

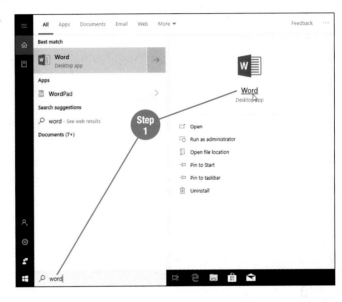

**Word 2019** is the application used to create, edit, and format text-based documents. When you start Word, the Start screen appears. The **Start screen** for Office applications shows the *Recent* option list used to open a document worked on recently.

**2** Click the Maximize button ☐ near the top right of the window if Word does not fill the screen. Skip this step if you see the Restore Down button ❐ since the window is already maximized.

**3** Compare your screen with the Word Start screen shown in Figure 3.1.

**4** Click the Start button and then click the Excel tile in the right pane on the Start menu. If an Excel tile is not available, scroll down the application list in the left pane, and then click *Excel*. Click the Maximize button if the Excel window does not fill the screen.

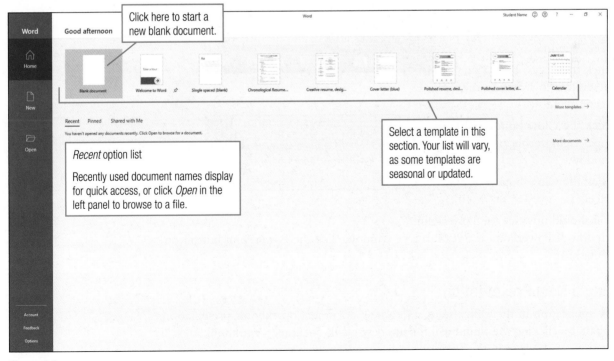

**Figure 3.1**

Word 2019 opens with the Word Start screen shown above.

Use **Excel 2019** when your focus is on entering, calculating, formatting, and analyzing numerical data.

⑤ Compare the Excel Start screen with the Word Start screen shown in Figure 3.1 on the previous page.

Each Office application offers templates in the template gallery on the Start screen. A template is a preformatted document, worksheet, presentation, or database. Click a thumbnail to preview a larger picture of the template. The first thumbnail is used to start a new blank document, workbook, presentation, or database.

⑥ Start PowerPoint by typing powerpoint in the search text box and then pressing Enter or by using the Start menu. If necessary, maximize the window.

Create slides for an oral or kiosk-style presentation that includes text, images, sound, video, or other multimedia in **PowerPoint 2019**.

⑦ Compare the PowerPoint Start screen with the Word Start screen shown in Figure 3.1.

⑧ Start Access using any method described in Step 1 or Step 4. If necessary, maximize the window and then compare the Access Start screen with the Word Start screen shown in Figure 3.1.

**Access 2019** is a database application in which you organize, store, and manage related data, such as information about customers or products.

**app tip**

Each Office application has its own color scheme. Word is blue, Excel is green, PowerPoint is orange, and Access is maroon.

## Switching between Office Applications

As learned in Chapter 1, you switch to another open application by clicking the application button on the taskbar. Point to a button on the taskbar to see a thumbnail appear above the button with a preview of the open document. If more than one document is open, the thumbnails make it easy to switch to the correct window.

⑨ Point to the Excel button on the taskbar to preview the thumbnail of the Excel Start screen and then click the button to switch to the Excel window.

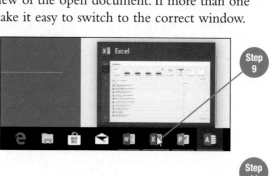

Step 9

⑩ Click the Access button on the taskbar to switch to the Access window.

⑪ Click the Close button (✕) at the upper right corner of the Access window.

Step 11

⑫ Click the Close button to close Excel.

⑬ You should now see the PowerPoint window. If PowerPoint is not the active window, click the PowerPoint button on the taskbar.

## Starting a New Presentation

For any application in the Office suite, start a new document, workbook, presentation, or database by clicking the thumbnail for the new blank document, workbook, presentation, or database in the Templates gallery on the Start screen.

14 Click *Blank Presentation* on the PowerPoint
Start screen.

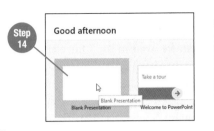

## Exploring the Ribbon

The ribbon appears along the top of each Office
application. Buttons within the ribbon are used to
access commands and features within the application.
The ribbon is split into individual tabs, with each tab divided into groups of related
buttons, as shown in Figure 3.2. Word, Excel, PowerPoint, and Access all have the File
and Home tabs as the first two tabs in the ribbon. The File tab is used to perform
document-level routines, such as saving, printing, and exporting. This tab is explored
in the next topic. The Home tab contains the most frequently-used features in each
application, such as formatting and editing buttons.

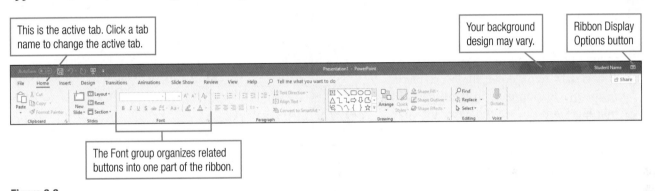

This is the active tab. Click a tab
name to change the active tab.

Your background
design may vary.

Ribbon Display
Options button

The Font group organizes related
buttons into one part of the ribbon.

**Figure 3.2**
The PowerPoint ribbon appears along the top of the application window, as it does for all Office applications.

Clicking the **Ribbon Display Options button** near the top right of the
window displays a drop-down menu to change the ribbon display from *Show Tabs
and Commands* to *Show Tabs*, which hides the command buttons until you click a tab,
or *Auto-hide Ribbon*, to display the ribbon only when you click along the top of the
window.

15 Click the Insert tab to view the groups
and buttons on the Insert tab ribbon for
PowerPoint.

16 Click the Word button on the taskbar
and then click *Blank document* on
the Start screen to start a new Word
document.

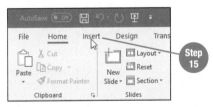

17 Click the Insert tab to view the groups and buttons on the Word Insert tab
ribbon.

18 Click the Design tab in Word and then review the buttons on the Word Design
tab ribbon.

19 Switch to PowerPoint and then click the Design tab.

20 Spend a few moments exploring the other tabs on the PowerPoint ribbon and
then close PowerPoint.

21 Spend a few moments exploring the other tabs on the Word ribbon and then
close Word.

## 3.2 Using the Backstage Area to Manage Documents

The File tab opens the **backstage area** in the Office applications. The backstage area is where you find file management options, such as *Open*, *Save*, *Save As*, *Print*, *Share*, *Export*, and *Close*. You also go to the backstage area within an application to start a new document. Other tasks performed in the backstage area include displaying information about a document, protecting a document, and managing the document properties and versions.

Each application in the Office suite provides options that can be personalized in the Options dialog box accessed from the left panel in the backstage area. Your Microsoft account and/or connected services and the background or theme for all the Office applications are managed in the Account backstage area.

*Note: If necessary, insert your storage medium into an empty port before starting this topic. Close the File Explorer window if it opens after you insert a removable drive.*

1 Start Word.

2 On the Word Start screen, click *Open* in the left panel.

3 Click the *Browse* option.

4 If necessary, scroll down the list of places in the Navigation pane in the Open dialog box until you can see the entry for your storage medium.

5 Click the entry in the Navigation pane for your storage medium.

6 Double-click *StudentDataFiles* in the Content pane.

7 Double-click *Ch3*.

The Word document **Cottage_rental_listing** is displayed in the Content pane. Within an Office application, the Open dialog box by default shows only file types that the active application can read.

8 Double-click *Cottage_rental_listing*.

Word document file names have a file extension *.docx*. Your system may be set to show the file extensions. In the next steps, you will use the Save As command to save a copy of the document in another folder.

9 If a yellow message bar displays indicating the document is shown in Protected view, click the Enable Editing button to turn off Protected view.

As a precaution, documents open in Protected view whenever the file originates from an online source, such as by downloading the file from a website or from a saved email attachment.

10 Click the File tab to display the backstage area.

When a document is open, the backstage area displays the Info panel with document properties for the active file and buttons to protect, inspect, and manage versions of the document.

**Watch & Learn**

**skills**

Open, Save As, Print, and Close a document

Export a document as a PDF file

**app tip**

The *Recent* option list on the Start screen displays by default the last 50 documents opened. If the file you want is in the *Recent* option list, click the file name to open the document.

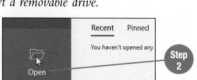

**Tutorial**

Opening a Document from the Recent Option List

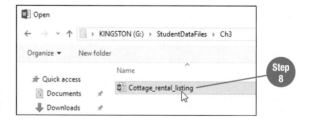

11  Click *Save As* in the left panel.

The Save As command is used to save a copy of a document in another location or to save a copy of the document in the same location but with a different file name.

12  With *This PC* already selected in the Save As backstage area, click the Up arrow next to the current file location G: > *StudentDataFiles* > Ch3 in the right panel.

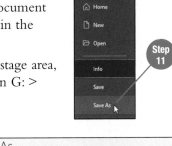

13  Click the Up arrow a second time.

14  Click the *ComputerCourse* folder name in the right panel.

15  Click the *CompletedTopics* folder name in the right panel and then click the More options hyperlink above the folder list.

16  In the Save As dialog box, click the New folder button on the Command bar, type **Ch3**, and then press Enter.

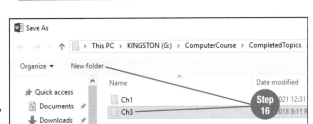

17  Double-click *Ch3*.

18  Click in the *File name* text box or drag to select the current file name, type CottageListing-YourName, and then press Enter or click the Save button.

19  Click in the document next to *List Date:* below the address *3587 Bluewater Road* (in the second column of the table) and then type the current date.

20  Click next to *Assigned Agent:* below the date (in the second column of the table) and then type your name.

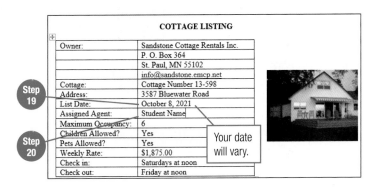

**quick steps**

**Open a Document**
1. On Word Start screen, click *Open* in the left panel.
2. Click *Browse*.
3. Navigate to drive and folder and then double-click file name.

**Use Save As to Save a Copy of a Document**
1. Click File tab.
2. Click *Save As*.
3. Navigate to desired location using folder names in right panel or by clicking Browse option.
4. If necessary, change file name.
5. Click Save.

**Tutorials**
Managing Documents
Managing Folders

**Oops!**
Typing mistake? Press Backspace to delete what you have typed if you make an error and then retype the text. You can also drag across text and then press Delete.

## Printing a Document

Display the Print backstage area when you want to preview and print a document. Before printing, review the document in the **Print Preview** panel of the backstage area shown in Step 23. The bottom of the Print Preview panel shows the number of pages needed to print the document, and navigation buttons are included to move to the next page and previous page in a multipage document.

In the Print panel, choose the printer at which to print the document in the *Printer* section, and modify the print settings and page layout options in the *Settings* section. When you are ready to print, click the Print button.

21 Click the File tab and then click *Print*.

22 Examine the document in the Print Preview panel, check the name of the default printer, and review the default options in the *Settings* section.

When print settings are changed, the options are stored with the document, so you do not need to change them again the next time you want to print.

23 Click the Print button to print the document.

The document is sent to the printer, and the backstage area closes.

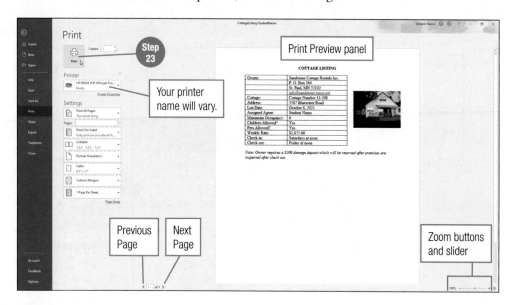

## Exporting a Document as a PDF File

Many people exchange documents as Portable Document Format (PDF) files via email or websites. A **PDF document** is a document saved in an open standard for exchanging files as electronic documents. The standard was developed by Adobe Systems. A PDF document can be viewed in any web browser.

The advantage of a PDF file is that the document looks and prints as it would in the application in which it was created but without having to open or install the source application. Export a Word document as a PDF if you need to send a document to someone who does not have Word installed on his or her computer.

24 Click the File tab and then click *Export*.

app tip

Change *Print All Pages* in the *Settings* section to print a selection of text only, the current page only, or specific pages or a range of pages. To print specific pages, change *Print All Pages* to *Custom Print* and then type the pages separated by commas in the *Pages* text box.

**Tutorials**
Previewing and Printing
Printing a Document

app tip

In Word, you can open a PDF document, and Word converts the PDF into an editable document.

**25** With *Create PDF/XPS Document* selected in the Export backstage area, click the Create PDF/XPS button.

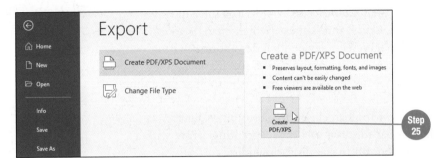

**26** Click the Publish button in the Publish as PDF or XPS dialog box.

By default, the PDF is created in the same drive and folder in which the Word document resides and with the same file name but with the file extension *.pdf*. The same name can be used for both the Word document and the PDF document because PDF files have a file extension different than a Word document.

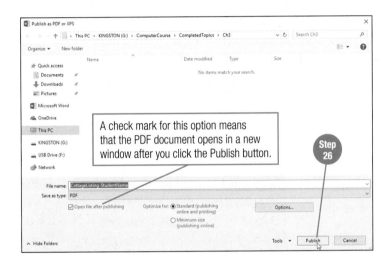

A check mark for this option means that the PDF document opens in a new window after you click the Publish button.

**27** The published PDF document opens in a reader app, such as Windows Reader or Adobe Reader, or in a browser window. If no application is associated with PDF documents, a message box similar to the one shown at the right opens with the prompt *How do you want to open this file?* In that case, click the OK button to open the PDF document in the selected option, *Microsoft Edge*.

**28** Close the window in which the PDF document opened to return to Word.

**29** Click the File tab and then click *Close*.

**30** Click the Save button when prompted in the message box asking if you want to save your changes. Leave Word open for the next topic.

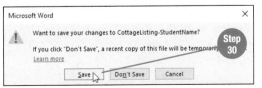

Close a document when you are finished editing, saving, printing, and publishing. A blank window displays when no documents are open.

**Tutorial**

Closing a Document and Closing Word

# 3.3 Customizing and Using the Quick Access Toolbar

The **Quick Access Toolbar** is at the upper left corner of each Office application window. With the default installation, the Quick Access Toolbar has the AutoSave option turned off and buttons to Save, Undo, and Repeat (which changes to Redo after Undo has been used). A touchscreen device includes a fourth button—the Touch/Mouse Mode button—that is used to optimize the spacing between commands for the mouse or for touch. Most people customize the toolbar by adding buttons that are used often.

To add a button to the Quick Access Toolbar, click the Customize Quick Access Toolbar button at the end of the toolbar (displays as a down-pointing arrow with a bar above) and then click the desired button in the drop-down list.

*Note: Skip this topic if the Quick Access Toolbar on the computer you are using already displays the New, Open, Quick Print, and Print Preview and Print buttons in Word, PowerPoint, and Excel. Skip any steps in which a drop-down list option already displays with a check mark, which means the button is already on the toolbar.*

**Watch & Learn**

**Hands On Activity**

— — — — — — Skills

Add buttons to the Quick Access Toolbar

Use buttons on the Quick Access Toolbar to perform commands

① With a blank Word document, click the Customize Quick Access Toolbar button.

② Click *New* in the drop-down list.

Options displayed with a check mark are already added to the toolbar. Buttons added to the Quick Access Toolbar are placed at the end of the toolbar.

③ Click the Customize Quick Access Toolbar button.

④ Click *Open* in the drop-down list.

⑤ Click the Customize Quick Access Toolbar button.

⑥ Click *Quick Print* in the drop-down list.

⑦ Click the Customize Quick Access Toolbar button.

⑧ Click *Print Preview and Print* in the drop-down list.

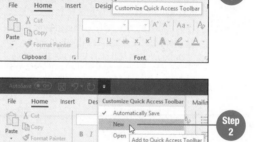
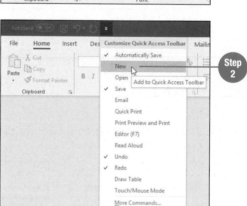

— — — — — app tip

Clicking a check-marked option removes the button from the Quick Access Toolbar.

— — — — — app tip

Right-click any button on the ribbon and then click *Add to Quick Access Toolbar* to add the button.

Touch-enabled devices also show a Touch/Mouse Mode button.

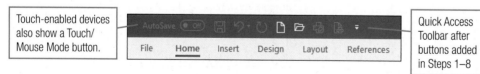

Quick Access Toolbar after buttons added in Steps 1–8

⑨ Click the Open button on the Quick Access Toolbar.

The Open backstage area opens with the *Recent* option list in the right panel.

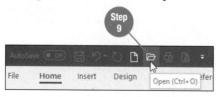

10 Click the file **CottageListing-StudentName** in the *Recent* option list.

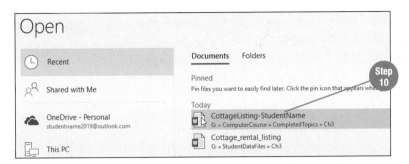

11 Click the Print Preview and Print button on the Quick Access Toolbar.

The Print backstage area opens.

12 Click the Back button (left-pointing arrow inside circle) to return to the document without printing.

13 Start PowerPoint.

14 Click *Blank Presentation* on the PowerPoint Start screen.

Notice the customized Word Quick Access Toolbar does not carry over to other Office applications.

15 Customize the Quick Access Toolbar in PowerPoint by adding the New, Open, Quick Print, and Print Preview and Print buttons.

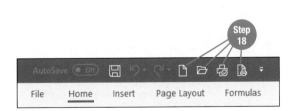

16 Close PowerPoint.

17 Start Excel and then click *Blank Workbook* on the Excel Start screen.

18 Customize the Excel Quick Access Toolbar to add the New, Open, Quick Print, and Print Preview and Print buttons.

19 Close Excel.

20 In the Word document, click the File tab and then click *Close*.

21 Click the New button on the Quick Access Toolbar.

A new blank document window opens. Leave this document open for the next topic.

---

**quick steps**

**Add a Button to the Quick Access Toolbar**
1. Click Customize Quick Access Toolbar button.
2. Click desired button in drop-down list.

**app tip**

Clicking the Quick Print button added to the Quick Access Toolbar automatically sends the current document to the printer using the active printer and print settings.

**app tip**

In all Office applications, the keyboard command Ctrl + F4 closes the current document and Alt + F4 closes the application.

**Tutorial**

Customizing the Quick Access Toolbar

## 3.4 Selecting Text or Objects, Using the Ribbon and Mini Toolbar, and Selecting Options in Dialog Boxes

Creating a document, worksheet, presentation, or database involves working with the ribbon to select options or perform commands. Some options involve using a button, list box, or gallery, and some commands cause a task pane or dialog box to open in which you select options.

In many instances, before you choose an option from the ribbon, you first select text or an object as the target for the action. Select text by clicking within a word, paragraph, cell, or placeholder or by dragging across the text you want to select. Select an object, such as an image or other graphic, by clicking on the object.

Selected objects display with a series of selection handles. A **selection handle** is a circle at the middle and/or corners of an object, or at the beginning and end of text on a touch-enabled device. Selection handles are used to manipulate the object or to define the selection area on a touch-enabled device. Table 3.3 provides instructions for selecting text using a mouse or touch and for selecting an object.

 **Watch & Learn**

 **Hands On Activity**

●────── Skills

Select text and objects

Perform commands using the ribbon and Mini toolbar

Display a task pane and dialog box

Choose options in a dialog box

**Table 3.3** Selecting Text and Objects Using the Mouse and Touch

| Selecting Text Using a Mouse | Selecting Text Using Touch | Selecting Objects |
|---|---|---|
| 1. Point at the beginning of the text or cell to be selected.<br><br>The pointer displays as I, called an *I-beam* in Word and PowerPoint, or as ⊕, called a *cell pointer* in Excel.<br><br>2. Hold down the left mouse button and drag to the end of the text or cells to be selected.<br><br>3. Release the mouse button.<br><br>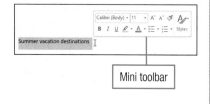<br><br>In some cases, a Mini toolbar displays when you release the mouse after selecting text. | 1. Tap at the beginning of the text to be selected. A selection handle appears below the text (displays as an empty circle).<br><br><br><br>2. Slide the selection handle right to move the insertion point to edit text. Alternatively, double-tap at the beginning of the text to select the first word and show a second selection handle, then slide the right selection handle to extend the selection.<br><br>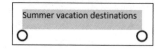<br><br>3. After releasing your finger, you can slide the left or right selection handle to redefine the area, if necessary.<br><br>4. To display the Mini toolbar, press and hold inside the selected text area. The toolbar appears when you release your finger and displays already optimized for touch.<br><br> | 1. Click anywhere on the object.<br><br>Selection handles appear at the ends, corners, and middle (depending on width and height) of each side of the object.<br><br>A Layout Options button also appears next to a selected object with options for aligning and moving the object with surrounding text.<br><br>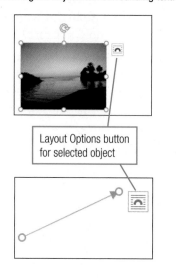 |

**1** In a blank Word document, type Summer vacation destinations and then press Enter.

**2** Type Explore the beaches of Florida and experience the Florida sunset with friends or family. and then press Enter.

**3** Select the title *Summer vacation destinations* to display the Mini toolbar.

The **Mini toolbar** appears near text after text is selected or with the shortcut menu when you right-click a selection. The toolbar contains frequently-used formatting commands for quicker access within the work area than using the ribbon.

If necessary, refer to the instructions in Table 3.3 for selecting text using the mouse or touch.

**4** Click the Bold button on the Mini toolbar.

**5** Click in the blank line below the sentence that begins with *Explore* to deselect the title.

**6** Click the Insert tab.

**7** Click the Pictures button in the Illustrations group.

This opens the Insert Picture dialog box.

**8** If necessary, scroll down the list of places in the Navigation pane. Click the entry in the Navigation pane for your storage medium.

**9** Double-click *StudentDataFiles*, double-click *Ch3*, and then double-click *FloridaSunset.jpg*.

The FloridaSunset image is inserted in the document and is automatically selected.

**10** Drag the selection handle at the lower right corner of the image until the image is resized to the approximate height and width shown below.

The mouse pointer changes shape to a double-headed diagonal arrow (⤢) when you point at the lower right corner of the image. Drag upward and to the left when you see this icon. The pointer changes shape to a crosshairs—a large, thin, black cross (✛)—while you drag the mouse. When you release the mouse, the selection handles reappear.

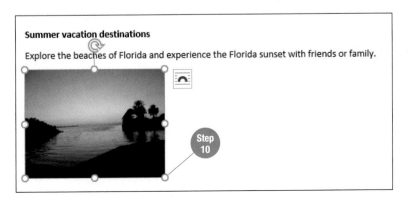

11  Click the Save button on the Quick Access Toolbar.

12  With the *Recent* option selected, click the Ch3 folder name associated with your storage medium and the folders » ComputerCourse » CompletedTopics.

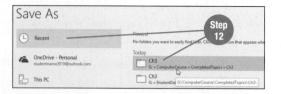

13  Click in the *File name* text box (displays the dimmed text *Enter file name here*), type VacationDestinations-YourName, and then press Enter or click the Save button.

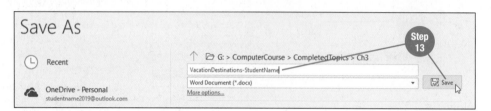

## Working with Objects, Contextual Tabs, and Dialog Boxes

An **object** is an image, shape, chart, or other item that can be manipulated separately from text or other objects around it. When an object is selected, a contextual ribbon tab appears. More than one contextual tab may appear. A **contextual tab** contains commands or options that are related to the type of object that is currently selected.

Some buttons in the ribbon display a drop-down gallery. A **gallery** displays visual representations of options for the selected item in a drop-down list or grid. Pointing to an option in a gallery displays a **live preview** of the text or object if the option is applied. Live previews let you see how formatting will look before applying an option. Use live preview to try out a few format options before making a choice, or click on the screen away from the list to leave the existing formatting intact.

14  With the image still selected, click the Picture Tools Format tab if the tab is not the active tab and then click the Corrections button in the Adjust group.

15  Point to the third option in the last row in the Corrections gallery. Notice the colors in the image deepen to show a more vibrant sunset when you point to the option.

16  Click the third option in the last row in the Corrections gallery to apply the *Brightness: 0% (Normal) Contrast: +40%* correction.

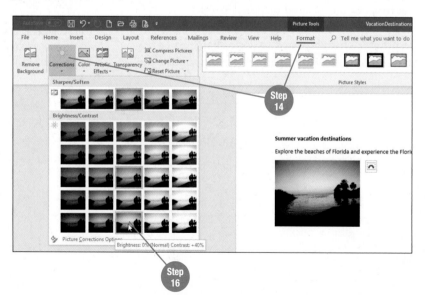

Some ribbon groups have a small button at the lower right corner of the group that is a diagonal downward-pointing arrow ( ). This button is called a **dialog box launcher**. Clicking the button causes a task pane or a dialog box to appear. A **task pane** appears at the left or right side of the window, whereas a **dialog box** opens in a separate window above (or, from the viewer's perspective, in front of) the document. Task panes and dialog boxes contain more options as buttons, lists, sliders, check boxes, text boxes, and option buttons.

**17** Click the dialog box launcher at the lower right corner of the Picture Styles group.

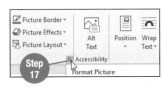

The Format Picture task pane opens at the right. You will work in a task pane in the next topic.

**18** Click the Close button (×) at the upper right corner of the Format Picture task pane to close the pane.

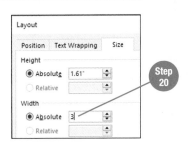

**19** Click the dialog box launcher at the lower right corner of the Size group.

The Layout dialog box opens with the Size tab active.

**20** Select the current value in the *Absolute* text box in the *Width* section and then type 3.

**21** Click the Text Wrapping tab and then click the *Square* option in the *Wrapping style* section.

**22** Click OK.

**23** With the picture still selected, position the pointer on top of the picture and then drag the picture up to the top of the document, releasing it when the green horizontal and vertical alignment guides show the picture aligned at the top and left margins.

An **alignment guide** is a colored horizontal or vertical line that appears when you are moving an object to help you align and place the object within the document boundaries or in relation to surrounding text or other nearby objects.

**24** Click the Save button on the Quick Access Toolbar. Leave the document open for the next topic.

Because the document has already been saved once, the Save button saves the changes using the existing file name and location.

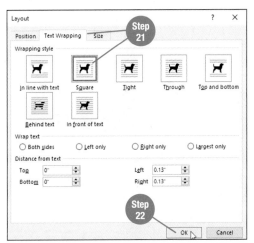

Alignment guides show the picture is aligned at the upper left margin.

## 3.5 Using the Office Clipboard and Formatting Using a Task Pane

The Clipboard group is standardized across Office applications. The buttons in the Clipboard group are Cut, Copy, Paste, and Format Painter. You used Cut, Copy, and Paste in Chapter 1 when you learned how to move and copy files and folders. Cut, Copy, and Paste are also used to move or copy text or objects.

**Format Painter** is used to copy formatting options from selected text or an object to other text or another object.

**1** With the **VacationDestinations–YourName** document still open, start PowerPoint and then click *Blank Presentation.*

**2** Click anywhere on *Click to add title* on the blank slide and then type Florida Sunset.

**3** Click the Word button on the taskbar.

**4** Select the sentence that begins with *Explore* below the title and then click the Copy button in the Clipboard group on the Home tab.

**5** Click the PowerPoint button on the taskbar to switch to PowerPoint and then click anywhere on *Click to add subtitle* on the slide to place an insertion point.

**6** Click the top part of the Paste button in the Clipboard group. Do *not* click the down-pointing arrow on the button.

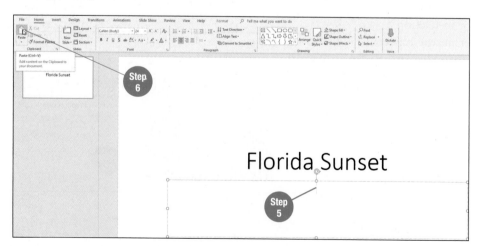

☁ **Watch & Learn**

●------ SKİLLS

Copy and paste text and an object

Copy and paste formatting options using Format Painter

Use a task pane to format a selected object

●------ app tip

Some buttons on the ribbon have two parts. Clicking the top or left of the button causes the default action to occur. Clicking the bottom or right of the button (arrow) displays a list of options to modify the action that occurs.

The selected text that was copied from Word is pasted on the slide in PowerPoint. Notice also the Paste Options button that appears below the pasted text.

**7**  Click the Paste Options button to display the Paste Options gallery.

Paste options vary depending on the pasted text or object. Buttons in the gallery allow you to change the appearance or behavior of the pasted text or object in the destination location.

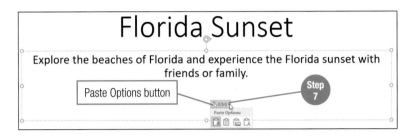

**8**  Click the Word button on the taskbar to switch to Word.

**9**  Click on the image to select it, click the Copy button, click the PowerPoint button on the taskbar, and then click the Paste button. (Remember not to click the down-pointing arrow on the Paste button.)

The pasted image is dropped onto the slide overlapping the text.

**10**  Click the Paste Options button. Notice that the paste options for a picture are different than the paste options for text.

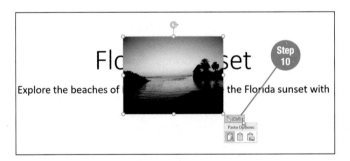

**11**  Click on white space away from the image to remove the Paste Options gallery.

**12**  If necessary, click to select the image object.

**13**  Drag the selected image below the text. Release the image when the orange guide shows the image is aligned with the middle of the text placeholders.

**14**  Click the Save button on the Quick Access Toolbar.

15 In the Save As backstage area, click *This PC* and then click *Browse*.

16 Navigate to the Ch3 subfolder in the folders › ComputerCourse › CompletedTopics on your storage medium.

17 Select the current text in the *File name* text box, type FloridaVacation-YourName, and then press Enter or click the Save button.

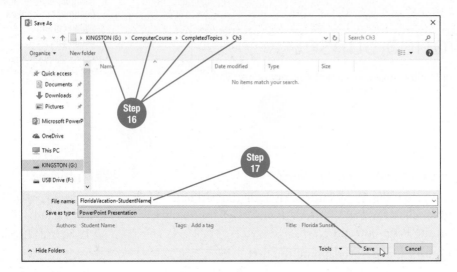

## Using Format Painter to Copy Formatting Options

Sometimes instead of copying text or an object, you want to copy formatting options. Format Painter copies to the Clipboard the formatting attributes for selected text or an object.

Click the Format Painter button to do a one-time copy of formatting options or double-click the button if you want to paste the formatting options multiple times. Double-clicking the Format Painter button turns the feature on until you click the button again to turn the feature off. A button that operates in an on or off state is called a **toggle button**.

18 Select the first occurrence of the word *Florida* in the subtitle on the slide.

19 Click the *Font Color* option box arrow (down-pointing arrow at the right of the Font Color button) in the Font group on the Home tab.

20 Click the *Purple* square (the last square in the *Standard Colors* section).

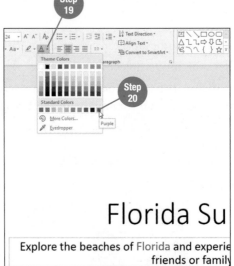

**Tutorials**

Applying Font Formatting Using the Font Group

Formatting with Format Painter

21  With the text still selected, click the Italic button in the Font group on the Home tab.

Step 21

22  With the text still selected, click the Format Painter button in the Clipboard group.

23  Drag the mouse pointer with the paintbrush icon across the second occurrence of the word *Florida* in the subtitle.

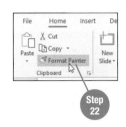

Step 22

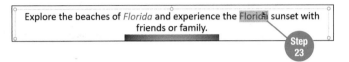

Explore the beaches of *Florida* and experience the Florida sunset with friends or family.

Step 23

**quick steps**

**Copy and Paste Formatting Options**
1. Select text or object with desired formatting applied.
2. Click Format Painter button.
3. Click object or drag to select desired text to which formatting is to be copied.

24  Click on white space away from the selected text to deselect the text.

25  Click the Save button on the Quick Access Toolbar.

26  Click on the image to select it and display the contextual tab.

27  Click the Picture Tools Format tab.

Step 27

28  Click the *Drop Shadow Rectangle* option in the Picture Styles gallery (fourth picture style option).

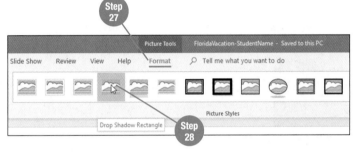

Drop Shadow Rectangle

Step 28

29  With the image still selected, click the Home tab, click the Format Painter button, and then click the title text *Florida Sunset* at the top of the slide.

30  With the Florida Sunset placeholder selected, click the dialog box launcher in the Drawing group on the Home tab to open the Format Shape task pane at the right side of the window.

31  Click the Size & Properties tab (last option) in the Format Shape task pane.

32  Click *Text Box* to reveal the options.

33  Click the *Vertical alignment* option box arrow (down-pointing arrow at the right of *Bottom*) and then click *Middle*.

34  Close the Format Shape task pane.

35  Save and then close PowerPoint. Leave the Word document open for the next topic.

**app tip**

You can also align text in a PowerPoint placeholder using the Align Text button in the Paragraph group on the Home tab.

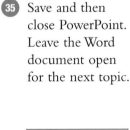

title aligned in the middle of the placeholder vertically at Steps 31–34

Step 31

Step 32

Step 33

## 3.6 Using the Tell Me Feature and Help Task Pane

 **Watch & Learn**

●------- Skills

A text box to the right of the Help tab on the ribbon containing the dimmed text *Search* is the **Tell Me** feature in Office 2019. Use the search text box to type an option and quickly access the command or feature directly from the *Tell Me* text box instead of navigating the ribbon tabs. You can also use the *Tell Me* text box to type a word or phrase and search resources in a Help task pane to learn how to use an option or feature. The **Smart Lookup** option in the Tell Me drop-down list opens a task pane at the right side of the window that lets you see a definition of the term or explore web resources related to the term.

Use Tell Me to locate an option

Use Tell Me to look up a definition

Use Help task pane to find information

1 With the **VacationDestinations-YourName** document still open, click to place the insertion point after the period in the sentence below the document title, press Enter, and then type Visit https://CA3.ParadigmEducation.com /FloridaTours for information on our Florida vacation packages.

2 Press Enter and then type Our exclusive Got2GoSunset package is our most popular booking.

3 Click to place the insertion point after the last *t* in *Got2GoSunset*.

Assume that Got2GoSunset is a trademarked name and you want to insert the trademark sign but do not know where the symbol feature is located.

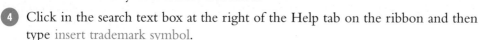

4 Click in the search text box at the right of the Help tab on the ribbon and then type insert trademark symbol.

5 Click *Insert a Symbol* in the drop-down list and then click the trademark symbol ™ in the Symbol palette that appears.

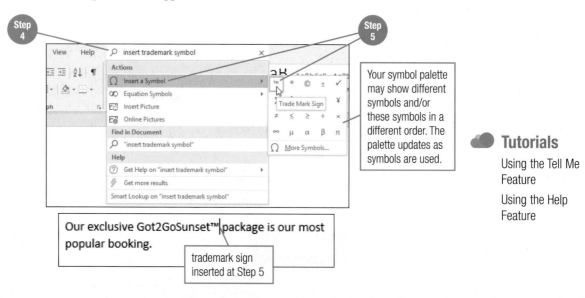

Your symbol palette may show different symbols and/or these symbols in a different order. The palette updates as symbols are used.

**Tutorials**

Using the Tell Me Feature

Using the Help Feature

Our exclusive Got2GoSunset™ package is our most popular booking.

trademark sign inserted at Step 5

6. Click in the search text box and then type *watermark*.

7. Click *Smart Lookup on "watermark"*.

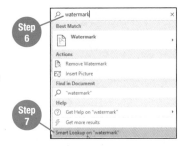

The Smart Lookup task pane opens at the right side of the window. On the Explore tab of the task pane are web links to resources related to watermark. On the Define tab, you can view a definition of the word *watermark*.

8. Click the Turn on button in the Smart Lookup task pane if a message displays to turn on intelligent services to let Office get web results for your highlighted text. Skip this step if no message about intelligent services displays in the task pane.

9. Click the Define tab in the Smart Lookup task pane and then read the definition for the word *Watermark* in the pane.

10. Close the Smart Lookup task pane.

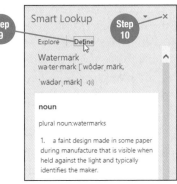

11. Click the Help tab and then click the Help button in the Help group.

A Help task pane opens at the right side of the window.

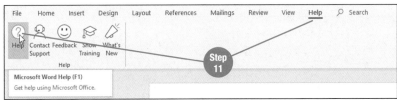

12. If necessary, click in the search text box in the Help task pane, type *watermark*, and then press Enter or click the search icon (magnifying glass).

13. Click the hyperlink to the <u>Add a watermark</u> help topic that displays in the Help task pane below the search text box.

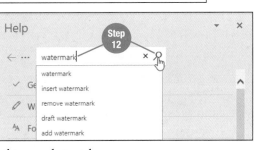

14. Read the steps for adding a watermark in the Help task pane and then close the Help task pane.

You can continue searching for other help information by typing a keyword or phrase in the search text box, or continue clicking hyperlinks in help topics displayed in the task pane.

15. Click in the search text box, type *watermark*, click *Watermark* in the drop-down list, and then click *SAMPLE 1* in the Watermark gallery.

16. View the SAMPLE watermark added to the document background, click the Save button on the Quick Access Toolbar, and then close the Word window. Click *No* when prompted to save the copied item.

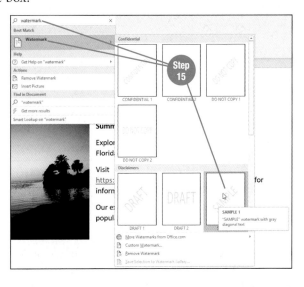

## 3.7 Using OneDrive for Storage, Scrolling in Documents, and Using Undo

 **Watch & Learn**

•———————— SKills

Save and open files to and from OneDrive

Navigate in longer documents

Use Undo

**OneDrive** is secure online storage available to individuals signed in with a Microsoft account (often referred to as *cloud storage*). You can save files to and open files from OneDrive, giving you the ability to access the files from any internet-connected device.

When working with longer documents, you will need to scroll the display or change the Zoom setting to see more or less text in the window. You will learn how to use the Zoom feature in the next topic. Undo reverses an action and is used often to restore a document when you make a mistake.

*Note: To complete this topic, you need to be signed in with a Microsoft account. If you do not have a Microsoft account, skip the OneDrive section in Steps 2 to 19 and proceed to Step 20 after opening the presentation.*

**Tutorial**
Using OneDrive for Storage

1. Start PowerPoint and then open the presentation **SpeechTechniques** from the Ch3 subfolder within the StudentDataFiles folder on your storage medium. If a yellow message bar displays indicating the presentation is open in Protected view, click the Enable Editing button to remove Protected view.

2. Click the File tab and then click *Save As*.

3. Click the *OneDrive - Personal* option in the Save As backstage area.

4. Click the <u>More options</u> hyperlink in the right panel.

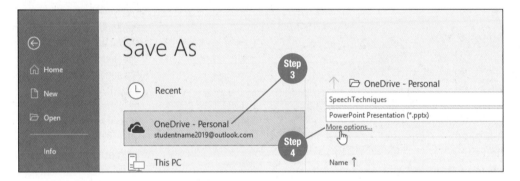

5. With the text in the *File name* text box in the Save As dialog box selected, type SpeechPres-YourName and then click the Save button. When the save is complete, if a message displays below the Quick Access Toolbar that AutoSave is On, click the Got it button to dismiss the message.

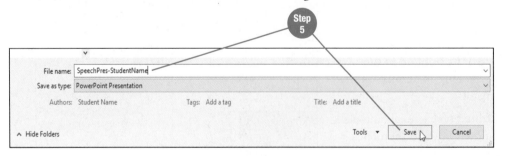

If you decide to use OneDrive for saving work, you should use folders to organize files. In the next steps, you will save the presentation to OneDrive a second time by creating a folder as you save the file.

⑥ Click the File tab and then click the *Save a Copy* option.

⑦ With your OneDrive account already selected in the Save As backstage area, click the *Browse* option.

⑧ In the Save As dialog box, click the New folder button on the Command bar.

Notice that the file saved in Step 5 appears in the Content pane.

⑨ Type CompletedTopics and then press Enter.

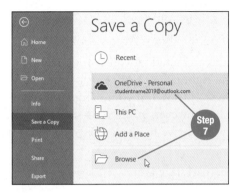

**quick steps**

**Save a File to OneDrive**
1. Click File tab.
2. Click *Save As*.
3. Click OneDrive account name.
4. Click folder in right panel or click <u>More options</u> hyperlink.
5. If necessary, navigate to desired folder in Save As dialog box.
6. Type name for file in *File name* text box.
7. Click Save.

⑩ Double-click *CompletedTopics*.

⑪ Click the New folder button on the Command bar, type Ch3, and then press Enter.

⑫ Double-click *Ch3*.

⑬ Click the Save button to save the presentation using the same name as before.

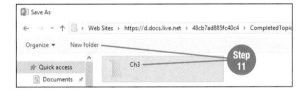

**app tip**

You may notice a one- or two-second delay when navigating folders in the Open or Save As dialog box when working on OneDrive. When that happens, wait a few seconds for the screen to update—do not click options or commands a second time.

You now have two copies of the presentation saved on OneDrive.

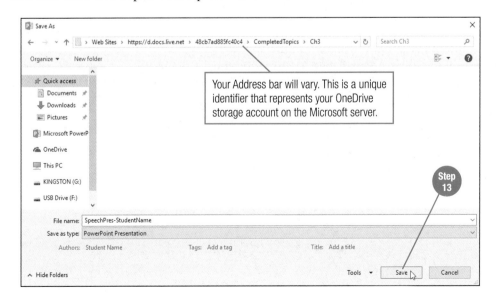

Your Address bar will vary. This is a unique identifier that represents your OneDrive storage account on the Microsoft server.

(14) Click the File tab and then click *Close*.

Depending on the speed of your internet connection, you may notice an Uploading to OneDrive message box while the file is uploaded to your OneDrive account.

(15) Click the Open button on the Quick Access Toolbar.

Because you just closed the presentation, you will notice the presentation appears twice in the *Recent* list in the right panel. Notice that your OneDrive account is associated with each of these entries. You could reopen the presentation from OneDrive using the *Recent* list; however, in the next steps, you will use Browse to practice opening a file from OneDrive in case a file you need is not in the *Recent* list for the Office application.

(16) Click *Browse* in the Open backstage area.

You will notice the file you want to reopen is already in the Content pane because the Quick Access feature shows recently opened presentations; however, you will practice navigating in case you ever need a file that has not been recently opened.

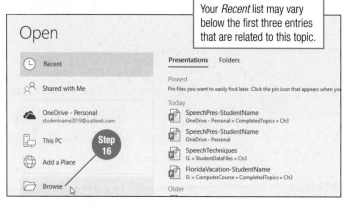

(17) Click *OneDrive* in the Navigation pane.

(18) In the Content pane, double-click the folders *CompletedTopics* and *Ch3*.

(19) Double-click **SpeechPres-YourName** in the Content pane.

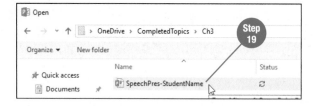

## Using the Scroll Bars

Larger files with content beyond the current view display in the application window with a horizontal and/or vertical scroll bar. A **scroll bar** has arrow buttons at the top and bottom or left and right ends used to scroll up, down, left, or right. A **scroll box** in the scroll bar between the two arrow buttons is also used to scroll by dragging the box up or down, or left or right. PowerPoint also has two buttons at the bottom of the vertical scroll bar that are used to navigate to the next or previous slide.

(20) Click the Next Slide button at the bottom of the vertical scroll bar to view the second slide in the presentation.

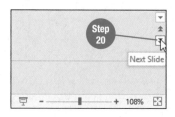

---

**21** Click the Next Slide button two more times to move to Slide 4.

**22** Click the up arrow button at the top of the vertical scroll bar repeatedly until you are returned to the first slide.

**23** Drag the scroll box at the top of the vertical scroll bar downward until you reach the end of the slides.

As you drag the scroll box downward, a ScreenTip displays the slide numbers and titles for the slides so you know when to release the mouse.

**24** Press Ctrl + Home.

Ctrl + Home is the universal keyboard shortcut for returning to the beginning of a file.

**25** Press Ctrl + End.

Ctrl + End is the universal keyboard shortcut for navigating to the end of a file.

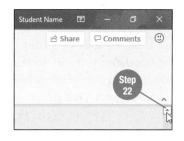

## Using Undo

The **Undo** command in all Office applications can be used to restore a document, presentation, worksheet, or Access object to its state before the last action that was performed. If you make a change to a file and do not like the results, immediately click the Undo button on the Quick Access Toolbar. Some actions, such as Save, cannot be reversed with Undo.

**26** Navigate to the first slide in the presentation.

**27** Select the title text *Speech Techniques*.

**28** Click the Bold button on the Mini toolbar or in the Font group on the Home tab.

**29** With the text still selected, click the Underline button on the Mini toolbar or in the Font group on the Home tab.

**30** Click on any part of the slide away from the selected text to deselect the title text.

**31** Click the Undo button on the Quick Access Toolbar. Do *not* click the down-pointing arrow on the button.

The underline is removed from the title text, and the text is selected.

**32** Click the Undo button a second time to remove the bold formatting.

**33** Click on the slide away from the selected text to deselect the text and then click the Save button on the Quick Access Toolbar.

**34** Close the presentation and then close PowerPoint.

# 3.8 Changing the Zoom Option and Screen Resolution

☁ **Watch & Learn**

Word, Excel, and PowerPoint display a **Zoom slider** bar near the bottom right corner of the window. Using the slider, you can zoom out or zoom in to view more or less of a document, worksheet, slide, or presentation. At each end of the slider bar is a **Zoom In** (plus symbol) and a **Zoom Out** (minus symbol) control that increases or decreases the magnification by 10 percent on each click.

● ━━━━━━━ Skills

Change the Zoom setting

View the screen resolution and change the setting if possible to match textbook illustrations

1. Start Excel and then open the workbook **CutRateRentals** in the Ch3 folder in StudentDataFiles. Click the Enable Editing button if a yellow message bar displays indicating the worksheet is opened in Protected view.

Notice the current Zoom setting near the bottom right corner of the Excel window, which is 100 percent if the setting is at the default option.

2. Click the Zoom In control (displays as a plus symbol).

3. Click the Zoom In control two more times.

The worksheet is now much larger in the display area with the magnification increased 30 percent.

4. Click the Zoom Out control (displays as a minus symbol).

5. Drag the Zoom slider left or right and watch magnification of the worksheet decrease or increase as you move the slider.

6. Drag the Zoom slider to the middle of the slider bar to change the zoom to 100 percent.

7. Click *100%* at the right of the Zoom In control.

● ━━━━━━━ Oops!

Having trouble using the Zoom slider on a touch device? Tap the controls or the percentage number to open the Zoom dialog box (Steps 7–8), or use the Zoom buttons on the View tab on the ribbon.

This opens the Zoom dialog box, where you can choose a predefined magnification, type a custom percentage value, or choose the *Fit selection* option to fit a group of selected cells to the window.

8. Click *75%* and then press Enter or click OK.

9. Click the View tab and then click the 100% button in the Zoom group.

Zoom In, Zoom Out, the Zoom slider, and the Zoom dialog box function the same in Word and PowerPoint.

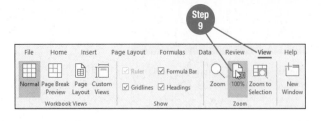

## Viewing and Changing Screen Resolution

The ribbon on your computer may show fewer or more buttons than the illustrations in this textbook, or some buttons may show icons only (no labels). The appearance of the ribbon is affected by the screen resolution (Figure 3.3 on the next page). **Screen resolution** refers to the number of picture elements, called *pixels*, that form the image

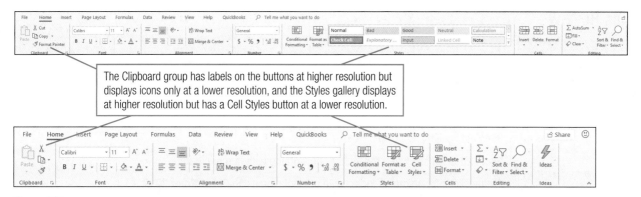

**Figure 3.3**

Shown is the Excel Home tab at 1920 x 1080 pixels (top) and at 1280 x 1024 pixels (bottom).

on the display. A pixel is a square with color values. Millions of pixels are used to render display images. Resolution is expressed as the number of horizontal pixels by the number of vertical pixels.

*Note: Check with your instructor before proceeding onward. Some schools do not allow the display properties to be changed. If necessary, perform Steps 10–17 on your home PC.*

10. Minimize the Excel window to display the desktop.

11. Right-click a blank area on the desktop and then click *Display settings*.

12. If the current setting for *Resolution* is *1920 x 1080*, skip to Step 14; otherwise, click the *Resolution* option box arrow and then click *1920 x 1080* if the option is available; otherwise, use the resolution option that displays with *(Recommended)*.

13. Click the Keep changes button.

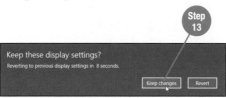

14. Close the Settings app window.

15. Click the Excel button on the taskbar to view the Excel window at the new screen resolution.

16. If you changed your screen resolution, examine the screen to see if you like the new setting. If the interface is not of good quality, repeat Steps 10 to 14 to restore the resolution option to its original setting. The setting that displays with *(Recommended)* is usually the best choice for your device.

You do not need to change the screen resolution to the same setting used for the images in this textbook. Just be aware that some illustrations may not match exactly what is on your display if your resolution is at a different setting.

17. Close the **CutRateRentals** worksheet and then close Excel.

---

**quick steps**

**Change Zoom Magnification**
Click Zoom In or Zoom Out button or drag Zoom slider to desired setting.
OR
1. Click zoom percentage.
2. Select desired zoom option in Zoom dialog box.
3. Click OK.

---

**app tip**

Screen resolution is an operating system setting. A higher resolution uses more pixels and means the image quality is sharper or clearer. It also means more content can be displayed in the viewing area.

---

**Tutorial**

Changing the Zoom

## Topics Review

| Topic | Key Concepts |
|---|---|
| **3.1 Starting and Switching Applications, Starting a New Presentation, and Exploring the Ribbon** | The Office suite is sold as a one-time purchase edition or as a subscription plan called **Office 365**.<br><br>**Word 2019** is the application used for text-based documents.<br><br>An Office application starts with the **Start screen**, which shows a list of recently opened files and a templates gallery.<br><br>**Excel 2019** is used when the focus is on working with numeric data.<br><br>Create slides for an oral or kiosk-style presentation using **PowerPoint 2019**.<br><br>**Access 2019** is a database application used to organize, store, and maintain related data, such as information about customers or products.<br><br>All applications display the ribbon along the top of the window, which contains buttons divided into tabs as well as groups for commands and features within the software.<br><br>The **Ribbon Display Options button** lets you control whether the ribbon shows tabs only, tabs with commands, or no ribbon. |
| **3.2 Using the Backstage Area to Manage Documents** | The **backstage area**, accessed from the File tab on the ribbon, is where you perform document-level routines, such as *Open*, *Save*, *Save As*, *Print*, *Share*, *Export*, and *Close*.<br><br>Use the *Open* option on the Word Start screen to navigate to a document not in the *Recent* option list.<br><br>Use the *Save As* option to save a copy of a document in another location or in the same location but with a different file name.<br><br>Use the Print backstage area to preview a document in the **Print Preview** panel to see how the document will look with the current print settings before printing.<br><br>A **PDF document** (Portable Document Format) is a document saved in an open standard for exchanging electronic documents that opens as it would appear if printed.<br><br>Use the *Export* option in the backstage area to publish a Word document in PDF format. |
| **3.3 Customizing and Using the Quick Access Toolbar** | The **Quick Access Toolbar** at the top left of each application window contains buttons for frequently used commands. The toolbar can be customized for each application.<br><br>Add to or remove buttons from the Quick Access Toolbar by clicking the Customize Quick Access Toolbar button and then clicking the desired option in the drop-down list. |
| **3.4 Selecting Text or Objects, Using the Ribbon and Mini Toolbar, and Selecting Objects in Dialog Boxes** | Each circle at the middle and/or corners of a selected object is called a **selection handle**.<br><br>The **Mini toolbar** appears near selected text or with the shortcut menu containing buttons for formatting the selected text or object.<br><br>Dragging a selection handle resizes a picture, shape, chart, or other item referred to as an **object**.<br><br>A **contextual tab** appears when an object is selected with command buttons related to the object.<br><br>A **gallery** is a drop-down list or grid with visual representations of options. A **live preview** shows the text or object as it would look if the option on which the mouse is resting is applied.<br><br>At the lower right of a group on the ribbon is a **dialog box launcher** that displays a task pane or **dialog box** when clicked.<br><br>Dialog boxes and task panes provide additional options for the related ribbon group as buttons, lists, sliders, check boxes, text boxes, and option buttons. A **task pane** opens at the left or right side of the document window, whereas a dialog box opens in a separate window on top of the document text.<br><br>A colored horizontal or vertical line called an **alignment guide** appears when moving an object to help you place and align the object with text, margins, or other nearby objects. |

*continued…*

| Topic | Key Concepts |
|---|---|
| **3.5 Using the Office Clipboard and Formatting Using a Task Pane** | The Clipboard group on the ribbon is standardized across all Office applications. |
| | Use Cut, Copy, and Paste buttons to move or copy text or objects. |
| | A Paste Options button appears when you paste text or an object with options for modifying the paste action. |
| | The **Format Painter** button in the Clipboard group is used to copy formatting options from selected text or an object to other text or object(s). |
| | A button, such as the Format Painter button, that operates in an on or off state is called a **toggle button**. |
| **3.6 Using the Tell Me Feature and Help Task Pane** | Click in the **Tell Me** text box that displays *Tell me what you want to do* and then type a term or option to locate the option directly from the Tell Me list or to look up other resources about the feature in a task pane. |
| | Click the **Smart Lookup** option in the Tell Me drop-down list to open a task pane with web links and definitions for the term. |
| | Click the Help tab on the ribbon and then click the Help button to open the Help task pane to find information about Office features. |
| **3.7 Using OneDrive for Storage, Scrolling in Documents, and Using Undo** | **OneDrive** is cloud storage where you can save files that can be accessed from any other internet-connected device by selecting your OneDrive account in the Open and Save As backstage areas. |
| | A horizontal or vertical bar called a **scroll bar** with arrow buttons and a **scroll box** is used to navigate documents, workbooks, presentations, or database tables that do not fit in the window. |
| | The **Undo** feature is used to reverse an action performed, restoring a document to its previous state. |
| **3.8 Changing the Zoom Option and Screen Resolution** | Use the **Zoom In**, **Zoom Out**, **Zoom slider**, and Zoom dialog box in Word, Excel, and PowerPoint to increase or decrease the magnification setting. |
| | Screen resolution refers to the number of horizontal and vertical pixels (squares with color values) used to render the display image. Millions of pixels are used for generating an image. |
| | The **screen resolution** setting for your PC or mobile device affects the display of the ribbon. Change the screen resolution using the *Display settings* option on the desktop shortcut menu. |

 **Topics Review**

# Chapter 4

# Using OneNote for Windows 10

OneNote is a note-taking application referred to as a *digital notebook*. OneNote is like an electronic binder with notes organized by dividers. Note-taking software can store, organize, search, and share notes of any type, including typed notes, handwritten notes, web pages, images, documents, presentations, worksheets, messages, appointments, contacts, and more. A OneNote notebook can collect everything you want to keep track of for a subject or topic in one place.

OneNote notebooks are stored on OneDrive so that you can access the notes from any internet-connected device. Another advantage to storing a notebook on OneDrive is that you can share it with others and allow each participant to edit notes. For group projects, OneNote is used to collaborate and share ideas, research, and content.

OneNote was originally available as a desktop application bundled with the Office suite with Word, Excel, PowerPoint, and Outlook. With the release of Office 2019, Microsoft converted OneNote to an app that is included with Windows 10.

In this chapter, you will learn how to use the OneNote app for Windows 10 to add sections, pages, and content to the default OneNote notebook. You'll also learn how to tag notes, create a new notebook, and share a notebook with others.

If the OneNote app is not on your Windows 10 device, download and install the app from the Windows Store before starting this chapter. If you are using Windows 7, you will need to use OneNote Online. Go to https://www.onenote.com and sign in with your Microsoft account. Windows apps can be updated at any time, and the screens shown and instructions provided may vary.

## Read & Learn

## Learning Objectives

**4.1** Add a section and a page in a notebook

**4.2** Add a note, apply color to a note, and add web content

**4.3** Insert an image, a document, and embed a PDF copy of a presentation into a notebook

**4.4** Tag a note and search tags

**4.5** Search notes

**4.6** Create a new notebook and share a notebook

## Content Online

The online course includes additional training and assessment resources.

## 4.1 Adding Sections and Pages in a Notebook

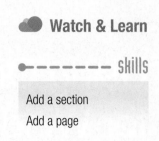

**Watch & Learn**

●------- Skills

Add a section

Add a page

**OneNote** is the note-taking app automatically included on a Windows 10 PC. The app is also available for download from the Apple or Google app stores for use on Mac or Android devices. A **OneNote notebook** is organized into sections, which are accessed by tabs. Think of a section like a divider you would use in a binder to organize notes by subject, topic, or category. Within each section you add pages. Notes and other content are added on a page. You can add as many sections and pages as you like to organize a notebook. Content can be placed anywhere on a page in a section.

1 Click the Start button and then click the OneNote tile or scroll down the list of applications in the left pane and then click *OneNote*. If necessary (for example, if this is the first time you have started OneNote), click the Get Started button. By default, OneNote opens automatically signed in with the Microsoft account associated with Windows 10. If an introduction screen appears, click *Skip*, or if a *What's New in OneNote* box appears, click the option to dismiss the message box.

The OneNote window (Figure 4.1) appears with a notebook open. By default a personal notebook for your Microsoft account is created and opened. The notebook is stored on OneDrive. The OneNote app works with notebooks stored on OneDrive.

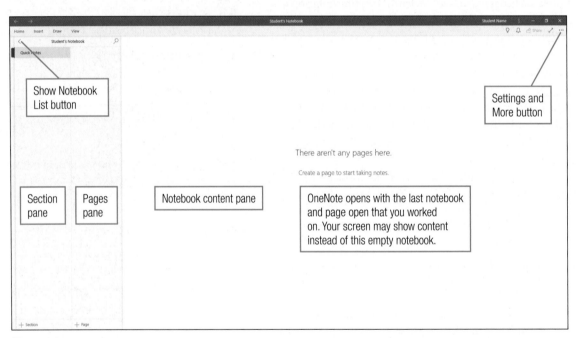

**Figure 4.1**

The OneNote window is shown with a notebook open and the Quick Notes tab active. If this is the first time you have used OneNote, the personal notebook is empty as shown here.

A section is like a divider in a binder. Sections organize content by category, topic, or subject. Each section can hold multiple pages. To create a new section in a notebook, click the **Section button** at the bottom of the Section pane, type a name for the section, and then press Enter. For each new section created, a blank page opens with an insertion point in the page title area. Type a title for the page and then press Enter.

Click the **Page button** at the bottom of the Page pane to add a new page within the active section. Type a title for the page and then press Enter. Starting to work with a digital notebook is like getting a new binder for a project and putting dividers and blank pages into it. You can add new sections and pages as you go along, but creating a few sections and pages lets you get started.

**2** Click the Section button (bottom of Section pane).

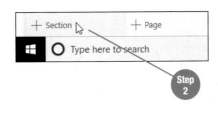

A new tab is added to the notebook with a section name added by default and with the default text selected. Type a name for the section and press Enter. In a digital notebook to keep track of school content, one way to organize the notebook is to create a section for each course.

**3** Type Computers and then press Enter.

**4** Click above the dimmed line in the in the page title area in the Notebook content pane, type Web Info and then press Enter.

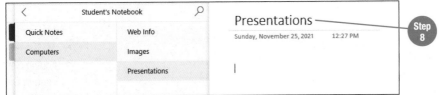

**5** Click the Page button (bottom of Pages pane).

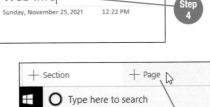

**6** Type Images as the page title and then press Enter.

**7** Click the Page button.

**8** Type Presentations as the page title and then press Enter.

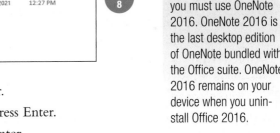

**9** Click the Section button, type Law, and then press Enter.

**10** Click in the page title area, type Family Law, and then press Enter.

**11** Click the Section button, type English, and then press Enter.

**12** Click in the page title area, type Assignments, and then press Enter.

To add a new page to an existing section, first make the desired section active by clicking the section tab, and then use the Page button to add a page.

**13** Click the Law tab to make Law the active section.

**14** Click the Page button.

**15** Type Civil Law and then press Enter.

Leave OneNote open for the next topic.

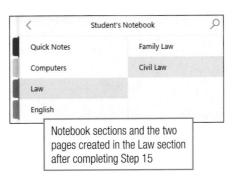

Notebook sections and the two pages created in the Law section after completing Step 15

# 4.2 Adding Notes to Pages and Inserting Web Content

**Watch & Learn**

To add a note to a page, make the page active and start typing to add text at the location of the insertion point. If you don't see an insertion point, click on the page. Click to start typing a new note anywhere on the current page. OneNote content is placed inside a box referred to as a **note container**. A selection handle, which can be used to select text that you want to format, appears on the left side of the note container when the pointer is resting on the note text. A four-headed white arrow pointer appears when you point to the top gray bar on the note container. To move a note somewhere else on the page, drag the note container when you see the four-headed white arrow.

**1** With OneNote open and the Civil Law page active, click the English tab in the Section pane.

**2** With Assignments the active page and an insertion point positioned below the page title, type Reflection paper due by end of week four.

**Skills**

Add a note to a page

Apply highlight color to text in a note

Add a web link in a note

Copy and paste web content into OneNote

**app tip**

If you have a touch-enabled device, you can handwrite notes or draw illustrations. Options on the Draw tab let you change ink color and convert ink to text or ink to a shape. The Math option recognizes and solves equations in a Math pane.

**3** Click beyond the right edge of the first note container on the same page and then type Child literature assignment is a group presentation.

**4** Point to the gray bar along the top of the note container. When you see the four-headed white arrow pointer, drag the note container to the top of the page below the first note as shown in the image at the right.

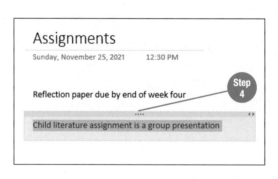

**Tutorial**

Adding Content to a Page in OneNote

**5** Click to place an insertion point within the note text in the first note container.

A selection handle appears left of the note container. You can use the selection handle to select text, or drag across text you want to select to apply a formatting option.

**6** Point to the selection handle that appears (  ) at the left of the note container for the first note and click when you see the mouse pointer change to the four-headed white arrow pointer.

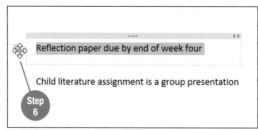

**app tip**

To delete a note, select the note text using the selection handle or the gray bar at the top of the note container and then press the Delete key on the keyboard.

The text Reflection paper due by end of week four is selected. A Highlight tool is used to apply color highlighting to notes just as you would use a highlighter to highlight important points while reading a textbook. Click the **Highlight button** to apply the default yellow highlighting to the selected text, or use the down-pointing arrow on the button to choose a different highlight color.

7 Click the Home tab if the ribbon is not currently visible and then click the Highlight button. Do not click the down-pointing arrow on the button.

One reason people like to use OneNote is that the app can be used as a receptacle for organizing information found online that you want to keep for later use. In the next steps, you will add a web link in a note and copy and paste content found on a web page into OneNote.

8 Click the Computers tab in the Section pane and then click the Web Info page in the Page pane.

9 With an insertion point active below the Web Info page title, type https://twitter.com/Techmeme and then press Enter twice.

OneNote automatically formats web addresses as hyperlinks.

10 Type Techmeme provides daily summaries of leading technology stories and tweets headlines with links throughout the day. and then click in a blank area away from the note container.

11 Move the mouse pointer over https://twitter.com/Techmeme and click the hyperlink to view the Twitter page in Microsoft Edge or other browser window.

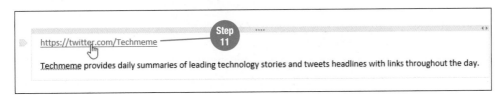

12 Scroll down to view the latest tweets by Techmeme.

13 Click in the Address bar in the browser window to select the web address or drag to select the web address, type https://en.wikipedia.org/wiki/Tim_Berners-Lee, and then press Enter.

Use the standard copy and paste tools to select and copy text from a web page and paste into a notebook. When you paste text copied from a web page, OneNote automatically includes the source URL below the pasted text.

14 Select the first three paragraphs of text below the title "Tim Berners-Lee" on the Wikipedia page and then press Ctrl + C.

Recall from Chapter 1 that the universal shortcut keys for copy and paste are Ctrl + C (copy) and Ctrl + V (paste).

15 Close the browser window to return to OneNote.

16 Click on the Web Info page away from the existing note container to start a new note and then press Ctrl + V to insert the copied text from the Wikipedia page.

17 Move the note container below the first note, if necessary. Leave OneNote open for the next topic.

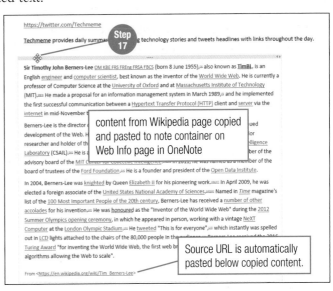

# 4.3 Inserting Files into a Notebook

Images, documents, presentations, workbooks, contacts, email messages, appointment details, and more can be added to a notebook as a link to the source content or as a PDF copy of the content. Some people use a OneNote notebook as a repository to collect all the data related to a course, subject, or other topic. The advantage to assembling content in one place is that you no longer need to keep track of web links or web pages separately from documents and other notes for a subject or project.

Items are inserted into OneNote using buttons on the Insert tab. A document can be inserted as an attachment that links to the source file, or a PDF printout of the content can be embedded into the notebook. Once inserted, you can add annotations with your own notes.

**1** With OneNote open and the Web Info page active, click the Images page in the Computers section.

**2** Click the Insert tab, click the Pictures button, and then click *From File*.

**3** In the Open dialog box, navigate to the Ch4 folder in StudentDataFiles on your storage medium.

**4** Double-click *AnalogCptr_1950s*.

**5** With an insertion point positioned in the note container below the image, type Analog computer from the 1950s.

**6** Click the Page button, type Documents, and then press Enter.

**7** Click the File button on the Insert tab.

**8** In the Open dialog box with the Ch4 folder in StudentDataFiles active, double-click the Word document *Tech_Wk1_SocialMedia*.

**9** With an insertion point positioned in the note container below the Word document icon and file name, type Week 1 Assignment and then click on the page away from the note container.

A link to the source document is inserted in the note. The document can be opened from OneNote by double-clicking the Word document icon.

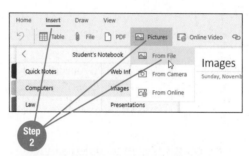

Step 2

Analog computer from the 1950s

Step 5

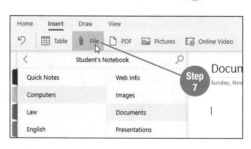

Step 7

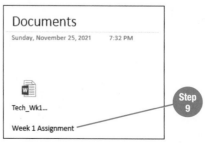

## Documents
Sunday, November 25, 2021    7:32 PM

Tech_Wk1...

Week 1 Assignment

Step 9

☁ **Watch & Learn**

●------ SKILLS

Insert an image

Insert a link to a document

Embed a copy of a presentation

☁ **Tutorials**

Inserting Files into a Notebook

Inserting a Picture in a OneNote Page

●------ app tip

You can also drag and drop an image onto a OneNote page from a File Explorer window.

●------ app tip

Use the Audio button on the Insert tab to have OneNote create a file by recording audio. Use this option to record a lecture in class or an interview as long as you are sitting close enough for the microphone on your device to pick up the voice. You could also use this at home to create audio notes rather than typing note text.

10  Double-click the document icon.

Word opens and the document linked to the OneNote page is opened in Protected view.

11  Scroll down to review the Word document and then close Word.

12  Click Presentations in the Page pane.

13  Click the PDF button on the Insert tab.

14  In the Open dialog box, navigate to the Ch4 folder in StudentDataFiles on your storage medium.

15  Double-click the PDF file *Tech_Wk1*.

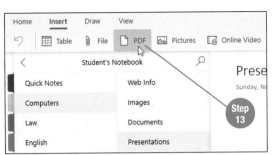

Use the **Insert Printout from PDF** option to embed a copy of content on the current page that has been saved in PDF format. This option allows you to add your own notes with the content. For example, you can save a PowerPoint presentation with lecture notes for a class as a PDF file to embed the slides in OneNote. Once the slides are inserted, type your own notes directly on or next to a slide.

16  Scroll up to the top of the Presentations page. Notice that an icon with a link to the source file is included above the embedded slides.

17  Scroll down to the last slide on the page.

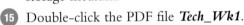

18  Double-click to start a new note on the slide below the last bulleted list entry and then type Jasmine described using an app that let her use virtual reality to visit the Great Wall of China! She talked about using virtual reality to visit places all over the world.

19  Select the text in the note container, click the Home tab, click the *Font Size* option arrow and then click *12*.

20  Click the Highlight button to apply yellow highlight to the text. Leave OneNote open for the next topic.

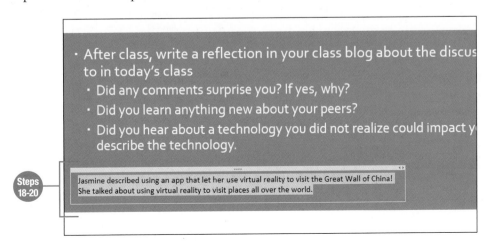

●━ ━ ━ ━ quick steps

**Insert an Image**
1. Activate desired page.
2. Click Insert tab.
3. Click Pictures button.
4. Click *From File*.
5. Navigate to drive and folder, then double-click image file name.
6. If necessary, type descriptive note.

**Insert a File as an Attachment**
1. Activate desired page.
2. Click Insert tab.
3. Click File button.
4. Navigate to drive and folder, then double-click file name.
5. If necessary, type descriptive note.

**Embed a PDF Copy of File Contents**
1. Activate desired page.
2. Click Insert tab.
3. Click PDF button.
4. Navigate to drive and folder, then double-click file name.
5. If necessary, type descriptive note.

## 4.4 Tagging Notes

A **tag** is a category assigned to a note. The tag allows you to identify the note later as an item flagged for a to-do list, as important, as a question, as critical content, as a contact, as an address, or as a phone number. OneNote includes a **Tag This Note button** on the Home tab with predefined tags in a drop-down list. Use the Create New Tag option to open the Create a Tag pane at the right side of the window to create your own tag.

1. With OneNote open and the Presentations page active with the last slide visible in the embedded presentation, double-click next to the slide title *REFLECTION BLOG* to start a new note.

2. Type *Blog post due by Friday.*

3. Click at the beginning of the note text *Blog post due by Friday* in the note container, click the Tag This Note button on the Home tab, and then click *To Do.*

OneNote inserts a check box left of the insertion point. The To-Do tag check box is interactive. When you complete the task, click the check box to insert a check mark to mark the activity completed.

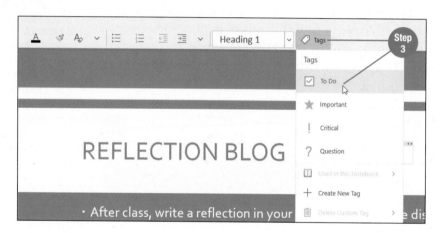

4. Click the Web Info page in the Page pane.

5. Double-click in white space near the top of the page next to the content pasted from Wikipedia to place an insertion point.

6. Click the Tag This Note button, and then click *Important.*

A gold star tag is inserted inside a new note container on the page.

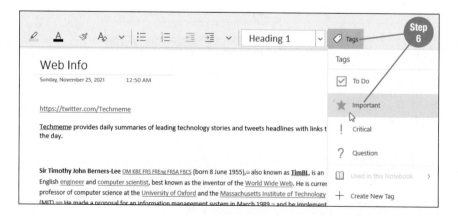

**Watch & Learn**

Skills

Tag a note

**Tutorial**

Tagging Notes

⑦ Type Use this information in Assignment 1. If necessary, drag the top gray bar to move the note container to the right side of the page as shown in the image below if it overlaps other text.

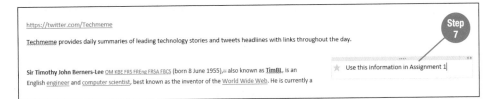

⟟╶╶╶ **quick steps**

**Add Tag to a Note**
1. Position insertion point at start of note text.
2. Click Tag This Note button.
3. Click desired tag.

⑧ Click the Images page in the Page pane.

⑨ Click at the beginning of the caption text below the photo of the analog computer to place an insertion point, click the Tag This Note button, and then click *Question*.

A purple question mark appears next to the caption text as shown in the image below.

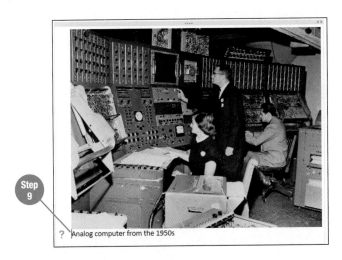

⑩ Click the Documents page in the Page pane.

⑪ Click to place an insertion point at the beginning of the text *Week 1 Assignment* in the note container, click the Tag This Note button, and then click *To Do*.

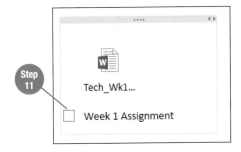

⑫ Click the English tab in the Section pane.

⑬ Click to place an insertion point at the beginning of the text *Reflection paper due by end of week four*, click the Tag This Note button, and then click *Critical*.

The Critical tag inserts a red exclamation mark next to the note text.

## 4.5 Searching Notes and Tags

An advantage to using an electronic versus a paper-based notebook is the ability to search all notebook pages for a keyword or phrase and instantly locate each occurrence. You can also search notebooks by a tag, such as the To-Do tag, to locate tagged notes. Click the magnifying glass icon next to the notebook name above the Page pane to open the search pane and then type the search keyword or phrase in the search text box, or click the tag you want to locate. OneNote begins displaying matches to the search text as you type. Matches are highlighted on the active page and other pages and in the search results list on the Pages tab or the Tags tab. Click the Pages tab or the Tags tab and then click an entry in the search results list to display the page where the text or tag appears. Click the Close button (displays as ×) to close the Search pane when finished.

**Watch & Learn**

—————— Skills

Search notes
Close a notebook

**Tutorial**

Searching Notes

1. With OneNote open, click the Law tab in the Section pane. If necessary, click to place an insertion point below the Family Law page title.

2. Click the Insert tab and then click the PDF button.

3. Double-click *FamilyAndLawAssgnt*. If you don't see the document, navigate to the Ch4 folder in StudentDataFiles on your storage medium.

4. With an insertion point below the embedded PDF printout, type Draft copy of first assignment, and then click in white space away from the note container.

5. Click the Search icon.

6. Type Berners-Lee in the search text box and then press Enter.

7. If necessary, click the Pages tab below the search text box and then click the Web Info page in the search results list to display the page and view the highlighted text entries.

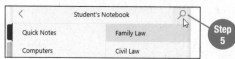

Step 5

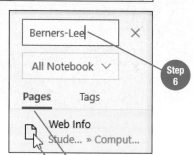

Step 6

Step 7

Sir Timothy John **Berners-Lee** OM KBE FRS FREng FRSA FBCS (born 8 June 1955), also known as **TimBL**, is an English engineer and computer scientist, best known as the inventor of the World Wide Web. He is currently a professor of Computer Science at the University of Oxford and at Massachusetts Institute of Technology (MIT). He made a proposal for an information management system in March 1989, and he implemented the first successful communication between a Hypertext Transfer Protocol (HTTP) client and server via the internet in mid-November the same year.

**Berners-Lee** is the director of the World Wide Web Consortium (W3C), which oversees the continued development of the Web. He is also the founder of the World Wide Web Foundation and is a senior researcher and holder of the 3Com founders chair at the MIT Computer Science and Artificial Intelligence Laboratory (CSAIL). He is a director of the Web Science Research Initiative (WSRI), and a member of the advisory board of the MIT Center for Collective Intelligence. In 2011, he was named as a member of the board of trustees of the Ford Foundation. He is a founder and president of the Open Data Institute.

In 2004, **Berners-Lee** was knighted by Queen Elizabeth II for his pioneering work. In April 2009, he was elected a foreign associate of the United States National Academy of Sciences. Named in *Time* magazine's list of the 100 Most Important People of the 20th century, **Berners-Lee** has received a number of other

"Inventor of the World Wide Web" during the 2012

OneNote highlights matches to the searched text on each page in the search results list.

peared in person, working with a vintage NeXT

ed "This is for everyone", which instantly was spelled

eople in the audience. Berners-Lee received the 2016

first web browser, and the fundamental protocols and

algorithms allowing the Web to scale".

From <https://en.wikipedia.org/wiki/Tim_Berners-Lee>

8  Select *Berners-Lee* in the search text box, type analog, and then press Enter.

9  Click *Images* on the Pages tab in the search results list to display the page with the photo of the analog computer.

10  Select *analog* in the search text box, type blog, and then press Enter.

11  Click *Presentations* on the Pages tab in the search results list to display the page with the embedded presentation and, if necessary, scroll down to the last slide with the blog note text.

12  Select *blog* in the search text box, type assign, and then press Enter

13  Click *Assignments* on the Pages tab in the search results list.

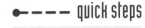

**quick steps**

**Search Notes**
1. Click Search icon (magnifying glass).
2. Type search keyword or phrase.
3. Click pages in search results list.
4. Click Close button when finished.

In this search, multiple pages have matches to the search text. Notice the search text is matched to part of the word *Assignments* in a page name as well as text in notes on the page. Typing a partial word is helpful if you think you may have notes with inconsistent spellings or forms of a word.

**app tip**

To search for a tag, delete search text in the search text box and click the tag you want to locate in the drop-down list that appears. The Tags tab becomes active automatically in the search results list with pages that contain the tag.

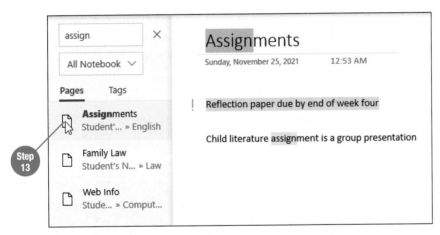

14  Click each of the pages in the search results list to view the matches to the search text.

By default, all notebooks are searched. Use the down-pointing arrow next to *All Notebooks* below the search text box to narrow the search scope to *Current Notebook*, *Current Section*, or *Current Page*.

15  Click the down-pointing arrow next to *All Notebooks* below the search text box.

16  Click *Current Section: Computers*.

Notice the search results list now displays two pages instead of four.

17  Click the Close button (displays as ✕) next to the search text box to close the Search pane and restore the Section pane and Page pane. Leave OneNote open for the next topic.

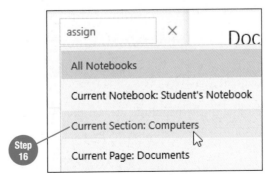

## 4.6 Creating a New Notebook and Sharing a Notebook

Some people like to organize all their notes for all purposes within one notebook (the default notebook file), using sections and pages to create an organizational structure. Others prefer to create separate notebooks to organize notes. For example, you may want to maintain separate notebooks for home, work, and school items. To create a new notebook, display the Notebooks list using the **Show Notebook List button**, click the **Notebook button** at the bottom of the Notebook pane, type a name for the new notebook, and then click the Create Notebook button.

A shared notebook is used to collaborate with others when working on a group project at school or work. You may also want to share a notebook with family members or friends to share ideas for an event or trip. In a shared notebook, members of the group can each post research, links, images, ideas, or other notes in one place. Click the **Share button** to open the Share pane and send an email invitation to each participant.

*Note: Check with your instructor for the name of the classmate with whom you will share the notebook created in this topic. In Step 12, you will need the Microsoft account email address for the classmate. If necessary, you can practice sharing the notebook with a friend or relative.*

1. With OneNote open, click the Show Notebooks List button (left-pointing arrow next to the notebook name below the ribbon).

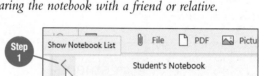

2. Click the Notebook button at the bottom of the Notebook pane.

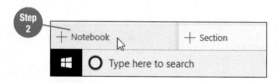

3. With the insertion point positioned in the *Notebook name* text box in the Create a notebook dialog box, type MyElectives-xx. Substitute your first and last initials for *xx*.

4. Click the Create notebook button.

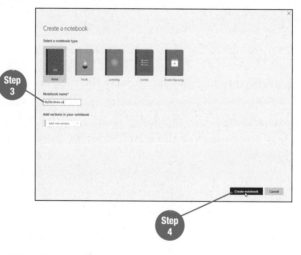

OneNote creates the new notebook and makes the notebook active.

5. Click the Section button to add a section in the new notebook.

6. Type Child Lit and then press Enter.

7. Click above the dimmed line in the page title area in the Notebook content pane, type Children's Literature, and then press Enter.

8  Insert the PowerPoint presentation **ChildLitPres** on the page using the File button on the Insert tab and type The Butter Battle Book by Dr. Seuss as note text below the file icon.

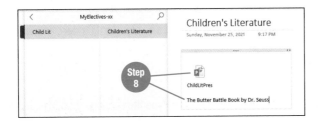

9  Add a second section named *Film* with the page title *Film Genres*. Insert the Word document **ApNowReflectionPaper** as a file and type Paper on Apocalypse Now movie as note text below the file icon.

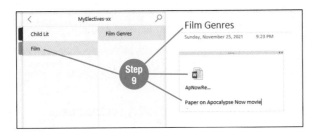

You decide to share the notebook with a classmate taking the same electives as you, so you can each add notes to the notebook.

10  Click the Share button near the upper right corner of the window below the Title bar.

The Share pane opens at the right side of the OneNote window.

11  Click in the *Type email addresses* text box and then type the email address for the classmate with whom you have been partnered for this topic. If necessary, use the email address for a friend or relative.

12  Click the Share button.

The *This notebook shared with* section of the Share pane is updated with the name(s) of people with whom the notebook was successfully shared.

13  Close OneNote.

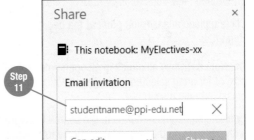

### Optional

14  Open a browser window, navigate to https://outlook.live.com, and sign in if you are not automatically signed in.

15  View the message with subject text similar to *MyElectives-xx has been shared with you*, and click the View in OneDrive button to open the notebook in OneNote Online. Dismiss any messages that appear about new features.

16  Add a note of your choosing to one of the pages and then close the browser window.

●━ ━ ━ ━ quick steps

**Create a New Notebook**
1. Click Show Notebook List button.
2. Click Notebook button.
3. Type name for new notebook.
4. Click Create notebook button.

**Share a Notebook**
1. Click Share button.
2. Type email address.
3. Click Share button.

☁ **Tutorial**
Sharing a Notebook

●━ ━ ━ ━ ━ ━ app tip

By default, each person with whom you share a notebook can make changes to the notebook. To share a notebook and only allow the other person to view the content, click the permissions option box arrow (displays *Can edit*) and then click *Can view*.

## Topics Review

| Topic | Key Concepts |
|---|---|
| **4.1 Adding Sections and Pages in a Notebook** | **OneNote** is a note-taking application referred to as a *digital notebook*.<br><br>A **OneNote notebook** is organized into sections (like dividers in a binder), with each section holding one or more pages.<br><br>When you start the OneNote app for Windows 10, the default notebook for the signed in user opens. All OneNote notebooks are stored on OneDrive.<br><br>To create a new section, click the **Section button**, type the section name, type a page title, and then press Enter.<br><br>To add a new page to a section, make it active if necessary, click the **Page button**, type the page title, and then press Enter. |
| **4.2 Adding Notes to Pages and Inserting Web Content** | Click on a page where you want a note to appear and then type the note text. Each note on a page is stored within a **note container**, which is a box that surrounds the note text.<br><br>Drag the gray bar along the top of the note container to move a note.<br><br>Select text in a note container using the selection handle at the left of the note container.<br><br>Use the **Highlight button** on the Home tab to add color to notes, similar to using a highlighter to emphasize text in a textbook.<br><br>A web address is automatically converted to a hyperlink in a note container.<br><br>A portion of a web page can be copied and pasted into a notebook. |
| **4.3 Inserting Files into a Notebook** | A OneNote notebook can be used as a repository to collect all the files related to a course, subject, or other topic.<br><br>Buttons on the Insert tab are used to insert an image, a file as an icon linked to the source document, or a file embedded on a page with the **Insert Printout from PDF** option. |
| **4.4 Tagging Notes** | A **tag** is a category assigned to a note. Tags are useful to identify notes that you want to flag for later review or follow-up. OneNote includes the **Tag This Note button** on the Home tab with a list of predefined tags, such as *Important* or *Question*.<br><br>Click to place an insertion point at the beginning of note text and then add the desired tag from the Home tab. |
| **4.5 Searching Notes and Tags** | OneNote searches all pages in all open notebooks for a keyword or phrase typed in the search text box, or for a tag selected in the Tags section of the search list. Click the Search icon (magnifying glass) to begin a search.<br><br>OneNote highlights matches to the search keyword or phrase on the active page and displays pages with matches on the Pages tab in the search results list in the Search pane.<br><br>Narrow the scope of a search by changing the *All Notebooks* option to *Current Notebook*, *Current Section*, or *Current Page*. |
| **4.6 Creating a New Notebook and Sharing a Notebook** | Display the Notebooks list by clicking the **Show Notebook List button**.<br><br>Click the **Notebook button**, type a name for a notebook, and then click the Create notebook button to create a new notebook for organizing notes.<br><br>Click the **Share button** to open the Share pane in which you type the email address for each person that you want to share the notebook with. |

Topics Review

# Communicating and Scheduling Using Outlook

Chapter **5**

Outlook is often referred to as a **personal information management (PIM)** application. A PIM application lets you organize items such as email messages, appointments or meetings, events, contacts, to-do lists, and notes. Reminders and flags help you remember and follow up on activities.

In the workplace, Outlook is often used with an Exchange server, which allows employees within the organization to easily share calendars, schedule meetings, and assign tasks with one another. You can use Outlook on your home PC or mobile device to connect to the mail server operated by your ISP and manage your messages. Outlook also lets you organize your time, activities, address book, and to-do list.

In this chapter, you will learn how to use Outlook for creating and sending email messages; scheduling appointments, events, and meetings; creating contacts; and maintaining a to-do list.

If you do not have access to the Outlook application on your desktop or laptop PC, you can use the free, web-based version by signing in with your Microsoft account at Outlook.com. You can use the same features on Outlook.com to practice the skills taught in this chapter; however, the screens shown will not match what you see in the browser, and the step-by-step instructions will vary because the web application uses a different interface. Where possible, instructions or marginal notes direct you to the method used at Outlook.com. In some cases, a feature is not available in the web-based edition, meaning you will have to skip some instructions.

 **Read & Learn**

## Learning Objectives

**5.1** Create, send, read, reply to, and forward email messages

**5.2** Attach a file to a message, delete a message, and empty the Deleted Items folder

**5.3** Preview, open, and save a file attachment

**5.4** Schedule an appointment and an event

**5.5** Schedule an appointment that repeats at fixed intervals, and edit an appointment

**5.6** Schedule and accept a meeting request

**5.7** Add and edit contact information

**5.8** Add, update, and delete a task

**5.9** Search Outlook messages, appointments, contacts, and tasks

### Content Online

The online course includes additional training and assessment resources.

## 5.1 Using Outlook to Send Email

**Electronic mail (email)** is communication between individuals by means of sending and receiving messages electronically. Email is the business standard for communication in the workplace. Individuals also use email to communicate with relatives and friends around the world. While text messaging is popular for brief messages between individuals, email is still used to send longer messages.

### Setting Up Outlook

The screen that you see when you start **Outlook** for the first time depends on whether a prior version of Outlook existed on the computer you are using. Outlook can transfer information from an older version of the application to a new data file. If no prior data file exists, Outlook will present a dialog box to set up a new email account. By default, Outlook inserts the Microsoft account email address you used to sign in on the device you are using. If you see the message shown in Figure 5.1a, type your email address if the email address shown is not the correct address. Click the Connect button to instruct Outlook to set up a new email profile. Type your password if a Windows Security dialog box appears and then click OK. When the message displays that the account was successfully added (Figure 5.1b), click Done.

*Note: The instructions in this chapter assume Outlook has already been set up and that you are connected to the internet with an always-on connection at school or at home. If necessary, connect to the internet and sign in to your email account before starting the topic activities. Check with your instructor for assistance if you are not sure how to proceed.*

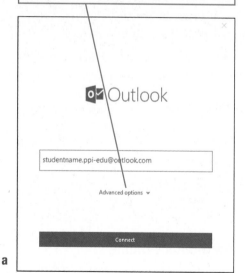

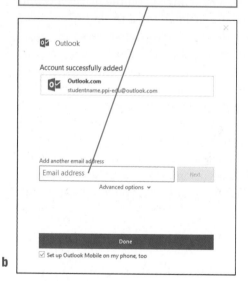

Click *Advanced options* if you want to set up your email account manually by entering the incoming and outgoing mail server addresses provided by your ISP for a non-Microsoft email account.

Type another email address here and click Next if you want to set up Outlook to show messages from more than one email account. Click Done when finished.

a    b

**Figure 5.1**

(a) The startup dialog box is shown when Outlook cannot locate an email profile from a prior version of Outlook on the device. (b) The dialog box after an account is successfully added is shown.

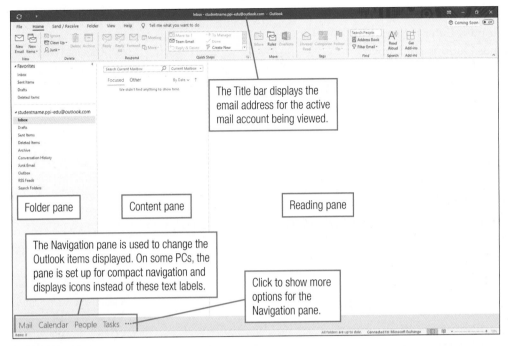

**Figure 5.2**
The Outlook window starts with the Inbox folder in Mail shown. Outlook checks for new messages automatically when the program is started.

Once Outlook has been set up with an email account, the Outlook window appears (Figure 5.2). By default, **Mail** is the active component. The mail folder shown when Outlook starts is **Inbox**. Messages in the Content pane are shown with the newest message received at the top of the message list. The **Focused Inbox** is turned on by default. In this view, Inbox messages are separated into two tabs: Focused and Other. The Focused tab displays the most important emails. All other messages are accessible on the Other tab. Messages such as newsletters or machine-generated emails are automatically placed on the Other tab. You can tailor the placement of messages from one tab to the other to meet your needs. For example, if a message appears in the Other list that you want in Focused, right-click the message header and then click *Move to Focused* so that future messages from the same sender will appear in the Focused list.

The left pane, called the *Folder pane*, is used to switch the display to another mail folder. At the bottom of the Folder pane is the Navigation pane, used to navigate to another Outlook component, such as Calendar. The right pane, called the *Reading pane*, displays the contents of the selected message.

## Creating and Sending a Message

Click the **New Email button** in the New group on the Home tab to start a new email message. Type the recipient email address in the *To* text box, type a brief description in the *Subject* text box, and then type your message text in the white message text box. Click the Send button when finished.

*Note: Check with your instructor for instructions on whom you should exchange messages and meeting requests with for this chapter. Your instructor may designate an email partner for each person or allow you to choose your email partner. If necessary, send messages to yourself.*

●------ app tip

Click the View tab, click the Folder Pane button in the Layout group, and choose *Normal, Minimized,* or *Off* to change the Folder pane. The Reading Pane button in the same group is used to display the Reading pane at the right or bottom of the screen or to turn off the Reading pane.

🌐 **Tutorial**

Creating and Sending a Message

**1**   Click in the search text box on the taskbar, type Outlook, and then press Enter to start Outlook. If you are using Outlook.com, open a browser window, navigate to https://outlook.live.com, and sign in to your Microsoft account if you are not automatically signed in.

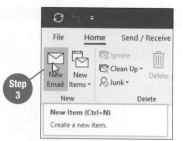

Step 3

**2**   Dismiss any messages that may pop up with information about new features in Outlook.

**3**   Click the New Email button in the New group on the Home tab. In Outlook.com, click <u>New message</u> in the browser window.

**4**   Type the email address for the recipient in the *To* text box.

For a message to be sent to multiple recipients, type a comma (,) or a semicolon (;) between email addresses in the *To* text box.

**5**   Press the Tab key twice or click in the *Subject* text box.

**6**   Type Social Media Project and then press Enter.

**7**   With the insertion point positioned at the top of the message text box, type the following text:

Hi [Recipient's Name], [press Enter twice]

I think we should do our project on Pinterest.com. Pinterest is a virtual pinboard where people pin pictures of things they have seen on the internet that they want to share with others. [press Enter twice]

What do you think?

**8**   Press Enter twice at the end of the message text and then type your name as the sender.

**9**   Click the Send button.

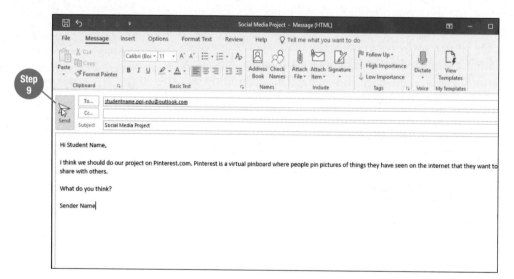

Step 9

## Replying to a Message

New messages appear at the top of the message list in the Content pane with message headers that show the sender, subject, time, and first line of message. Click to select a message header and reply directly from the Reading pane using the **Reply button**. Replying from the Reading pane is called an **inline reply**. You can also double-click the message header in the Content pane to open the message in a Message window.

**Tutorial**

Replying to a Message

**10** Click the Send/Receive All Folders button on the Quick Access toolbar to update the Content pane. Skip this step if you can already see the message sent to you by a classmate or yourself from Step 9.

**11** Click to select the message header for the message received from Step 9, if the message is not already selected, and then read the message text in the Reading pane.

**12** Click the Reply button in the Reading pane.

**13** Type the reply message text shown in the image below and then click the Send button.

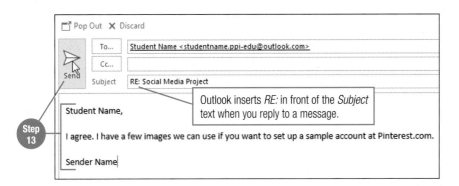

Outlook inserts *RE:* in front of the *Subject* text when you reply to a message.

## Forwarding a Message

Forward a message if you want someone else to receive a copy of a message you have received. Click the **Forward button**, type the email address for the person to whom you want to forward the message, type a brief explanation, and then click Send. Think carefully before forwarding a message to be certain that the original sender would not object to another person reading the message without his or her permission. If in doubt, do not forward a message.

**14** With the message header for the message selected in Step 11 still active, click the Forward button in the Reading pane.

**15** Type the email address for another classmate in the *To* text box.

**16** Click in the message text box above the original message text, type the text shown in the image below, and then click the Send button.

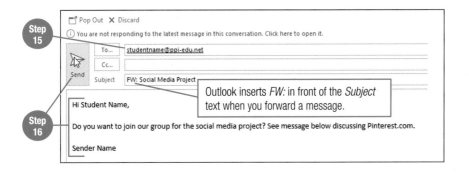

Outlook inserts *FW:* in front of the *Subject* text when you forward a message.

## 5.2 Attaching a File to a Message and Deleting a Message

Files are often exchanged between individuals via email. To attach a file to an email message, use the **Attach File button** in the Include group on the Message tab. The recipient of an email message with a file attachment can choose to open the file from the mail server or save it to a storage medium to open later.

Messages that are no longer needed should be deleted to keep your mail folders a manageable size. You can delete a message in the Inbox folder if you replied to the message because you can view the original text with your reply in the Sent Items folder. Set aside a time each day or week to clean up your Inbox by deleting messages.

**1** With Outlook open and Inbox the active folder, click the New Email button.

**2** Type the email address for the recipient in the *To* text box.

As you begin typing an email address, Outlook suggests email addresses that match what you type based on people you have messaged recently in the *Recent People* list or at some other time in the *Other Suggestions* list. This feature is referred to as **AutoComplete**. Rather than type the entire email address, you can click the correct recipient in the list. People in the workplace who use Outlook connected to a mail server called an *Exchange server* can send a message to someone within the organization by typing the name of the recipient instead of the full email address.

**3** Press the Tab key twice or click in the *Subject* text box.

**4** Type Picture for Pinterest and then press Enter.

**5** With the insertion point positioned in the message text box, type the following text:

Hi [Recipient's Name], [press Enter twice]

Attached is a picture we can put on Pinterest.

**6** Press Enter twice at the end of the message text and then type your name as the sender.

**7** Click the Attach File button in the Include group on the Message tab in the message window.

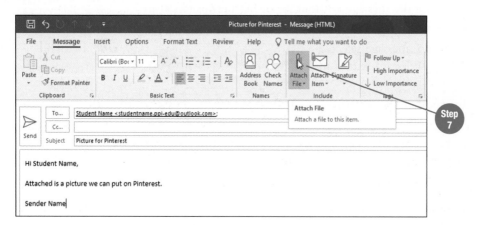

**8** Click *Browse This PC*.

The *Recent Items* list from the Attach File button displays file names for files you have worked on recently that you can attach to a message with one click.

**9** In the Insert File dialog box, navigate to the StudentDataFiles folder on your storage medium and then double-click *Ch5*.

**Watch & Learn**

● ━ ━ ━ ━ ━ ━ ━ **Skills**

Attach a file to a message

Delete a message

Empty Deleted Items folder

**Tutorial**

Attaching a File to a Message

**10** Double-click the image file *MurresOnIce.jpg*.

Outlook adds the file to an *Attached* area below the *Subject* text box.

**11** Click the Send button.

Over time, your mail folders (Inbox and Sent Items) become filled with messages that are no longer needed. To delete a message, select the message header in the Content pane and then click the Delete button in the Delete group on the Home tab. Deleted messages are moved to the **Deleted Items** folder where the message can be restored if needed. Periodically empty the Deleted Items folder to permanently delete the messages.

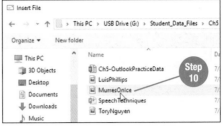

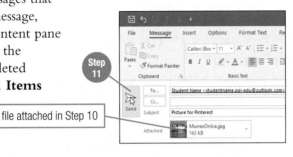

file attached in Step 10

**12** Click the Send/Receive All Folders button on the Quick Access Toolbar to update the Inbox. Skip this step if you can already see the message sent to you by a classmate or yourself in Step 11.

**13** Click *Sent Items* in the Folder pane.

**14** If necessary, click to select the message header for the message with the picture attached that you sent to a classmate or yourself in this topic.

**15** Click the Delete button in the Delete group on the Home tab.

**16** Click *Deleted Items* in the Folder pane.

Notice the message you deleted appears in the Content pane.

**17** Right-click *Deleted Items* in the Folder pane and then click *Empty Folder*.

**18** Click Yes in the Microsoft Outlook message box asking whether you want to continue to permanently delete everything in the Deleted Items folder.

**19** Click *Inbox* in the Folder pane.

# 5.3 Previewing File Attachments and Using File Attachment Tools

When you receive an email message with a file attached, you can preview, open, print, save, remove, or copy the file attachment from the Reading pane or from a message window. Click the down-pointing arrow next to the file name to select an action in the option list, or click the file name in the Reading pane to preview the attachment. During preview, the message text disappears, replaced with the contents of the attached file, and the Attachment Tools Attachments tab becomes active. Some files cannot be viewed within the Reading pane. In those instances, double-click the file name to open the file attached to the message.

●――――――― skills

Preview a file attachment

Open a file attachment

Save a file attachment

1. With Outlook open and Inbox the active folder, click the message header for the message with the subject *Picture for Pinterest*.

2. Click the file **MurresOnIce.jpg** in the Reading pane. Do *not* click the down-pointing arrow at the right of the file name.

●――――――― app tip

If you are using Outlook .com, clicking the file name in the message pane previews the file in a separate pane. Click the Close button (displays as ×) in the preview pane when finished viewing the image. An Attachment Tools ribbon tab is not available in Outlook.com. Instead, you have two options: Download and Save to OneDrive.

Outlook removes the message text and displays in the Reading pane the picture attached to the message. Notice also that the Attachment Tools Attachments tab is active on the ribbon.

3. Click *Back to message* at the top of the Reading pane to restore the message text.

When an attached message is previewed in the Reading pane, the message text is removed.

4. Click the New Email button and then type the email address for the recipient in the *To* text box.

5. Click in the *Subject* text box and then type Presentation for Business Class.

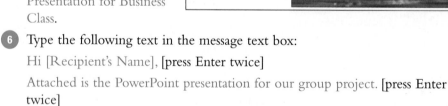

6. Type the following text in the message text box:

   Hi [Recipient's Name], [press Enter twice]

   Attached is the PowerPoint presentation for our group project. [press Enter twice]

   [Sender Name]

7. Click the Attach File button in the Include group on the Message tab and then click *Browse This PC*.

8. In the Insert File dialog box, with the Ch5 folder in StudentDataFiles the active folder, double-click the file **SpeechTechniques.pptx**.

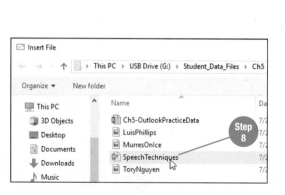

9. Click the Send button.

10  Click the Send/Receive All Folders button on the Quick Access Toolbar to update the Inbox folder if the message sent to you in Step 9 is not in your Inbox.

11  Click the message header for the message with the subject *Presentation for Business Class*.

12  Click the down-pointing arrow at the right of the file **SpeechTechniques.pptx** in the Reading pane and then click *Preview*.

13  Scroll down the Reading pane to the last slide in the presentation.

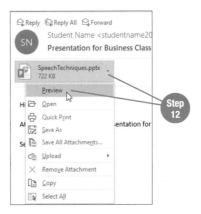

14  Click the Open button in the Actions group on the Attachment Tools Attachments tab.

PowerPoint starts with the attached file open. The presentation opens in **Protected View**, which allows you to read the file content in the source application; however, editing is not permitted until you click the Enable Editing button on the Message bar.

15  Close PowerPoint to return to Outlook.

16  Click the Save As button in the Actions group on the Attachment Tools Attachments tab.

17  In the Save Attachment dialog box, navigate to the CompletedTopics folder on your storage medium and then create a new folder named *Ch5*.

18  Double-click the *Ch5* folder and then click the Save button.

19  Click the Show Message button in the Message group on the Attachment Tools Attachments tab to close the presentation preview in the Reading pane.

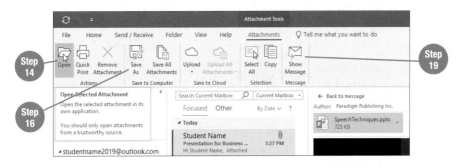

# 5.4 Scheduling Appointments and Events in Calendar

The **Calendar** component in Outlook is used to schedule appointments and events, such as meetings or conferences. An **appointment** is any activity where you want to track the day or time that the activity begins and ends in your schedule or when you want to be reminded to be somewhere. An appointment in Calendar can be a class, a meeting, a medical test, or a lunch date.

*Note: In this topic and the next two topics, you will schedule appointments, an event, and a meeting in October 2021. Check with your instructor for alternate instructions.*

**1** With Outlook open and Inbox the active folder, click *Calendar* on the Navigation pane.

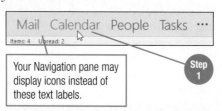

Your Navigation pane may display icons instead of these text labels.

Outlook displays the current date, week, or month in the Content pane in Day, Week, or Month view. A **Date Navigator** displays at the top of the Folder pane with the current month and next month, along with directional arrows to browse forward or back to upcoming or previous months. Open the **Go To Date** dialog box to display a date further out.

**2** Click the Day button in the Arrange group on the Home tab if Day is not the current view, and then click the Go to Date dialog box launcher at the lower right of the Go To group.

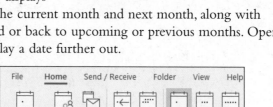

**3** In the Go To Date dialog box, type 10/11/2021, and then press Enter or click OK.

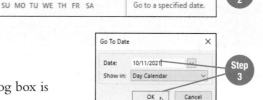

In Outlook.com, the Go To Date dialog box is not available. Click the down-pointing arrow next to the date displayed at the top of the Appointment area to jump to a specific date. Click the up or down arrow to change the year, click the month button, and then click the day within the month.

**4** Click next to 9:00 a.m. in the Appointment area, type Meet with program adviser, and then press Enter or click in another time slot outside the appointment box.

By default, Outlook schedules an appointment for a half hour.

**5** Drag the bottom boundary of the appointment box to 10:00 a.m.

**6** Click next to 11:00 a.m. in the Appointment area.

**7** Click the New Appointment button in the New group on the Home tab.

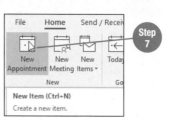

In Outlook.com click *New Event* in the browser window. Use the **New Appointment button** to open an Appointment window, where you can provide details about the appointment or select options to apply to the appointment.

 **Watch & Learn**

---------- skills

Schedule an appointment

Schedule an event

---------- app tip

A feature called *peek* displays a pop-up with current information when you point to an item. For example, peek at the appointments for the day without leaving Mail by pointing to *Calendar* in the Navigation pane.

**Tutorial**

Displaying the Calendar

**8** Type Intern Interview in the *Subject* text box.

**9** Press Tab or click in the *Location* text box and then type Room 3001.

**10** Click the *End time* option box arrow and then click *12:00 PM (1 hour)*.

**11** Click the Save & Close button in the Actions group on the Appointment tab.

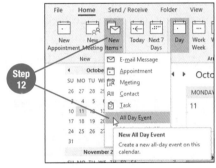

An **event** differs from an appointment in that it is an activity that lasts an entire day or longer. Examples of events include conferences, trade shows, or vacations. An event does not occupy a time slot in the Calendar.

**12** Click the New Items button in the New group on the Home tab and then click *All Day Event* to open an Event window.

In Outlook.com, click the *All day* option to change an appointment to an all day event.

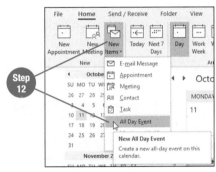

**13** Type Career Fair in the *Subject* text box.

**14** Press Tab or click in the *Location* text box and then type Student Center.

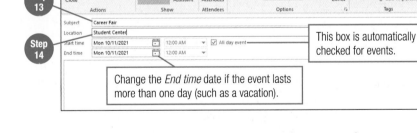

This box is automatically checked for events.

Change the *End time* date if the event lasts more than one day (such as a vacation).

**15** Click the Save & Close button in the Actions group on the Event tab.

Notice the event appears at the top of the Appointment area (Figure 5.3).

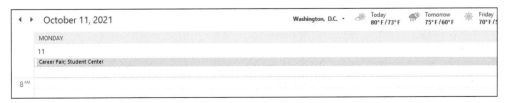

**Figure 5.3**

Event information appears in a banner along the top of the day in the Appointment area.

# 5.5 Scheduling a Recurring Appointment and Editing an Appointment

An appointment that occurs on a regular basis at fixed intervals need only be entered once and set up as a recurring appointment. To do this, open the Appointment Recurrence dialog box by clicking the **Recurrence button** in the Options group on the Appointment tab or the Calendar Tools Appointment tab. Enter the recurrence pattern for a repeating appointment, click OK, and then save and close the appointment.

1. With Outlook open and Calendar active for October 11, 2021, click the Forward button to display October 12, 2021, in the Appointment area.

2. Click next to 3:00 p.m. in the Appointment area, type Math Extra Help Sessions, and then press Enter.

3. With the Math Extra Help Sessions appointment box selected in the Appointment area, click the Recurrence button in the Options group on the Calendar Tools Appointment tab.

By default, Outlook sets the *Recurrence pattern* details for the appointment to recur *Weekly* at the same day and time as the appointment.

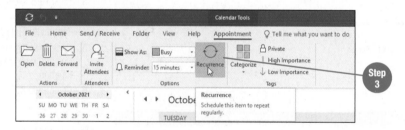

4. In the Appointment Recurrence dialog box, select *25* in the *End after* text box in the *Range of recurrence* section and then type *6*.

5. Click OK.

A recurring icon displays at the right end of the appointment box in the Appointment area for any appointment that repeats.

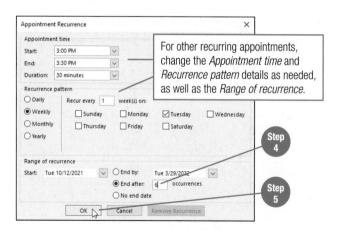

6. Click *19* in the October 2021 calendar in the Date Navigator to display the appointments for the next Tuesday and view the recurring appointment.

☁ **Watch & Learn**

●------- Skills

Schedule a recurring appointment

Edit an appointment

☁ **Tutorials**

Scheduling a Recurring Appointment

●------- Oops!

Using Outlook.com? Set the recurrence option by changing the *Repeat* option to *Every week* and changing the end date to November 16, 2021.

●------- app tip

Consider entering your class schedule in the Outlook calendar for the current semester as recurring appointments.

**7** Click *26* in the October 2021 calendar in the Date Navigator to view the Math Extra Help Sessions appointment in the Appointment area.

**8** Click the Go to Date dialog box launcher in the Go To group, type 11/23/2021 in the Go To Date dialog box, and then press Enter or click OK. Notice the Math Extra Help Sessions appointment does not appear in the Appointment area because the range of recurrence has ended.

Assign options or tags to an existing appointment by selecting the appointment and using the buttons on the Calendar Tools Appointment tab. Change the subject, location, day, or time of an appointment by double-clicking an appointment to open the Appointment window.

**9** Display October 11, 2021, in the Calendar.

**10** Click to select the appointment scheduled at 11:00 a.m.

A selected appointment box displays with a black outline. A pop-out opens at the left with the appointment details when you point at an appointment.

**11** Click the High Importance button in the Tags group on the Calendar Tools Appointment tab.

Skip Step 11 if you are using Outlook.com. The High Importance tag is not available.

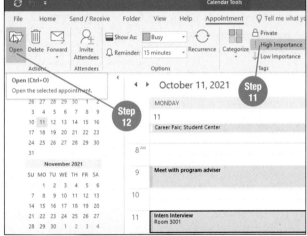

**12** Click the Open button in the Actions group.

If you are using Outlook .com, click the *Edit* option.

**13** Click the calendar icon at the right end of the *Start time* text box.

**14** Click *12* in the calendar.

**15** Click Save & Close in the Actions group on the Appointment tab.

**16** Display October 12, 2021, in the Appointment area.

Notice the revised Intern Interview appointment appears next to 11:00 a.m. You can also move an

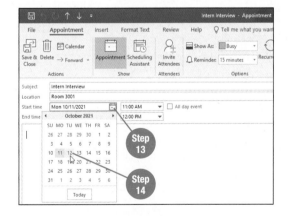

appointment using a drag-and-drop technique. In Day view, drag an appointment box to another time slot on the same day. Display the Calendar in Week view and drag an appointment box from one day to another day in the same week, and change the start time if needed. In Month view, drag an appointment box from one day to another. The appointment is moved with the same start time.

## 5.6 Scheduling a Meeting

Schedule a meeting by displaying the day in the Appointment area, selecting the time, and then clicking the **New Meeting button** in the New group on the Home tab. In the Meeting window *To* text box, enter the email addresses for the individuals in your invite list. Enter the meeting subject, location, and other details as needed and then click the Send button. Meeting attendees receive a **meeting request** email message. Responses to the meeting request are sent back to the meeting organizer via buttons in the email message window or Reading pane.

1. With Outlook open and Calendar active with October 12, 2021, displayed in the Appointment area, click next to 1:00 p.m.

2. Click the New Meeting button in the New group on the Home tab.

3. Type the email address for the classmate with whom you have been exchanging email messages in the *To* text box.

*Note: If you have been sending email messages to yourself in this chapter, send the meeting request message to a friend or relative, or use an email address for yourself that is not your Microsoft account address. You cannot send a meeting request to the same email sending the request. You will not be able to complete Step 8 to Step 16 if you do not receive a meeting request message from someone else.*

4. Press Tab or click in the *Subject* text box and then type Fundraising Planning Meeting.

5. Press Tab or click in the *Location* text box and then type Room 1010.

6. Click the *End time* option box arrow and then click *2:30 PM (1.5 hours)*.

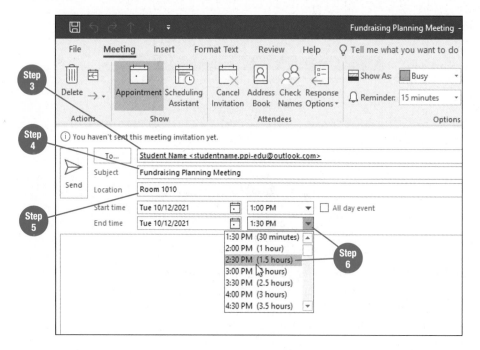

7. Click the Send button.

8  Click *Mail* on the Navigation pane.

9  Click the Send/Receive All Folders button on the Quick Access Toolbar to update your Inbox, if the meeting request message sent to you by a classmate at Step 7 has not been received.

Step 8

Mail  Calendar  People  Tasks  ⋯

10  Click the message header for the meeting request to view the message details in the Reading pane.

Buttons along the top of the Reading pane or in the Respond group on the Meeting tab in a message text box for a meeting request are used to respond to the meeting organizer. Click the **Accept button** to send a reply that you will attend the meeting. Use the Tentative, Decline, or Propose New Time buttons if you are not required or unable to attend a scheduled meeting. Outlook also provides a Calendar button at the top of the Reading pane, which is used to view your appointments for the meeting day and time to see if you are available before you send a response to the meeting organizer.

11  Double-click the message header to open the meeting request message with the Meeting tab active.

Note that you will see two entries for the meeting in the message window. One entry is the meeting that you created in Step 1 to Step 7. You are referred to as the *meeting organizer* for this meeting. The other entry is a result of the meeting request sent to you by someone else for which you are a *meeting invitee*.

12  Click the Accept button in the Respond group on the Meeting tab.

13  Click *Send the Response Now*.

A meeting request email message is deleted from the Inbox after you respond to the meeting invitation.

**quick steps**

**Schedule a Meeting**
1. Display Calendar.
2. Navigate to meeting date.
3. Click next to meeting time.
4. Click New Meeting button.
5. Enter meeting details.
6. Click Send.

**Accept a Meeting Request**
1. Display Inbox.
2. Select meeting request message header OR open Meeting window.
3. Click Accept button.
4. Click *Send the Response Now*.

**Tutorial**

Scheduling a Meeting

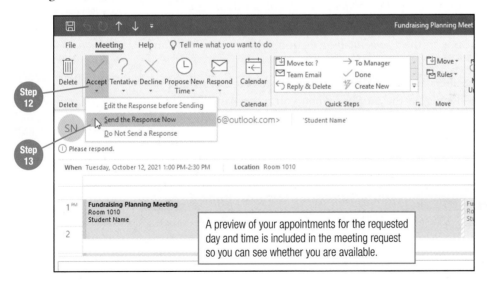

Step 12

Step 13

A preview of your appointments for the requested day and time is included in the meeting request so you can see whether you are available.

14  Click *Sent Items* in the Folder pane.

15  Click the message header for the message sent to the meeting organizer with your Accepted reply and then read the message in the Reading pane.

This is the default reply sent to a meeting organizer when a meeting invitation is accepted. Use the *Edit the Response before Sending* option to type your own message.

16  Click *Inbox* in the Folder pane.

## 5.7 Adding and Editing Contacts in People

☁ **Watch & Learn**

The **People** component in Outlook is used to store and organize contact information, such as email addresses, mailing addresses, telephone numbers, and other information about the people with whom you communicate. Think of People as an electronic address book. To add information about a person, click the **New Contact button** in the New group on the Home tab. Contacts display alphabetically in the *People* list in the Content pane. Click a contact name to display the information about the person in the Reading pane.

**●————— Skills**

Add a Contact

Edit a Contact

1. With Outlook open and Inbox the active folder, click *People* on the Navigation pane.

2. Click the New Contact button in the New group.

3. Type Tory Nguyen in the *Full Name* text box in the Contact window.

4. Press Tab or click in the *Company* text box.

Notice the *File as* text box automatically updates when you move past the *Full Name* entry, with the last name of the person followed by the first name. The *File as* entry is used to organize the *People* list alphabetically by last names.

5. Type NuWave Personnel in the *Company* text box.

6. Press Tab or click in the *Job title* text box and then type Recruitment Specialist.

7. Click in the *Email* text box and then type tory@ppi-edu.net.

8. Click in the *Business* text box in the *Phone numbers* section and then type 8885559840.

9. Click in the *Mobile* text box in the *Phone numbers* section, type 8885553256, and then press Tab.

Notice that the phone numbers automatically format to show brackets around the area code, a space, and a hyphen in the number when you move past the field. In Outlook.com, the formatting of telephone numbers does not occur. Type the brackets and hyphens if desired.

10. Click the Add Contact Picture control that displays as a silhouette in a box between the name and business card sections in the Contact window.

**●————— Oops!**

Using Outlook.com? Click *New contact* to enter the information in the browser window. Outlook.com uses a different order and input grouping. For example, type the first name and last name in separate areas. Use the *Add more* option at the bottom of the window to find additional entry options not shown. For example, click *Add More* and point to the *Work* option to find the *Job title* entry.

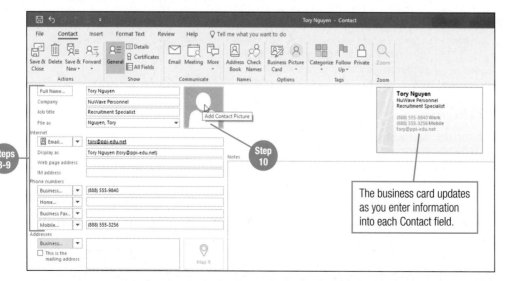

The business card updates as you enter information into each Contact field.

11  In the Add Contact Picture dialog box, navigate to the Ch5 folder in StudentDataFiles and then double-click the image file ***ToryNguyen.jpg***.

12  Click the Save & Close button in the Actions group on the Contact tab.

Contact information for the selected person displays in the Reading pane with options for communicating with the person, such as sending an email message.

13  Double-click *Tory Nguyen* in the *People* list. In Outlook.com, click the *Edit contact* option in the Reading pane to reopen the Contact window.

14  Click at the end of the telephone number *(888) 555-9840* in the *Business* text box, press the spacebar, and then type extension 3115.

15  Click in the *Notes* text box and then type the following text:

NuWave Personnel was recommended by Dana. Tory is also the regional director.

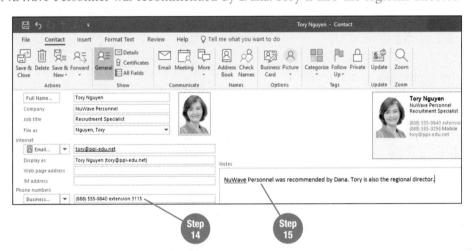

16  Click the Save & Close button. The updated information displays in the Reading pane.

17  Double-click in any white space below the last entry in the *People* list to open a new Contact window.

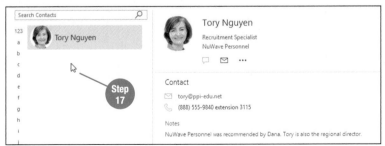

18  Enter the name of your instructor in the *Full Name* text box, the name of your school in the *Company* text box, and the email address for your instructor in the *Email* text box, and then click the Save & Close button.

## 5.8 Adding and Editing Tasks

The **Tasks** component in Outlook maintains a to-do list. You can track information about each task, such as how much of the task is completed, how much time has been spent on the task, and the priority of the task. In **To-Do List** view, uncompleted tasks are shown grouped in descending order by the due date.

**1** With Outlook open and People active, click *Tasks* on the Navigation pane.

**2** Click *To-Do List* in the Current View group on the Home tab if To-Do List is not the active view.

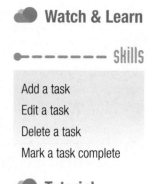

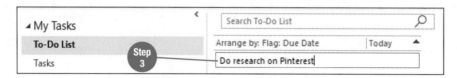

**3** Click in the text box at the top of the To-Do List that displays the dimmed text *Type a new task*, type Do research on Pinterest, and then press Enter.

**4** Type Gather images to create Pinterest pinboard and then press Enter.

**5** Type Create resume for Career Fair and then press Enter.

Adding a task using the *Type a new task* text box adds the entry in the To-Do List with a Today follow-up flag by default. In the next steps, you will open a Task window using the **New Task button** to add a task with more details and options.

**6** Click the New Task button in the New group on the Home tab.

**7** Type Prepare study notes for exams in the *Subject* text box.

**8** Click the *Priority* option box arrow and then click *High*.

**9** Click the calendar icon at the right end of the *Due date* text box, navigate to the last month of the current semester, and then click the Monday one week before the last week of your semester.

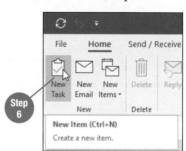

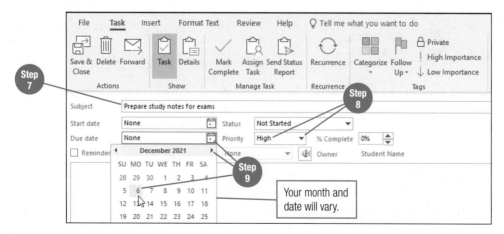

Your month and date will vary.

**10** Click the Save & Close button in the Actions group on the Task tab.

Editing or updating a task can include activities such as assigning or changing a due date, assigning a priority, entering the percentage of completion, or changing the status of a task. When a task is completed, use the **Remove from List button** in the Manage Task group on the Home tab to remove the follow-up flag on the task, or the **Mark Complete button** to remove the task from the To-Do List.

**11** Click *Do research on Pinterest* to select the task. Notice the task details in the Reading pane.

**12** Click the Remove from List button in the Manage Task group on the Home tab.

Notice the task is removed from the To-Do List. You can also use the Delete button in the Delete group on the Home tab to remove a task you have completed.

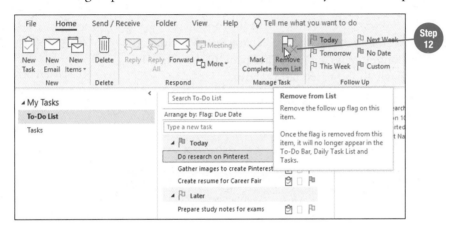

**13** Double-click the task *Create resume for Career Fair* to open the Task window.

**14** Click the Mark Complete button in the Manage Task group on the Task tab.

The Task window closes, and the entry is removed from the To-Do List.

**15** Double-click the task entry *Gather images to create Pinterest pinboard*.

**16** Click the *Status* option box arrow and then click *Waiting on someone else*.

**17** Click in the text box with white space below the *Reminder* options and then type Waiting for Leslie to send me images from her renovation clients.

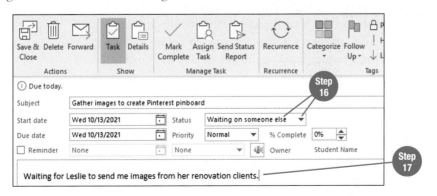

**18** Click the Save & Close button. Notice the updated task details appear in the Reading pane.

**19** Display the Inbox folder in Mail.

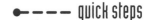

## quick steps

**Add a Task**
1. Display Tasks.
2. Click in *Type a new task* text box, type task description, and press Enter.

OR

1. Display Tasks.
2. Click New Task button.
3. Enter task details.
4. Click Save & Close.

 **Tutorial**
Editing Tasks

## 5.9 Searching Outlook Items

The search feature in Outlook, referred to as **Instant Search**, is used to locate a message, appointment, contact, or task. A search text box located between the ribbon and content is used to find items. Outlook begins a search as soon as you begin typing an entry in the search text box. Matched items are highlighted in the search results. Once located, you can open an item to view or edit the information by double-clicking the entry in the filtered search results list.

**1** With Outlook open and Inbox the active folder in Mail, click in the search text box at the top of the Content pane (contains the dimmed text *Search Current Mailbox*).

**2** Type *pinterest*.

Outlook immediately begins matching items as you type characters. Matched words are highlighted in both the Content pane and Reading pane in the filtered search results list. Notice also the Search Tools Search tab is active.

**3** Click the Current Folder button in the Scope group on the Search Tools Search tab.

**Watch & Learn**

— — — — — — **skills**

Search messages

Search appointments

Search contacts

Search tasks

— — — — — — **app tip**

The keyboard shortcut Ctrl + E opens the search text box.

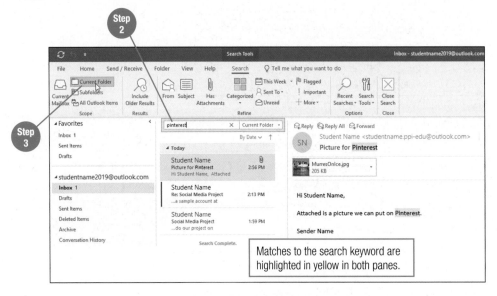

Matches to the search keyword are highlighted in yellow in both panes.

**4** Click each message in the search results list to view the message in the Reading pane.

Notice that for each message, *Pinterest* is highlighted in the Reading pane. Use buttons in the Scope and Refine groups on the Search Tools Search tab to further eliminate messages that do not meet the criterion for content you are trying to find. You can search for a message by typing a sender name, a subject, or any keyword or phrase that exists in the message.

**5** Click *Close Search* in the search text box to close the search results list and return the Inbox to the full message list.

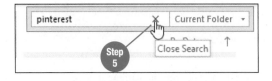

6  Display the Calendar.

7  Click in the search text box at the upper right of the Appointment area (contains the dimmed text *Search Calendar*) and then type career fair.

Outlook displays the appointment found with the search keywords in a filtered list. The search keywords are highlighted in yellow within the entry.

8  Double-click the entry in the filtered list to view the details in the Event window and then close the window.

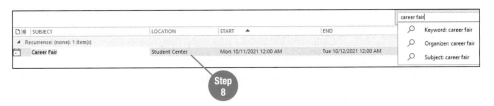

9  Click *Close Search* to restore the Calendar to the appointments for the current day.

10  Display People.

11  Click in the search text box at the top of the *People* list (contains the dimmed text *Search Contacts*), type nuwave, and then press Enter.

Outlook displays the contact for Tory Nguyen, who works at NuWave Personnel. You can use the search feature to find any Outlook item by any entry in the message, appointment, contact, or task. For example, you could find a contact by name, job title, company name, or telephone number.

12  Click *Close Search* to restore the *People* list.

13  Display Tasks.

14  Click in the search text box at the top of the task list (contains the dimmed text *Search To-Do List* or *Search Tasks*, depending on the active selection in the Folder pane) and then type leslie.

15  Click to select the task entry *Gather images to create Pinterest pinboard* and then view the task details with *Leslie* highlighted in yellow in the Reading pane.

Outlook can locate a task by searching text in any task option.

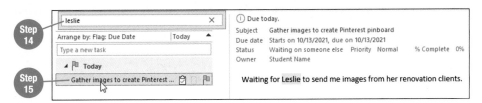

16  Click *Close Search* to restore the To-Do List.

17  Display Mail with the Inbox folder.

18  Close Outlook.

**quick steps**

**Search for an Outlook Item**
1. Display Mail, Calendar, People, or Tasks.
2. Click in search text box.
3. Type search keyword.

**Tutorial**

Searching for People and Appointments

| Topic | Key Concepts |
|---|---|
| **5.1 Using Outlook to Send Email** | **Outlook** is an application for organizing messages, appointments, contacts, and tasks referred to as a **personal information management (PIM)** application. |
| | **Electronic mail (email)** is the exchange of messages between individuals electronically. |
| | When Outlook is started, **Mail** is the active component within Outlook, with the Inbox folder shown by default. |
| | The **Inbox** displays messages with the newest message received on the **Focused Inbox** tab at the top of the Content pane. The Focused tab shows your important messages—those that are not newsletters or machine-generated emails, which appear on the Other tab. |
| | Create and send a new email message using the **New Email button** in the New group on the Home tab. |
| | Click the message header for a new message received to read the message contents in the Reading pane. |
| | Reply directly to a message by clicking the **Reply button** in the Reading pane (referred to as an **inline reply**). |
| | Send a copy of a message you have received to someone else using the **Forward button**. |
| **5.2 Attaching a File to a Message and Deleting a Message** | Attach a file to a message using the **Attach File button** in the Include group on the Message tab. |
| | As you type an email address in the *To* text box, the **AutoComplete** feature shows email addresses in a *Recent People* list and an *Other Suggestions* list that match what you are typing. As you type, click an entry in the list if the correct recipient appears before you finish typing. |
| | Delete messages from mail folders that are no longer needed to keep folders a manageable size. |
| | Deleted messages are moved to the **Deleted Items** folder. |
| | Empty the Deleted Items folder to permanently delete messages. |
| **5.3 Previewing File Attachments and Using File Attachment Tools** | Preview a file attached to a message by clicking the file name in the Reading pane. |
| | While a file is being previewed, message text is temporarily removed from the Reading pane. |
| | Some files cannot be viewed in the Reading pane and must be viewed by double-clicking the file name to open the file in the source application. |
| | Open or save a file using buttons in the Actions group on the Attachment Tools Attachments tab or with the options accessed by clicking the down-pointing arrow at the right of the file name in the Reading pane. |
| | A file opened from a message is opened in **Protected View** in the source application, which allows you to read the contents. You cannot edit the file until you click the Enable Editing button on the Message bar. |

*continued…*

| Topic | Key Concepts |
|---|---|
| **5.4 Scheduling Appointments and Events in Calendar** | The **Calendar** component is used to schedule appointments and events. |
| | An **appointment** is any activity for which you want to record the occurrence by day and time. |
| | A **Date Navigator** at the top of the Folder pane displays the current month and next month with directional arrows to browse to the previous or next month. |
| | Use the **Go To Date** dialog box to navigate directly to a specific date you want to display in the Appointment area. |
| | Click next to the time in the Appointment area and then type a description to enter a new appointment. |
| | Click the **New Appointment button** in the New group on the Home tab to enter details for a new appointment in an Appointment window. |
| | An **event** is an appointment that lasts an entire day or longer. |
| | Click the New Items button in the New group on the Home tab and then click *All day event* to create an event in an Event window. |
| **5.5 Scheduling a Recurring Appointment and Editing an Appointment** | An appointment that occurs at fixed intervals on a regular basis can be entered once, and Outlook schedules the remaining appointments automatically. |
| | Click the **Recurrence button** in the Options group on the Appointment tab to set the recurrence pattern and range of recurrence details for a recurring appointment. |
| | Click to select an appointment in the Appointment area to assign options or tags to the appointment using buttons on the Calendar Tools Appointment tab. |
| | Open the Appointment window to make changes to the subject, location, day, or time of a scheduled appointment. |
| | You can also move an appointment by dragging the appointment box to another day or time in Day view or Week view. In Month view, drag an appointment box to another day at the same appointment start time. |
| **5.6 Scheduling a Meeting** | A meeting is an appointment to which you invite people. |
| | Click the **New Meeting button** in the New group on the Home tab to create a new meeting by entering the email addresses for meeting attendees, as well as and the meeting particulars. |
| | Information about a meeting is sent to people via a **meeting request**, which is an email message sent to meeting participants. |
| | A meeting attendee responds to a meeting request from the Reading pane or the message window by clicking a response button, such as the **Accept button**. |
| **5.7 Adding and Editing Contacts in People** | Use the **People** component to store and organize contact information for people with whom you communicate. |
| | Click the **New Contact button** or double-click in a blank area in the *People* list to open a Contact window and add contact information. |
| | Click the Add Contact Picture control in the Contact window to select an image file for a contact in the Add Contact Picture dialog box and display the image next to the name in the *People* list. |
| | Double-click a name in the *People* list to edit contact information in the Contact window. |

*continued…*

| Topic | Key Concepts |
|---|---|
| **5.8 Adding and Editing Tasks** | Use **Tasks** in Outlook to maintain a to-do list. |
| | Click in the *Type a new task* text box to add a task to the **To-Do List**, or click the **New Task button** to enter a new task in a Task window. |
| | Open a Task window to add a due date or to add other task information, such as a priority or status. |
| | Select a task and click the **Remove from List button** to remove the follow-up flag on the task, which also removes the task from the To-Do List. |
| | Select a task or open a task in the Task window and use the **Mark Complete button** to indicate a task is completed. Completed tasks are removed from the To-Do List view. |
| **5.9 Searching Outlook Items** | The **Instant Search** feature is a search text box at the top of Mail, Calendar, People, and Tasks in which you search for an Outlook item by typing a keyword or phrase. |
| | Matched items start to appear in a filtered list as soon as you type characters in the search text box. Continue typing the search keyword or phrase to continue filtering the list. |
| | Matches to the search keyword or phrase are highlighted in the filtered lists and in content in the Reading pane for a selected item. |
| | Use buttons on the Search Tools Search tab to further refine a search. |
| | Click *Close Search* in the search text box to remove the filtered list and redisplay all items. |

 **Topics Review**

# Creating, Editing, and Formatting Word Documents

Microsoft Word (referred to as Word) is a **word processing application** used to create documents that are mostly text. Word documents can also include images, charts, tables, or other visual objects to make the document more interesting and the content easier to understand. Letters, essays, reports, invitations, recipes, agendas, contracts, and resumes are examples of documents that can be created using Word.

Word automatically corrects some errors as you type and indicates other potential spelling and grammar errors for you to consider. Other features provide tools to format and enhance a document. In this chapter, you will learn how to create, edit, and format documents. You will create new documents starting from a blank page and other documents by selecting from the Template gallery in Word.

If necessary, review the Office essentials you learned in Chapter 3 before starting to do the work in this chapter. It is assumed that you have the skills to start an application; save, open, and print a document; and perform commands using the Quick Access Toolbar, ribbon, dialog boxes, and task panes.

## Learning Objectives

**6.1** Create a new document and insert, delete, and edit text

**6.2** Insert symbols and check spelling and grammar

**6.3** Find and replace text

**6.4** Move text and create bulleted and numbered lists

**6.5** Format text using font options, and change paragraph alignment

**6.6** Indent paragraphs, change line spacing, and change spacing before and after paragraphs

**6.7** Apply a style to text and change the style set

**6.8** Create a new document from a template

### Read & Learn

### Content Online

The online course includes additional training and assessment resources.

## 6.1 Creating and Editing a New Document

The Word Start screen appears when Word is initially opened. You can choose to open an existing document, create a new blank document, or search for and select a template to create a new document. Creating a new document includes typing the document text, editing the text, and correcting errors. As you type, features in Word help you fix common typing errors and apply formatting to characters or paragraphs.

**1** Start Word.

**2** On the Word Start screen, click *Blank document* in the Templates gallery and compare your screen with the one shown in Figure 6.1.

Table 6.1 on the next page describes the elements shown in Figure 6.1.

**3** Type Social Bookmarking and press Enter.

Notice that extra space is automatically added below the text before the next line.

**4** Type the text on the next page, allowing the lines to end automatically; press Enter only where indicated.

Word moves text to a new line automatically when you reach the end of the current line. This feature is called **wordwrap**. Word displays a red wavy line below a word not recognized by the software. These red wavy lines indicate possible spelling errors. If a red wavy line appears below a word spelled correctly, you can ignore it. If the red wavy line appears for a typing error, correct the typing mistake by pressing the Backspace key to delete the character(s) and then type the correct character(s). You will learn other editing methods later in this topic.

**Watch & Learn**

**Hands On Activity**

●———————— Skills

Enter text

Describe AutoCorrect actions

Describe AutoFormat actions

Edit, insert, and delete text

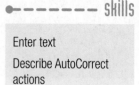

**Tutorials**

Exploring the Word Start Screen

Entering Text

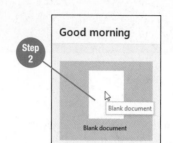

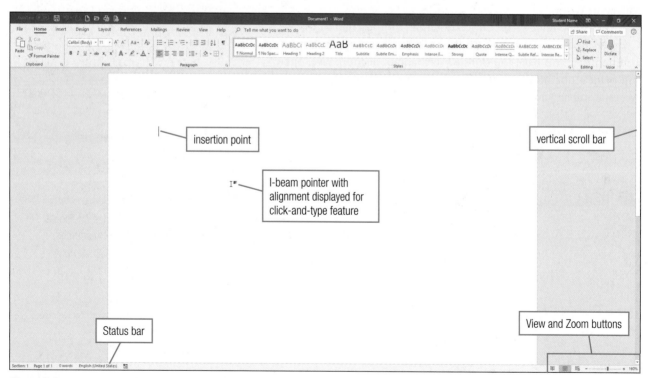

**Figure 6.1**

A new blank document screen is shown. The default settings in Word display a new document in Print Layout view. Your display may vary if settings have been changed on the computer you are using. See Table 6.1 on the next page for a description of screen elements.

**Table 6.1** Word Features

| Feature | Description |
|---|---|
| Insertion point | A blinking vertical bar indicates where the next character typed will appear. |
| I-beam pointer | This pointer appears for text entry or selection when you move the pointer using a mouse or trackpad.<br>     The I-beam pointer displays with a paragraph alignment option (left, center, or right) depending on the location of the I-beam within the current line. You can double-click and type text anywhere on a page, and the alignment option will be left-aligned, center-aligned, or right-aligned. This feature is called **click and type**. |
| Status bar | The status bar displays page numbers with total number of pages and number of words in the current document. The right end of the Status bar has view and zoom options. The default view is Print Layout view, which displays how a page will look when printed with the current print options. |
| Vertical scroll bar | Use the scroll bar to view parts of a document not displayed in the current window. |
| View and Zoom buttons | By default, Word opens in Print Layout view. Other view buttons include Read Mode and Web Layout. Read Mode view maximizes reading space and removes editing tools, providing a more natural environment for reading. In Web Layout view, the document displays the page width, text, and formatting options as the document would look in a browser. Zoom buttons, as you learned in Chapter 3, are used to enlarge or shrink the display. |

Social bookmarking websites are used to organize, save, and share web content. Links are called bookmarks and include tags, which are keywords you assign to the content when you create the bookmark. [press Enter]

Many websites now include icons for popular social bookmarking sites that capture the page references for bookmarking. Another way to bookmark a page is to add the bookmarklet for the social bookmarking site you use to your browser's toolbar. Bookmarklets add a bookmark instantly when clicked. [press Enter]

5  Type teh and press the spacebar. Notice that Word changes the text to *The*.

A feature called **AutoCorrect** changes commonly misspelled words as soon as you press the spacebar and capitalizes the first character in a new sentence if the first character is typed lowercase.

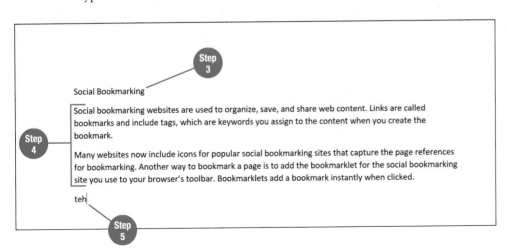

 app tip

AutoCorrect also fixes common capitalization errors, such as two initial capitals and no capital in the name of a day. AutoCorrect also turns off the Caps Lock key and corrects text when a new sentence is started with the key left on. Use Undo if AutoCorrect changes text that you don't want changed.

**Tutorial**

Undoing an AutoCorrect Correction

**6**    **Type** popular social bookmarking site Pinterest.com is used to pin pictures found on the Web to virtual pinboards. **and press Enter.**

**7**    **Type** a study by a marketing company found that 1/2 of frequent web surfers use a social bookmarking site, with Pinterest the 1st choice for most females. **and press Enter.**

Notice that Word automatically corrects the capitalization of the first word in the sentence, *1/2* is changed to a fraction character (½), and *1st* is automatically formatted as an ordinal, with the *st* displayed as superscript text (superscript characters are smaller text placed at the top of the line). The **AutoFormat** feature automatically changes some fractions, ordinals, quotes, hyphens, and hyperlinks as you type. AutoFormat also converts straight apostrophes (') or quotation marks (") to smart quotes ('smart quotes'), also called *curly quotes* ("curly quotes").

●------- app tip

AutoFormat does not recognize all fractions. For example, typing *1/3* will not format to the one-third fraction character. You will learn about inserting symbols for these characters in the next topic.

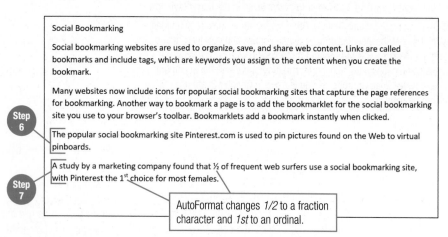

Social Bookmarking

Social bookmarking websites are used to organize, save, and share web content. Links are called bookmarks and include tags, which are keywords you assign to the content when you create the bookmark.

Many websites now include icons for popular social bookmarking sites that capture the page references for bookmarking. Another way to bookmark a page is to add the bookmarklet for the social bookmarking site you use to your browser's toolbar. Bookmarklets add a bookmark instantly when clicked.

**Step 6**
The popular social bookmarking site Pinterest.com is used to pin pictures found on the Web to virtual pinboards.

**Step 7**
A study by a marketing company found that ½ of frequent web surfers use a social bookmarking site, with Pinterest the 1st choice for most females.

AutoFormat changes *1/2* to a fraction character and *1st* to an ordinal.

**8**    Click the Save button on the Quick Access Toolbar.

The Save As backstage area appears because this is the first time the document has been saved.

**9**    Click *This PC* and then click *Browse,* or click the option in the Save As backstage area for your storage medium if you are not saving to a USB flash drive.

**10**    Navigate in the Save As dialog box to the CompletedTopics folder on your storage medium and then create a new folder, *Ch6*.

**11**    Double-click the new folder *Ch6*.

**12**    Select the current text in the *File name* text box, type SocialMediaProject -YourName, and then press Enter or click the Save button.

☁ **Tutorials**

Saving with the Same Name

Saving with a New Name

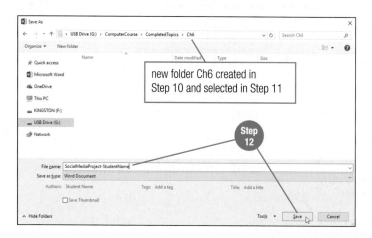

new folder Ch6 created in Step 10 and selected in Step 11

**Step 12**

When creating a new document, changes often need to be made to the text after the text has been typed. You may need to correct typing errors, change a word or phrase to some other text, add new text, or remove text. Making changes to a document after the document has been typed is called **editing**. The first step to edit text is to move the insertion point to the location of a change.

13 Click to position the insertion point at the beginning of the last paragraph that begins with the text *A study by a marketing company* (insertion point will be blinking just left of *A*), type Experian Hitwise, and then press the spacebar.

Word automatically inserts the new text and moves existing text to the right.

Step 13

> Experian Hitwise A study by a marketing company found that ½ of frequent web surfers use a social bookmarking site, with Pinterest the $1^{st}$ choice for most females.

14 With the insertion point still positioned at the left of *A* in *A study*, press the Delete key until you have removed *A study by*, type (, click to position the insertion point just after the *y* in *company*, and then type ) conducted a survey of frequent web surfers and.

15 Position the insertion point just left of ½, press the Delete key until you have removed *½ of frequent web surfers*, and then type one-half.

16 Position the insertion point at the left of $1^{st}$, delete $1^{st}$, and then type first.

17 Position the insertion point below the last paragraph and then type your first and last name.

18 Check your text with the document shown in Figure 6.2. If necessary, make further corrections by moving the insertion point and inserting and deleting characters as needed.

> Social Bookmarking
>
> Social bookmarking websites are used to organize, save, and share web content. Links are called bookmarks and include tags, which are keywords you assign to the content when you create the bookmark.
>
> Many websites now include icons for popular social bookmarking sites that capture the page references for bookmarking. Another way to bookmark a page is to add the bookmarklet for the social bookmarking site you use to your browser's toolbar. Bookmarklets add a bookmark instantly when clicked.
>
> The popular social bookmarking site Pinterest.com is used to pin pictures found on the Web to virtual pinboards.
>
> Experian Hitwise (a marketing company) conducted a survey of frequent web surfers and found that one-half use a social bookmarking site, with Pinterest the first choice for most females.
>
> Student Name

**Figure 6.2**

The document text for **SocialMediaProject-StudentName** is displayed.

19 Click the Save button on the Quick Access Toolbar. Leave the document open for the next topic.

Because the document has already been assigned a file name in Step 12, the Save button saves the document changes using the same name.

## quick steps

**Create a New Document**
1. Start Word.
2. Click *Blank document*.
3. Type text.

**Save a New Document**
1. Click Save button on the Quick Access Toolbar.
2. Navigate to drive and folder.
3. Enter file name.
4. Click Save.

**Save a Document Using the Existing Name**
Click Save button on the Quick Access Toolbar.

**Edit a Document**
1. Position insertion point at location of change.
2. Type new text or delete text as needed.
3. Save changes.

## Tutorial

Moving the Insertion Point and Inserting and Deleting Text

## Oops!

Insertion point will not go below last paragraph? Click to place the insertion point at the end of the last sentence and then press Enter to move down to a blank line.

## app tip

You can move the insertion point by pressing the Up, Down, Left, or Right Arrow keys to move within a document. Hold down an arrow key to rapidly move the insertion point.

## 6.2 Inserting Symbols and Completing a Spelling and Grammar Check

In some documents, you need to insert a symbol or special character, such as a copyright symbol (©), registered trademark (™), or a fraction character for a fraction that AutoCorrect does not recognize, such as one-third (⅓). Symbols and special characters are inserted using the **Symbol gallery** or the Symbol dialog box.

The **Check Document button** on the Review tab opens the Editor task pane. The Result(s) button near the top of the pane displays the number of errors detected with the *Corrections* list showing the number of errors categorized as *Spelling* or *Grammar*. The Spelling feature matches words in the document with words in a dictionary and flags a word not found as a possible error. When you click the Result(s) button to review potential errors, the Editor task pane changes to display suggestions for a word not found and give options to ignore, to change, or to add the word to the dictionary. The feature also checks for duplicate words, with an option to remove the repeated word. The Check Document feature helps you correct many errors; however, you still need to proofread each document. For example, the errors in the following sentence escaped detection: The plain fair was expensive!

1. With the **SocialMediaProject-YourName** document open, position the insertion point after *n* in *Experian* in the last paragraph.

2. Click the Insert tab.

3. Click the Symbol button in the Symbols group.

4. Click *Trade Mark Sign*. Note that the symbol may appear in a different position than shown in the image at the right.

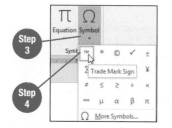

5. Position the insertion point after the period at the end of the paragraph that begins *Experian* and press the spacebar to insert a space.

6. Type The survey sample size of 1,000 interviews provides a standard error at 95% confidence of and then press the spacebar.

7. Click the Symbol button and then click *More Symbols*.

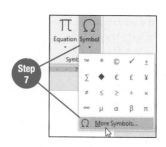

8. In the Symbol dialog box, with *Font* set to *(normal text)* and *Subset* set to *Letterlike Symbols*, scroll up the symbol list until the *Subset* changes to *Latin-1 Supplement*, click ±, and then click the Insert button.

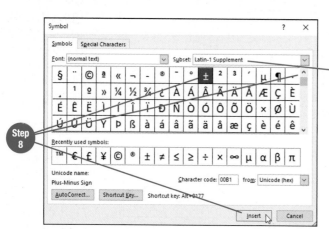

Scroll up almost to the top of the dialog box in Step 8 until the *Subset* changes to *Latin-1 Supplement* and you can see the plus-minus symbol.

**9**   Click the Close button to return to the document and then type 3%.

You may have noticed the Plus–Minus Sign symbol in the Symbol gallery. In Steps 7 to 9, you practiced using the Symbol dialog box so that you will know how to find a symbol that is not displayed in the Symbol gallery.

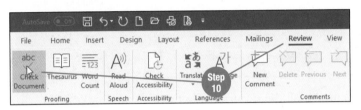

> Experian™ Hitwise (a marketing company) conducted a survey of frequent web surfers and found that one-half use a social bookmarking site, with Pinterest the first choice for most females. The survey sample size of 1,000 interviews provides a standard error at 95% confidence of ±3%.
>
> Student Name

Step 9

**10**   Click the Review tab and then click the Check Document button in the Proofing group.

**11**   Click the 1 Result button near the top of the Editor task pane to check the error detected by Word.

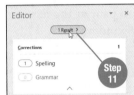

**12**   When the word *Hitwise* is selected, click *Ignore All* in the Editor task pane.

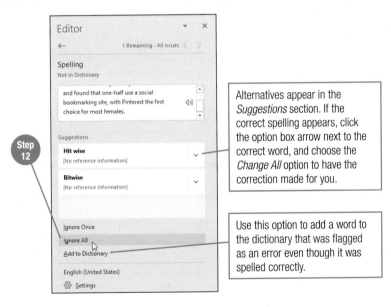

Step 12

Alternatives appear in the *Suggestions* section. If the correct spelling appears, click the option box arrow next to the correct word, and choose the *Change All* option to have the correction made for you.

Use this option to add a word to the dictionary that was flagged as an error even though it was spelled correctly.

**13**   If the spelling feature stops at other typing errors in your document, click the option box arrow next to the correct spelling entry in the *Suggestions* section, and then click *Change All*. If the correct spelling is not offered, click in the document at the end of the typing mistake, press the Backspace key to delete the characters, type the correct text, and then click the Resume button in the Proofing task pane.

**14**   Click OK at the message that the spelling and grammar check is complete and then close the Editor task pane.

**15**   Save the document using the same name. Leave the document open for the next topic.

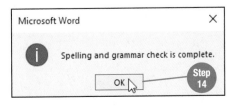

Microsoft Word

Spelling and grammar check is complete.

OK

Step 14

# 6.3 Finding and Replacing Text

The **Find** command moves the insertion point to each occurrence of a word or phrase. Find is helpful if you think you have overused a particular term and want to review how many times it appears in a document, or if you want to move the insertion point to a specific location in the document very quickly and a unique word or phrase exists near the location. **Replace** looks for each occurrence of a word or phrase and automatically changes the text to another word or phrase that you specify. Use Replace to make a global change throughout the document, such as changing a person's name in a will or legal contract.

1. With the **SocialMediaProject-YourName** document open, click at the top left of the document to position the insertion point at the beginning of the title text.

2. Click the Home tab.

3. Click the Find button in the Editing group. Do *not* click the down-pointing arrow on the button.

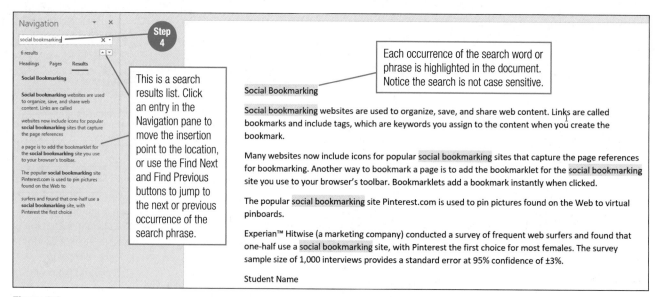

This opens the Navigation pane at the left side of the Word document window.

4. Type social bookmarking in the search text box.

When you finish typing, Word highlights in the document all the occurrences of the search word or phrase and displays the search results below the search text box. The total number of occurrences appears at the top of the *Results* list. Each entry in the search results list is a link that moves to the search word location in the document when clicked (Figure 6.3).

**Figure 6.3**

Search results for *social bookmarking* are highlighted in the **SocialMediaProject-YourName** document.

5. Click each entry one at a time in the *Results* list in the Navigation pane.

Notice that each occurrence of the search word is selected as you move to the phrase location.

 **Watch & Learn**

 **Hands On Activity**

●– – – – – – – **Skills**

Use Find Command

Use Replace Command

 **Tutorials**

Finding Text

Navigating Using the Navigation Pane

●– – – – – – **App tip**

You can find text using a partial word search. For example, enter *exp* to find *Experian*, *expert*, and *experience*.

**6** Click the Close button at the top right of the Navigation pane to close the pane.

**7** Press Ctrl + Home to position the insertion point at the beginning of the document.

Find and Replace searches begin from the location of the insertion point in the document. Move the insertion point when necessary before a Replace command.

**8** Click the Replace button in the Editing group on the Home tab.

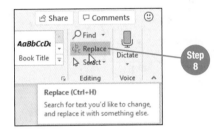

**9** In the Find and Replace dialog box, with *social bookmarking* selected in the *Find what* text box, type bookmarklet.

**10** Press Tab or click in the *Replace with* text box and then type bookmark button.

**11** Click the Replace All button.

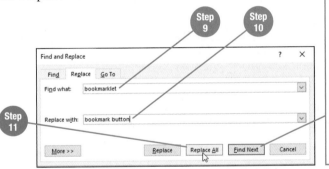

**12** Click OK at the message that 2 replacements were made.

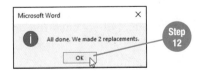

**13** Click the Close button to close the Find and Replace dialog box.

Notice that Word matches the correct case of a word when the word is replaced at the beginning of a sentence, as seen in the last sentence of the paragraph that begins with *Many websites now include.*

**14** Save the document using the same name. Leave the document open for the next topic.

**quick steps**

**Find Text**
1. Position insertion point at beginning of document.
2. Click Find button.
3. Type search text.
4. Review *Results* list.

**Replace Text**
1. Position insertion point at beginning of document.
2. Click Replace button.
3. Type *Find what* text.
4. Press Tab or click in *Replace with* text box.
5. Type replacement text.
6. Click Replace All.
7. Click OK.
8. Click Close button.

**Tutorial**

Finding and Replacing Text

**Oops!**

No replacements made at Step 12? Click OK and check your spelling in the *Find what* text box. Correct the text and try Replace All again. Still have 0 replacements? Cancel the Replace command and check the spelling of *bookmarklet* within your document. Correct the text in the document and repeat Steps 8 to 12.

# 6.4 Moving Text and Inserting Bullets and Numbering

The Cut button in the Clipboard group on the Home tab is used to move a selection of text from one location in the document to another location. Bulleted and numbered lists set apart information that is structured in short phrases or sentences. A bulleted list is text set apart from a paragraph with a list of items that are in no particular sequence. The **Bullets button** in the Paragraph group is used to create this type of list. A numbered list is used for a sequential list of tasks, items, or other text and is created using the **Numbering button** in the Paragraph group.

① With the **SocialMediaProject–YourName** document open, position the insertion point at the beginning of the paragraph that begins with *The popular social bookmarking site.*

② Click the Show/Hide button in the Paragraph group on the Home tab.

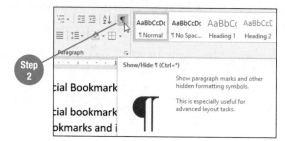

Show/Hide turns on the display of hidden formatting symbols. For example, each time you press Enter, a paragraph symbol (¶) is inserted in the document. Each time you press the spacebar, a dot is inserted (·). Revealing these symbols is helpful when you are preparing to move or copy text because you often want to make sure you move or copy the paragraph symbol with the paragraph.

③ Select the paragraph *The popular social bookmarking site Pinterest.com is used to pin pictures found on the Web to virtual pinboards.* Make sure to include the paragraph formatting symbol at the end of the text in the selection as shown in the image below.

④ Click the Cut button in the Clipboard group on the Home tab.

The text is removed from the current location and placed in the Clipboard.

⑤ Position the insertion point at the beginning of the paragraph that begins with *Many websites now include.*

⑥ Click the top of the Paste button in the Clipboard group. Do *not* click the down-pointing arrow on the button.

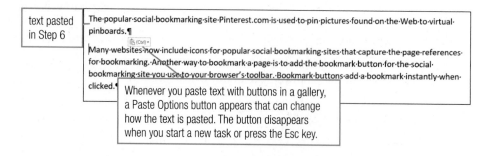

text pasted in Step 6

Whenever you paste text with buttons in a gallery, a Paste Options button appears that can change how the text is pasted. The button disappears when you start a new task or press the Esc key.

 **Watch & Learn**

 **Hands On Activity**

— — — — — Skills

Move text
Create a bulleted list
Create a numbered list

 Oops!

Forgot how to select text? Refer to Topic 3.4 in Chapter 3 for help with selecting text and objects.

7 Click the Show/Hide button to turn off the display of hidden formatting symbols.

8 Position the insertion point after the period that ends the sentence *The popular social networking site Pinterest.com is used to pin pictures found on the Web to virtual pinboards*, press the spacebar, type Other social bookmarking sites include:, and then press Enter.

9 Click the left part of the Bullets button in the Paragraph group. Do *not* click the down-pointing arrow on the button.

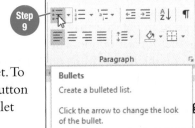

Step 9

Paragraph

**Bullets**

Create a bulleted list.

Click the arrow to change the look of the bullet.

This action indents and then inserts the default bullet character, which is a solid round bullet. To choose a different bullet symbol, click the Bullets button arrow to choose another bullet character in the Bullet Library.

10 Type Mix.com and press Enter.

11 Type Pinboard.in and press Enter.

12 Type Digg.com and press Enter.

13 Type Folkd.com.

> The popular social bookmarking site Pinterest.com is used to pin pictures found on the Web to virtual pinboards. Other social bookmarking sites include:
>
> Steps 10-13
> - Mix.com
> - Pinboard.in
> - Digg.com
> - Folkd.com

14 Position the insertion point after the period that ends the sentence *Bookmark buttons add a bookmark instantly when clicked*, press the spacebar, type To add a bookmark button:, and then press Enter.

15 Click the left part of the Numbering button in the Paragraph group. Do *not* click the down-pointing arrow on the button.

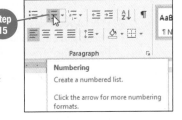

Step 15

Paragraph

**Numbering**

Create a numbered list.

Click the arrow for more numbering formats.

This indents and inserts *1.* followed by a tab. To choose another number style, such as *1)* or *A.*, click the Numbering button arrow to choose another number option in the Numbering Library.

16 Type Display the browser's Favorites toolbar or Bookmarks bar. and then press Enter.

17 Type Right-click the bookmark button and then choose the Add to favorites option. and then press Enter.

18 Type Choose the Add button at the dialog box that appears.

> Many websites now include icons for popular social bookmarking sites that capture the page references for bookmarking. Another way to bookmark a page is to add the bookmark button for the social bookmarking site you use to your browser's toolbar. Bookmark buttons add a bookmark instantly when clicked.
>
> Steps 16-18
> 1. Display the browser's Favorites toolbar or Bookmarks bar.
> 2. Right-click the bookmark button and then choose the Add to favorites option.
> 3. Choose the Add button at the dialog box that appears.

19 Save the document using the same name. Leave the document open for the next topic.

# 6.5 Formatting Text with Font and Paragraph Alignment Options

Changing the appearance of text is called **formatting**. Changing the appearance of characters is called **character formatting**. Changing the appearance of a paragraph is called **paragraph formatting**. Some people format text as they type the document by changing the character or paragraph format option before typing the text. Depending on the format option in use, you may have to turn off the option after typing text, or change to another format option. For example, if you want to apply bold to text as you type, turn on the bold feature, type the text, and then turn off the bold feature. Other people type the document text first and then apply formatting options. For many formatting options, the first step in formatting text that has already been typed is to select the characters or paragraphs to be changed.

## Applying Character Formatting

The Font group on the Home tab contains buttons to change character formatting. A **font** includes the design and shape of the letters, numbers, and special characters of a particular typeface. A large collection of fonts is available, from simple to artistic character design. The font size is set in points, with one point equal to approximately 1/72 of an inch in height. The default font in a new document is 11-point Calibri.

The Font group includes buttons to increase or decrease the font size; change the case; change the font style (bold, italic, or underline) or font color; highlight text; add the effects strikethrough, subscript, or superscript; or add an artistic look using the Text Effects and Typography button (outline, shadow, glow, and reflection accents).

① With the **SocialMediaProject-YourName** document open, select the title text, *Social Bookmarking*.

② Click the *Font* option box arrow on the Mini toolbar.

③ Scroll down and then click *Century Gothic*.

④ With the title text still selected, click twice the Increase Font Size button on the Mini toolbar or in the Font group on the Home tab.

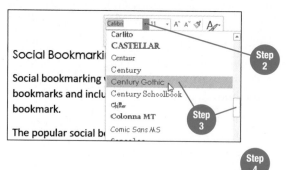

The first time you increase the font size, the size changes to 12 points. The second time, the size changes to 14 points and continues to increase 2 point sizes each time until you reach 28. After 28 points, the size changes to 36, 48, and then 72.

⑤ With the title text still selected, click the Bold button on the Mini toolbar or in the Font group on the Home tab.

⑥ With the title text still selected, click the Font Color button (*not* the Font Color button arrow) on the Mini toolbar or in the Font group.

The default Font Color is Red.

**7** Click in the document away from the selected title to deselect the text.

## Applying Paragraph Formatting

The Paragraph group on the Home tab contains the buttons used to change paragraph formatting. The bottom row of buttons in the group contains the buttons for changing the alignment of paragraphs from the default **Align Left** to **Center**, **Align Right**, or **Justify**. Justified text adds space within a line so that the text is distributed evenly between the left and right margins. You will explore other buttons in the Paragraph group in the next topic.

**8** Click to place the insertion point anywhere within the title text *Social Bookmarking*.

To format a single paragraph, you do not need to select the paragraph text, because paragraph formatting applies to all text within the paragraph to the point where the Enter key was pressed.

**9** Click the Center button in the Paragraph group.

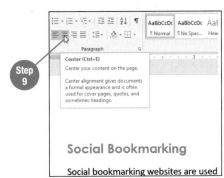

**10** Click to place the insertion point anywhere within the first paragraph of text and then click the Justify button in the Paragraph group.

Justified text spreads the lines out so that the text ends evenly at the right margin.

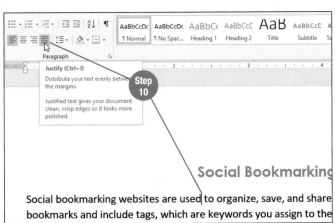

**11** With the insertion point still positioned in the first paragraph, click the Align Left button in the Paragraph group.

To change the paragraph alignment for more than one paragraph in a continuous sequence, select the paragraphs first and then click the desired alignment option on the ribbon.

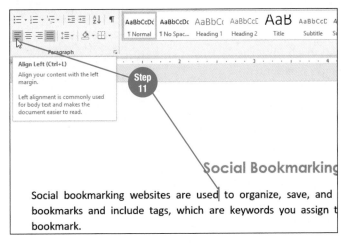

**12** Save the document using the same name. Leave the document open for the next topic.

## 6.6 Indenting Text and Changing Line and Paragraph Spacing

Watch & Learn

Hands On Activity

————— Skills

Indent text

Change line spacing

Change spacing after paragraphs

Paragraphs are indented to set the paragraph apart from the rest of the document. In reports, essays, or research papers, long quotes are indented. A paragraph can be indented for the first line only or for all lines in the paragraph. Paragraphs can also be indented from the right margin. A paragraph where the first line remains at the left margin but subsequent lines are indented is called a **hanging indent**. Hanging indents are used in bulleted lists, numbered lists, bibliographies, and lists of cited works.

Use the **Line and Paragraph Spacing button** in the Paragraph group on the Home tab to change the spacing between lines of text within a paragraph and to change the spacing before and after paragraphs.

① With the **SocialMediaProject-YourName** document open, position the insertion point at the left margin of the first paragraph (begins with the text *Social bookmarking websites*).

② Press the Tab key.

Pressing the Tab key indents the first line of a paragraph 0.5 inch and is referred to as a *first line indent*. The AutoCorrect Options button also appears. Use the button when it appears to change the first line indent back to a Tab, to stop setting indents when you press Tab, or to change other AutoFormat options.

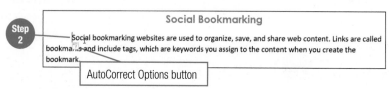

Tutorial

Indenting Text

③ Position the insertion point anywhere within the second paragraph (begins with the text *The popular*).

④ Click the Increase Indent button in the Paragraph group on the Home tab.

The **Increase Indent button** indents all lines of a paragraph 0.5 inch. Click the button more than once if you want to move the paragraph farther away from the margin. Each time the button is clicked, the paragraph is indented another 0.5 inch.

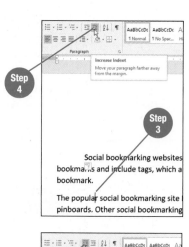

⑤ With the insertion point still positioned in the second paragraph, click the Decrease Indent button in the Paragraph group.

**Decrease Indent button** moves the paragraph back toward the left margin. When a paragraph has been indented more than one position, clicking the Decrease Indent button moves the paragraph back towards the left margin 0.5 inch each time the button is clicked.

⑥ Position the insertion point anywhere within the third paragraph (begins with the text *Many websites now include*).

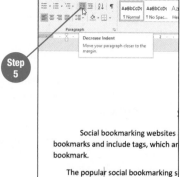

————— app tip

Use the Clear All Formatting button in the Font group to remove all formatting from selected text.

7 Click the Line and Paragraph Spacing button in the Paragraph group and then click *Line Spacing Options*.

8 Select the current entry in the *Left* text box and then type 0.5.

9 Select the current entry in the *Right* text box and then type 0.5.

10 Click OK.

The paragraph is indented from both margins by 0.5 inch.

11 With the insertion point still positioned in the paragraph that begins with the text *Many websites*, click the Line and Paragraph Spacing button and then click *Line Spacing Options*.

12 Change the entry in the *Left* and the *Right* text boxes to 0.

13 Click the *Special* option box arrow and then click *First line*.

14 Click OK.

15 Format the second paragraph, which begins with the text *The popular*, and the last paragraph, which begins with the text *Experian*, with a first line indent by opening the Paragraph dialog box and then changing the *Special* option to *First line*.

16 Click the Select button in the Editing group on the Home tab and then click *Select All*.

Another way to select all the text in a document is to use the keyboard shortcut Ctrl + A.

17 Click the Line and Paragraph Spacing button and then click *1.5*.

The line spacing is changed to 1.5 lines for the entire document. Notice the other line spacing options are *1.0, 1.15, 2.0, 2.5,* and *3.0.*

18 With the entire document still selected, click the Line and Paragraph Spacing button and then click *Remove Space After Paragraph*.

19 Click in any text in the document to deselect the text.

20 Save the document using the same name. Leave the document open for the next topic.

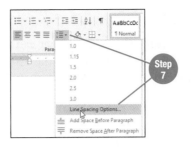

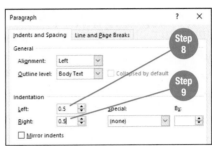

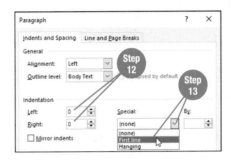

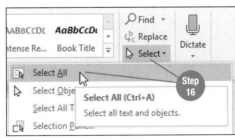

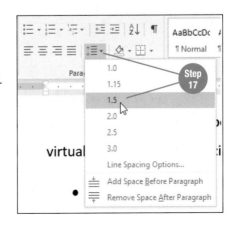

## 6.7 Formatting Using Styles

A **style** is a set of predefined formatting options that can be applied to selected text or paragraphs with one step. The Styles gallery on the Home tab displays style options. You can also create your own style. Use the More button at the bottom of the scroll bar in the Styles gallery to access the options *Create a Style*, *Clear Formatting*, and *Apply Styles*.

Once styles have been applied to text in the document, buttons in the Document Formatting group on the Design tab change the **style set**. A style set changes the look of a document by applying a group of formatting options to the active style. Each style set has different formatting options associated with it.

**Watch & Learn**

**Hands On Activity**

●━━━━━━ SKILLS

Apply styles
Change style set

1. With the **SocialMediaProject-YourName** document open, click the File tab and then click *Save As*.

2. Click the <u>More options</u> hyperlink in the current file name section in the right panel of the backstage area.

3. Click in the *File name* text box in the Save As dialog box to deselect the file name, and then click a second time to place the insertion point at the end of the word *Project* and before -*YourName*.

4. Type WithStyles.

The file name is now **SocialMediaProjectWithStyles-YourName**.

5. Click the Save button.

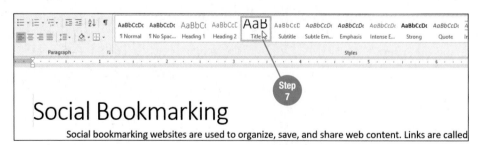

6. Position the insertion point anywhere within the title text *Social Bookmarking*.

7. Click the *Title* style in the Styles gallery on the Home tab.

**Tutorial**

Applying Styles and Style Sets

8. Position the insertion point anywhere within the first paragraph of text below the title.

9. Click the *Quote* style option in the Styles gallery. If the *Quote* style option is not visible, click the More button (⌄) at the bottom of the scroll bar at the right of the Styles gallery and then click the *Quote* style option.

Step 9

first paragraph with *Quote* style preview

10 Select *Pinterest.com* in the second paragraph and then click the *Intense Reference* style.

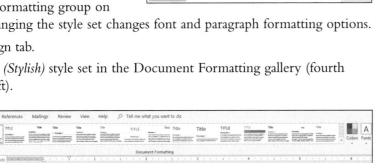

Step 10

11 Deselect *Pinterest.com*.

Once styles have been applied to text, you can experiment with various style sets in the Document Formatting group on the Design tab. Changing the style set changes font and paragraph formatting options.

12 Click the Design tab.

13 Click the *Basic (Stylish)* style set in the Document Formatting gallery (fourth option from left).

Step 12   Step 13

The *Basic (Stylish)* style set causes the look of the document to change.

14 Position the insertion point anywhere within the title text *Social Bookmarking*, click the Home tab, and then click the *Heading 1* style.

 **Tutorial**

Applying and Modifying a Theme

The title text formats with the options for the Heading 1 style in the new style set. The formatting applied by a style is determined by the **theme**. Each theme has a color scheme and a font scheme that affect style options. You can change the theme by choosing another theme option in the Themes gallery in the Document Formatting group on the Design tab.

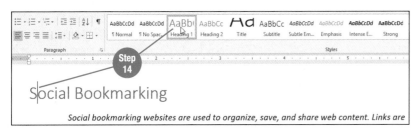

Step 14

15 Save the document using the same name (**SocialMediaProjectWithStyles -YourName**).

16 Click the File tab and then click *Close* to close the document, leaving Word open for the next topic.

## 6.8 Creating a New Document from a Template

A **template** is a document that has been created with formatting options applied. Several professional-quality templates for various types of documents are available that you can use rather than creating a new document from scratch. At the Word Start screen, browse and preview available templates by category, or type a keyword for the type of document you are looking for in the search text box (contains the text *Search for online templates*) and browse the templates in the search results. The New backstage area is used to browse for a template.

1  Click the File tab and then click *New*.

2  In the New backstage area, click *Business* in the *Suggested searches* section.

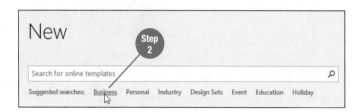

3  Scroll down and review the various types of business templates available, then click the Back button (left arrow) at the top of the New backstage area.

4  Click in the search text box, type time sheet, and then press Enter or click the Start searching icon (magnifying glass at the end of the search text box).

5  Click the *Time sheet* template in the Templates gallery that looks like the one in the image shown below.

A preview of the template opens with a description that provides information on the template design.

6  Click the Create button.

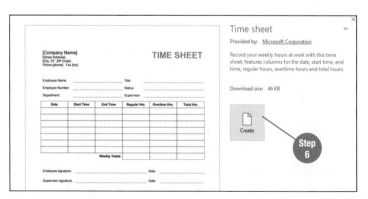

**Watch & Learn**

**Hands On Activity**

— — — — — — — Skills

Create a new document from a template

**Tutorial**

Creating a Document Using a Template

— — — — — — — Oops!

Your *Time sheet* template looks different? Available templates are changed often. Look for the template closest to the one displayed. If necessary, adjust the remaining instructions to suit the available template.

7  Click *[Company Name]* and then type A+ Tutoring Advantage.

8  Click *[Street Address]* and then type 1015 Montreal Way.

9  Click *[City, ST ZIP Code]* and then type St. Paul, MN 55102.

10  Click *[phone]* and then type 888-555-3125.

11  Click *[fax]* and type 888-555-3445.

12  Click next to *Employee Name* and then type your name.

13  Complete the remainder of the time sheet document using the text shown in Figure 6.4.

**A+Tutoring Advantage**
1015 Montreal Way
St. Paul, MN 55102
Phone 888-555-3125   Fax 888-555-3445

Steps 7-11

**TIME SHEET**

Step 12

| Employee Name: | Student Name | Title: | Computer Tutor |
| Employee Number: | 101 | Status: | Part-time |
| Department: | Computers | Supervisor: | Dayna Summerton |

| Date | Start Time | End Time | Regular Hrs. | Overtime Hrs. | Total Hrs. |
|---|---|---|---|---|---|
| Oct 12 | 9:00 am | 12:00 pm | 3.0 | | 3.0 |
| Oct 13 | 1:00 pm | 4:30 pm | 3.5 | | 3.5 |
| Oct 14 | 7:00 pm | 9:30 pm | 2.5 | | 2.5 |
| Oct 15 | 10:00 am | 1:00 pm | 3.0 | | 3.0 |
| | | | | | |
| | | | | | |
| | | | | | |
| | **Weekly Totals** | | 12.0 | | 12.0 |

Employee signature: _____   Date: _____

Supervisor signature: _____   Date: _____

**Figure 6.4**
Shown is a completed **Oct12to15TimeSheet-YourName** document.

14  Save the completed time sheet in the Ch6 folder in the CompletedTopics folder on your storage medium as **Oct12to15TimeSheet-YourName**. Click OK when a message displays that the document will be upgraded to the newest file format.

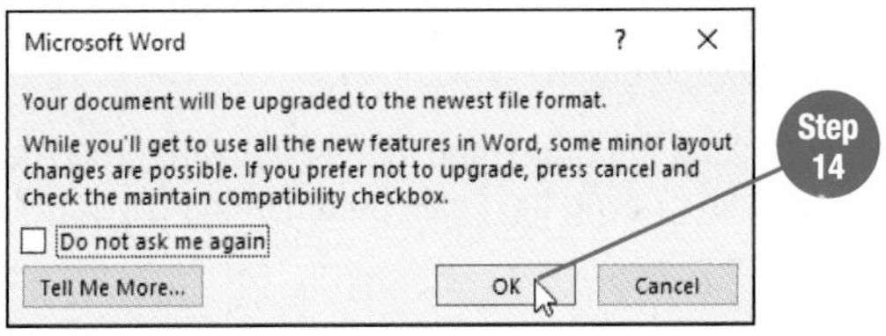

| Microsoft Word | ? | X |
|---|---|---|

Your document will be upgraded to the newest file format.

While you'll get to use all the new features in Word, some minor layout changes are possible. If you prefer not to upgrade, press cancel and check the maintain compatibility checkbox.

☐ Do not ask me again

[ Tell Me More... ]   [ OK ]   [ Cancel ]

Step 14

15  Close the document.

## Topics Review

| Topic | Key Concepts |
|---|---|
| **6.1 Creating and Editing a New Document** | A **word processing application** is software used to create documents that are composed mostly of text. |
| | Start a new blank document from the Word Start screen. |
| | Creating a new document includes typing text, editing text, and correcting errors. |
| | The term **wordwrap** describes Word moving text to the next line automatically when you reach the right margin. |
| | Double-clicking on the page in blank space and typing is referred to as **click and type**. Text is automatically aligned left, center, or right depending on the location in the line at which click and type occurs. |
| | As you type new text, **AutoCorrect** fixes common misspellings, and **AutoFormat** automatically converts some text to fractions, ordinals, quotes, hyphens, and hyperlinks. |
| | A change made to text that has already been typed is referred to as **editing** and involves inserting, deleting, and replacing characters. |
| | The first step in editing is to position the insertion point at the location of the change. |
| **6.2 Inserting Symbols and Completing a Spelling and Grammar Check** | Symbols or special characters, such as a copyright symbol or registered trademark, are entered using the **Symbol gallery** or Symbol dialog box. |
| | The **Check Document** button is used to match words in the document with words in the dictionary; words not found are flagged as potential errors. Click the Result(s) button near the top of the Editor task pane to review potential errors found. |
| | A word not found in the dictionary is highlighted, and suggestions for replacement appear in the Editor task pane. |
| | *Ignore Once*, *Ignore All*, and *Add to Dictionary* are options in the Editor task pane to respond to each potential error. Alternative words appear in the *Suggestions* section. A suggested word can be substituted for the error using the option box arrow next to the correctly spelled alternative. |
| **6.3 Finding and Replacing Text** | The **Find** command highlights all occurrences of a keyword or phrase and provides in the Navigation pane a link to each location in the document. |
| | Use the **Replace** command if you want Word to automatically change each occurrence of a keyword or phrase with another word or phrase. |
| | Word matches the case of the word when the word is replaced at the beginning of a sentence. |
| **6.4 Moving Text and Inserting Bullets and Numbering** | A bulleted list is set apart from the paragraph and is entered in no particular sequence. The **Bullets button** in the Paragraph group on the Home tab is used to type a bulleted list. |
| | A numbered list is a sequential series of tasks or other items that are each preceded by a number created using the **Numbering button** in the Paragraph group. |
| | Displaying hidden formatting symbols is helpful when moving text to make sure the paragraph symbol is selected before cutting the text. Turn on the display of hidden formatting symbols using the Show/Hide button in the Paragraph group. |
| | Selected text is moved by clicking the Cut button in the Clipboard group, moving the insertion point to the desired location, and then clicking the Paste button in the Clipboard group. |

*continued…*

| Topic | Key Concepts |
|---|---|
| **6.5 Formatting Text with Font and Paragraph Alignment Options** | Changing the appearance of text is called **formatting**. **Character formatting** involves applying changes to the appearance of characters, whereas **paragraph formatting** changes the appearance of an entire paragraph. |
| | A **font** is also called a *typeface* and refers to the design and shape of letters, numbers, and special characters. Change a font, font size, case, font style, and font color; highlight text; and add font effects to change character formats. |
| | Change a paragraph's alignment from the default **Align Left** to **Center**, **Align Right**, or **Justify** using the buttons in the Paragraph group on the Home tab. |
| | Justified text has extra space within a line so that the left and right margins are even. |
| **6.6 Indenting Text and Changing Line and Paragraph Spacing** | A paragraph in which all lines are indented except the first line is called a **hanging indent**. |
| | Change line spacing and the amount of space before and after a paragraph using options from the **Line and Paragraph Spacing button** in the Paragraph group or in the Paragraph dialog box. |
| | Press Tab at the beginning of a paragraph to indent only the first line or change *Special* to *First line* in the Paragraph dialog box. |
| | Indent all lines of a paragraph using the **Increase Indent button** or change the *Left* text box entry in the Paragraph dialog box. |
| | A paragraph indents 0.5 inch each time the Increase Indent button is clicked. |
| | Use the **Decrease Indent button** to move a paragraph closer to the left margin; the paragraph moves left 0.5 inch each time the button is clicked. |
| | Indent a paragraph from both margins using the *Left* and *Right* text boxes in the Paragraph dialog box. |
| **6.7 Formatting Using Styles** | Format text by applying a **style**, which is a set of predefined formatting options. |
| | Change the **style set** using buttons in the Document Formatting group on the Design tab. Each style set applies a different set of formatting options for the styles on the Home tab, meaning you can change the appearance of a document by changing the style set. |
| | A **theme** is a set of colors, fonts, and font effects that alter the appearance of a document. |
| **6.8 Creating a New Document from a Template** | A **template** is a document that is already set up with text and/or formatting options. |
| | Browse available templates in the Template gallery on the Word Start screen or in the New backstage area. |
| | Find a template by browsing the gallery by a category or by typing a keyword in the search text box. |
| | Click a template design to preview the template and create a new document based upon the template. |
| | Within a template, text placeholders or instructional text is included to help you personalize the document. |

## Topics Review

# Enhancing a Word Document with Special Features

Several features in Word allow you to add visual appeal, organize information, or format a document for a special purpose, such as a research paper. Word provides different views to work in, making navigating a document easy. Word integrates collaborative tools, such as comments, to help you effortlessly work on a document with multiple people. Word also provides several templates to help you build professional documents without spending a lot of time on design. For example, Word offers résumé and cover letter templates to help you build job search documents.

In this chapter, you will enhance documents already typed by inserting images, adding borders and shading to text, and inserting a text box. You will add text to an existing document in a column-and-row format by inserting a table. To finalize an academic research paper, you will learn to format an essay according to the MLA style guide, insert citations, and create a Works Cited page. Lastly, you will create a résumé and cover letter using templates.

Throughout the step-by-step instructions, you will notice prompts to save the document you're working on. These prompts are simply reminders. You are encouraged to save more frequently to avoid losing your work.

## Learning Objectives

**7.1** Insert, edit, and label images in a document

**7.2** Add borders and shading to text and insert a text box in a document

**7.3** Insert a table in a document

**7.4** Format and modify a table

**7.5** Change layout options

**7.6** Add text and page numbers in a header for a research paper

**7.7** Insert and edit citations in a research paper

**7.8** Create a Works Cited page for a research paper and display a document in different views

**7.9** Insert and reply to comments in a document

**7.10** Create a résumé and cover letter from templates

### Read & Learn

### Content Online

The online course includes additional training and assessment resources.

# 7.1 Inserting, Editing, and Labeling Images in a Document

Adding a graphic element, such as an image, not only adds visual appeal to a document but can also help a reader understand content. You can insert an image from a web resource using the **Online Pictures button** in the Illustrations group on the Insert tab. Once an image has been inserted, you can change the way text wraps around the sides of the image using the **Layout Options button** that appears. Resize the image with the selection handles or drag the image to move it to another location in the document. Use buttons on the contextual Picture Tools Format tab to edit the image.

## Inserting an Image from an Online Source

You can search for and insert an image from the web using the search text box in the Online Pictures dialog box. By default, Bing search results apply a copyright filter displaying images tagged with a Creative Commons license. This means that to use the image in a document, you need to provide attribution (credit to the image author). Word provides a hyperlink to the source website of an image inserted from the Online Pictures dialog box.

1. Start Word and open **InsulaSummary** from the Ch7 folder in StudentDataFiles. If necessary, click the Enable Editing button to close Protected view.

2. Click the File tab and then click *Save As*. Navigate to the CompletedTopics folder on your storage medium and create a new folder, *Ch7*. Double-click *Ch7*, click twice in the *File name* text box after the *y* in *InsulaSummary* and before the period, type – followed by your name, and then click the Save button. The file name is now **InsulaSummary-YourName.docx**.

3. Click to place the insertion point at the beginning of the first paragraph.

4. Click the Insert tab and then click the Online Pictures button in the Illustrations group.

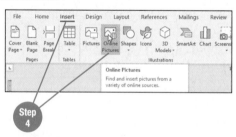

Step 4

5. With the insertion point positioned in the search text box, type brain and then press Enter.

Step 5

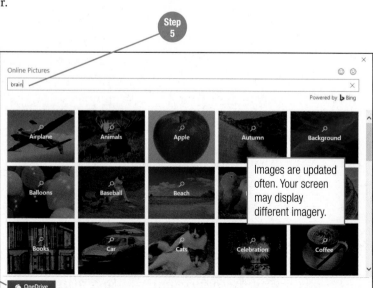

Images are updated often. Your screen may display different imagery.

Images you have stored on OneDrive can be seen by clicking the OneDrive button.

●------- Skills

Insert an image from a web resource

Insert an image from a file on your computer

Edit an image

Insert a caption

●------- app tip

The Online Video button in the Media group on the Insert tab is used to add videos to a Word document.

**6** If necessary, scroll down the search results list to the checked image shown in the Step 6 illustration, click to select the image, and then click the Insert button. If you cannot locate the image shown, close the Online Pictures dialog box and then insert the image using the Pictures button in the Illustrations group (see Steps 13 to 14). Use the image **Brain.jpg** in the Ch7 folder in StudentDataFiles.

**7** Click the Layout Options button located at the upper right of the inserted image.

**8** Click *Square*, the first option in the *With Text Wrapping* section in the Layout Options gallery, and then click the Close button.

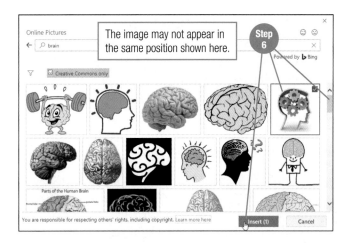

The Layout Options gallery allows you to control how text wraps around the image and to control whether the image should remain fixed at its current position or move with the text. Notice also that the Picture Tools Format tab becomes active when an image is selected. You will work with buttons on this tab later in this topic.

**9** If necessary, scroll down to see the selection handles along the bottom of the image. Drag the bottom right corner selection handle on the image up and left until the image is resized to the approximate size shown in the Step 9 illustration. (You do not need to match the size exactly.)

**10** Move the mouse pointer anywhere over the image and then drag the image right until the right edge of the image is aligned at the right margin. Use the green alignment guides to help you position the image as shown in the Step 10 illustration.

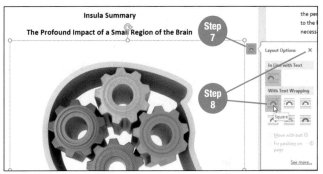

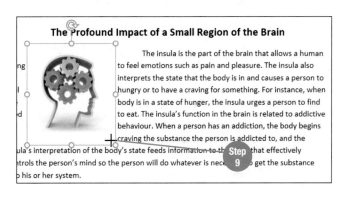

When the mouse pointer is positioned over an image, the pointer displays with a four-headed arrow attached. The four-headed arrow indicates a move command will occur when you drag the mouse.

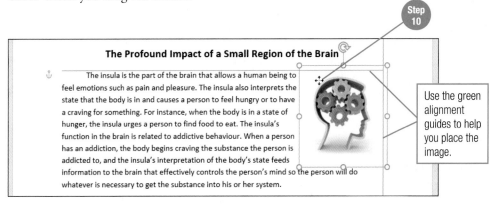

**11** Save the revised document using the same name (**InsulaSummary-Your Name**).

## Inserting an Image from a File

Images you have scanned or imported from a digital camera to a storage device, or other images you have saved on any storage medium, can be inserted into a document using the **Pictures button** in the Illustrations group. The Layout Options button also appears for an image inserted from a file.

12  Click to place the insertion point at the beginning of the second paragraph.

13  Click the Insert tab and then click the Pictures button in the Illustrations group.

14  Navigate to the Ch7 folder in the StudentDataFiles folder on your storage medium and then double-click *CampusBldg.jpg*.

15  Click the Layout Options button, click *Square*, and then close the Layout Options gallery.

16  Resize the image to the approximate size shown in the Step 16 illustration below and align the photo at the left margin.

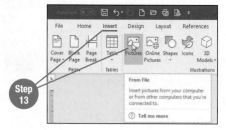

Step 13

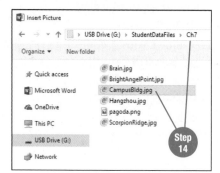

Step 14

Step 16

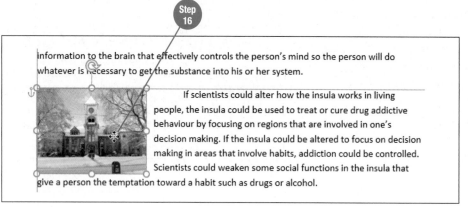

## Editing a Picture

Buttons on the Picture Tools Format tab are used to edit an image. Use buttons in the Adjust group to modify the appearance of an image, such as the sharpness, contrast, or color tone, or to apply an artistic effect. Add a border or image effect with options in the Picture Styles group. Change the image position, text wrapping option, order, alignment, or rotation with buttons in the Arrange group. Crop or specify exact measurements for the image height and width with buttons in the Size group.

17  Click to select the image at the right margin in the first paragraph.

18  Click *Drop Shadow Rectangle* (fourth option) in the Picture Styles group on the Picture Tools Format tab.

Step 18

19  Click to select the image at the left margin in the second paragraph.

20  Click the Corrections button in the Adjust group and then click *Sharpen: 50%* (last option) in the *Sharpen/Soften* section.

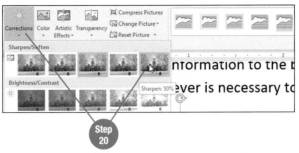

Notice details in the image are more pronounced. The Corrections gallery contains two options to show more softness and two options to show more sharpness in the image. Use the thumbnails in the *Brightness/Contrast* section to show more or less brightness or contrast.

## Inserting a Caption with an Image

Adding a caption below an image helps a reader understand the context of the image, or captions can serve as figure numbers to chronologically order images; e.g. in a formal report. With the **Insert Caption** feature, Word automatically numbers images, inserting the number after the label *Figure*.

21  With the image in the second paragraph still selected, click the References tab.

22  Click the Insert Caption button in the Captions group.

23  With the insertion point positioned in the *Caption* text box, press the spacebar and then type Insula research is done at many universities.

24  Press Enter or click OK.

25  Click in the document away from the caption box to deselect the caption.

Use the Caption feature in a formal report, such as a research paper with several illustrations, to automatically generate a Table of Figures from the caption text added to graphical objects. You can learn how to generate a Table of Figures using the Tell Me feature or Help task pane (see Chapter 3, Topic 3.6).

26  Save the revised document using the same name (**InsulaSummary -YourName**). Leave the document open for the next topic.

**Tutorial**

Creating and Customizing Captions

## 7.2 Adding Borders and Shading, and Inserting a Text Box

Add a border and/or add color behind text (called **shading**) to make text stand out from the rest of a document. You can add borders and shading to a single paragraph or to selected paragraphs. With the **Page Borders button**, you can add a border around the edges of the page. Add a line that spans the entire page width above the insertion point location using the *Horizontal Line* option in the **Borders gallery**.

A text box is used to set a short passage of text apart from the rest of a document. Word has several built-in text box styles that can be used for this purpose. Inserting text inside a box is a way to draw the reader's attention to an important quote or point in a document. A quote inside a text box is called a **pull quote**. Use the **Text Box button** in the Text group on the Insert tab to create a text box.

**1** With the **InsulaSummary-YourName** document open, click the Home tab.

**2** Select the title and subtitle text (first two lines) and then click the Borders button arrow in the Paragraph group.

**3** Click *Outside Borders*.

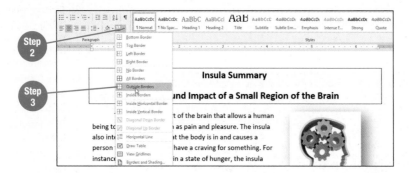

**4** With the first two lines of text still selected, click the Shading button arrow in the Paragraph group.

**5** Click *Orange, Accent 6, Lighter 80%* (last color in second row of *Theme Colors* section).

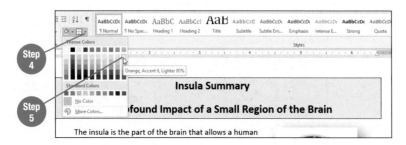

**6** Click in any paragraph to deselect the text.

**7** Click the Design tab.

**8** Click the Page Borders button in the Page Background group.

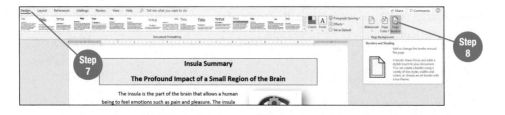

**Watch & Learn**

**Hands On Activity**

●————— Skills

Add a paragraph border

Add shading in a paragraph

Add a border to a page

Insert a text box

**Tutorials**

Applying Borders

Applying Shading

●————— app tip

Use *Borders and Shading* in the Borders gallery to create a custom border in the Borders and Shading dialog box by changing the border style, color, and width options.

⑨ Click the page thumbnail next to *Shadow* in the *Setting* section in the Borders and Shading dialog box with the Page Border tab active.

⑩ Click the *Width* option box arrow and then click *1 ½ pt* in the drop-down list.

⑪ Click OK.

⑫ Click the Insert tab.

⑬ Click the Text Box button in the Text group, scroll down, and then click *Whisp Quote*.

Word places a text box overlapping other text in the document and with default text already selected inside the text box.

⑭ Type Some people with damage to the insula were able to quit smoking instantly!

⑮ Right-click *[Cite your source here.]* and then click *Remove Content Control*.

⑯ Point to the border of the box. When the mouse pointer displays with the four-headed move icon, drag the text box to the bottom of the page to the approximate location shown in the Step 16 illustration below.

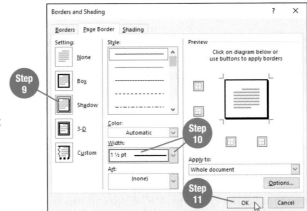

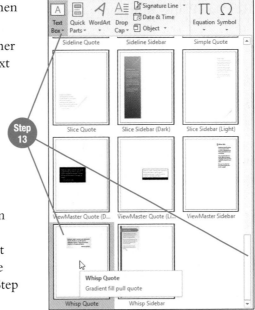

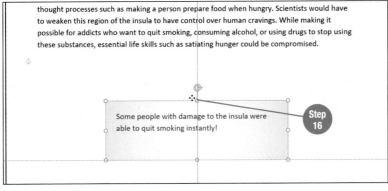

⑰ Drag the top middle selection handle down to reduce the height of the text box to the approximate height shown in the Step 17 illustration at the right.

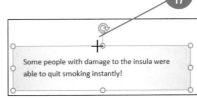

⑱ Click in any paragraph to deselect the text box and then save the revised document using the same name (**InsulaSummary-YourName**).

⑲ Close the document.

---

**quick steps**

**Add a Paragraph Border**
1. Select paragraph(s).
2. Click Borders button arrow.
3. Click desired border style.

**Add Shading to a Paragraph**
1. Select paragraph(s).
2. Click Shading button arrow.
3. Click desired color.

**Add a Page Border**
1. Click Design tab.
2. Click Page Borders button.
3. Select desired *Setting* and *Style*, *Color*, *Width*, or *Art* options.
4. Click OK.

**Insert a Text Box**
1. Click Insert tab.
2. Click Text Box button.
3. Click desired text box style.
4. Type text.
5. Move and/or resize box as needed.

**Tutorials**

Inserting a Page Border

Inserting a Text Box

Formatting a Text Box

## 7.3 Inserting a Table

A table organizes and presents data in columns and rows. Text is typed within a **table cell**, which is a rectangular box that is the intersection of a column with a row. When you create a table, you specify the number of columns and rows the table will hold, and Word creates a blank grid into which you type the table text. Text that you want to place side by side in columns, or in rows, is ideal for a table. For example, a price list or a catalog with items, descriptions, and prices is best typed in a three-column table.

You can also create a table using the Quick Tables feature. A **Quick Table** is a predefined and formatted table with sample data that you can replace with your own text.

1. Open the document **RezMealPlans** from the Ch7 folder in StudentDataFiles. If necessary, click the Enable Editing button to close Protected view.

2. Save the document as **RezMealPlans-YourName** in the Ch7 folder in CompletedTopics.

3. Click to place the insertion point at the left margin in the blank line below the subheading *Meal Plans with Descriptions*.

4. Click the Insert tab and then click the Table button in the Tables group.

5. Click the square in the grid that is three columns to the right and two rows down (*3x2 Table* displays above grid).

6. With the insertion point positioned in the first table cell, type Meal Plan Name and then press Tab or click in the next table cell.

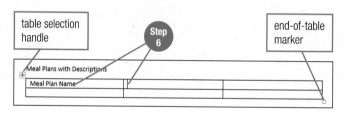

7. Type Cost and then press Tab or click in the next table cell.

8. Type Description and then press Tab or click in the first table cell in the second row.

9. Type the second row of data as specified below. When you finish typing the text in the last column, press Tab to add a new row to the table automatically.

   Minimum     $2,000     Suitable for students with small appetites who plan to be away from residence most weekends.

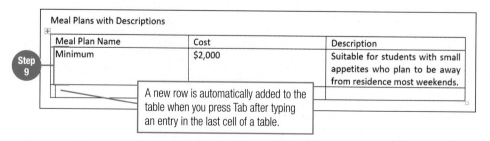

10. Type the remainder of the table (starting with column 1, "Light") as shown in Figure 7.1 on the next page by completing steps similar to those in Steps 6 to 9, except do not press Tab after typing the last table cell entry.

### Watch & Learn

### Hands On Activity

● - - - - - - - Skills

Insert a table

Type text in a new table grid

### Tutorial

Creating a Table

● - - - - - - - app tip

The advantage to using a table versus typing information in tabbed columns is that information can wrap within a table cell.

Meal Plans with Descriptions

| Meal Plan Name | Cost | Description |
|---|---|---|
| Minimum | $2,000 | Suitable for students with small appetites who plan to be away from residence most weekends. |
| Light | $2,200 | Best plan for students with a lighter appetite who spend occasional weekends on campus. |
| Full | $2,400 | Full is the most popular plan. This plan is for students with an average appetite who will stay on campus most weekends. |
| Plus | $2,600 | Students with a hearty appetite who will stay on campus most weekends choose the Plus plan. |

**Figure 7.1**

Text for the first table in the **RezMealPlans-YourName** document is shown.

**quick steps**

**Insert a Table**
1. Click Insert tab.
2. Click Table button.
3. Click square in drop-down grid that represents the desired number of columns and rows.
4. Type table text.
OR
1. Click Insert tab.
2. Click Table button.
3. Click *Insert Table*.
4. Type number of columns.
5. Type number of rows.
6. Click OK.
7. Type table text.

⑪ Click to place the insertion point at the left margin in the blank line below the subheading *Meal Plan Fund Allocations*.

⑫ Click the Insert tab, click the Table button, and then click *Insert Table*.

You can also insert a new table using a dialog box in which you specify the number of columns and rows.

⑬ With the value in the *Number of columns* text box already set to *5*, press Tab or select the value in the *Number of rows* text box, type *5*, and then press Enter or click OK.

⑭ Type the text in the new table as shown in Figure 7.2. Click in the paragraph below the table after typing the text in the last table cell.

⑮ Save the document using the same name (**RezMealPlans-YourName**). Leave the document open for the next topic.

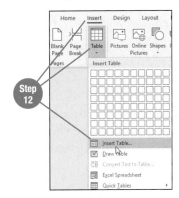

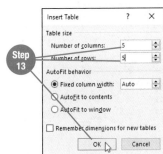

Click outside the table grid after typing the last table cell entry to avoid adding a new row to the table.

Meal Plan Fund Allocations

| Meal Plan Name | Total Cost | Operating Fund | Basic Fund | Flex Fund |
|---|---|---|---|---|
| Minimum | $2,000 | $200 | $1,575 | $225 |
| Light | $2,200 | $200 | $1,725 | $275 |
| Full | $2,400 | $200 | $1,775 | $425 |
| Plus | $2,600 | $200 | $1,850 | $550 |

Note that the Basic fund is tax exempt and is designed for use at all on-campus restaurants. Flex fund purchases are taxable.

**Figure 7.2**

Text for the second table in the **RezMealPlans-YourName** document is shown.

# 7.4 Formatting and Modifying a Table

Once a table has been created, use buttons on the Table Tools Design and Table Tools Layout tabs to format the table and to add or delete rows and columns. Choose from a predesigned collection of formatting options to add borders, shading, and color to a table from the **Table Styles** gallery.

1. With the **RezMealPlans–YourName** document open, click to place the insertion point in any table cell within the first table.

2. Click the Table Tools Design tab.

3. Click the More button located at the bottom of the scroll bar in the Table Styles gallery.

4. Click *Grid Table 4 – Accent 2* (tenth option in second row of *Grid Tables* section; the option's location may vary depending on your screen size and resolution setting).

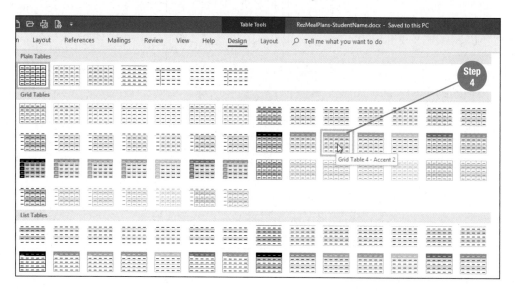

Notice the formatting applied to the column headings and text in the first column. A row with shading and other formatting applied uniformly to every other row to make the data easier to read is referred to as a **banded row**. The border around each cell is now colored orange. The check boxes in the Table Style Options group, the Shading button in the Table Styles group, and the buttons in the Borders group are used to further modify the table formatting.

5. Click the *First Column* check box in the Table Style Options group to clear the check mark.

Notice the bold formatting is removed from the text in the first column.

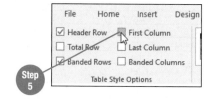

 **Watch & Learn**

 **Hands On Activity**

●───────── SHILLS

Apply and customize a table style

Insert and delete rows and columns

Change column width

Change cell alignment

Merge cells

●──── quick steps

**Format a Table**
1. Position insertion point within table cell.
2. If necessary, click Table Tools Design tab.
3. Click desired Table Style option.
4. Customize Table Style Options, Shading, or Borders as desired.

**Insert Columns or Rows**
1. Position insertion point in desired table cell.
2. If necessary, click Table Tools Layout tab.
3. Click Insert Above, Insert Below, Insert Left, or Insert Right button.

**Delete Columns or Rows**
1. Position insertion point in desired table cell.
2. If necessary, click Table Tools Layout tab.
3. Click Delete button.
4. Click *Delete Cells*, *Delete Columns*, *Delete Rows*, or *Delete Table*.

**Tutorial**

Changing the Table Design

**6** Select the column headings in the first row of the table, click the Shading button arrow on the Mini toolbar or in the Table Styles group, and then click *Orange, Accent 2, Darker 50%* (sixth option in last row of *Theme Colors* section).

| Meal Plan Name | Cost | Description |
|---|---|---|
| Minimum | $2,000 | Suitable for students with sm appetites who plan to be awa from residence most weeken |
| Light | $2,200 | Best plan for students with a lighter appetite who spend occasional weekends on campus. |
| Full | $2,400 | Full is the most popular plan. This plan is for students with an average appetite who will stay on campus most weekends. |
| Plus | $2,600 | Students with a hearty appetite who will stay on campus most weekends choose the Plus plan. |

## Inserting and Deleting Columns and Rows

Buttons in the Rows & Columns group on the Table Tools Layout tab are used to insert or delete columns or rows. Position the insertion point within a table row and click the Insert Above or Insert Below button to add a new row to the table. The Insert Left and Insert Right buttons are used to add a new column to the table.

Position the insertion point within a table cell, select multiple rows or columns or select the entire table, and then click the Delete button to delete cells, a column, a row, or the entire table.

**7** Click to place the insertion point within any table cell in the third row of the first table (begins with *Light*).

**8** Click the Table Tools Layout tab.

**9** Click the Delete button in the Rows & Columns group.

**10** Click *Delete Rows*.

**11** Click to place the insertion point within any table cell in the third row of the second table (begins with *Light*), click the Delete button, and then click *Delete Rows*.

**12** Click to place the insertion point within any table cell in the last column of the first table.

**13** Click the Insert Left button in the Rows & Columns group.

A new column is created between the *Cost* and *Description* columns. Notice that Word adjusts each column width. New rows are inserted by following a similar process.

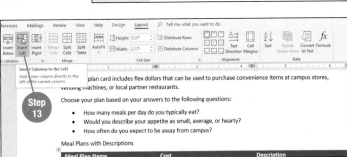

**14** Click to place the insertion point within the table cell in the first row of the new column (between *Cost* and *Description*) and then type Daily Spending.

**15** Type the values below the column heading in rows 2, 3, and 4 as follows:

$18.35

$22.00

$23.85

## Modifying Column Width and Alignment and Merging Cells

☁ **Tutorial**

Changing the Table Layout

Adjust the width of a column by dragging the border line between columns left or right. You can also enter precise width measurements in the *Width* text box in the Cell Size group on the Table Tools Layout tab. Use the buttons in the Alignment group to align text within cells horizontally and/or vertically. Combine two or more cells into one cell using the Merge Cells button, or divide a cell into two or more cells using the Split Cells button in the Merge group.

16 Click to place the insertion point within any table cell in the second column of the first table (column heading is *Cost*).

17 Click the *Table Column Width* option box down arrow in the Cell Size group until the value is *1"*.

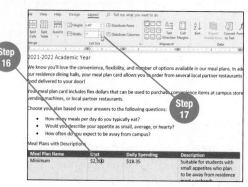

18 Click to place the insertion point within any table cell in the last column of the first table (column heading is *Description*) and then click the *Table Column Width* option box up arrow until the value is *2.3"*.

19 With the insertion point still positioned within the last column of the first table, click the Select button in the Table group and then click *Select Column*.

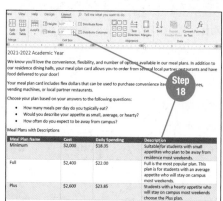

20 Click the Align Top Center button in the Alignment group (second button).

21 Select the first column in the first table (column heading is *Meal Plan Name*) and then click the Align Center button in the Alignment group (second button in second row).

22 Repeat Step 21 to Align Center the second and third columns in the first table.

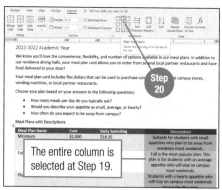

The entire column is selected at Step 19.

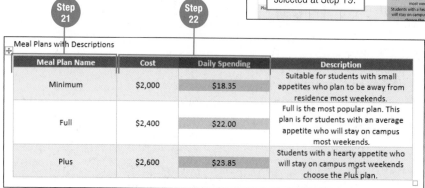

| Meal Plan Name | Cost | Daily Spending | Description |
|---|---|---|---|
| Minimum | $2,000 | $18.35 | Suitable for students with small appetites who plan to be away from residence most weekends. |
| Full | $2,400 | $22.00 | Full is the most popular plan. This plan is for students with an average appetite who will stay on campus most weekends. |
| Plus | $2,600 | $23.85 | Students with a hearty appetite who will stay on campus most weekends choose the Plus plan. |

**23** Click to place the insertion point within any table cell in the first row of the second table and then click the Insert Above button in the Rows & Columns group.

**24** With the new row already selected, click the Merge Cells button in the Merge group.

**25** With the new row still selected, type Breakdown of Meal Plan Cost by Fund and then click the Align Top Center button in the Alignment group.

---- **quick steps**

**Modify Column Width**
1. Position insertion point within table cell in desired column.
2. If necessary, click Table Tools Layout tab.
3. Change *Width* measurement value to desired width.

**Modify Cell Alignment**
1. Select column, row, or table cells.
2. If necessary, click Table Tools Layout tab.
3. Click desired button in Alignment group.

**Merge or Split Cells**
1. Select column, row, or table cells.
2. If necessary, click Table Tools Layout tab.
3. Click Merge Cells or Split Cells button.

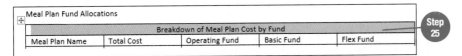

**26** Point in the second table at the beginning of the text *Total Cost* (second column, second row) until the pointer displays as a black diagonal arrow (↗), then drag the mouse right and down to select the column headings and all the values in columns 2, 3, 4, and 5. Click the Align Center button.

You cannot use the Select Column option from the Select button to select the columns, because this option includes the first merged row.

**27** Click to place the insertion point within any table cell in the second table, click the Select button and then click *Select Table*.

You can also select a table by clicking the Table selection handle at the upper left of the table.

**28** Click the Table Tools Design tab.

**29** Click the Borders button arrow in the Borders group and then click *No Border*.

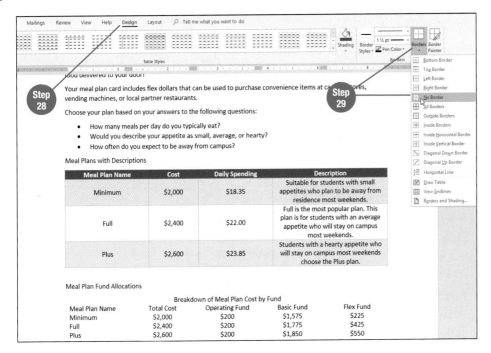

**30** Click in the paragraph below the table to deselect the table.

**31** Apply the Heading 2 style to the headings *Meal Plans with Descriptions* and *Meal Plan Fund Allocations* above the first and second tables.

**32** Save the document using the same name (**RezMealPlans-YourName**) and then close the document.

------ **app tip**

Create a table for any type of columnar text instead of setting tabs. Tables are simpler to create and have several formatting options.

**Tutorial**

Customizing Cells in a Table

# 7.5 Changing Layout Options

By default, new documents in Word are set up for a letter-size page (8.5 × 11 inches) in portrait orientation with 1-inch margins at the left, right, top, and bottom. When the page is vertical (taller than it is wide) it is in **portrait** orientation. In portrait orientation, a page has a standard 6.5-inch line length (8.5 inches minus 2 inches for the left and right margins). This is the orientation commonly used for documents and books. You can change to **landscape** orientation (the page is rotated to make it wider than it is tall). In landscape orientation, the text has a 9-inch line length (11 inches minus 2 inches for the left and right margins).

The Page Setup group on the Layout tab contains buttons to change the margins and page orientation. A Size button allows you to change to a legal size page (8.5 × 14 inches), another predefined page size, or set a custom page width and height. The Columns button formats a document into two or more newspaper-style columns.

**Watch & Learn**

**Hands On Activity**

------------- Skills

Change page orientation
Change margins
Insert a page break

1 Open the document **ChildLitBookRpt** from the Ch7 folder in StudentDataFiles. If necessary, click the Enable Editing button to close Protected view.

2 Save the document as **ChildLitBookRpt-YourName** in the Ch7 folder in CompletedTopics.

3 Click the Layout tab.

4 Click the Orientation button in the Page Setup group and then click *Landscape*.

Notice the width of the page is extended, and the page is now wider than it is tall.

**Tutorials**

Changing Page Orientation

Changing Margins

5 Scroll down to view the document in landscape orientation.

6 With the insertion point positioned at the top of the document, click the Margins button in the Page Setup group.

7 Click *Custom Margins*.

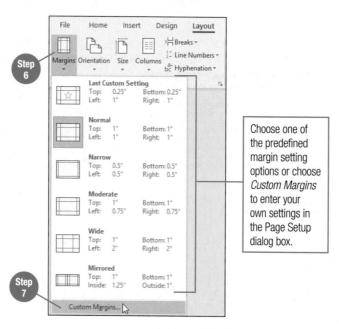

Choose one of the predefined margin setting options or choose *Custom Margins* to enter your own settings in the Page Setup dialog box.

8. With the insertion point positioned in the *Top* text box in the *Margins* section in the Page Setup dialog box, press Tab twice or select the current value in the *Left* text box, and then type 1.2.

9. Press Tab or select the current value in the *Right* text box, type 1.2, and then press Enter or click OK.

10. Scroll down to view the document with the new left and right margin settings.

Sometimes you want to end a page before the point at which Word automatically ends a page and starts a new page (referred to as a **soft page break**). Soft page breaks occur when the maximum number of lines that can fit within the current page size and margins has been reached. A page break that you insert at a different location is called a **hard page break**. Insert hard page breaks as your last step in preparing a document, because hard page breaks do not adjust if you add or delete text.

11. Click to place the insertion point at the left margin next to the subtitle *The Allegories* near the bottom of page 1.

12. Click the Insert tab.

13. Click the Page Break button in the Pages group.

Notice that all the text from the insertion point onward is moved to page 2.

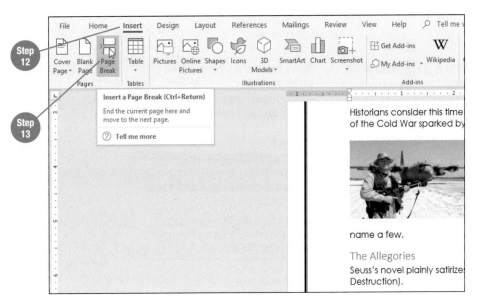

14. Scroll up and down to view the book report with the new page break.

15. Save the document using the same name (**ChildLitBookRpt-YourName**).

16. Close the document.

## 7.6 Adding Text and Page Numbers in a Header for a Research Paper

During the course of your education, chances are you will have to submit a research paper or essay that is formatted for a specific **style guide** (a set of rules for paper formatting and referencing). Style guides are used in academic and professional writing; MLA (Modern Language Association) and APA (American Psychological Association) are two organizations that publish widely used guides. See Table 7.1 for general MLA and APA guidelines. Always check an instructor's assignment instructions for additional requirements that may be unique to the guidelines set forth in Table 7.1.

 **Watch & Learn**

 **Hands On Activity**

● ─ ─ ─ ─ ─ ─ ─ ─ Skills

Add text in a header

Insert page numbers in a header in MLA style

**Table 7.1** Formatting and Page Layout Guidelines for MLA and APA

| Layout Item | MLA | APA |
| --- | --- | --- |
| Paper size and margins | 8.5 × 11" with 1-inch margins | 8.5 × 11" with 1-inch margins |
| Font size | 12-point; typeface is not specified but should be easily readable | 12-point, with Times New Roman the preferred typeface |
| Line and paragraph spacing | 2.0 with no spacing between paragraphs | 2.0 with no spacing between paragraphs |
| Paragraph indent | Indent first line 0.5 inch | Indent first line 0.5 inch |
| Page numbering | Top right of each page with one space after your last name | Top right of each page with the title of the paper all uppercase at the left margin on the same line |
| Title page | No (unless specifically requested by your instructor) | Yes<br>Running head: title of the paper at the left margin all uppercase with page number at the right margin 1 inch from the top; in the upper half of the page centered horizontally include:<br>    Title of the paper<br>    Your name<br>    School name |
| First page | Top left corner (double-spaced):<br>Your name<br>Instructor's name<br>Course title<br>Date<br>A double space after date<br>Title (title case), centered<br>Begin the paper (left-aligned) | Center the word *Abstract* at the top of the page.<br>    Type a brief summary of the paper in a single paragraph in block format (no indents). Limit yourself to approximately 150 words.<br>    Start a new page after the abstract page, center the paper title (title case) at the top of the new page, and then begin typing the paper. |
| Bibliography | Create a separate Works Cited page at the end of the document organized alphabetically by author. | Create a separate References page at the end of the document organized alphabetically by author. |

 **Tutorials**

Formatting a Report in APA Style

Formatting a Report in MLA Style

A **header** is text that appears at the top of each page, and a **footer** is text that appears at the bottom of each page. Word provides several predefined headers and footers or you can create your own. Page numbering is added within a header or footer using the **Page Number button** in the Header & Footer group on the Header & Footer Tools Design tab.

● ─ ─ ─ ─ ─ go online

**https://CA3.Paradigm Education.com/MLA Guide**

Go here for a comprehensive MLA guide.

① Open the document **China&TibetEssay** from the Ch7 folder in StudentDataFiles. If necessary, click the Enable Editing button to close Protected view.

② Save the document as **China&TibetEssay-YourName** in the Ch7 folder in CompletedTopics.

③ Scroll down and review the formatting in the essay. Notice the paper size, font, margins, line and paragraph spacing, and first line indents are already formatted.

④ Position the insertion point at the beginning of the document and replace the text *Michael Seguin* with your first and last name.

The first four lines of this report are set up in MLA format for a first page; however, you need to add the page numbering for an MLA report.

⑤ Click the Insert tab and then click the Header button in the Header & Footer group.

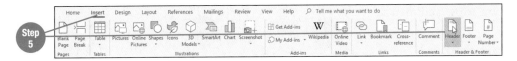

⑥ Click *Edit Header.*

The document text is dimmed and a Header pane appears above a dashed line. In the Header pane, you type the text that you want to appear at the top of each page and/or insert page numbering. A Header or Footer pane is divided into three sections: left, center, and right.

⑦ Press Tab twice to move the insertion point to the right section, type your last name, and then press the spacebar.

⑧ Click the Page Number button in the Header & Footer group on the Header & Footer Tools Design tab.

⑨ Point to *Current Position* and then click *Plain Number.*

⑩ Select your last name and the page number in the Header pane.

⑪ Click the *Font* option box arrow on the Mini toolbar, scroll down the font list, and then click *Times New Roman.*

⑫ Click the *Font Size* option box arrow on the Mini toolbar and then click *12.*

⑬ Click within the Header pane to deselect the text.

⑭ Click the Close Header and Footer button in the Close group on the Header & Footer Tools Design tab.

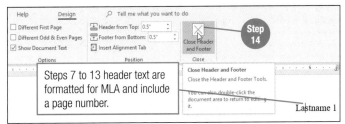

Steps 7 to 13 header text are formatted for MLA and include a page number.

⑮ Scroll down through the document to view the page number on each page.

⑯ Save the document using the same name (**China&TibetEssay-YourName**). Leave the document open for the next topic.

**quick steps**

**Insert a Header or Footer**
1. Click Insert tab.
2. Click Header or Footer button.
3. Click built-in option or choose *Edit Header* or *Edit Footer.*
4. Type text and/or add options as needed.
5. Click Close Header and Footer button.

**Insert Page Numbering**
1. Click within Header or Footer pane at desired location.
2. Click Page Number button.
3. Point to *Current Position.*
4. Click desired page number option.
5. Click Close Header and Footer button.

**Tutorials**

Editing a Header and Footer

Inserting and Removing Page Numbers

Creating a Different First Page Header and Footer

## 7.7 Inserting and Editing Citations

In an academic paper, direct quotations copied from sources or content you have paraphrased from another source need to be referenced in a **citation**. A citation tells the reader the source for the information you are quoting or paraphrasing. The first time you insert a citation for a source, add the information about the source. The next time you need to cite the same source, insert the citation by choosing the source entry in the *Insert Citation* list. To add a page number in a citation, click in the citation text and click the Citation Options arrow to open the Edit Citation dialog box.

 **Watch & Learn**

 **Hands On Activity**

● ─ ─ ─ ─ ─ ─ ─ ─ skills

Insert a citation

Edit a citation

1. With the **China&TibetEssay–YourName** document open, click the References tab.

2. Look at the *Style* option in the Citations & Bibliography group. If the *Style* is not *MLA*, click the *Style* option box arrow and then click *MLA*.

3. Scroll down the first page, click to place the insertion point left of the period at the end of the quotation in the first sentence in the second paragraph that begins *China currently claims . . .* and then press the spacebar.

4. Click the Insert Citation button in the Citations & Bibliography group.

5. Click *Add New Source*.

6. Click the *Type of Source* option box arrow, scroll down the list box, and then click *Web site* in the Create Source dialog box.

7. Click in the *Author* text box and then type Bajoria, Jayshree.

8. Click in the *Name of Web Page* text box and then type The Question of Tibet.

9. Press Tab or click in the *Year* text box and then type 2008.

10. Continue to press Tab or click in the designated text boxes and type the information as follows:

| | | | |
|---|---|---|---|
| *Month* | December | *Month Accessed* | November |
| *Day* | 5 | *Day Accessed* | 15 |
| *Year Accessed* | 2021 | *Medium* | Web |

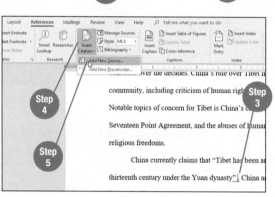

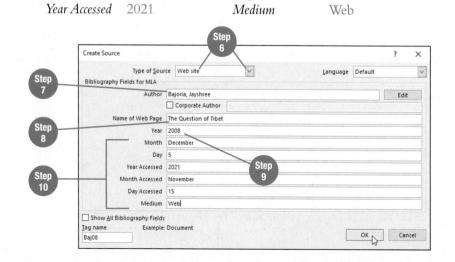

● ─ ─ ─ ─ in the know

In the seventh edition of the MLA handbook, URLs are no longer required. MLA advises writers to include URLs only if a reader is unlikely to find the source without the web address.

11  Click OK. The citation is inserted using the correct formatting as per MLA style.

12  Scroll down if necessary, click to place the insertion point left of the period at the end of the quoted text in the fifth sentence in the second paragraph that begins *Not until the 18th century . . .*, press the spacebar, click the Insert Citation button, and then click *Add New Source*.

13  Click the *Type of Source* option box arrow, scroll up, and then click *Book*.

14  Enter the information in the designated text boxes in the Create Source dialog box as follows and click OK when finished.

| | |
|---|---|
| *Author* | Praag, Michael C. |
| *Title* | The Historical Status of Tibet: History, Rights, and Prospects in International Law |
| *Year* | 1987 |
| *City* | Boulder |
| *Publisher* | Westview Press |
| *Medium* | Print |

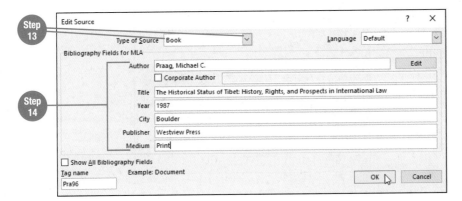

15  Click to place the insertion point in the *(Praag)* citation and then click the Citation Options arrow that appears.

16  Click *Edit Citation*.

17  Type 3 in the *Pages* text box in the Edit Citation dialog box and then press Enter or click OK.

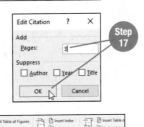

For citations that refer to a print-based source, open the Edit Citation dialog box using the Citation Options arrow to enter the page number quoted or paraphrased.

18  Click to place the insertion point left of the period at the end of the fourth sentence in the third paragraph on page 2, which begins *While independence was . . .* and then press the spacebar.

19  Click the Insert Citation button and then click *Bajoria, Jayshree*.

20  Save the document using the same name (**China&TibetEssay-YourName**). Leave the document open for the next topic.

---

◆ - - - - **quick steps**

**Insert a Citation with a New Source**
1. Position insertion point.
2. Click References tab.
3. Click Insert Citation button.
4. Click *Add New Source*.
5. If necessary, change *Type of Source*.
6. Enter required information.
7. Click OK.

**Insert a Citation with an Existing Source**
1. Position insertion point.
2. Click References tab.
3. Click Insert Citation button.
4. Click required source.

**Edit a Citation**
1. Position insertion point in citation text.
2. Click Citation Options arrow.
3. Click *Edit Citation*.
4. Type page or other reference.
5. Click OK.

☁ **Tutorials**

Inserting Sources and Citations

Editing a Citation and Source

## 7.8 Creating a Works Cited Page and Using Word Views

The **Bibliography button** in the Citations & Bibliography group on the References tab is used to generate a **Works Cited** page for MLA papers or a References page for APA papers. The MLA style guide requires a Works Cited page to be on a separate page at the end of the document and organized alphabetically by author's name, or by title when an author's name is absent.

Word provides various ways to view a document, including Read Mode view, which provides maximum screen space for reading longer documents.

1. With the **China&TibetEssay-YourName** document open, press Ctrl + End to move the insertion point to the end of the document.

2. Click the Insert tab and then click the Page Break button in the Pages group to start a new page.

3. Click the References tab.

4. Click the Bibliography button in the Citations & Bibliography group.

5. Click *Works Cited*.

Word automatically generates the Works Cited page. In the next steps, you will format the text to match the font, size, and spacing of the rest of the document.

6. If necessary, scroll up to view the Works Cited page.

7. Select all the text on the Works Cited page.

Word surrounds the text on the page with a border and displays a Bibliographies button and an Update Citations and Bibliography button at the top of the placeholder.

8. Click the Home tab and then make the following changes to the selected text:

   a. Change the font to 12-point Times New Roman.

   b. Change the line spacing to 2.0.

   c. Open the Paragraph dialog box and change *Before* and *After* in the *Spacing* section to *0 pt*.

9. Select the title text *Works Cited*, change the font color to Automatic (black), and center the title.

**Watch & Learn**

**Hands On Activity**

— — — — — — Skills

Create a Works Cited page

Browse a document in different views

**Tutorial**

Inserting a Works Cited Page

— — — — — app tip

Do not change formatting until you are sure your Works Cited page is complete, because the page will revert to predefined formats if you make changes to sources and then update the Works Cited page.

Use the Bibliographies button to change to a different bibliography style or to convert the Works Cited page to static text that can be edited.

Regenerate the Works Cited page using this button if you make a change to any of the source information.

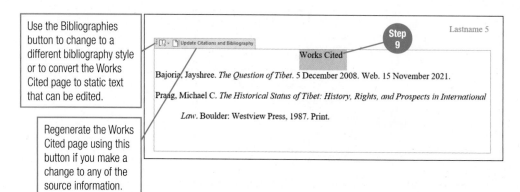

The default view for new documents is **Print Layout view**, which displays the document as it will appear when printed. **Read Mode view** displays a document full screen in columns, allowing you to read longer documents more easily without screen elements such as the Quick Access Toolbar and ribbon. **Draft view** hides print elements, such as headers and footers. **Web Layout view** displays a document as it would appear in a browser, and **Outline view** displays content as bulleted points.

**10** Press Ctrl + Home to move the insertion point to the beginning of the document.

**11** Click the View tab and then click the Read Mode button in the Views group.

Use Read Mode to view a document without editing.

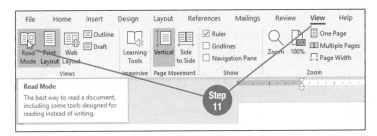

**12** Click the View tab, point to *Layout*, and then click *Paper Layout* to display the document as single pages. Skip this step if your screen is already displaying the document as single pages.

**13** Scroll down to the end of the document to view the document in Read Mode, click the View tab, point to *Layout*, and then click *Column Layout*.

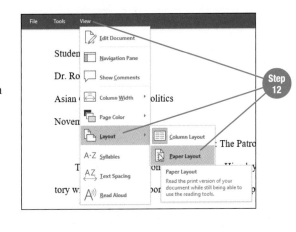

**14** Click the Previous Screen button (left-pointing arrow inside circle at the middle left of the screen) to move back to the previous screen until you have returned to the beginning of the document.

**15** Click the View tab and then click *Edit Document* to return to Print Layout view.

**16** Click the Draft button in the Views group on the View tab and then scroll through the document to view the document without the header or margins.

**17** Click the Print Layout button near the right end of the Status bar.

Read Mode and Print Layout buttons are also at the right end of the Status bar.

**18** Save the document using the same name (**China&TibetEssay-YourName**). Leave the document open for the next topic.

quick steps

**Generate a Works Cited Page**
1. Position insertion point at end of document.
2. Insert page break.
3. Click References tab.
4. Click Bibliography button.
5. Click *Works Cited*.
6. Format as required.

**Change Document View**
1. Click View tab.
2. Click desired view button.

Tutorial

Changing Document Views

app tip

In Print Layout view, turn on the Navigation pane (View tab, *Navigation Pane* check box in the App tip on page 167) and click *Pages* at the top of the pane to move through a document by clicking miniature page thumbnails.

# 7.9 Inserting and Replying to Comments

A **comment** is a short note associated with text that provides explanatory information to a reader. Comments are useful for individuals when working on a document with another person or team to give explanations, feedback, or to pose a question to other reviewers of the document. You can reply to a comment made by someone else and mark a comment resolved when a question is answered.

A comment and all replies added to a document are tagged with the user name of the comment writer and the date and the time the comment text was inserted.

1. With the **China&TibetEssay-YourName** document open, click to place the insertion point at the beginning of the first paragraph in the document.

2. Select the words *Sakya Lama* in the second sentence in the first paragraph.

3. Click the Review tab.

4. Click the New Comment button in the Comments group.

Word opens a new comment box (referred to as a *comment balloon*) in the Markup Area. The **Markup Area** is at the right side of the screen, where comments and other document changes, such as insertions and deletions, are displayed when the Track Changes feature is turned on.

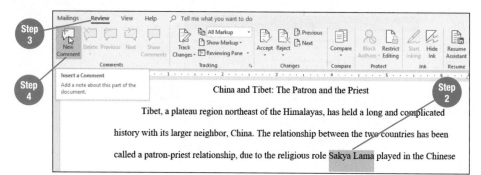

5. Type Consider adding more explanation about Sakya Lama and the Yuan dynasty and then click in the document outside the comment box.

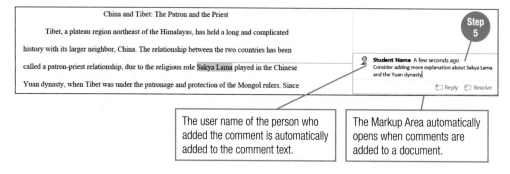

The user name of the person who added the comment is automatically added to the comment text.

The Markup Area automatically opens when comments are added to a document.

6. Select *modern historians* in the third sentence in the second paragraph.

7. Click the New Comment button.

8. Type Provide a few names of modern historians? and then click in the document.

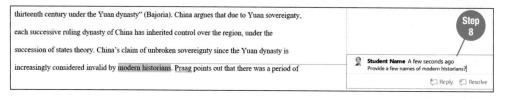

**Watch & Learn**

**Hands On Activity**

**— — — — — — Skills**

Insert comments
Change the markup view
Reply to a comment
Resolve a comment

**Tutorial**

Inserting Comments

**— — — — — — app tip**

When a document has been shared with others, the Real Time Presence feature lets you see who else is editing the document you have open and the changes each person is making in real time.

⑨ Position the insertion point at the beginning of the document.

⑩ Click the Display for Review option box arrow in the Tracking group (currently displays *All Markup* or *Simple Markup*) and then click *No Markup*.

Notice the two comments on page 1 are removed from the document display.

⑪ Click the Display for Review option box arrow and then click *Simple Markup*.

Simple Markup view displays a red line from the document text to the associated comment when you click inside a comment in the Markup Area. Other edited text displays a vertical red bar in the margin area next to insertions or deletions to highlight areas in the document where changes have been made.

⑫ Click in the first comment balloon on page 1.

Notice Word displays a red line pointing to the text for which the comment is associated. The comment box also displays a Reply option and Resolve option.

⑬ Click *Reply* in the first comment box.

⑭ Type Asked Dr. Smith and he said this is not necessary and then click in the document.

Notice the reply comment text is indented below the original comment in a conversation-style dialogue.

⑮ Scroll down if necessary, point to the second comment balloon on page 1, and then click *Reply*.

⑯ Type I'll check with Jose to see if he wants to include a few names.

⑰ Scroll up if necessary and then click *Resolve* in the first comment balloon.

⑱ Click in the document outside the comment box.

Notice the comment text is dimmed for the resolved comment. You can also delete a comment instead of marking the comment resolved.

⑲ Save the document using the same name (**China&TibetEssay-YourName**).

⑳ Close the document.

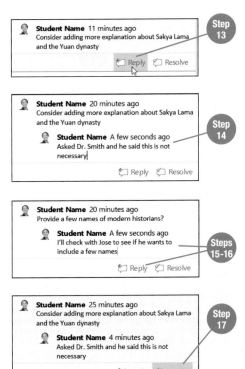

● – – – **quick steps**

**Insert a Comment**
1. Select text.
2. Click Review tab.
3. Click New Comment button.
4. Type comment text.
5. Click in document.

**Reply to a Comment**
1. Point to comment balloon.
2. Click *Reply*.
3. Type reply text.
4. Click in document.

**Resolve a Comment**
1. Point to comment balloon.
2. Click *Resolve*.

☁ **Tutorial**

Managing Comments

## 7.10 Creating a Résumé and Cover Letter from Templates

Word provides several professionally designed and formatted résumé and cover letter templates that take the work out of designing and formatting these two crucial documents, letting you focus your efforts on writing documents that will win you a job interview!

1. Click the File tab and then click *New*.

2. In the New backstage area, click in the search text box, type entry level resume, and then press Enter or click the Start searching icon.

3. Click the thumbnail for *Entry-level resume* in the Templates gallery.

4. Click the Create button in the preview window.

5. Double-click *YOUR NAME* in the First Page Header pane and then type Dana Jelic. Note that the text will appear all uppercase.

6. Select *Sreet Address, City, ST ZIP Code* and press Delete.

7. Select *Phone* and then type 800-555-4577.

8. Select *Email* and then type jelic@ppi-edu.net.

9. Select *Website*, press Delete, and then click the Close Header and Footer button in the Close group on the Header & Footer Tools Design tab.

10. Type the remainder of the document by selecting text below a heading and typing text, by deleting template text, and by formatting text to match Figure 7.3.

11. Save the document as **JelicResume-YourName** in the Ch7 folder in CompletedTopics.

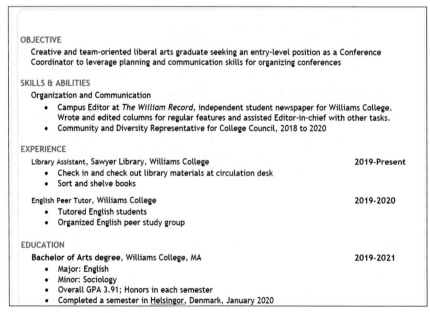

**OBJECTIVE**
Creative and team-oriented liberal arts graduate seeking an entry-level position as a Conference Coordinator to leverage planning and communication skills for organizing conferences

**SKILLS & ABILITIES**
Organization and Communication
- Campus Editor at *The William Record*, independent student newspaper for Williams College. Wrote and edited columns for regular features and assisted Editor-in-chief with other tasks.
- Community and Diversity Representative for College Council, 2018 to 2020

**EXPERIENCE**
Library Assistant, Sawyer Library, Williams College                    2019-Present
- Check in and check out library materials at circulation desk
- Sort and shelve books

English Peer Tutor, Williams College                    2019-2020
- Tutored English students
- Organized English peer study group

**EDUCATION**
Bachelor of Arts degree, Williams College, MA                    2019-2021
- Major: English
- Minor: Sociology
- Overall GPA 3.91; Honors in each semester
- Completed a semester in Helsingor, Denmark, January 2020

**Figure 7.3**

Shown is the résumé text for Topic 7.10.

**Watch & Learn**

**Hands On Activity**

●------- Skills

Create a résumé

Create a cover letter

**Tutorial**

Changing Vertical Alignment

**12** Close the résumé document and then display the New backstage area.

**13** Type entry level cover letter in the search text box and then press Enter or click the Start searching icon.

**14** Select and create a document using the template *Cover letter for entry-level resume*.

**15** Create the letter as shown in Figure 7.4 by completing steps similar to those in Steps 5 through 10. (Note that the letter closing in the template is not shown in Figure 7.4. Use the default text in this section.)

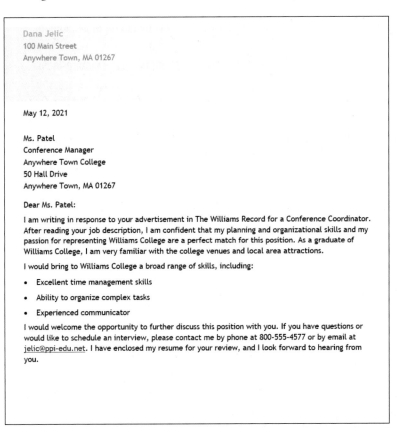

Dana Jelic
100 Main Street
Anywhere Town, MA 01267

May 12, 2021

Ms. Patel
Conference Manager
Anywhere Town College
50 Hall Drive
Anywhere Town, MA 01267

Dear Ms. Patel:

I am writing in response to your advertisement in The Williams Record for a Conference Coordinator. After reading your job description, I am confident that my planning and organizational skills and my passion for representing Williams College are a perfect match for this position. As a graduate of Williams College, I am very familiar with the college venues and local area attractions.

I would bring to Williams College a broad range of skills, including:

- Excellent time management skills
- Ability to organize complex tasks
- Experienced communicator

I would welcome the opportunity to further discuss this position with you. If you have questions or would like to schedule an interview, please contact me by phone at 800-555-4577 or by email at jelic@ppi-edu.net. I have enclosed my resume for your review, and I look forward to hearing from you.

**Figure 7.4**

Use this cover letter text for Topic 7.10.

**16** Click the Layout tab and then click the Page Setup dialog box launcher.

**17** Click the Layout tab in the Page Setup dialog box.

**18** Click the *Vertical alignment* option box arrow, click *Center*, and then click OK.

**19** Save the document as **JelicCoverLetter –YourName** in the Ch7 folder in CompletedTopics and then close the document.

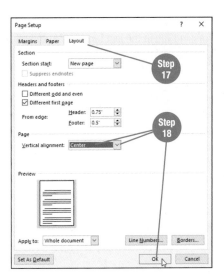

## Topics Review

| Topic | Key Concepts |
|---|---|
| **7.1 Inserting, Editing, and Labeling Images in a Document** | Graphic elements assist with comprehension and/or add visual appeal to documents. |
| | Insert an image from a web resource or OneDrive using the **Online Pictures button** in the Illustrations group on the Insert tab. |
| | Click the **Layout Options button** to control how text wraps around the image and to change the position of the image on the page. |
| | Use the **Pictures button** to insert an image from a file stored on your computer or other storage medium. |
| | Edit the appearance of an image and/or add special effects using buttons on the Picture Tools Format tab. |
| | A caption is explanatory text above or below an image that is added using the **Insert Caption** feature on the References tab. |
| | Word automatically numbers images as Figures and can generate a Table of Figures automatically. |
| **7.2 Adding Borders and Shading, and Inserting a Text Box** | Add a border or **shading** to paragraphs to make text stand out on a page. Shading is color applied to the page behind the text. |
| | A page border surrounds the entire page and is added using the **Page Borders button** on the Design tab. |
| | Apply a border to selected text using the **Borders gallery** from the Borders button arrow in the Paragraph group on the Home tab. |
| | Shading is added using the Shading button arrow in the Paragraph group. |
| | A **pull quote** is a quote typed inside a text box. |
| | Insert text inside a box using the **Text Box button** in the Text group on the Insert tab. |
| **7.3 Inserting a Table** | A table is a grid of columns and rows in which you type text and is used when you want to arrange text side by side or in rows. |
| | A **table cell** is a rectangular-shaped box in the table grid that is the intersection of a column and a row. Text in the table is typed inside table cells. |
| | A **Quick Table** is a predesigned table with sample data, such as a calendar or a tabular list. |
| | Create a table by clicking a square in the table grid accessed from the Insert tab that represents the number of columns and rows or by entering the number of columns and rows in the Insert Table dialog box. |
| | Pressing Tab in the last table cell automatically adds a new row to the table. |

*continued…*

| Topic | Key Concepts |
|---|---|
| **7.4 Formatting and Modifying a Table** | Apply a predesigned collection of borders, shading, and color to a table using an option from the **Table Styles** gallery. |
| | Shading or other formatting applied to every other row to make the table data easier to read is called a **banded row**. |
| | Check boxes in the Table Style Options group on the Table Tools Design tab are used to customize the formatting applied from a Table Style. |
| | Apply shading or borders using buttons in the Table Styles group and Borders group on the Table Tools Design tab. |
| | New rows and columns are inserted above, below, left, or right of the active table cell using buttons in the Rows & Columns group on the Table Tools Layout tab. |
| | Remove selected table cells, rows, or columns or the entire table using options from the Delete button in the Rows & Columns group. |
| | Adjust the width of a column by changing the *Width* text box value in the Cell Size group on the Table Tools Layout tab or by dragging the column border. |
| | Buttons to change alignment options for selected table cells are found in the Alignment group on the Table Tools Layout tab. |
| | Cells in a table can be merged or split using buttons in the Merge group on the Table Tools Layout tab. |
| **7.5 Changing Layout Options** | **Portrait** orientation means the text on the page is oriented to the taller side (8.5-inch width), while **landscape** orientation rotates the text to the wider side of the page (11-inch measurement becomes the page width). |
| | Change the margins by choosing one of the predefined margin options or by entering measurements for the top, bottom, left, and right margins at the Page Setup dialog box. |
| | A **soft page break** is a page break that Word inserts automatically when the maximum number of lines that can fit on a page has been reached. |
| | A **hard page break** is a page break inserted by you in a different location than where the soft page break occurred. |
| **7.6 Adding Text and Page Numbers in a Header for a Research Paper** | A **style guide** is a set of rules for formatting and referencing academic papers. |
| | A **header** is text that appears at the top of each page, while a **footer** is text that appears at the bottom of each page. |
| | Page numbers are added to a document at the top or bottom of a page within a Header or Footer pane using the **Page Number button** on the Header & Footer Tools Design tab. |
| | Click the Insert tab and choose the Header or Footer button to create a header or footer in the Header or Footer pane. |

*continued...*

| Topic | Key Concepts |
|---|---|
| **7.7 Inserting and Editing Citations** | A **citation** provides a reader with the reference for information quoted or paraphrased within an academic paper. |
| | Check the *Style* option in the Citations & Bibliography group on the References tab for the correct style guide and change the option, if necessary, before inserting a citation or creating sources. |
| | Position the insertion point where a citation is needed and use the Insert Citation button on the References tab to create a reference. |
| | Click *Add New Source* in the *Insert Citation* list to enter information for a new reference for a citation in the Create Source dialog box. |
| | Click in citation text, click the Citation Options arrow that appears, and then click *Edit Citation* to add a page number to the reference. |
| | Add a citation for a source that has already been defined by positioning the insertion point, clicking the Insert Citation button, and then clicking the source entry in the list. |
| **7.8 Creating a Works Cited Page and Using Word Views** | A Works Cited page is a separate page at the end of the document with the references used for the paper. |
| | Use the **Bibliography button** in the Citations & Bibliography group on the References tab to generate a **Works Cited** page formatted for the MLA style guide. |
| | **Print Layout view** displays the document as it will appear when printed. |
| | **Read Mode view** displays a document full screen in columns or pages without editing tools. |
| | **Draft view** hides print elements, such as headers, footers, and page numbering. |
| | **Web Layout view** displays the document as a web page. |
| | **Outline view** displays content as bullet points. |
| **7.9 Inserting and Replying to Comments** | A **comment** is a short note associated with text that provides explanatory information or poses a question to a reader. |
| | Select text that you want to associate with a comment and type the comment text inside a comment balloon by clicking the New Comment button on the Review tab. |
| | Comment balloons display in the **Markup Area**, which is a pane that opens at the right side of the document when comments are added. |
| | A document with comments can be displayed with *No Markup*, *Simple Markup*, or *All Markup* view, which refers to the way comment boxes and other editing changes are displayed. |
| | Point at a comment balloon and use the *Reply* option to enter reply text that responds to a comment. |
| | Mark a comment as complete to retain the comment text but display the comment dimmed in the Markup Area by pointing at the comment balloon and clicking the *Resolve* option. |
| **7.10 Creating a Résumé and Cover Letter from Templates** | In the New backstage area, search for a résumé template and a cover letter template in various styles, themes, and purposes. |
| | Personalize a résumé and cover letter template by selecting existing text and typing new text or by selecting and deleting text not needed. |

 **Topics Review**

# Creating, Editing, and Formatting Excel Worksheets

## Chapter 8

Microsoft Excel (referred to as Excel) is a **spreadsheet application** used to create, process, and present information that is organized in a grid of columns and rows. In Excel, data can be calculated, analyzed, and graphed in a chart. The ability to do "what-if" analysis is a popular feature of the application. In this type of analysis, one or more values are changed to view the effect on other values. Excel is used for budget, income, expense, investment, loan, schedule, grading, attendance, inventory, and research data. Any information that can be set up in a grid-like structure is suited to Excel.

A file that you save in Excel is called a **workbook**. A workbook contains a collection of worksheets. A **worksheet** is the structure into which you enter, edit, and manipulate data. Think of a workbook as a binder and a worksheet as a page within the binder. Initially, a workbook has only one worksheet (page), but you can add more as needed.

Many of the features that you learned about in Word operate the same or similarly in Excel, which will make learning Excel faster and easier. You will begin by creating new worksheets in a blank workbook and then open other worksheets to practice navigating, editing, sorting, and formatting tasks.

## Learning Objectives

**8.1** Create and edit a new worksheet

**8.2** Format cells with font, number, border, and merge and center options

**8.3** Adjust column width and row height, and change cell alignment

**8.4** Use the Fill feature to enter and copy data, and use AutoSum to add a column or row of values

**8.5** Insert and delete rows and columns

**8.6** Sort data and apply cell styles

**8.7** Change page orientation, scale a worksheet, and display formulas in cells

**8.8** Insert and rename a worksheet, copy cells, and indent data within a cell

**8.9** Use Go To to move to a cell, freeze panes, apply shading options, and wrap text and rotate cell entries

### Read & Learn

### Content Online

The online course includes additional training and assessment resources.

# 8.1 Creating and Editing a New Worksheet

When you create a new blank workbook, you begin with a blank worksheet, which is divided into columns and rows. The intersection of a column and a row is called a **cell**, and in each cell you can type text, a value, or a formula. The cell with the green border is called the **active cell**. The active cell is the location in which the next entry you type is stored. The active cell is also the cell affected by the next command (action) you perform. Each cell is identified with the letter of the column and the number of the row that intersect to form the cell. For example, A1 refers to the cell in column A, row 1. A new workbook starts with one worksheet labeled *Sheet1* that has columns labeled A to Z, AA to AZ, BA to BZ, and so on to the last column, which is labeled XFD. Rows are numbered 1, 2, 3, up to 1,048,576.

## Creating a New Worksheet

Begin a new worksheet by entering titles, column headings, and row headings to give the worksheet an organizational layout and provide context for the reader. Next, enter values in the columns and rows. Complete the worksheet by inserting formulas that perform calculations or otherwise summarize data.

1. Start Excel.

2. At the Excel Start screen, click *Blank workbook* in the Templates gallery. Compare your screen with the one shown in Figure 8.1 and read the descriptions of screen elements in Table 8.1 on the next page.

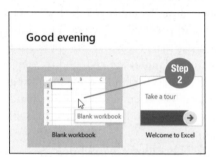

**Watch & Learn**

**Hands On Activity**

● - - - - - - - - SKILLS

Start a new workbook
Enter text and values
Create formulas
Edit cells

**Tutorial**
Opening a Blank Workbook

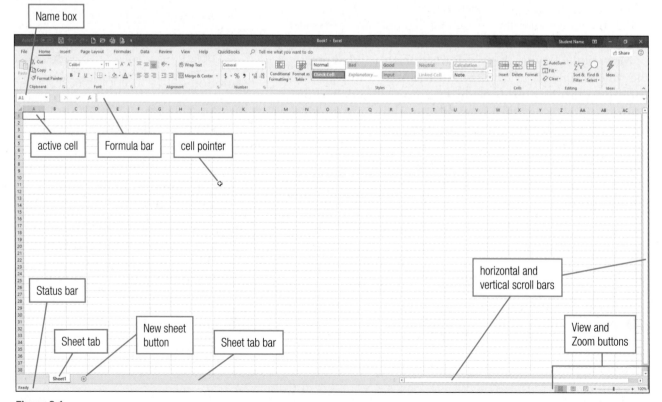

**Figure 8.1**

A new blank Excel worksheet is shown here. The worksheet area below the ribbon is divided into columns and rows, creating cells into which data and formulas are typed and stored.

**Table 8.1** Excel Features

| Feature | Description |
|---------|-------------|
| Active cell | This location is where the next typed data will be stored and that will be affected by the next command. Make a cell active by clicking it or by moving to it using the arrow keys. |
| Cell pointer | This icon (⊕) displays when you are able to select cells with the mouse by clicking or dragging. On a touch device with no mouse attached, tap a cell to display selection handles (round circles) at the upper left and bottom right corners. |
| Formula bar | Bar that displays contents stored in the active cell and is also used to create formulas |
| Horizontal and vertical scroll bars | Tools used to view parts of a worksheet not displayed in the current viewing area |
| Name box | Box that displays the address or name of the active cell |
| New sheet button | Button on the Sheet tab bar used to insert a new worksheet |
| Sheet tab | This tab displays the name of the active worksheet. By default, new sheets are named *Sheet#* where *#* is the number of the sheet in the workbook. |
| Sheet tab bar | Bar that displays sheet tabs and is used to navigate between worksheets |
| Status bar | Bar that displays messages indicating the current mode of operation; e.g., *Ready* indicates the worksheet is ready to accept new data |
| View and Zoom buttons | These buttons are used to change the appearance of the worksheet. Excel opens in Normal view. Other view buttons include Page Layout and Page Break Preview. Zoom buttons are used to enlarge or shrink the display. |

## Entering Text and Values

When you start a new worksheet, the active cell is A1 at the upper left corner of the worksheet. Entries are created by activating a cell and typing text, a value, or a formula.

**3** With A1 the active cell, type Car Purchase Cost and then press Enter or click cell A2 to make A2 the active cell.

**4** Type Preowned Ford Focus Sedan and then press Enter twice or click cell A4.

**5** Type Total Purchase Price and then click cell A6.

**6** Type Loan Details: and then click cell B7.

**7** Type the remaining row headings by moving the active cell and typing the text shown in the image at the right.

**8** Click cell F4, type 16700.00, and then click cell E7.

By default, text entries align at the left edge of a cell. Numeric entries align at the right edge of a cell, and zeros to the right of a decimal are not stored or displayed. You will learn how to format a cell to display the decimal place values in Topic 8.2.

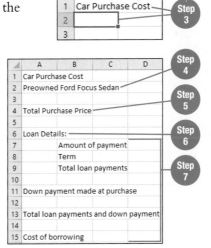

● – – – – – – **app tip**

You can use arrow keys to move up, down, left, or right from cell to cell.

☁ **Tutorial**
Entering Data

⑨ Type the remaining values by activating the cell and typing the numbers shown in the image below.

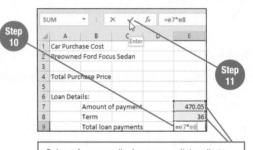

## Creating Formulas to Perform Calculations

A **formula** is used to perform mathematical operations on values. A formula entry begins with the equals sign (=) to indicate to Excel the entry that follows is a calculation. Following the equals sign, type the first cell address that contains a value you want to use, type a mathematical operator, and then type the second cell address. Continue typing mathematical operators and cell addresses until finished. The mathematical operators are + (addition), – (subtraction), * (multiplication), / (division), and ^ (exponentiation).

⑩ Click cell E9 to make it the active cell and then type =e7*e8.

⑪ Click the Enter button on the Formula bar.

Excel calculates the result and displays the value in E9. Notice that the cell in the worksheet area displays the formula result, while the Formula bar displays the formula used to calculate the result. Notice also that Excel capitalizes column letters in cell addresses within formulas.

Color references display as you click cells to build a formula. These colors make it easy to check the values that are part of the formula or to help find errors.

Another way to enter a formula is to use the pointing method. In this method, you click the desired cells instead of typing their cell addresses.

⑫ Make F13 the active cell and then type =.

⑬ Click cell E9.

A moving dashed border surrounds E9, the cell is color coded, and the address E9 is inserted in both cell F13 and the Formula bar.

⑭ Type +.

⑮ Click cell E11 and then click the Enter button on the Formula bar or press Enter.

⑯ Make F15 the active cell, type the formula =f13-f4 or enter the formula using the pointing method, and then click the Enter button on the Formula bar or press Enter.

The result, 1921.8, displays in the cell.

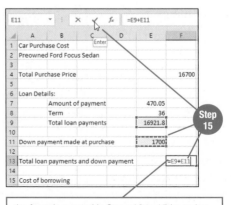

the formula created in Steps 12 to 15 by typing the equals sign (Step 12), clicking cells (Steps 13 and 15), and typing the operator (Step 14)

**Tutorials**

Entering Formulas Using the Keyboard

Entering Formulas Using the Mouse

●– – – – – – – **Oops!**

Clicked the wrong cell? Just click the correct cell—the cell reference can be changed until you type an operator. You can also press the Esc key to start over.

**Tutorial**

Determining the Order of Operations

## Editing Cells

An entire cell entry can be changed by making the cell active and typing a new entry to replace the existing contents. Double-click a cell to open it for editing in the worksheet area; this will allow you to change the entry rather than replace it. You can also edit the content in a cell by making the cell active and then inserting or deleting characters or spaces in the Formula bar.

To delete the contents of an active cell, click the cell and then press the Delete key or click the **Clear button** in the Editing group on the Home tab and then click *Clear All*.

**17**  Make E7 the active cell, type 480.95, and then press Enter.

Notice that the new payment amount causes the values in E9, F13, and F15 to update.

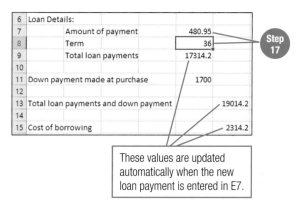

These values are updated automatically when the new loan payment is entered in E7.

**18**  Double-click cell F4, position the insertion point between *6* and *7*, press Backspace to remove *6*, type *5*, and then click any other cell.

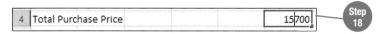

**19**  Make E11 the active cell, click in the Formula bar, position the insertion point between *1* and *7*, press Backspace to remove *1*, and then click any other cell.

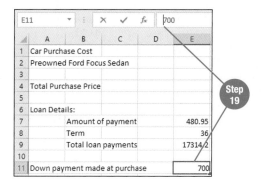

**20**  Save the new workbook as **CarCost-YourName** in a new folder *Ch8* in the CompletedTopics folder on your storage medium. Leave the workbook open for the next topic.

## 8.2 Formatting Cells

 **Watch & Learn**

 **Hands On Activity**

Much like the Home tab in Word, the Home tab in Excel contains the formatting options for changing the appearance of text, values, or formula results. The Font group contains buttons to change the font, font size, and font color and to apply bold, italic, underline, borders, and shading. The Alignment group contains buttons to align text or values within the cell edges.

●━━━━━━ skills

Select cells
Change the font
Apply bold formatting
Format values
Add borders
Merge cells

### Selecting Cells and Applying Font Formatting

**To select cells using a mouse:** Select adjacent cells with a mouse by positioning the cell pointer (✛) in the starting cell and dragging in the required direction until all the desired cells have been included in a shaded selection rectangle. Select nonadjacent cells by holding down the Ctrl key while you click the mouse in each desired cell. The Quick Analysis button displays when a group of cells has been selected. You will learn about the options available from this button in Chapter 9.

**To select cells using touch:** Selecting cells on a touch device in Excel is similar to selecting text in Word, with the exception that two selection handles display in the active cell, as shown in Figure 8.2. Tap inside the selection area to display the Mini toolbar.

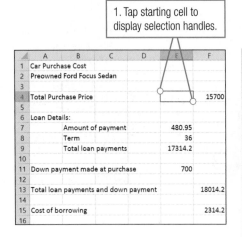

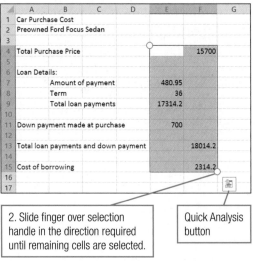

1. Tap starting cell to display selection handles.

2. Slide finger over selection handle in the direction required until remaining cells are selected.

Quick Analysis button

 **Tutorial**

Selecting Cells

**Figure 8.2**
Selecting cells on a touch device is not much different from selecting cells using a mouse.

●━━━━━━ app tip

Select cells using the keyboard by holding down the Shift key while you press an arrow key (Up, Down, Left, or Right).

① With the **CarCost-YourName** workbook open, starting at cell A1, select all the cells down and right to F15.

A rectangular-shaped group of cells is referred to as a **range**. A range is referenced with the address of the cell at the upper left corner, a colon (:), and the address of the cell at the lower right corner. For example, the reference for the range selected in Step 1 is A1:F15.

**2** Click the *Font* option box arrow in the Font group on the Home tab, scroll down, and then click *Century Gothic*.

**3** Click any cell to deselect the range.

**4** Select A1:A2 and then click the Bold button in the Font group.

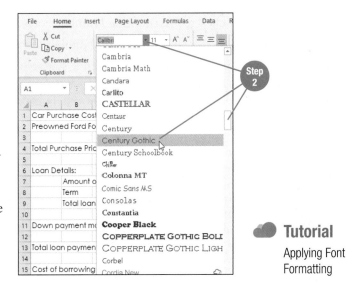

Notice the entire text in A1 and A2 is bold, including the characters that spill over the edge of column A into columns B, C, and D. This is because the entire text entry is stored in the cell that was active when the text was typed. Overflow text that displays in adjacent columns is not problematic when the adjacent columns are empty. You will learn how to widen a column to fit overflow text in the next topic.

**Tutorial**

Applying Font Formatting

**5** Click cell F15 and then click the Bold button.

## Formatting Numbers

Cells in a new worksheet are in the General format, which has no specific appearance options. Buttons in the Number group on the Home tab are used to format the appearance of values in a worksheet. Add a dollar symbol, insert commas to indicate thousands, and/or adjust the number of decimal places to improve the appearance of values. Use the Percent Style format to convert decimal values to percentages and include the percent symbol. The *Number Format* option box (displays *General*) offers other formats for dates, times, fractions, or scientific values and is used to open the Format Cells dialog box from the *More Number Formats* option (Figure 8.3).

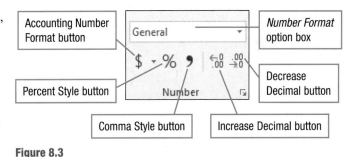

**Figure 8.3**

Shown are the buttons in the Number group on the Home tab. Click the *Number Format* option box arrow for more options.

**6** Select E4:F15.

**7** Click the Comma Style button in the Number group.

**Comma Style** formats values with a comma in thousands and two decimal places.

**8** Click cell F4 and then click the Accounting Number Format button in the Number group.

**Accounting Number Format** adds a currency symbol ($ for the United States and Canada), commas in numbers above 1,000, and two decimal places. Use the *Accounting Number Format* option box arrow to choose a different currency symbol (such as €, which stands for Euro).

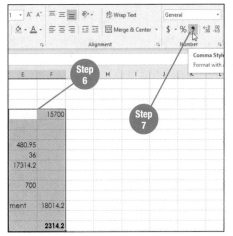

**app tip**

Excel automatically widens columns as needed when you apply a format that adds more characters to a column, such as Comma Style.

**9** Click cell E7 and then click the Accounting Number Format button.

**10** Select F13:F15 and then click the Accounting Number Format button.

The Accounting Number format aligns the currency symbol at the left edge of the cell. The *Currency* option (an option in the *Number Format* drop-down list), places the currency symbol immediately left of the value in the cell. Formatting values using the *Currency* option means that the currency symbols do not align at the same position in the cell if the length of the values varies. Use the Accounting Number Format option if you want all dollar symbols to align at the same position.

**11** Click cell E8 and then click the Decrease Decimal button in the Number group twice.

One decimal place is removed from the active cell or range each time you click the **Decrease Decimal button**. Click the **Increase Decimal button** to add one decimal place.

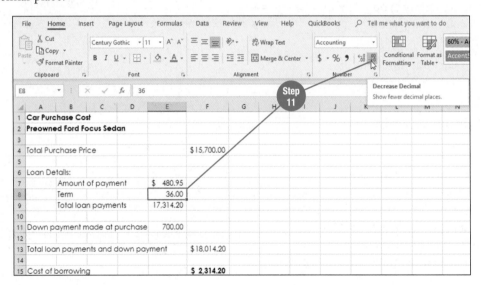

**Tutorials**

Applying Formatting Using the Format Cells Dialog Box

Applying Number Formatting

## Adding Borders

Borders in various styles and colors can be added to the top, left, bottom, or right edge of a cell. Borders are used to underscore column headings or totals or to otherwise emphasize cells.

**12** Click cell F13.

**13** Click the *Bottom Border* button arrow in the Font group.

**14** Click *Top and Bottom Border*.

**15** Click cell F15.

**16** Click the *Top and Bottom Border* button arrow and then click *Bottom Double Border*.

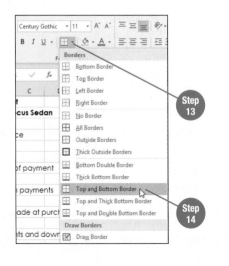

**Tutorial**

Adding Borders to Cells

**17** Click any other cell to view the border style applied to F15.

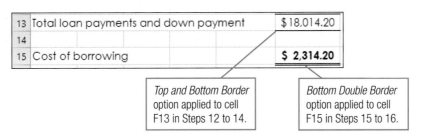

| 13 | Total loan payments and down payment | | | | $18,014.20 |
| 14 | | | | | |
| 15 | Cost of borrowing | | | | $ 2,314.20 |

*Top and Bottom Border* option applied to cell F13 in Steps 12 to 14.

*Bottom Double Border* option applied to cell F15 in Steps 15 to 16.

## Merging and Centering Cells

A worksheet title is often centered across the columns used in the worksheet. The **Merge & Center button** in the Alignment group on the Home tab is used to combine a group of cells into one large cell and center its contents. The Merge & Center button arrow on the Merge & Center button lets the user alter the Merge & Center option (to merge without centering, or to unmerge a merged cell).

**18** Select A1:F1.

**19** Click the Merge & Center button in the Alignment group.

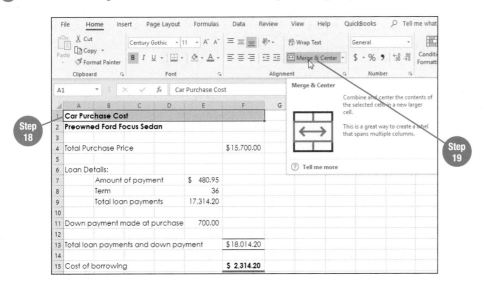

**20** Select A2:F2 and then click the Merge & Center button.

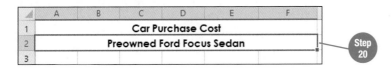

Step 20

**21** Save the workbook using the same name (**CarCost–YourName**). Leave the workbook open for the next topic.

**quick steps**

**Change a Font**
1. Select cell(s).
2. Click *Font* option box arrow.
3. Click desired font.

**Change Number Format**
1. Select cell(s).
2. Click Accounting Number Format button, Percent Style button, or Comma Style button.

OR

1. Select cell(s).
2. Click *Number Format* option box arrow.
3. Click desired format option.

**Adjust Decimal Places**
1. Select cell(s).
2. Click Increase Decimal or Decrease Decimal button.

**Add Borders**
1. Select cell(s).
2. Click Borders button arrow.
3. Click border style.

**Merge Cells**
1. Select cells.
2. Click Merge & Center button.

**Tutorial**

Merging and Centering Cells

## 8.3 Adjusting Column Width and Row Height and Changing Cell Alignment

In a new worksheet, each column width is 8.43 and each row height is 15. The column width value is the number of characters at the default font that can be displayed in the column. The row height value is a points measurement, with 1 point being approximately 1/72 of an inch. Make cells larger by widening the width of a column or by increasing the height of a row. In many instances, Excel automatically makes columns wider and rows taller to accommodate the cell entry, formula result, or format that you apply. Manually changing the column width or the row height is a technique used to add more space between cells to improve readability or emphasize a section of the worksheet.

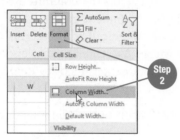

1 With the **CarCost-YourName** workbook open, make active any cell in column E.

2 Click the Format button in the Cells group on the Home tab and then click *Column Width*.

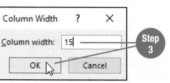

3 Type 15 in the *Column width* text box in the Column Width dialog box and then press Enter or click OK.

4 Make active any cell in column F, click the Format button, and then click *Column Width*.

5 Type 10 in the Column Width dialog box and then press Enter or click OK.

   Notice the cells with values in column F have been replaced with a series of pound symbols (#####). This occurs when the column width is too narrow to display all the characters.

6 Make F4 the active cell.

7 Click the Format button and then click *AutoFit Column Width*.

   **AutoFit** changes the width of the column to fit the contents of the active cell. F4 was made active in Step 6 because this cell has one of the longest numbers in the column. Notice the pound symbols have disappeared, and the values are redisplayed now that the column is wide enough.

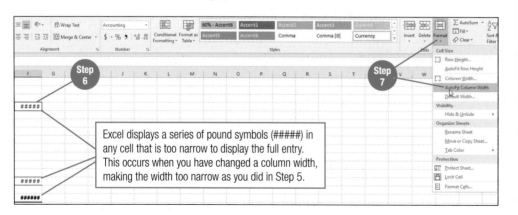

Excel displays a series of pound symbols (#####) in any cell that is too narrow to display the full entry. This occurs when you have changed a column width, making the width too narrow as you did in Step 5.

**Watch & Learn**

**Hands On Activity**

— — — — — — — Skills

Adjust column width

Adjust row height

Change cell alignment

**Tutorials**

Adjusting Column Width and Row Height

Applying Alignment Formatting

— — — — — — app tip

You can also change the column width or row height using the mouse by dragging the column boundary at the right of the column letter or the row boundary below the row number. Drag the boundary line right or left to increase or decrease the column width or down or up to increase or decrease the row height. Double-click the boundary line to AutoFit the width or height.

⑧ Select A1:A2.

To change the height or width of more than one row or column at the same time, you first select cells in the rows or columns to be changed.

⑨ Click the Format button and then click *Row Height*.

⑩ Type 26 in the *Row height* text box in the Row Height dialog box, and then press Enter or click OK.

The Alignment group on the Home tab contains buttons to align the entry of a cell horizontally and/or vertically. You can align cell content at the left, center, or right horizontally or at the top, middle, or bottom vertically within the cell boundaries. The **Middle Align button** centers text vertically between the top and bottom edges within a cell.

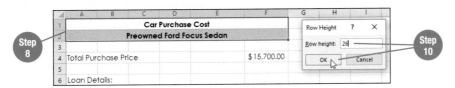

⑪ With A1:A2 still selected, click the Middle Align button in the Alignment group.

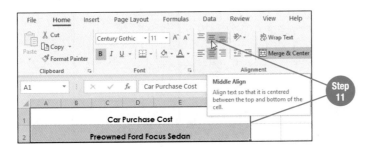

⑫ Make active any cell in row 15.

⑬ Click the Format button, click *Row Height*, type 26, and then press Enter or click OK.

⑭ Select A15:F15 and then click the Middle Align button.

⑮ Click any cell to deselect the range and then compare your worksheet to the one shown in the image above.

⑯ Save the workbook using the same name (**CarCost-YourName**).

⑰ Click the File tab and then click *Close* to close the workbook.

## 8.4 Entering or Copying Data with the Fill Command and Using AutoSum

The **Auto Fill** feature is used to enter data automatically based on a pattern or series that exists in an adjacent cell or range. For example, if *Monday* is entered in cell A1, Auto Fill can enter *Tuesday*, *Wednesday*, and so on automatically in the cells immediately to the right of or below A1. Excel fills many common text or number series, and also detects patterns for other data when you select the first few entries in a list. When no pattern or series applies, the **Fill** feature is used to copy an entry or formula across or down to other cells.

The **Flash Fill** feature automatically fills data as soon as a pattern is recognized. When Flash Fill presents a suggested list in dimmed text, press Enter to accept the suggestions, or ignore the suggestions and continue typing. A Flash Fill Options button with options to undo Flash Fill, accept the suggestions, or select changed cells appears when a list is presented.

### Using Auto Fill and Fill Right

1. Click the File tab, click *New*, and then click *Blank workbook*.
2. Type the text entries in A2:A13 as shown in the image at the right.
3. Change the width of column A to 18.
4. Make B1 the active cell, type Sep, and then click the Enter button on the Formula bar.
5. Select B1:I1.
6. Click the Fill button in the Editing group, and then click *Series*.
7. Click *AutoFill* in the *Type* section in the Series dialog box and then click OK.

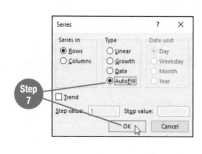

| | A | B |
|---|---|---|
| 1 | | |
| 2 | Expenses | |
| 3 | Housing | |
| 4 | Food | |
| 5 | Transportation | |
| 6 | Smartphone | |
| 7 | Internet | |
| 8 | Entertainment | |
| 9 | Total Expenses | |
| 10 | | |
| 11 | Income | |
| 12 | | |
| 13 | Cash left over | |

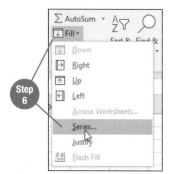

Auto Fill enters the column headings *Oct* through *Apr* in the selected range.

8. Make B3 the active cell, type 875, and then click the Enter button.
9. Select B3:I3, click the Fill button, and then click *Right*.

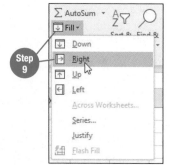

**Fill Right** copies the entry in the first cell to the other cells within the selected range.

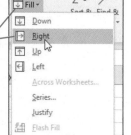

**10** Enter the remaining values as shown in the Step 10 illustration at the right. In rows 5, 6, and 7, use Fill Right to enter the data.

| | A | B | C | D | E | F | G | H | I |
|---|---|---|---|---|---|---|---|---|---|
| 1 | | Sep | Oct | Nov | Dec | Jan | Feb | Mar | Apr |
| 2 | Expenses | | | | | | | | |
| 3 | Housing | 875 | 875 | 875 | 875 | 875 | 875 | 875 | 875 |
| 4 | Food | 260 | 340 | 310 | 295 | 320 | 280 | 300 | 345 |
| 5 | Transportation | 88 | 88 | 88 | 88 | 88 | 88 | 88 | 88 |
| 6 | Smartphone | 48 | 48 | 48 | 48 | 48 | 48 | 48 | 48 |
| 7 | Internet | 42 | 42 | 42 | 42 | 42 | 42 | 42 | 42 |
| 8 | Entertainment | 150 | 110 | 95 | 175 | 100 | 85 | 95 | 120 |
| 9 | Total Expenses | | | | | | | | |

Step 10

## Using the Fill Handle to Copy Cells

The **fill handle** is a small, green square at the lower right corner of the active cell or range. The fill handle is used to copy data from a cell or range to adjacent cells.

**To use the fill handle with a mouse:** When you point at the fill handle with a mouse, the cell pointer changes appearance from the large white cross to a thin, black crosshairs (**+**). Drag right or down when you see the thin, black crosshairs to copy data or a formula to the cells right or below the active cell or to extend a series, such as September or Monday, from the active cell to adjacent cells.

**To use the fill handle on a touch device:** Tapping a cell on a touch device displays the active cell with two selection handles instead of the fill handle. See Figure 8.4 for instructions on how to use the fill handle on a touch device.

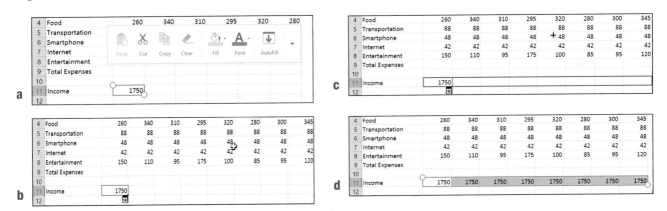

**Figure 8.4**

The steps to using the fill handle on a touch device are shown here. First, (a) tap inside the cell and tap the AutoFill button, resulting in (b) the selection handles disappearing and the fill handle and fill icon displaying. Next, (c) slide right or down to highlight the cells you want filled, resulting in (d) Excel copying the entry to the selected cells.

**11** Make B11 the active cell, type *1750*, and then click the Enter button.

**12** Point to the fill handle in B11 and then drag the fill handle right to I11 when you see the pointer change to a thin, black crosshairs.

The value *1750* is copied to the cells in the selected range when you release the mouse.

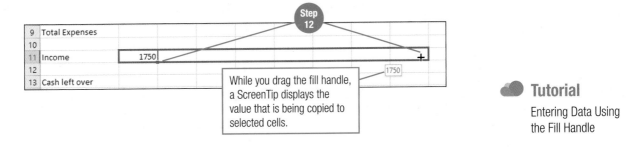

Step 12

While you drag the fill handle, a ScreenTip displays the value that is being copied to selected cells.

☁ **Tutorial**

Entering Data Using the Fill Handle

## Using the SUM function and AutoSum

To add the expenses in B9, you could type the formula =*B3+B4+B5+B6+B7+B8*; however, Excel includes a built-in preprogrammed function called SUM that can be used to add a column or row of numbers. The SUM function is faster and easier to use. To add the expenses in B9 using SUM, you would type the formula =*SUM(B3:B8)*. Notice after =*SUM* you need to provide only the range of cells to add enclosed in parentheses, rather than all the individual cell references and addition symbols. Because the SUM function is used frequently, an **AutoSum button** is included in the Editing group on the Home tab and the AutoSum feature automatically detects the range to be added.

(13)  Make B9 the active cell.

(14)  Click the AutoSum button in the Editing group on the Home tab.

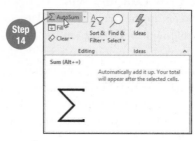

Step 14

**Tutorial**

Entering Formulas Using the AutoSum Button

Excel inserts the formula =*SUM(B3:B8)* in B9 with the suggested range B3:B8 selected. In most cases, the suggested range is correct; however, always check the entry before completing the formula because sometimes AutoSum assumes you want to add values above the active cell when you intend to add values to the left. If necessary, you can drag the mouse to select a different range before completing the formula.

| SUM | | ▼ | ⋮ | × | ✓ | ƒx | =SUM(B3:B8) | | |
|---|---|---|---|---|---|---|---|---|---|
| ◢ | A | B | C | D | E | F | G | H | I |
| 1 | | Sep | Oct | Nov | Dec | Jan | Feb | Mar | Apr |
| 2 | Expenses | | | | | | | | |
| 3 | Housing | 875 | 875 | 875 | | | | 875 | 875 |
| 4 | Food | 260 | 340 | 310 | With the active cell B9 when | | | 300 | 345 |
| 5 | Transportation | 88 | 88 | 88 | the AutoSum button was | | | 88 | 88 |
| 6 | Smartphone | 48 | 48 | 48 | clicked, Excel suggests the | | | 48 | 48 |
| 7 | Internet | 42 | 42 | 42 | values above the active cell | | | 42 | 42 |
| 8 | Entertainment | 150 | 110 | 95 | as the range to be added. | | | 95 | 120 |
| 9 | Total Expenses | =SUM(B3:B8) | | | | | | | |
| 10 | | SUM(**number1**, [number2], ...) | | | | | | | |

(15)  Press Ctrl + Enter.

Ctrl + Enter is the keyboard shortcut to complete the entry and keep the current cell active—the equivalent of clicking the Enter button on the Formula bar.

(16)  With B9 the active cell, point to the fill handle and then drag the fill handle right to I9 when you see the pointer change to a thin, black crosshairs.

**Tutorial**

Copying Formulas

In this instance, using the fill handle copies the formula in B9 to the selected cells. The **Auto Fill Options button** appears when you release the mouse after dragging the fill handle to copy a cell. Use the button to fill the cells with the formatting options only from the copied cell (no data is copied) or to copy the data with no formatting options (only data is copied). The button disappears when you start typing data in a new cell.

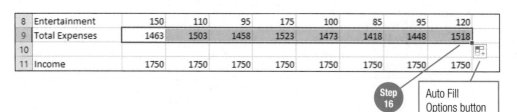

| 8 | Entertainment | 150 | 110 | 95 | 175 | 100 | 85 | 95 | 120 |
|---|---|---|---|---|---|---|---|---|---|
| 9 | Total Expenses | 1463 | 1503 | 1458 | 1523 | 1473 | 1418 | 1448 | 1518 |
| 10 | | | | | | | | | |
| 11 | Income | 1750 | 1750 | 1750 | 1750 | 1750 | 1750 | 1750 | 1750 |

Step 16

Auto Fill Options button

17. Make B13 the active cell, type =b11-b9, and then click the Enter button or press Ctrl + Enter.

18. With B13 the active cell, drag the fill handle right to I13.

| 11 | Income | 1750 | 1750 | 1750 | 1750 | 1750 | 1750 | 1750 | 1750 |
| 12 | | | | | | | | | |
| 13 | Cash left over | 287 | 247 | 292 | 227 | 277 | 332 | 302 | 232 |
| 14 | | | | | | | | | |

Step 18

19. Make J1 the active cell, type Total, and then press Enter.

20. Make J3 the active cell and then click the AutoSum button.

In this instance, Excel suggests the range B3:I3 in the SUM function. Excel looks for values immediately above or to the left of the active cell. Because no value exists above J3, Excel correctly suggests adding the values to the left in the same row.

21. Click the Enter button to accept the formula.

Step 20

Step 21

| SUM | ▾ | : | × | ✓ | fx | =SUM(B3:I3) |

| ⊿ | A | B | C | D | E | F | G | H | I | J | K | L |
|---|---|---|---|---|---|---|---|---|---|---|---|---|
| 1 | | Sep | Oct | Nov | Dec | Jan | Feb | Mar | Apr | Total | | |
| 2 | Expenses | | | | | | | | | | | |
| 3 | Housing | 875 | 875 | 875 | 875 | 875 | 875 | 875 | 875 | =SUM(B3:I3) | | |
| 4 | Food | 260 | 340 | 310 | 295 | 320 | 280 | 300 | 345 | SUM(number1, [number2], …) | | |
| 5 | Transportation | 88 | 88 | 88 | 88 | 88 | 88 | 88 | 88 | | | |
| 6 | Smartphone | 48 | 48 | 48 | 48 | 48 | 48 | 48 | 48 | | | |

22. With J3 the active cell, drag the fill handle down to J13.

23. Make J10 the active cell, and then press the Delete key.

24. Make J12 the active cell and then press the Delete key.

| ⊿ | A | B | C | D | E | F | G | H | I | J |
|---|---|---|---|---|---|---|---|---|---|---|
| 1 | | Sep | Oct | Nov | Dec | Jan | Feb | Mar | Apr | Total |
| 2 | Expenses | | | | | | | | | |
| 3 | Housing | 875 | 875 | 875 | 875 | 875 | 875 | 875 | 875 | 7000 |
| 4 | Food | 260 | 340 | 310 | 295 | 320 | 280 | 300 | 345 | 2450 |
| 5 | Transportation | 88 | 88 | 88 | 88 | 88 | 88 | 88 | 88 | 704 |
| 6 | Smartphone | 48 | 48 | 48 | 48 | 48 | 48 | 48 | 48 | 384 |
| 7 | Internet | 42 | 42 | 42 | 42 | 42 | 42 | 42 | 42 | 336 |
| 8 | Entertainment | 150 | 110 | 95 | 175 | 100 | 85 | 95 | 120 | 930 |
| 9 | Total Expenses | 1463 | 1503 | 1458 | 1523 | 1473 | 1418 | 1448 | 1518 | 11804 |
| 10 | | | | | | | | | | |
| 11 | Income | 1750 | 1750 | 1750 | 1750 | 1750 | 1750 | 1750 | 1750 | 14000 |
| 12 | | | | | | | | | | |
| 13 | Cash left over | 287 | 247 | 292 | 227 | 277 | 332 | 302 | 232 | 2196 |

Step 24

25. Save the new workbook as **SchoolBudget-YourName** in the Ch8 folder in CompletedTopics on your storage medium. Leave the workbook open for the next topic.

**quick steps**

**AutoFill Series**
1. Select range.
2. Click Fill button.
3. Click *Series*.
4. Click *AutoFill*.
5. Click OK.

**Fill Right**
1. Select range.
2. Click Fill button.
3. Click *Right*.

**Copy Using Fill Handle**
1. Make cell active.
2. Drag fill handle as required.

**Add with the SUM Function**
1. Activate formula cell.
2. Click AutoSum button.
3. Click Enter button.

OR

Select correct range and then click Enter button.

## 8.5 Inserting and Deleting Rows and Columns

New rows or columns are inserted or deleted using the Insert or Delete buttons in the Cells group on the Home tab. New rows are inserted above the row in which the active cell is positioned, and new columns are inserted to the left. Cell references within formulas and formula results are automatically updated when new rows or columns with data are added to or removed from a worksheet.

To insert a new row, activate any cell in the row below where a new row is required and click *Insert Sheet Rows* from the **Insert button** option list.

**1** With the **SchoolBudget-YourName** workbook open, make any cell in row 4 active.

| | A | B | C | D | E | F | G | H | I |
|---|---|---|---|---|---|---|---|---|---|
| 1 | | Sep | Oct | Nov | Dec | Jan | Feb | Mar | Apr |
| 2 | Expenses | | | | | | | | |
| 3 | Housing | 875 | 875 | 875 | 875 | 875 | 875 | 875 | 87 |
| 4 | Food | 260 | 340 | 310 | 295 | 320 | 280 | 300 | 34 |
| 5 | Transportation | 88 | 88 | 88 | 88 | 88 | 88 | 88 | 8 |

*Step 1*

**2** Click the Insert button arrow in the Cells group on the Home tab.

**3** Click *Insert Sheet Rows*.

A new blank row is inserted between *Housing* and *Food*.

*Step 2*

*Step 3*

**4** Type the following entries in the cells indicated:

| | | | |
|---|---|---|---|
| A4 | Utilities | F4 | 128 |
| B4 | 110 | G4 | 106 |
| C4 | 115 | H4 | 118 |
| D4 | 132 | I4 | 112 |
| E4 | 147 | | |

**5** Make J3 the active cell and then drag the fill handle down to J4 to copy the SUM formula to the new row.

| | A | B | C | D | E | F | G | H | I | J |
|---|---|---|---|---|---|---|---|---|---|---|
| 1 | | Sep | Oct | Nov | Dec | Jan | Feb | Mar | Apr | Total |
| 2 | Expenses | | | | | | | | | |
| 3 | Housing | 875 | 875 | 875 | 875 | 875 | 875 | 875 | 875 | 7000 |
| 4 | Utilities | 110 | 115 | 132 | 147 | 128 | 106 | 118 | 112 | 968 |
| 5 | Food | 260 | 340 | 310 | 295 | 320 | 280 | 300 | 345 | 2450 |

*Step 4*   *Step 5*

**6** Select A1:A2, click the Insert button arrow, and then click *Insert Sheet Rows*.

Two rows are inserted above A1. You can also insert multiple rows by selecting the row numbers along the left edge of the worksheet area and displaying the contextual shortcut menu. To do this, point to the first row number until the pointer displays as a right-pointing black arrow (→), hold down the left mouse button, drag down to the last row number and release the mouse. Multiple rows will be highlighted in the worksheet area. Right-click within the selection area and then click *Insert*.

| | A | B | C | D | E | F | G | H | I | J |
|---|---|---|---|---|---|---|---|---|---|---|
| 1 | | | | | | | | | | |
| 2 | | | | | | | | | | |
| 3 | | Sep | Oct | Nov | Dec | Jan | Feb | Mar | Apr | Total |
| 4 | Expenses | | | | | | | | | |
| 5 | Housing | 875 | 875 | 875 | 875 | 875 | 875 | 875 | 875 | 7000 |

*Step 6*

7 Make the new cell A1 the active cell and then type Proposed School Budget.

8 Make the new cell A2 the active cell and then type First Year of Program.

9 Select A1:J1 and then click the Merge & Center button in the Alignment group.

10 Select A2:J2 and then click the Merge & Center button.

**Insert Rows or Columns**
1. Activate cell or select range.
2. Click Insert button arrow.
3. Click *Insert Sheet Rows* or *Insert Sheet Columns.*

**Delete Rows or Columns**
1. Activate cell or select range.
2. Click Delete button arrow.
3. Click *Delete Sheet Rows* or *Delete Sheet Columns.*

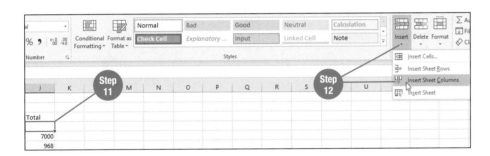

11 Make any cell in column J active.

12 Click the Insert button arrow and then click *Insert Sheet Columns.*

A new column is inserted between *Apr* and *Total.* Notice also an **Insert Options button** (🖉) appears below and right of the active cell. Options from this button are used to format the new column with the same formatting options as the column at its left or right, or to clear formatting in the new column.

To delete a single row or column, position the active cell within that row or column, and then click *Delete Sheet Rows* or *Delete Sheet Columns* in the **Delete button** option list. To remove multiple rows or columns from the worksheet, begin by selecting the range of rows or columns to be deleted before clicking the Delete button arrow.

13 Make any cell in row 10 active.

14 Click the Delete button arrow and then click *Delete Sheet Rows.*

Row 10 is removed from the worksheet, and existing rows below are shifted up to fill in the space. Delete a column by performing actions similar to those in Steps 13 to 14 except click *Delete Sheet Columns* in the Delete option list.

15 Save the workbook using the same name (**SchoolBudget-YourName**). Leave the workbook open for the next topic.

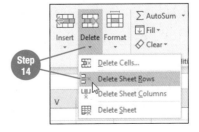

☁ **Tutorial**

Deleting Columns and Rows

# 8.6 Sorting Data and Applying Cell Styles

Data in cells within a range in Excel can be reordered by sorting in either ascending or descending order on one or more columns. For example, you can sort a list of names and cities first by the city and then by the last name. To sort the entire worksheet by one column, make active any cell in the column by which to order the data, click the **Sort & Filter button** in the Editing group on the Home tab and then click *Sort A to Z* or *Sort Z to A*. To sort a portion of the cells within a worksheet or to sort by more than one column, select the range and open the Sort dialog box from the Sort & Filter button option list.

**Watch & Learn**

**Hands On Activity**

- - - - - - - - — Skills

Sort rows of data

Apply cell styles

**Tutorial**
Sorting Data

1. With the **SchoolBudget-YourName** workbook open, select A5:K10.

To sort the list of expenses alphabetically in ascending order, notice you do not include the heading (in A4) or the totals (in A11:K11) in the sort range.

| 4 | Expenses | | | | | | | | | |
|---|---|---|---|---|---|---|---|---|---|---|
| 5 | Housing | 875 | 875 | 875 | 875 | 875 | 875 | 875 | 875 | 7000 |
| 6 | Utilities | 110 | 115 | 132 | 147 | 128 | 106 | 118 | 112 | 968 |
| 7 | Food | 260 | 340 | 310 | 295 | 320 | 280 | 300 | 345 | 2450 |
| 8 | Transportation | 88 | 88 | 88 | 88 | 88 | 88 | 88 | 88 | 704 |
| 9 | Smartphone | 48 | 48 | 48 | 48 | 48 | 48 | 48 | 48 | 384 |
| 10 | Entertainment | 150 | 110 | 95 | 175 | 100 | 85 | 95 | 120 | 930 |
| 11 | Total Expenses | 1531 | 1576 | 1548 | 1628 | 1559 | 1482 | 1524 | 1588 | 12436 |
| 12 | | | | | | | | | | |

*Step 1*

2. Click the Sort & Filter button in the Editing group on the Home tab.

3. Click *Sort A to Z*.

4. Click any cell to deselect the range, and review the new order of the expenses.

5. Select A5:K10.

6. Click the Sort & Filter button and then click *Custom Sort*.

7. Click the *Sort by* option box arrow in the *Column* section (currently displays *Column A*) in the Sort dialog box, and then click *Column K*.

8. Click the *Order* option box arrow (currently displays *Smallest to Largest*), and then click *Largest to Smallest*.

9. Click OK.

The data within the range A5:K10 is rearranged in descending order from highest expense total to lowest.

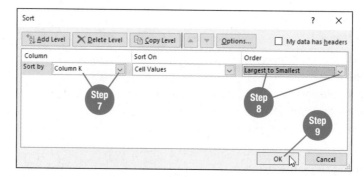

10. Click any cell to deselect the range, and review the new order of expenses.

Similar to the Styles feature in Word, **Cell Styles** in Excel offers a set of predefined formatting options that can be applied to a single cell or a range. Using Cell Styles to format a worksheet is fast and promotes consistency. The Cell Styles gallery groups styles by the sections *Good, Bad and Neutral; Data and Model; Titles and Headings; Themed Cell Styles;* and *Number Format.*

⑪ Click cell A1 and then the More button in the Styles group on the Home tab.

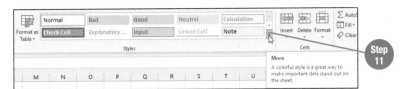

⑫ Click *Heading 1* in the *Titles and Headings* section of the Cell Styles gallery.

Live Preview displays how the cell will look with the style applied.

Step 12

⑬ Click cell A2, click the More button in the Cell Styles gallery, and then click *Heading 4* in the *Titles and Headings* section.

⑭ Click cell A4, hold down the Ctrl key, click cell A13, click cell A15, and then release the Ctrl key. With the three cells selected, apply the Heading 4 cell style.

⑮ Select B3:K3, apply the Accent1 style (first option in last row) in the *Themed Cell Styles* section of the Cell Styles gallery, and then center the content of those cells.

⑯ Select B5:K15 and apply the Comma [0] style in the *Number Format* section of the Cell Styles gallery.

⑰ Select B15:K15 and apply the Total style in the *Titles and Headings* section of the Cell Styles gallery.

⑱ Click any cell to deselect the range and compare your worksheet with the one shown in Figure 8.5.

**quick steps**

**Sort a Range by the First Column**
1. Select range.
2. Click Sort & Filter button.
3. Click *Sort A to Z* or *Sort Z to A*.

**Custom Sort**
1. Select range.
2. Click Sort & Filter button.
3. Click *Custom Sort*.
4. Change options and/or add levels as needed.
5. Click OK.

**Apply Cell Styles**
1. Select cell or range.
2. Click desired cell style in Cell Styles gallery.

**Oops!**

No Cell Styles gallery? On smaller displays, the gallery is accessed by clicking the Cell Styles button in the Styles group on the Home tab.

|  | A | B | C | D | E | F | G | H | I | J | K |
|---|---|---|---|---|---|---|---|---|---|---|---|
| 1 | Proposed School Budget | | | | | | | | | | |
| 2 | First Year of Program | | | | | | | | | | |
| 3 | | Sep | Oct | Nov | Dec | Jan | Feb | Mar | Apr | | Total |
| 4 | Expenses | | | | | | | | | | |
| 5 | Housing | 875 | 875 | 875 | 875 | 875 | 875 | 875 | 875 | | 7,000 |
| 6 | Food | 260 | 340 | 310 | 295 | 320 | 280 | 300 | 345 | | 2,450 |
| 7 | Utilities | 110 | 115 | 132 | 147 | 128 | 106 | 118 | 112 | | 968 |
| 8 | Entertainment | 150 | 110 | 95 | 175 | 100 | 85 | 95 | 120 | | 930 |
| 9 | Transportation | 88 | 88 | 88 | 88 | 88 | 88 | 88 | 88 | | 704 |
| 10 | Smartphone | 48 | 48 | 48 | 48 | 48 | 48 | 48 | 48 | | 384 |
| 11 | Total Expenses | 1,531 | 1,576 | 1,548 | 1,628 | 1,559 | 1,482 | 1,524 | 1,588 | | 12,436 |
| 12 | | | | | | | | | | | |
| 13 | Income | 1,750 | 1,750 | 1,750 | 1,750 | 1,750 | 1,750 | 1,750 | 1,750 | | 14,000 |
| 14 | | | | | | | | | | | |
| 15 | Cash left over | 219 | 174 | 202 | 122 | 191 | 268 | 226 | 162 | | 1,564 |

**Figure 8.5**
The **SchoolBudget-YourName** worksheet will display in sorted order and with cell styles applied from this topic.

⑲ Save the workbook using the same name (**SchoolBudget-YourName**). Leave the workbook open for the next topic.

**Tutorials**

Applying Cell Styles

Applying and Modifying Themes

## 8.7 Changing Orientation and Scaling, Displaying Cell Formulas, and Exporting a Worksheet

By default, new Excel workbooks have print options set to print the active worksheet on a letter-size page (8.5 × 11 inches), in portrait orientation, with 0.75-inch top and bottom margins, and 0.7-inch left and right margins. Preview a new worksheet before printing to determine whether these print options are appropriate.

Workbooks are often circulated electronically by email or by other means of electronic file transfer. In that case, a workbook is often distributed as a PDF file, which is essentially an electronic view of a printed worksheet. When a workbook is exported in PDF format, the PDF is created using the print options in effect. Preview a worksheet and change print options as needed before exporting in PDF file format.

1 With the **SchoolBudget-YourName** workbook open, click the File tab, click *Print*, and then compare your screen with the one shown in Figure 8.6.

2 Click the Next Page button at the bottom of the preview panel to view the second page of the worksheet.

3 Click the *Orientation* button arrow (currently displays *Portrait Orientation*) in the *Settings* section and then click *Landscape Orientation*.

Notice the worksheet now fits on one page. Landscape is a common layout used for wide worksheets. Print options are saved with the workbook content.

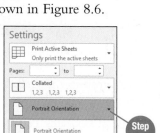

**Watch & Learn**

**Hands On Activity**

------- Skills

Preview a worksheet
Change orientation
Display cell formulas
Change scaling
Export worksheet as PDF

**Tutorials**
Displaying Formulas
Printing a Worksheet

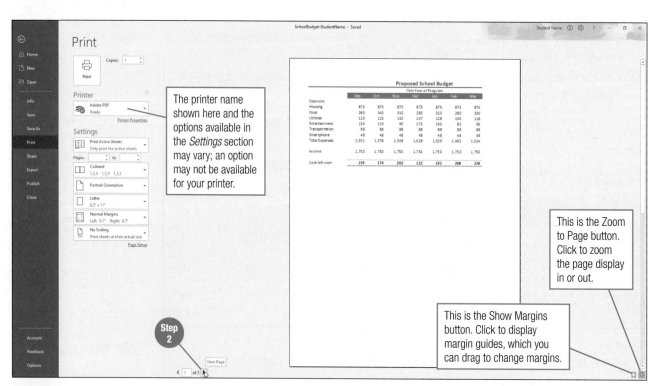

**Figure 8.6**
The Print backstage area is shown with the first page displayed for the **SchoolBudget-StudentName** worksheet.

④ Click Back to return to the worksheet and then save the revised workbook using the same name (**SchoolBudget-YourName**).

On a printed copy of the worksheet, only the formula result is printed. You may want to print a second copy of the worksheet with the formulas displayed in the cell as a backup or documentation strategy for a complex or otherwise important worksheet.

⑤ Click the Formulas tab.

⑥ Click the Show Formulas button in the Formula Auditing group.

The **Show Formulas button** is a toggle button that switches the display between showing the formula in each cell and showing the result in each cell. Ctrl + ` (grave accent, usually positioned on the keyboard above the Tab key), is the keyboard command to turn on or turn off the display of formulas.

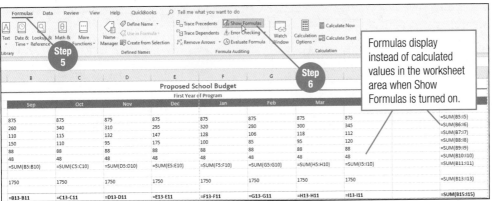

Formulas display instead of calculated values in the worksheet area when Show Formulas is turned on.

⑦ Scroll right if necessary to review the worksheet with formulas displayed.

⑧ Display the Print backstage area.

⑨ Click the *Scaling* option box arrow (currently displays *No Scaling*) in the *Settings* section and then click *Fit Sheet on One Page*.

**Fit Sheet on One Page** shrinks the size of text on the printout to fit all columns and rows on one page.

⑩ Click Back to return to the worksheet.

⑪ Use Save As to save a copy of the worksheet with the formulas displayed, named **SchoolBudgetFormulas-YourName** in the Ch8 folder in CompletedTopics.

⑫ Click the File tab and then click *Export*.

⑬ Click the Create PDF/XPS button on the right panel of the Export backstage area.

⑭ Click the Publish button in the Publish as PDF or XPS dialog box, accepting the default file name and storage location options.

⑮ If the PDF document opens in a new window, close the window to return to Excel.

⑯ Close the workbook.

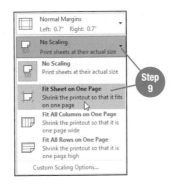

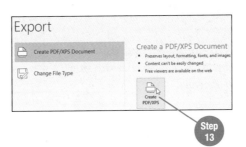

## 8.8 Inserting and Renaming a Worksheet, Copying Cells, and Indenting Cell Contents

A workbook can contain more than one worksheet. Use multiple worksheets as a method to organize or group data into manageable units. For example, a homeowner might have one household finance workbook and keep track of bills and loans in one worksheet, savings and investments in a second worksheet, and a household budget in a third worksheet. Insert, rename, and navigate between worksheets using the sheet tabs on the **Sheet tab bar**. Use the **New sheet button** on the Sheet tab bar to add a new worksheet to the workbook.

**1** Open the **SchoolBudget-YourName** workbook.

**2** Click the New sheet button on the Sheet tab bar.

A new worksheet with the name Sheet2 is added to the workbook to the right of the Sheet1 worksheet and made the active worksheet.

The left- and right-pointing arrows are used to scroll sheet tabs.

**3** Click the Sheet1 tab to make Sheet1 the active worksheet.

**4** Right-click the Sheet1 tab and then click *Rename*.

**5** Type First Year and then press Enter.

**6** Right-click the Sheet2 tab, click *Rename*, type Second Year, and then press Enter.

**7** Click the First Year tab to make First Year the active worksheet.

**8** Select A4:A15 and then click the Copy button in the Clipboard group on the Home tab.

**9** Click the Second Year tab, click cell A4, and then click the Paste button. Do *not* click the down-pointing arrow.

**10** Click the Paste Options button and then click *Keep Source Column Widths* (second option in second row in *Paste* section).

**11** Click the First Year tab, select A1:K3, and then click the Copy button.

**12** Click the Second Year tab, click cell A1, and then click the Paste button.

**13** Edit the content in cell A2 in the Second Year worksheet to change *First* to *Second* so that the title now reads *Second Year of Program*.

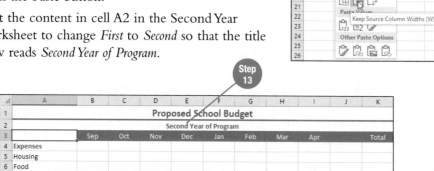

 **Watch & Learn**

 **Hands On Activity**

**Skills**

Insert a new worksheet
Rename worksheets
Copy cells
Indent cells

**app tip**

Other options on the sheet tab shortcut menu are used to delete, move, copy, hide, or protect entire sheets and to change the color of the background in the sheet tab.

**Oops!**

Pasted to the wrong starting cell? Drag the border of the selected range to the correct starting point, or use Cut and Paste to move cells.

**Tutorials**

Inserting and Renaming Worksheets

Copying and Pasting Cells between Worksheets

14 Enter the data in the Second Year worksheet as shown in Figure 8.7. Complete the worksheet by entering and copying formulas for the cells in row 11, row 15, and column K. Refer to the entries in the First Year worksheet if you need help with the formulas. Apply the Comma [0] cell style to all the values and the Total cell style to the cells in row 15.

| | A | B | C | D | E | F | G | H | I | J | K |
|---|---|---|---|---|---|---|---|---|---|---|---|
| 1 | | | | | Proposed School Budget | | | | | | |
| 2 | | | | | Second Year of Program | | | | | | |
| 3 | | Sep | Oct | Nov | Dec | Jan | Feb | Mar | Apr | | Total |
| 4 | Expenses | | | | | | | | | | |
| 5 | Housing | 910 | 910 | 910 | 910 | 910 | 910 | 910 | 910 | | |
| 6 | Food | 245 | 330 | 298 | 285 | 308 | 275 | 295 | 355 | | |
| 7 | Utilities | 112 | 118 | 140 | 151 | 131 | 118 | 122 | 124 | | |
| 8 | Entertainment | 160 | 95 | 100 | 185 | 110 | 95 | 90 | 125 | | |
| 9 | Transportation | 90 | 90 | 90 | 90 | 90 | 90 | 90 | 90 | | |
| 10 | Smartphone | 50 | 50 | 50 | 50 | 50 | 50 | 50 | 50 | | |
| 11 | Total Expenses | | | | | | | | | | |
| 12 | | | | | | | | | | | |
| 13 | Income | 1,800 | 1,800 | 1,800 | 1,800 | 1,800 | 1,800 | 1,800 | 1,800 | | |
| 14 | | | | | | | | | | | |
| 15 | Cash left over | | | | | | | | | | |

**Figure 8.7**
The completed Second Year worksheet is shown here.

The **Increase Indent button** in the Alignment group on the Home tab moves an entry approximately one character width inward from the left edge of a cell each time the button is clicked. Use this feature to indent entries in a list below a subheading. The **Decrease Indent button** moves an entry approximately one character width closer to the left edge of the cell each time the button is clicked.

15 With Second Year still the active worksheet, select A5:A10 and then click the Increase Indent button in the Alignment group.

16 Click cell A11 and then click the Increase Indent button twice.

17 Click the Page Layout tab.

18 Click the Orientation button in the Page Setup group and then click *Landscape*.

19 Make First Year the active worksheet. If necessary, press the Esc key to remove the moving marquee from the copied cells in A1:A2.

20 Select A5:A10, click the Home tab, and then click the Increase Indent button.

21 Click cell A11 and then click the Increase Indent button twice.

22 Save the workbook using the same name (**SchoolBudget-Your Name**) and then close the workbook.

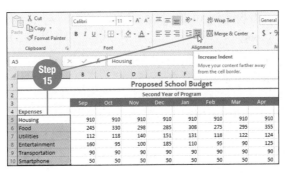

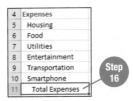

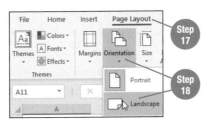

# 8.9 Using Go To; Freezing Panes; and Shading, Wrapping, and Rotating Cell Entries

🌩 **Watch & Learn**

🌩 **Hands On Activity**

In large worksheets where you cannot see all cells at once, the **Go To** and **Go To Special** commands accessed from the **Find & Select button** in the Editing group on the Home tab are used to move the active cell to a specific location in a worksheet. Column or row headings not visible when you scroll right or down beyond the viewing area can be fixed in place using the **Freeze Panes** buton. The position of the active cell when the *Freeze Panes* option is used determines which rows and columns remain fixed in place—all columns left and all rows above the active cell are frozen.

● ─ ─ ─ ─ ─ ─ ─ skills

Use Go To

Freeze panes

Add fill color

Wrap text

Rotate text

1. Open the **NSCSuppliesInventory** workbook from the Ch8 folder in StudentDataFiles. If necessary, click the Enable Editing button to close Protected view.

2. Save the workbook as **NSCSuppliesInv-YourName** in the Ch8 folder in CompletedTopics.

3. Scroll down the worksheet area until the titles and column headings are no longer visible.

4. Click the Find & Select button in the Editing group on the Home tab and then click *Go To*.

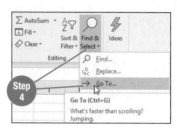

5. Type a4 in the *Reference* text box in the Go To dialog box and then press Enter or click OK.

6. Click the Find & Select button and then click *Go To Special*.

7. Click *Last cell* in the Go To Special dialog box and then click OK.

Use the *Last cell* option in the Go To Special dialog box in a large worksheet to quickly move the active cell to the bottom right of the worksheet.

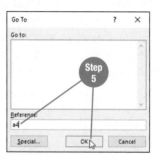

8. Use the Go To command to move the active cell back to A4.

9. Scroll up until you can see the first three rows, containing titles and column headings.

10. With A4 still the active cell, click the View tab and then click the Freeze Panes button in the Window group.

11. Click *Freeze Panes*.

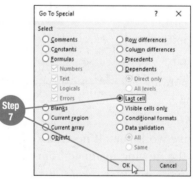

● ─ ─ ─ ─ ─ ─ ─ app tip

Ctrl + G is the keyboard command for Go To.

🌩 **Tutorials**

Navigating and Scrolling

Freezing and Unfreezing Panes

● ─ ─ ─ ─ ─ ─ ─ app tip

Once cells have been frozen, the *Freeze Panes* option from the Freeze Panes button changes to *Unfreeze Panes*.

12. Scroll down several rows past all data. Notice that rows 1 to 3 do not scroll out of the viewing area.

Column headings can be formatted to stand out from the rest of the worksheet by shading the background of the cell using the **Fill Color button** in the Font group on the Home tab or by rotating the cell entries. Cells with long entries can be housed in narrower columns by formatting the text to automatically wrap within the width of the cell using the **Wrap Text button** in the Alignment group. In Steps 13 to 20, you will format worksheet headings to match the headings shown in Figure 8.8.

**quick steps**

**Freeze Panes**
1. Activate cell below and right of cells to freeze.
2. Click View tab.
3. Click Freeze Panes button.
4. Click *Freeze Panes*.

**Shade Cell Background**
1. Select cell or range.
2. Click Fill Color button arrow.
3. Click desired color.

**Rotate Cells**
1. Select cell or range.
2. Click Orientation option in Alignment group.
3. Click desired rotate option.

13. Scroll to the top of the worksheet and then click cell A1.

14. Click the Home tab, click the Fill Color button arrow in the Font group, and then click *Orange, Accent 6* (last color in first row of *Theme Colors* section).

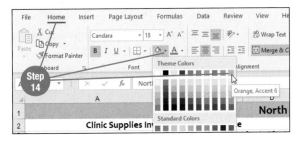

15. Select A2:M2, click the Fill Color button arrow and then click *Orange, Accent 6, Lighter 60%* (last option in third row of *Theme Colors* section).

16. Select A3:M3, click the Fill Color button arrow, and then click *Orange, Accent 6, Lighter 80%* (last option in second row of *Theme Colors* section).

17. Make M3 the active cell, change the row height to 30, and change the column width to 10.

18. With M3 still the active cell, click the Wrap Text button in the Alignment group.

The column heading is visible again, with the text wrapping within the cell.

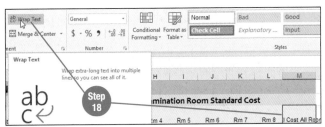

19. Make D3 the active cell, change the column width to 9, and then click the Wrap Text button.

20. Select E3:L3, click the Orientation button in the Alignment group, and then click *Angle Counterclockwise*.

**Angle Counterclockwise** rotates the text within the cell boundaries 45 degrees.

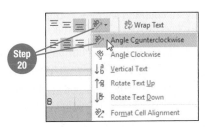

🌐 **Tutorials**

Adding Fill Color to Cells

Using Format Painter and the Repeat Command

21. Click any cell to deselect the range and then display the Print backstage area.

22. Change settings to *Landscape Orientation* and *Fit All Columns on One Page* and then go back to the worksheet display.

23. Save the workbook using the same name (**NSCSuppliesInv-YourName**) and then close the workbook.

| | A | B | C | D | E | F | G | H | I | J | K | L | M |
|---|---|---|---|---|---|---|---|---|---|---|---|---|---|
| 1 | | | | North Side Medical Clinic | | | | | | | | | |
| 2 | Clinic Supplies Inventory Units and Price | | | | | | Patient Examination Room Standard Cost | | | | | | |
| 3 | Item | | Unit | Price | Standard Stock Qty | Rm 1 | Rm 2 | Rm 3 | Rm 4 | Rm 5 | Rm 6 | Rm 7 | Rm 8 | Total Cost All Rooms |

**Figure 8.8**
The **NSCSuppliesInv-StudentName** worksheet with the formatted headings is shown.

## Topics Review

| Topic | Key Concepts |
|---|---|
| **8.1 Creating and Editing a New Worksheet** | A **spreadsheet application** is software in which data is created, analyzed, and presented in a grid-like structure of columns and rows. |
| | A **workbook** is an Excel file that consists of a collection of individual **worksheets**. |
| | A new workbook opens with a blank worksheet into which you add text, insert values, and create formulas. |
| | The intersection of a column and a row is called a **cell**. |
| | The **active cell** is indicated with a green border and is the place where the next data typed will be stored or the next command will be acted upon. |
| | Create a worksheet by making a cell active and typing text, a value, or a formula. |
| | A **formula** is used to perform mathematical operations on values. |
| | Formula entries begin with an equals sign and are followed by cell references with operators between the references. |
| | Edit a cell by typing new data to overwrite existing data, by double-clicking to open the cell for editing, or by inserting or deleting characters in the Formula bar. |
| | Press the Delete key or use the **Clear button** in the Editing group on the Home tab to delete the contents in the active cell. |
| **8.2 Formatting Cells** | The Font group on the Home tab contains buttons to change the font, font size, font color, and font style of selected cells. |
| | Select cells with the mouse by positioning the cell pointer over the starting cell and dragging in the required direction. |
| | Select cells using touch by tapping the starting cell and then sliding your finger over a selection handle until the remaining cells are inside the selection rectangle. |
| | A rectangular-shaped group of cells is called a **range** and is referenced with the starting cell address, a colon, and the ending cell address (e.g., A1:F15). |
| | By default, cells in a new worksheet have the General format, which has no specific formatting options. |
| | The **Comma Style** format option adds a comma in thousands and two decimal places to numbers. |
| | The **Accounting Number Format** option adds a dollar symbol, a comma in the thousands place, and displays two digits after the decimal point for numbers. |
| | The **Decrease Decimal button** and **Increase Decimal button** in the Number group on the Home tab remove or add one decimal place each time the button is clicked. |
| | Borders in various styles and colors can be added to the edges of a cell using the Bottom Border button arrow in the Font group on the Home tab. |
| | The **Merge & Center button** in the Alignment group on the Home tab is often used to center a worksheet title over multiple columns. |

*continued...*

| Topic | Key Concepts |
|---|---|
| **8.3 Adjusting Column Width and Row Height, and Changing Cell Alignment** | A technique to add more space between cells to improve readability or emphasize a section is to widen a column or increase the height of a row. |
| | Open the Column Width dialog box from the Format button in the Cells group on the Home tab to enter a new value for the width of the column in which the active cell is positioned. |
| | Excel displays a series of pound symbols when a column width is too narrow to display all the cell contents. |
| | **AutoFit** changes the width of a column to fit the contents of the active cell. |
| | Open the Row Height dialog box from the Format button in the Cells group on the Home tab to enter a new value for the height of the row in which the active cell is positioned. |
| | Align the content in a cell within the cell boundaries at the left, center, or right horizontally, or top, middle, or bottom vertically using buttons in the Alignment group. |
| | The **Middle Align button** in the Alignment group on the Home tab centers cell contents vertically. |
| **8.4 Entering or Copying Data with the Fill Command and Using AutoSum** | **Auto Fill** can be used to automatically enter data in a series or pattern based upon an entry in an adjacent cell. When no pattern is detected, the **Fill** feature copies an entry or formula to other cells. |
| | The **Flash Fill** feature automatically suggests entries when a pattern is detected. |
| | Select a range and use the Fill button to open the Series dialog box or choose the *Down, Right, Up,* or *Left* option to copy an entry. For example, **Fill Right** copies the entry in the first cell to the cells that are right of the starting cell within the selected range. |
| | The small, green square at the lower right of an active cell is the **fill handle** and can be used to copy data and formulas or to enter a series by dragging the fill handle to adjacent cells. |
| | The **AutoSum button** in the Editing group on the Home tab enters a SUM function to add a column or row of numbers in the active cell. |
| | The **Auto Fill Options button** appears after dragging the fill handle with options for copying the data. |
| **8.5 Inserting and Deleting Rows and Columns** | A new row is inserted above the active cell or selected range. |
| | A new column is inserted left of the active cell or selected range. |
| | Use the **Insert button** in the Cells group on the Home tab to insert new rows or columns. |
| | Options for formatting new rows or columns are available from the **Insert Options button** that appears when rows or columns are inserted. |
| | Delete rows or columns using options from the **Delete button** in the Cells group on the Home tab. |
| **8.6 Sorting Data and Applying Cell Styles** | Select a range and choose the sort order option from the **Sort & Filter button** in the Editing group on the Home tab to arrange the rows by the entries in the first column. |
| | Open the Sort dialog box to sort by more than one column or to choose a different column in the range by which to sort. |
| | **Cell Styles** are a set of predefined formatting options applied to selected cells using options in the Cell Styles gallery in the Styles group on the Home tab. |
| | Use Cell Styles to format cells with a group of options faster and/or to promote consistency among worksheets. |

*continued...*

| Topic | Key Concepts |
|---|---|
| **8.7 Changing Orientation and Scaling, Displaying Cell Formulas, and Exporting a Worksheet** | New workbooks print on a letter-size page, in portrait orientation, with top and bottom margins of 0.75 inches, and left and right margins of 0.7 inches. |
| | Change print options even if you are only exporting the workbook as a PDF because PDFs are generated using the print settings. |
| | Change to landscape orientation using the Orientation option box in the Print backstage area. |
| | Landscape is a common layout used for wide worksheets. |
| | Turn on or turn off the display of formulas in cells by clicking the Formulas tab and then clicking the **Show Formulas button** in the Formula Auditing group. |
| | The **Fit Sheet on One Page** scaling option scales text on a printout so that all columns and rows print on one page. |
| | Export a worksheet as a PDF document by clicking the File tab, *Export*, and then the Create PDF/XPS button in the right panel of the Export backstage area. |
| **8.8 Inserting and Renaming a Worksheet, Copying Cells, and Indenting Cell Contents** | The **Sheet tab bar** near the bottom left of the window is used to insert, rename, and navigate among sheets in a workbook. |
| | Click the **New sheet button** on the Sheet tab bar to insert a new worksheet. |
| | Right-click a sheet tab and click *Rename* to type a new name for a worksheet. |
| | Copy and paste cells between worksheets using the Copy and Paste buttons in the Clipboard group. |
| | *Keep Source Column Widths* from the Paste Options button lets you paste new cells with the same column width as the source cell. |
| | The **Increase Indent button** moves an entry approximately one character width inward from the left edge of the cell each time the button is clicked. |
| | The **Decrease Indent button** moves an entry approximately one character width closer to the left edge of the cell each time the button is clicked. |
| **8.9 Using Go To; Freezing Panes; and Shading, Wrapping, and Rotating Cell Entries** | The **Go To** dialog box and **Go To Special** dialog box from the **Find & Select button** in the Editing group on the Home tab are used to move to a specific cell in a large worksheet. |
| | **Freeze Panes** fixes rows and/or columns in place for scrolling in large worksheets so that column and row headings do not scroll out of the viewing area. |
| | All rows above and all columns left of the active cell are frozen when Freeze Panes is turned on. |
| | Cells are shaded with color using the **Fill Color button** in the Font group on the Home tab. |
| | Long text entries in cells can be displayed in narrow columns by wrapping text within the cell column width using the **Wrap Text button** in the Alignment group on the Home tab. |
| | **Angle Counterclockwise** is an option from the Orientation button in the Alignment group on the Home tab that rotates text within a cell 45 degrees. |

**Topics Review**

# Inserting Functions and Enhancing an Excel Worksheet

## Chapter 9

Several hundred preprogrammed formulas grouped by category in the function library are used to perform data analysis, decision making, and data modeling. With each application update, the function library is updated and expanded. To enter references for some formulas correctly, you need to understand the difference between an absolute and a relative cell address. Creating a range name to reference a cell or a block of cells makes creating and understanding a formula easier.

There are all sorts of creative ways to enhance the look of an Excel worksheet. Charts present data in a visual snapshot. Data organized in a list can be defined as a table. Excel offers various design styles with formatting options, and the ability to sort or filter lists of data. As you design your data into a more visually appealing presentation, Page Layout view is useful for previewing page layout and print options and for adding headers or footers to worksheets that will be printed. You can then use comments to collaborate with multiple people that are editing or reviewing a worksheet.

In this chapter, you continue working with formulas by learning the types of references used in formulas and how to use functions to perform statistical, date, financial, and logical analysis. Next, you explore strategies for presenting data using charts, comments, and tables as well as formatting a worksheet using page layout options and print options.

## Learning Objectives

**9.1** Create formulas with absolute addresses and range names

**9.2** Create formulas with AVERAGE, MAX, and MIN statistical functions

**9.3** Enter and format dates and use the TODAY function

**9.4** Perform decision making using the logical IF function

**9.5** Use the PMT function to calculate a loan payment

**9.6** Create and edit a pie chart

**9.7** Create and edit a column chart

**9.8** Create and edit a line chart

**9.9** Use Page Layout view, insert a header, change margins, and center a worksheet

**9.10** Create and edit sparklines and insert comments into a worksheet

**9.11** Sort and filter a range defined as a table

**Read & Learn**

**Content Online**

The online course includes additional training and assessment resources.

## 9.1 Using Absolute Addresses and Range Names in Formulas

The formulas created in the previous chapter used a type of cell reference called a **relative address**, where each column letter and row number changes to the destination column or row when the formula is copied. For example, the formula =SUM(A4:A10) becomes =SUM(B4:B10) when copied from a cell in column A to a cell in column B. Relative addressing is the most common addressing method.

Sometimes you need a formula in which one or more addresses should not change when the formula is copied. In these formulas, use a cell reference that is an **absolute address**. A dollar symbol precedes a column letter and/or row number in an absolute address, for example, =$A$10. Some formulas have both a relative and an absolute reference; referred to as a **mixed address**. See Table 9.1 for formula addressing examples. A range name is a text label to describe the value stored in a cell.

**Table 9.1** Cell Addressing and Copying Examples

| Formula | Type of Reference | Action if Formula Is Copied |
|---|---|---|
| =B4*B2 | relative | Both addresses will update. |
| =$B$4*$B$2 | absolute | Neither address will update. |
| =B4*$B$2 | mixed | The address B4 will update; the address $B$2 will not update. |
| =B4*$B2 | mixed | The address B4 will update; the row number in the second address will update, but the column letter will not. |
| =B4*B$2 | mixed | The address B4 will update; the column letter in the second address will update, but the row number will not. |

① Start Excel and open the workbook **FinancialPlanner** from the Ch9 folder in StudentDataFiles on your storage medium. If necessary, click the Enable Editing button to close Protected view.

② Use Save As to save a copy of the workbook as **FinancialPlanner-YourName** in a new folder, Ch9, in the CompletedTopics folder on your storage medium.

③ Review the worksheet, noticing the three rates in row 2; these rates will be used to calculate gross pay, payroll deductions, and savings.

④ Click cell C4, type =b4*b2, and then click the Enter button on the Formula bar or press Ctrl + Enter.

⑤ Use the fill handle in C4 to copy the formula to C5:C30.

Notice the entry #VALUE! in cell C5, which is an **error value**. An error value appears when Excel cannot calculate a formula.

⑥ Click cell C5 and read the formula in the Formula bar.

The #VALUE! error occurred because the content stored in B3 is a label (Estimated Hours) and has no mathematical value. Although you see results in

● Watch & Learn

● Hands On Activity

●─────── Skills

Create a formula with an absolute address

Create a range name

Create a formula with a range name

● Tutorial

Absolute Addressing

●────── app tip

Other error values you might see when Excel encounters a formula that cannot be calculated are #NAME? or #DIV/0. Use the Tell Me feature or Help task pane to find explanations for an error value and tips for correcting the errors.

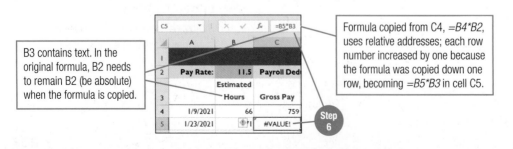

B3 contains text. In the original formula, B2 needs to remain B2 (be absolute) when the formula is copied.

Formula copied from C4, =B4*B2, uses relative addresses; each row number increased by one because the formula was copied down one row, becoming =B5*B3 in cell C5.

the cells C6:C30, the calculated values are not the correct gross pay amounts. For example, the formula in C6 is *=B6\*B4*, which is multiplying 58 × 66, returning the result *3828*. The correct calculation should be 58 × 11.5, returning the result 667.

**7** Select C5:C30 and press the Delete key, or click the Clear button in the Editing group on the Home tab and then click *Clear All*.

**8** Click cell C4, click at the end of the formula in the Formula bar, press Backspace twice to delete B2 content, and then type $B$2 so that the formula now reads *=B4\*$B$2*.

**9** Click the Enter button on the Formula bar or press Ctrl + Enter, and then use the fill handle in C4 to copy the formula to C5:C30.

**10** Click cell D4, type the formula =c4*$e$2, and then press Ctrl + Enter.

You can also press function key F4 to make the cell address absolute. F4 cycles through variations of addressing for the address in which the insertion point is positioned each time the key is pressed. The first variation is an absolute reference.

**11** Click cell E4, type the formula =c4–d4, and then press Ctrl + Enter.

**12** Select D4:E4 and then use the fill handle to copy the formulas to D5:E30.

A cell or a range in a formula can be referenced by a descriptive label, making the formula easier to understand. For example, the formula *=Hours\*PayRate* is readily understood. Names are also used when a formula needs an absolute reference, because the cell referenced in a range name is automatically an absolute address. Assign a range name using the **Name box** (left of the Formula bar).

**13** Click cell G2, click in the Name box, type SaveRate, and then press Enter.

A range name can use letters, numbers, and some symbols. Spaces are not valid in a range name, and the first character in a name must be a letter, an underscore, or a backslash (\).

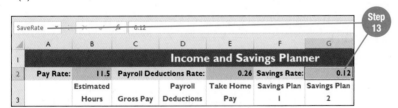

**14** Click cell F4, type =e4*SaveRate, and then press Ctrl + Enter.

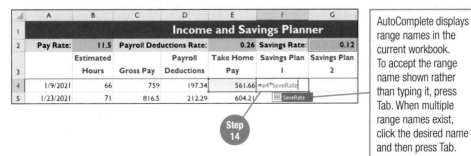

**15** Use the fill handle in F4 to copy the formula to F5:F30.

**16** Save the revised workbook using the same name (**FinancialPlanner-YourName**). Leave the workbook open for the next topic.

**quick steps**

**Make Cell Reference Absolute**
Type dollar symbol before column letter and/or row number, or press F4 to cycle through variations of addressing.

**Name a Cell or Range**
1. Select cell or range.
2. Type name in Name box.
3. Press Enter.

**app tip**

Use the Name Manager button on the Formulas tab to change a range name or the cell(s) the name references, if you want to make a change after creating the range name.

**Tutorials**

Naming and Using a Range

Managing Range Names

AutoComplete displays range names in the current workbook. To accept the range name shown rather than typing it, press Tab. When multiple range names exist, click the desired name and then press Tab.

# 9.2 Entering Formulas Using Statistical Functions

A preprogrammed formula in Excel is called a **function**. In the previous chapter, you learned how to use the AutoSum function, which is the formula to add a series of values. The function library in Excel contains more than 400 preprogrammed formulas grouped into 13 categories. All formulas based on functions begin with the name of the function followed by the function parameters in parentheses. The parameters for a function (referred to as an **argument**) vary depending on the formula chosen and can include a value, a cell reference, a range, multiple ranges, or a combination of values with cell references.

🔵 Watch & Learn

🔵 Hands On
   Activity

●────── skills

Enter AVERAGE formula
Enter MAX formula
Enter MIN formula

**1** With the **FinancialPlanner-YourName** workbook open, click cell I4.

**2** Click the AutoSum button arrow in the Editing group on the Home tab and then click *Average*.

Excel inserts *=AVERAGE(B4:H4)* in the cell with the range B4:H4 selected. In this instance, Excel suggests the wrong range.

**3** With the range B4:H4 highlighted in the formula cell, select B4:B30, and then press Enter or click the Enter button on the Formula bar.

Excel displays the result *68.96296296* in the formula cell, which is the arithmetic mean of the hours in column B. If empty cells or cells containing text are included in the formula argument, they are ignored.

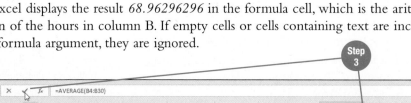

Excel provides the format for the function argument in a ScreenTip.

**4** Click cell I5, click the AutoSum button arrow, and then click *Max*.

**5** Type *=b4:b30* and then press Enter or click the Enter button.

Excel returns the value *80* in the formula cell. MAX returns the largest value found in the range included in the argument.

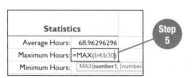

**6** Click cell I6 if I6 is not the active cell, type *=min(b4:b30)*, and then press Enter or click the Enter button.

The result *48* is displayed in I6. MIN returns the smallest value within the range included in the argument.

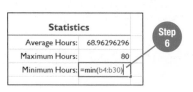

Excel's Insert Function dialog box assists with finding and entering functions and their arguments into a formula cell. A variety of methods can be used to open the Insert Function dialog box, including clicking the Insert Function button on the Formula bar, or clicking the Formulas tab and then clicking the Insert Function button in the Function Library group.

●────── app tip

Shift + F3 is the keyboard shortcut to open the Insert Function dialog box.

7  Click cell I8 and then click the Insert Function button on the Formula bar (button at the right of the Enter button).

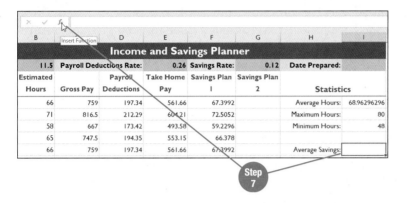

●━━━━ **quick steps**

**AVERAGE, MAX, or MIN Functions**
1. Activate formula cell.
2. Click AutoSum button arrow.
3. Click required function.
4. Type or select argument range.
5. Press Enter.

8  Click the *Or select a category* option box arrow and then click *Statistical*.

9  Click *AVERAGE* in the *Select a function* list box and then click OK.

10  Type f4:f30 in the *Number1* text box in the Function Arguments dialog box and then press Enter or click OK.

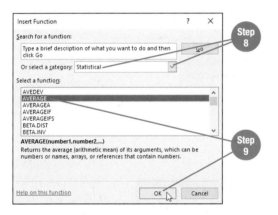
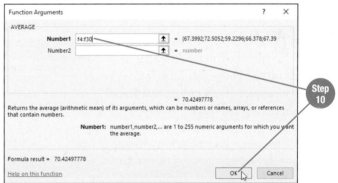

11  Click cell I9, click the AutoSum button arrow, and then click *More Functions*.

12  With the current text already selected in the *Search for a function* text box, type max and then click the Go button.

13  With MAX selected in the *Select a function* list box, click OK.

14  Type f4:f30 in the *Number1* text box in the Function Arguments dialog box, and then press Enter or click OK.

15  Click cell I10, click the AutoSum button arrow, and then click *Min*.

16  Type f4:f30 and then press Enter.

17  Save the revised workbook using the same name (**FinancialPlanner-YourName**). Leave the workbook open for the next topic.

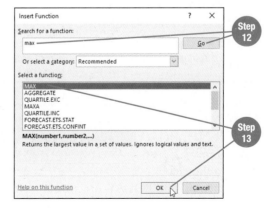

☁ **Tutorial**

Using Statistical Functions

●━━━━━━ **app tip**

The *Count Numbers* option in the AutoSum drop-down list returns the number of cells within a range that have values. Use this function to return a count of values in a list.

## 9.3 Entering, Formatting, and Calculating Dates

A date typed into a cell in normal date format, such as *May 1, 2021* or *5/1/2021,* is stored as a value. For example, entering May 1, 2021 in cell A1 displays *1-May-21* in the worksheet area and *5/1/2001* in the Formula bar. Excel assigns a value to a cell containing a date. For a date of May 1, 2021, the value assigned to cell A1 would be 44317. The value assigned to a date represents the number of days that have passed since January 1, 1900. (January 1, 1900, is assigned the value 1.)

A time entered into a cell is stored as a decimal value representing a fraction of a day. For example, entering 12:00 pm into cell A1 displays *12:00 PM* in the worksheet area, *12:00:00 PM* in the Formula bar, and Excel assigns the value 0.5 to the cell (one-half of the day has passed).

A formula can reference a cell that contains a date or a time, since the entry is considered a value by Excel. A variety of date and time formats can be applied to cells containing dates. Examples of worksheets that use dates or times in formulas include worksheets to calculate elapsed time such as scheduling, hours worked, memberships, vacation durations, seniority, or employee service.

1 With the **FinancialPlanner-YourName** workbook open, click cell I2.

2 Type *12/20/2021* and then press Ctrl + Enter.

3 Notice that *Date* appears in the *Number Format* option box in the Number group on the Home tab. Table 9.2 gives examples of cell entries that Excel recognizes as valid dates.

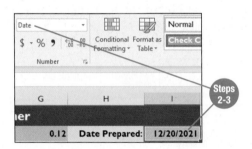

Watch & Learn

Hands On Activity

Skills

Enter a valid date

Enter the current date using a function

Create a formula using a date

Format dates

Oops!

*General* instead of *Date* appears? Excel did not recognize your entry as a valid date. Generally, this is because of a typing error. Try Step 2 again. Still *General*? You may need to check the Region in the Control Panel.

**Table 9.2** Entries Excel Recognizes as Valid Dates or Times

| Dates | Times |
| --- | --- |
| 12/20/21; 12-20-21 | 4:45 (stored as 4:45:00 AM) |
| Dec 20, 2021 | 4:45 PM (stored as 4:45:00 PM) |
| 20-Dec-21 or 20 Dec 21 or 20/Dec/21 | 16:45 (stored as 4:45:00 PM) |

*Note: The year can be entered as two digits or four digits and the month as three characters or spelled in full. Times are generally entered as hh:mm, but in situations that require a higher level of accuracy, they are entered as hh:mm:ss.*

4 With I2 still the active cell, press the Delete key to clear the contents, type =today(), and then press Ctrl + Enter.

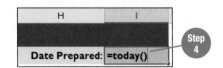

Excel enters the current date into the cell. No argument is required for this function. The TODAY function updates the cell entry to the current date whenever the worksheet is opened or printed. Do not use *=TODAY()* if you want the date to stay the same. Another DATE function that returns the current date is the NOW function. Entering *=NOW()* into a cell enters the current date along with the current time.

5 Click cell B3, click the Insert button arrow in the Cells group on the Home tab, and then click *Insert Sheet Columns.*

6  With B3 still the active cell, type Pay Date and then click A3.

7  Type End Date in A3 and then press Ctrl + Enter.

8  Select A3:B3 and then click the Center button in the Alignment group.

9  Click cell B4, type =a4+7, and then press Ctrl + Enter.

Excel returns *1/16/2021* in B4, which is seven days from January 9, 2021.

10  Use the fill handle in B4 to copy the formula to the range B5:B30.

11  Select A4:B30.

12  Click the *Number Format* option box arrow (the arrow next to *Date*) and then click *More Number Formats*.

13  With *Date* selected in the *Category* list box in the Format Cells dialog box, scroll down the *Type* list box and then click *14-Mar-12* (the Sample preview box will display *9-Jan-21*).

14  Click OK.

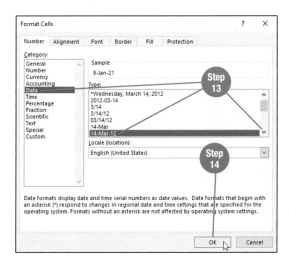

15  Click any cell to deselect the range.

16  Scroll down and review the dates in column A and column B.

You can also enter a date into a cell using a DATE function. A DATE function uses the argument *(Year,Month,Day)*. For example, the entry *=DATE(2021,12,20)* enters the date December 20, 2021 in a cell.

17  Save the revised workbook using the same name (**FinancialPlanner-YourName**). Leave the workbook open for the next topic.

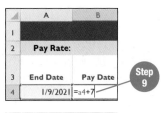

**quick steps**

**TODAY Function**
1. Activate formula cell.
2. Type =today().
3. Press Enter.

**in the know**

Many businesses that operate globally have adopted the International Standards Organization (ISO) date format YYYY-MM-DD to avoid confusion with a date written as *02/04/03*, which could mean February 4, 2003 (US), or April 2, 2003 (UK). Another strategy is to format a date with the month spelled out, like the format used in Step 13.

**Tutorial**

Using Date and Time Functions

## 9.4 Using the IF Function

**Watch & Learn**

**Hands On Activity**

●------- Skills

Enter IF function

Logical functions are used when you need a formula to perform a calculation based on a condition or comparison of a cell with a value or the contents of another cell. For example, in column G of the Income and Savings Planner worksheet, you calculated a savings value based on the take-home pay amounts in column F. Suppose you decide that you cannot afford to contribute to your savings plan unless your take-home pay is more than $500. The formula in column G does not accommodate this restriction; however, an IF formula can compare each take-home pay value with a set value and calculate the savings for those payroll earnings over the minimum.

**1**   With the **FinancialPlanner-YourName** workbook open, click cell H4.

**2**   Click the Formulas tab.

Functions are grouped on the Formulas tab in the Function Library group by category buttons. Click a button to view a list of functions in a drop-down list.

**3**   Click the Logical button in the Function Library group and then click *IF*.

Excel opens the Function Arguments dialog box for the IF statement. An IF statement has three arguments: *Logical_test*, *Value_if_true*, and *Value_if_false*.

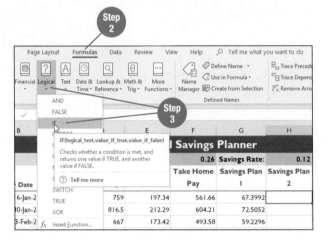

**4**   Type f4>500 in the *Logical_test* argument box and then press Tab or click in the *Value_if_true* argument box.

A logical test is a statement to evaluate a comparison and perform one of two actions. The statement *f4>500* tells Excel to determine whether the value in F4 is greater than 500. Either the value is greater than 500 *(true)* or the value is not greater than 500 *(false)*. See Table 9.3 on the next page for more examples of logical tests.

**5**   Type f4*SaveRate and then press Tab or click in the *Value_if_false* argument box.

Enter in the *Value_if_true* argument box the formula to calculate when the logical test proves true. In this instance, if the value in F4 is greater than 500, Excel will multiply the value in F4 times the value in the cell named *SaveRate (.12)*.

**6**   Type 0 and then click OK.

●------- app tip

Formulas, values, or text are all valid entries for the *Value_if_true* and *Value_if_false* argument boxes.

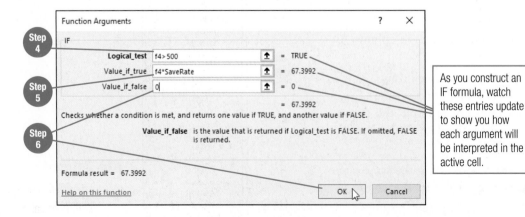

In the *Value_if_false* argument box, enter the formula to calculate when the logical test proves false. In this formula, if the value in F4 is 500 or less, zero is entered in the cell because you have decided that you cannot afford to contribute to your savings plan for earnings below the set value.

**7** Look in the Formula bar at the IF statement entered into cell H4: =IF(F4>500,F4*SaveRate,0).

Using the Function Arguments dialog box to build an IF statement is a good idea because the commas and parentheses are inserted automatically in the correct positions within the formula.

**8** Use the fill handle in H4 to copy the formula to the range H5:H30.

**9** Click any cell to deselect the range and then scroll down to view the results in H5:H30. Notice the cells with 0 occur in a row where the take-home pay value in column F is 500 or less.

**10** Select C4:H30 and then click the Quick Analysis button that appears below the selection.

**11** Click the Totals tab and then click the Sum button in the Totals gallery (first button).

Excel creates SUM functions in row 31 for each column in the selected range. Using the Quick Analysis button saves time when you need to add totals to multiple columns.

**12** Select D4:H31, click the Home tab, and then click the Comma Style button in the Number group.

**13** Apply the Comma Style format to C2 and then J4:J10.

**14** Apply the Percent Style format to F2 and H2.

**15** Select C31:H31 and then add a Top and Double Bottom Border.

**16** Save the revised workbook using the same name (**FinancialPlanner-YourName**). Leave the workbook open for the next topic.

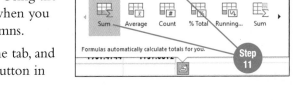

First 10 rows of worksheet with formatting applied in Steps 12 to 14

**Table 9.3** IF Statement Logical Test Examples

| Logical Test | Condition Evaluated | IF Statement Example |
|---|---|---|
| F4>=500 | Is the value in F4 greater than or equal to 500? | =IF(F4>=500,F4*SaveRate,0) |
| F4<500 | Is the value in F4 less than 500? | =IF(F4<500,0,F4*SaveRate) |
| F4<=500 | Is the value in F4 less than or equal to 500? | =IF(F4<=500,0,F4*SaveRate) |
| F4=$K$2 | Is the value in F4 equal to the value in K2? Assume value in K2 is the take-home pay value for which you will set aside savings. | =IF(F4=$K$2,F4*SaveRate,0) |
| Hours<>0 | Is the value in the cell named *Hours* not equal to 0? | =IF(Hours<>0,Hours*PayRate,0) Calculates Gross Pay when hours have been logged |

**Tutorial**

Using Logical IF Functions

## 9.5 Using the PMT Function

Watch & Learn

Hands On
Activity

●———————— Skills

Enter PMT function

Financial functions in Excel can be used for a variety of tasks that involve saving or borrowing money, such as calculating the future value of an investment, calculating the present value of an investment, or calculating borrowing criteria, such as interest rates, terms, or payments. If you are considering a loan or mortgage, use the Excel PMT function to determine an estimated loan payment. The PMT function uses a specified interest rate, number of payments, and loan amount to calculate a regular payment. Once a payment is displayed, you can change the interest rate, term, or loan amount to find a payment with which you are comfortable.

1 With the **FinancialPlanner–YourName** workbook open, click the LoanPlanner sheet tab.

2 Click cell B7.

3 Click the Formulas tab, click the Financial button in the Function Library group, scroll down the list, and then click *PMT*.

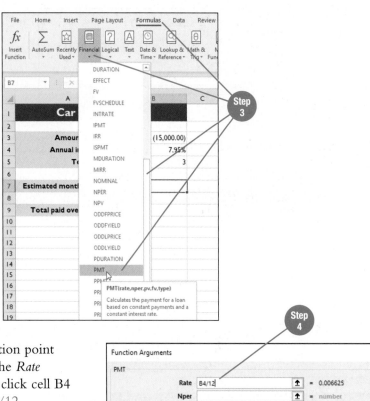

4 With the insertion point positioned in the *Rate* argument box, click cell B4 and then type /12.

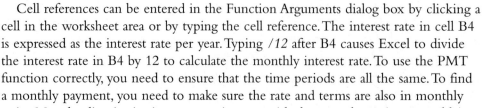

●———————— Oops!

Can't see B4? Drag the Function Arguments Title bar right until the dialog box is no longer obscuring your view of columns A and B.

Cell references can be entered in the Function Arguments dialog box by clicking a cell in the worksheet area or by typing the cell reference. The interest rate in cell B4 is expressed as the interest rate per year. Typing */12* after B4 causes Excel to divide the interest rate in B4 by 12 to calculate the monthly interest rate. To use the PMT function correctly, you need to ensure that the time periods are all the same. To find a monthly payment, you need to make sure the rate and terms are also in monthly units. Most lending institutions express interest with the annual rate (not monthly) but compound the interest monthly.

5  Click in the *Nper* argument box, click cell B5, and then type *12.

The value in B5 is the number of years you will take to pay back the loan. Multiplying the value times 12 will convert the value to the number of months to repay the loan. Most lending institutions express the repayment term in years (not months).

6  Click in the *Pv* argument box and then click B3.

Pv stands for *present value* and represents the amount you want to borrow (referred to as the *principal*). Notice the amount borrowed is entered as a negative value in this worksheet. By default, Excel considers payments as negative values because money is subtracted from your bank balance when you make a loan payment. By entering a negative number for the amount borrowed, the PMT formula will return a positive value for the calculated loan payment. Whether you prefer to display a negative value for the amount borrowed or for the estimated monthly loan payment is a matter of personal preference; both options are acceptable.

7  Click OK.

Excel returns the payment *$469.70* in B7. The PMT formula assumes a constant payment and a constant interest rate. Do not use the PMT function to estimate a loan payment where the payment or the interest rate is variable.

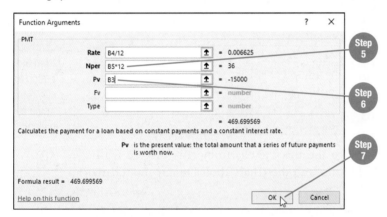

8  Look in the Formula bar at the PMT statement entered into the active cell =PMT(B4/12,B5*12,B3).

9  Click cell B9, type =b7*b5*12, and then press Ctrl + Enter.

Excel calculates the total cost for the loan to be $16,909.18.

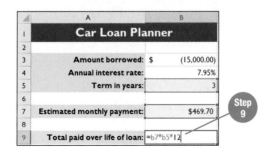

10  Change the value in B5 from *3* to *4*.

Notice that increasing the term one more year reduces your monthly payment to 365.84; however, the total cost of the loan increases to $17,560.41 because you are making 12 more payments and are being charged interest for an additional year.

11  Save the revised workbook using the same name (**FinancialPlanner-YourName**) and then close the workbook.

**Tutorial**

Using Financial Functions

## 9.6 Creating and Modifying a Pie Chart

Charts provide a visual snapshot of data. Charts illustrate trends, proportions, and comparisons more distinctly than numbers alone. Excel provides 15 categories of charts with multiple styles in each category. The Quick Analysis button recommends charts based on the type of data you are analyzing and allows you to preview the chart style with selected data. A **pie chart** is a circular graph with each data point (pie slice) sized to show its proportion to the total of all points within the selected range. A frequent use of pie charts is by governments to illustrate how tax dollars are allocated across programs and services.

**1** Open the workbook **SocialMediaStats**. If necessary, click the Enable Editing button to close Protected view.

**2** Use Save As to save a copy of the workbook as **SocialMediaStats-YourName** in the Ch9 folder in CompletedTopics on your storage medium.

**3** Select A5:B9 in the SocialMediaWebsites worksheet.

Before you can insert a chart, you first need to select the data that you want Excel to represent in a chart.

**4** Click the Quick Analysis button that appears below the selection area and then click the Charts tab.

**5** Click *Pie*.

Excel graphs the data in a pie chart and places the chart overlapping the cells within a chart object window. Notice also that the chart object is selected with selection handles, three chart editing buttons, and the Chart Tools Design and Chart Tools Format tabs in the ribbon.

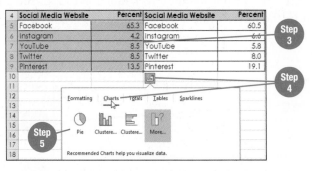

**6** Position the mouse pointer over any white unused area inside the chart borders and then drag the chart to the approximate location shown below.

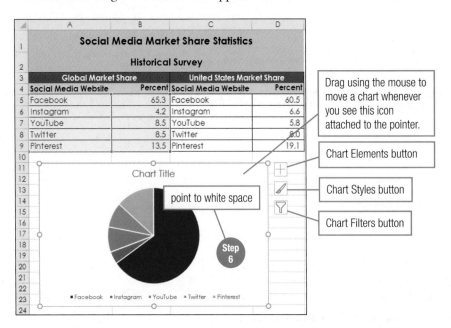

Drag using the mouse to move a chart whenever you see this icon attached to the pointer.

Chart Elements button

Chart Styles button

Chart Filters button

☁ **Watch & Learn**

☁ **Hands On Activity**

●‑‑‑‑‑‑‑ Skills

Create a pie chart

Add data labels

Modify the legend position

Edit the chart title

●‑‑‑‑‑‑‑ app tip

Roll the mouse over the chart options offered from the Quick Analysis button to see a live preview of the data graphed in the chart above the gallery.

**7**  Click the Chart Elements button (displays as a plus symbol next to the chart).

**8**  Click the *Data Labels* check box to insert a check mark, click the right-pointing arrow that appears at the right end of the *Data Labels* option, and then click *Outside End*.

Although pie charts show proportions well, adding data labels allows a reader to include context with the size of each pie slice.

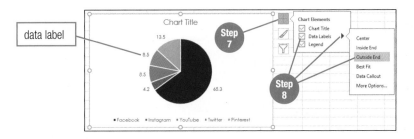

**9**  With Chart Elements still displayed, point to the *Legend* option, click the right-pointing arrow that appears, and then click *Right*.

**10**  Click to select the *Chart Title* object inside the chart window, select the text *Chart Title*, type Global Market Share, and then click in any white, unused area within the chart to deselect the title.

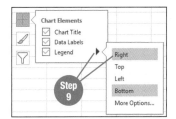

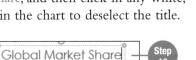

**11**  Select C5:D9 and insert a second pie chart, as shown in Figure 9.1, by completing steps similar to those in Steps 4 to 10. In Step 10, type United States Market Share as the chart title.

**12**  Save the revised workbook using the same name (**SocialMediaStats-YourName**). Leave the workbook open for the next topic.

| | Social Media Market Share Statistics | | | | | | | | |
|---|---|---|---|---|---|---|---|---|---|
| | Historical Survey | | | | | | | | |
| | Global Market Share | | United States Market Share | | | | | | |
| | Social Media Website | Percent | Social Media Website | Percent | | | | | |
| Facebook | | 65.3 | Facebook | 60.5 | | | | | |
| Instagram | | 4.2 | Instagram | 6.6 | | | | | |
| YouTube | | 8.5 | YouTube | 5.8 | | | | | |
| Twitter | | 8.5 | Twitter | 8.0 | | | | | |
| Pinterest | | 13.5 | Pinterest | 19.1 | | | | | |

**Figure 9.1**

The completed side-by-side pie charts are shown for the **SocialMediaStats-YourName** workbook.

## 9.7 Creating and Modifying a Column Chart

In a **column chart**, each data point is a colored bar that extends up from the **category axis** (horizontal axis, also called *x-axis*) with the bar height representing the value of the data point on the **value axis** (vertical axis, also called *y-* or *z-axis*). Use a column chart to compare one or more series of data side by side. Column charts are often used to identify trends or illustrate comparisons over time or by categories.

① With the **SocialMediaStats-YourName** workbook open, click the Facebook sheet tab.

② Select A7:B14.

③ Click the Quick Analysis button, click the Charts tab, and then click Clustered Column (first button).

④ Position the mouse pointer over any white unused area inside the chart borders and then drag the chart to position the top left corner in row 1 under column letter C.

⑤ Point to the bottom right-corner selection handle. When the mouse pointer changes to a diagonal two-headed white arrow, drag the pointer down and right to resize the chart, releasing the mouse when the bottom right corner is at the lower right border of J16.

Watch & Learn

Hands On Activity

━━━━━━━━━ SKILLS

Create a column chart
Change the chart style
Change the chart color scheme
Add axis titles

Tutorial
Changing Chart Design

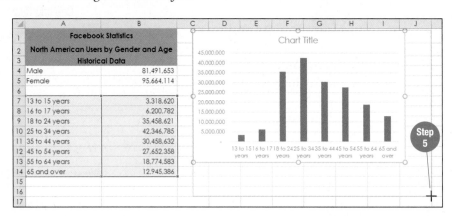

⑥ Click the Chart Styles button (displays as a paintbrush next to the chart).

⑦ Scroll down to the bottom of the Style list and then click the last option in the gallery (*Style 16*).

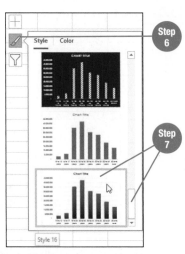

8.  With the Chart Styles gallery still open, click the Color tab and then click the third row in the *Colorful* section (*Colorful Palette 3*).

9.  Click the Chart Elements button.

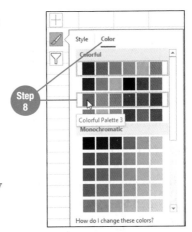

More Chart Elements options appear for a column chart than a pie chart because data points are graphed on two axes. For example, you can add titles to the horizontal and vertical axes on the chart to describe the values and add context for each bar. You can also add a data table, add error bars, show or hide gridlines, or add a trendline on the chart. The chart title, data labels, and legend options for a column chart are the same chart elements used for a pie chart that you learned about in the previous topic.

10. Click the *Axis Titles* check box to insert a check mark.

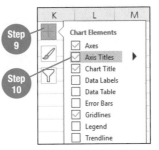

Excel adds an Axis Title object to the vertical axis and to the horizontal axis.

11. Right-click inside the Axis Title object along the vertical axis (displays *Axis Title* rotated 90 degrees), and then click *Edit Text*.

12. Press Delete until the text *Axis Title* is removed, and then type North American Users.

13. Click to select the *Axis Title* object along the horizontal axis and then press Delete to remove the object.

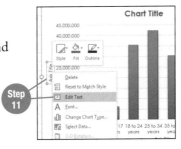

14. Click to select the *Chart Title* object, select the text *Chart Title*, type Facebook Audience by Age Group, and then click in the worksheet area outside the chart to deselect the chart.

15. Compare your chart to the one shown in Figure 9.2. If necessary, redo an action in Steps 4 to 14.

16. Save the revised workbook using the same name (**SocialMediaStats-YourName**). Leave the workbook open for the next topic.

**quick steps**

**Create a Column Chart**
1.  Select range.
2.  Click the Quick Analysis button.
3.  Click Charts tab.
4.  Click *Clustered Column*.
5.  Move and/or modify chart elements as required.

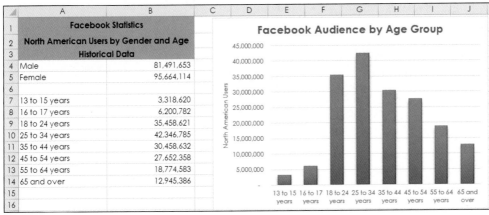

**Figure 9.2**
Shown is the completed column chart for the Facebook worksheet in the **SocialMediaStats-YourName** workbook.

## 9.8 Creating and Modifying a Line Chart

Line charts are best suited for data where you want to illustrate trends and changes in values over a period of time. With a **line chart**, a reader can easily spot a trend or identify growth spurts, dips, or unusual points in the series. Line charts are also often used to help predict future values based on the direction of the line.

1. With the **SocialMediaStats-YourName** workbook open, click the FacebookUserTimeline sheet tab.

2. Select A4:B18.

3. Click the Quick Analysis button, click the Charts tab, and then click *Line*.

4. Click the Move Chart button in the Location group on the Chart Tools Design tab.

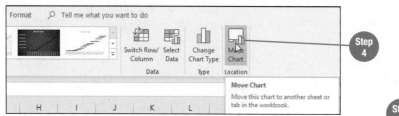

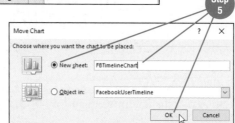

5. Click the *New sheet* option, type FBTimelineChart, and then press Enter or click OK.

   Click the **Move Chart button** to move a chart to its own chart sheet. Excel automatically scales the chart to fit a letter-size page in landscape orientation.

6. Click the second option in the Chart Styles gallery (*Style 2*) on the Chart Tools Design tab.

7. Click the Change Colors button in the Chart Styles group and then click the third row in the *Colorful* section (*Colorful Palette 3*).

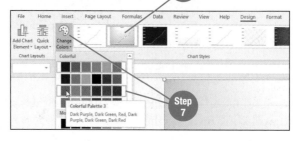

8. Click to select the *Chart Title* object, select the text *Chart Title*, and then type Facebook Active Users Historical Timeline.

Notice the dates in column A were incorrectly converted when the chart was created, changing *Dec* to *Jan*. You will correct the dates in Steps 9 to 12.

9. Click any date along the bottom of the chart to select the category axis. Make sure you see a border and selection handles around the axis labels.

10. Double-click inside the border of the selected axis labels to open the Format Axis task pane at the right side of the window.

11. Click *Text axis* in the *Axis Options* section.

    Notice the axis labels change to display the December dates as they appeared in the worksheet.

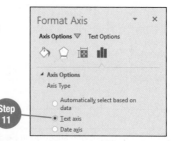

12. Close the Format Axis task pane.

**Watch & Learn**

**Hands On Activity**

●─────── Skills

Create a line chart

Move a chart to a new sheet

Format an axis

Format data labels

●─────── Oops!

No border around dates? Click the axis labels a second time. Sometimes the chart is selected the first time you click.

⑬ Click any data value on a data point to select the entire series of data labels.

⑭ Right-click any selected data value and then click *Format Data Labels*.

The Format Data Labels task pane displays.

⑮ Click *Above* in *Label Position* in the *Label Options* section.

⑯ Click the Text Options tab and then click *Text Fill* to expand the options list.

⑰ Click the Color button and then click *Black, Text 1* (second color in first row).

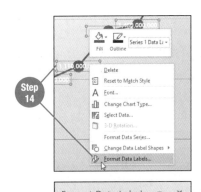

Step 14

Step 15

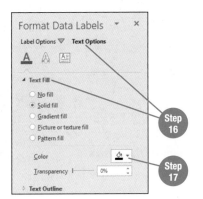

Step 16

Step 17

**quick steps**

**Create a Line Chart**
1. Select range.
2. Click Quick Analysis button.
3. Click Charts tab.
4. Click *Line*.
5. Move and/or modify chart elements as required.

**Tutorial**

Changing Chart Formatting

⑱ Close the Format Data Labels task pane and then click in the window outside the chart borders to deselect the data labels.

⑲ Compare your chart with the chart shown in Figure 9.3 and make corrections if necessary.

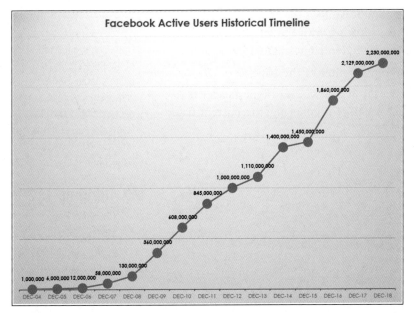

**Figure 9.3**

Shown is the line chart in a new chart sheet in the **SocialMediaStats-YourName** workbook.

⑳ Save the revised workbook using the same name (**SocialMediaStats-YourName**). Leave the workbook open for the next topic.

# 9.9 Using Page Layout View, Adding a Header, and Changing Margins

In **Page Layout view**, you can preview page layout options similarly to Print Preview; however, you also have the advantage of being able to edit the worksheet. The worksheet is divided into pages, with white space around the edges of each page displaying the size of the margins and a ruler along the top and left of the column letters and row numbers. Pages and cells outside the active worksheet are grayed out; however, you can click any page or cell to add new data.

1. With the **SocialMediaStats-YourName** workbook open, click the SocialMediaWebsites sheet tab.

2. Click the View tab and then click the Page Layout button in the Workbook Views group.

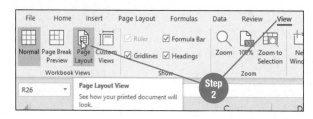

Notice in Page Layout view you can see the right pie chart is split over two pages.

3. Click the Page Layout tab, click the Orientation button in the Page Setup group, and then click *Landscape*.

Notice that changing the page orientation to print on the wider edge of the paper did not resolve fitting the entire worksheet on one page.

4. Click the *Width* option box arrow in the Scale to Fit group (displays *Automatic*) and then click *1 page*.

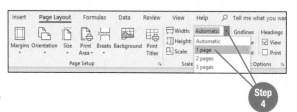

## Inserting a Header or Footer

A header prints at the top of each page and a footer prints at the bottom of each page. Headers and footers are divided into three sections, with the left section left-aligned, the center section centered, and the right section right-aligned by default.

5. Click the dimmed *Add header* text near the top center of the page.

The Header pane opens with three text boxes in which you can type header text and add header and footer options, such as images, page numbering, the current date or time, and file or sheet names.

6. Type your first and last names.

7. Click in the right box in the Header pane.

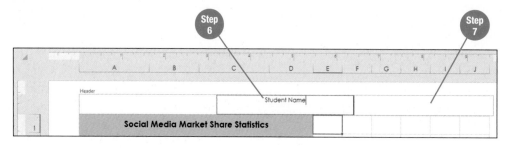

**8** Click the Sheet Name button in the Header & Footer Elements group on the Header & Footer Tools Design tab.

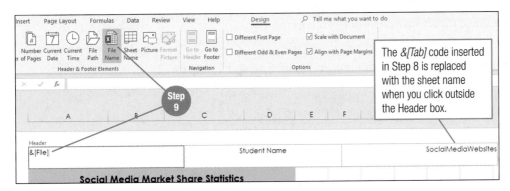

Excel inserts the code &[Tab], which is replaced with the sheet tab name when you click outside the right text box.

**9** Click the left box in the Header pane and then click the File Name button in the Header & Footer Elements group.

Excel inserts the code *&[File]*, which is replaced with the file name when you click outside the left section text box.

The *&[Tab]* code inserted in Step 8 is replaced with the sheet name when you click outside the Header box.

**10** Click any cell in the worksheet area.

## Changing Margins

Worksheet margins are 0.75 inch at the top and bottom of the page and 0.7 inch at the left and right of the page, with the header or footer printing 0.3 inch from the top or bottom of the page. Adjust margins to add more space around the edges of a page, or between the header and footer text and the worksheet. Center a smaller worksheet horizontally and/or vertically to improve the page appearance.

**11** Click the Page Layout tab, click the Margins button in the Page Setup group, and then click *Wide*.

The *Wide* preset margin option changes the top, bottom, left, and right margins to 1 inch and the header and footer margins to 0.5 inch.

**12** Click the Margins button and then click *Custom Margins*.

**13** Click the *Horizontally* and the *Vertically* check boxes in the *Center on page* section in the Page Setup dialog box to insert a check mark in each box and then click OK.

**14** Click the Facebook sheet tab, change to Page Layout view, and modify print options by completing steps similar to those in Steps 3 to 13 to improve the appearance of the printed worksheet.

**15** Save the revised workbook using the same name (**SocialMediaStats-YourName**) and then close the workbook.

# 9.10 Creating and Modifying Sparklines and Inserting Comments

A **sparkline chart** is a miniature chart inserted into an individual cell within a worksheet. Sparkline charts are used to draw attention to trends or variations in data on a smaller scale than a column or line chart. Excel offers three types of sparkline charts: Line, Column, and Win/Loss.

A comment attached to a cell pops up when the reader points or clicks the cell. Comments are used to add explanatory information, pose questions, or provide other feedback to readers when a workbook is shared.

1. Open the workbook **SchoolBudget**. If necessary, click the Enable Editing button to close Protected view.

2. Use Save As to save a copy of the workbook as **SchoolBudget-YourName** in the Ch9 folder in CompletedTopics on your storage medium.

3. Click cell K3.

4. Click the Insert tab and then click the Column button in the Sparklines group.

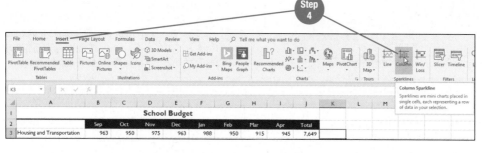

5. Type b3:i3 in the *Data Range* text box in the Create Sparklines dialog box and then press Enter or click OK.

Excel embeds a column chart within the cell.

6. Use the fill handle in K3 to copy the sparklines column chart from K3 to K4:K11.

7. Click the *High Point* check box in the Show group on the Sparkline Tools Design tab to insert a check mark.

Excel highlights the bar in the column chart with the highest value by coloring it red. Other options on the tab are used to change the chart type or style, emphasize other points, display markers, or edit the data source.

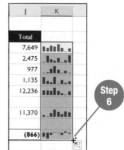

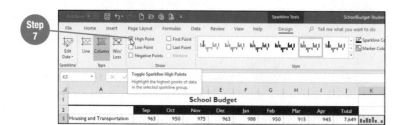

8. Click cell K2, type Trend, and then click cell E9.

---

☁ **Watch & Learn**

☁ **Hands On Activity**

●-------- skills

Insert sparkline charts

Insert a comment

Edit a comment

Print comments

☁ **Tutorial**

Summarizing Data with Sparklines

●------- app tip

Increase the row height and/or column width to enlarge sparkline charts.

9  Click the Review tab and then click the New Comment button in the Comments group.

10  Type Assuming extra hours during Christmas break. and then click I3.

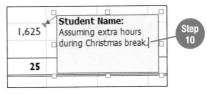

Excel inserts a red triangle in the upper right corner of a cell to indicate a comment exists for the cell.

11  Click the New Comment button, type May be able to use last month's rent., and then click any cell.

12  Point to I3 with the mouse to display the comment in a pop-up box.

13  Click I3 to activate the cell.

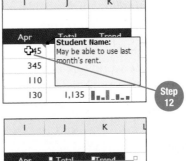

14  Click the Edit Comment button in the Comments group and edit the comment text to *May be able to use last month's rent*, which lowers this value to 55.

15  Click any cell to finish editing the Comment box.

16  Click the Show All Comments button in the Comments group to display both comment boxes in the worksheet.

17  Click the Show All Comments button to turn off the display of comment boxes.

By default, comments do not print with the worksheet. To print a version of the worksheet with comments shown, use the Show All Comments button to turn on the display of comments and print the worksheet. Alternatively, you can print comments on a separate page after the worksheet prints, which you will set up in Steps 18 to 20.

18  Click the File tab, click *Print*, and then click the Page Setup hyperlink at the bottom of the *Settings* section.

19  Click the Sheet tab in the Page Setup dialog box.

20  Click the *Comments* option box arrow, click *At end of sheet*, and then click OK.

21  Click the Next Page arrow in the Print Preview panel to view the separate page with the comments and then click Back to return to the worksheet.

22  Save the revised workbook using the same name (**SchoolBudget-YourName**) and then close the workbook.

**quick steps**

**Insert Sparklines**
1.  Activate cell.
2.  Click Insert tab.
3.  Click required sparklines chart button.
4.  Type source data range.
5.  Click OK.

**Insert a Comment**
1.  Activate cell.
2.  Click Review tab.
3.  Click New Comment button.
4.  Type comment text.
5.  Click any cell.

**Tutorials**

Inserting, Posting, and Printing Comments

Showing, Editing, and Deleting Comments

Page Setup dialog box with tabs Page, Margins, Header/Footer, Sheet:

Print area:

Print titles

Rows to repeat at top:

Columns to repeat at left:

Print

☐ Gridlines    Comments: At end of sheet

☐ Black and white    Cell errors as: displayed

☐ Draft quality

☐ Row and column headings

Page order

◉ Down, then over

○ Over, then down

## 9.11 Working with Tables

A range of cells in a worksheet can for formatted as a table to analyze, sort, and filter data as an independent unit. A worksheet can have more than one table, which means you can isolate and analyze data in groups. A table also allows you to choose from a variety of preformatted table styles, which is faster than manually formatting a range. Use tables for any block of data organized in a list format.

A **filter** temporarily hides any data that does not meet a criterion. Use filters to look at subsets of data without deleting rows in the table.

**Watch & Learn**

**Hands On Activity**

●------------ Skills

Format a range as a table

Sort a table

Filter a table

**Tutorial**

Formatting Data as a Table

1. Open the workbook **CalorieActivityTable**. If necessary, click the Enable Editing button to turn off Protected view.

2. Use Save As to save a copy of the workbook as **CalorieActivityTable-YourName** in the Ch9 folder in CompletedTopics on your storage medium.

3. Select A3:D23 and then click the Quick Analysis button.

4. Click the Tables tab and then click the Table button.

5. Select A1:A2, click the Fill Color button arrow, and then click *White, Background 1, Darker 5%* (first color in second row).

6. Click cell A4.

7. Click the Table Tools Design tab and then click the *White, Table Style Medium 1* option in the Table Styles gallery (option left of the active style).

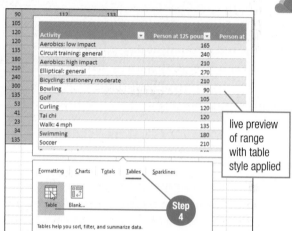

live preview of range with table style applied

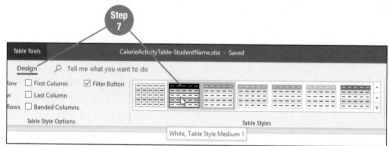

8. Click the *Activity* column filter arrow (down-pointing arrow at the top of the *Activity* column).

9. Click *Sort A to Z*.

The table rows are sorted in ascending order by the activity descriptions.

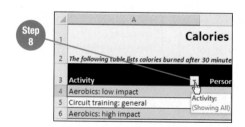

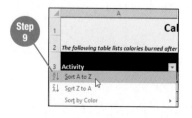

⑩ Click the *Person at 155 pounds* column filter arrow.

A check box is included for each unique value within the column. You can filter a table by clearing check boxes for values or items you do not want to see in a filtered list, or use the *Filter by Color* and *Number Filters* options to specify a filter condition.

⑪ Point to *Number Filters* and then click *Greater Than*.

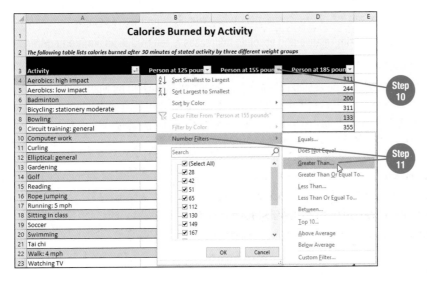

⑫ Type 200 with the insertion point already positioned in the text box next to *is greater than* in the Custom AutoFilter dialog box and then click OK.

Excel filters the table and displays only those activities in which the calories burned are more than 200 for 30 minutes of the activity by a person at 155 pounds.

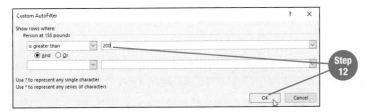

⑬ Point to the *Person at 155 pounds* filter arrow. Notice the funnel icon displayed on the button and the ScreenTip that pops up telling you how the current worksheet data has been filtered.

Other indicators that the worksheet data is filtered are the blue row numbers (row numbers for hidden rows do not display) and the message on the Status bar that reads *9 of 20 records found* (next to *Ready*).

⑭ Click the *Person at 155 pounds* filter arrow and then click *Clear Filter From "Person at 155 pounds"*.

Clearing a filter redisplays the entire table.

⑮ Change the orientation to landscape and center the worksheet horizontally.

⑯ Save the revised workbook using the same name (**CalorieActivityTable –YourName**) and then close the workbook.

**Tutorials**

Sorting a Table

Filtering a Table

## Topics Review

| Topic | Key Concepts |
|---|---|
| **9.1 Using Absolute Addresses and Range Names in Formulas** | By default, a cell address in a formula is a **relative address**, which means the column letter or row number will update as the formula is copied relative to the destination column and row. |
| | A dollar symbol in front of a column letter or row number makes the cell reference an **absolute address**, and means the reference will not update when the formula is copied. |
| | A formula that has both relative and absolute addresses is referred to as a **mixed address**. |
| | Excel displays an **error value** in a cell when the formula cannot produce a calculated result. |
| | A descriptive label can be assigned to a cell or range and is created by typing the label in the **Name box**. Each cell reference in a range name is automatically made an absolute address. |
| | A range name can be used in a formula in place of the cell address(es). |
| **9.2 Entering Formulas Using Statistical Functions** | A preprogrammed formula in Excel is called a **function**. |
| | A formula that uses a function begins with the function name followed by the parameters for the formula (called the **argument**) within parentheses. |
| | The AVERAGE function returns the arithmetic mean from the range used in the formula. |
| | The MAX function returns the largest value from the range. |
| | The MIN function returns the smallest value from the range. |
| | The Insert Function dialog box accessed from the Insert Function button provides tools to find and enter a function and argument. |
| **9.3 Entering, Formatting, and Calculating Dates** | A valid date or time entered into a cell is stored as a numerical value and can be used in formulas. |
| | The TODAY function enters the current date into the cell and updates the date whenever the worksheet is opened or printed. |
| | Date and time cells can be formatted to a variety of month, day, and year combinations in the Format Cells dialog box with the *Date* category selected. |
| **9.4 Using the IF Function** | The IF function compares the contents in a cell with a set value and and performs one of two calculations depending on whether the comparison proves true or false. |
| | Use the Insert Function dialog box to assist with entering an IF statement's arguments. |
| | The *logical_test* argument is the statement you want Excel to evaluate to determine which calculation to perform. |
| | The *value_if_true* argument is the value or formula to return if the logical test proves true. |
| | The *value_if_false* argument is the value or formula to return if the logical test proves false. |
| **9.5 Using the PMT Function** | Financial functions can be used for a variety of calculations that involve saving or borrowing money. |
| | The PMT function calculates a regular loan payment from a specified interest rate, term, and amount borrowed. |
| | Make sure the interest rate and terms are in the same units as the payment you want calculated. For example, divide the interest rate by 12 and/or multiply the term times 12 to calculate a monthly payment from an annual rate or terms. |
| | In the PMT argument, *Rate* means the interest rate, *Nper* means the term, and *Pv* means the amount borrowed. |

*continued...*

| Topic | Key Concepts |
|---|---|
| **9.6 Creating and Modifying a Pie Chart** | Charts are often used to portray a visual snapshot of data. |
| | A **pie chart** shows each data point as a pie slice. |
| | The size of each slice in the pie chart represents the value of the data point in proportion to the total of all the values. |
| | Use the Charts tab in the Quick Analysis gallery to create a pie chart from a selected range. |
| | The Chart Elements button is used to add or modify a chart title, data labels, or legend. |
| **9.7 Creating and Modifying a Column Chart** | A **column chart** shows one bar for each data point extending upward from a horizontal axis with the height of the bar representing its value. |
| | The horizontal axis in a column chart is the **category axis**, also called the *x-axis*, and shows the labels for each bar. |
| | The vertical axis in a column chart is called the **value axis**, also known as the *y-* or *z-axis*, and is scaled to the values of the bars graphed. |
| | A column chart is often used to illustrate trends or comparisons over time or by category. |
| | The Chart Styles button is used to choose a preformatted style for a column chart or to change the color scheme. |
| | The *Axis Titles* option from the Chart Elements button is used to add titles to each axis in a column chart. |
| **9.8 Creating and Modifying a Line Chart** | A **line chart** helps a reader identify trends, growth spurts, dips, or unusual points in a data series. |
| | Use the **Move Chart button** in the Location group on the Chart Tools Design tab to move a chart from the worksheet into a chart sheet. |
| | A chart in a chart sheet is automatically scaled to fill a letter-size page in landscape orientation. |
| | Change axis options in the Format Axis task pane, or data label options in the Format Data Labels task pane. |
| **9.9 Using Page Layout View, Adding a Header, and Changing Margins** | In **Page Layout view** the worksheet is divided into pages, with white space depicting the size of the margins and a ruler along the top and left edges. |
| | You can see page layout and print options in Page Layout view while viewing and editing the worksheet. |
| | Add a header in Page Layout view by clicking the dimmed text *Add header*. |
| | Use buttons on the Header & Footer Tools Design tab to add options to a header or footer, such as a picture, page numbering, date or time, or file or sheet names. |
| | Change to a preset set of margins from the Margins button on the Page Layout tab, or choose *Custom Margins* to enter your own margin settings in the Page Setup dialog box. |
| | Open the Page Setup dialog box with the Margins tab active to center a worksheet horizontally and/or vertically. |

*continued...*

| Topic | Key Concepts |
|---|---|
| **9.10 Creating and Modifying Sparklines and Inserting Comments** | A miniature chart embedded into a cell is called a **sparkline chart**. |
| | Sparkline charts emphasize trends or variations in data on a smaller scale. |
| | Activate a cell and choose a Line, Column, or Win/Loss sparkline chart from the Sparklines group on the Insert tab. |
| | Once created, add or modify sparkline options using buttons on the Sparkline Tools Design tab. |
| | A comment appears in a pop-up box when you point to or click a cell with an attached comment. |
| | Excel displays a red triangle in a cell containing a comment. |
| | Use the New Comment button on the Review tab to add a comment in the active cell. |
| | To change the text in an existing comment, activate the cell containing the comment and use the Edit Comment button. |
| | By default, comments do not print with the worksheet. Use the Show All Comments button to print a copy of the worksheet with the comments displayed, or print the comments on a separate page after the worksheet data by opening the Page Setup dialog box and changing the *Comments* option on the Sheet tab to *At end of sheet*. |
| **9.11 Working with Tables** | A block of data set up in list format can be formatted as a table for formatting, analyzing, sorting, or filtering purposes. |
| | A **filter** temporarily hides data that does not meet a criterion. |
| | Use a filter to review subsets of data without deleting rows. |
| | Click the filter arrow at the top of a column to sort or filter a table. |
| | Click the filter arrow at the top of a filtered column and then use the *Clear Filter From (column title)* option to redisplay the hidden rows in the table. |

## Topics Review

# Creating, Editing, and Formatting a PowerPoint Presentation

Presentations occur in meetings, seminars, and classrooms for a variety of purposes. Some presentations are informational, while others are designed to persuade you to buy a product or service. Some people use presentations at weddings, anniversaries, or family reunions to entertain an audience. In some organizations, presentations are used to provide information at a kiosk where a slide show runs continuously for individuals to view as they walk by or enter a booth. For example, at a trade show, a company might provide a slide show with information about a product. At school, you may have used a presentation as a study guide to prepare for an exam.

A presentation is made up of a collection of slides (referred to as a **slide deck**) containing text and multimedia. PowerPoint is a **presentation application** used to create a slide deck for presentations. In this chapter, you will learn how to create, edit, and format a presentation. You will create a presentation with a variety of text-based slide layouts; edit content and placeholders; move, duplicate, and delete slides; format slides using a variety of techniques; add notes and comments; and preview the presentation as a slide show. Finally, you will preview options for audience and speaker handouts.

## Learning Objectives

**10.1** Create a new presentation based on a theme, insert slides, and add content to slides

**10.2** Change the design theme and theme variant, and insert a table

**10.3** Format text using font and paragraph options

**10.4** Create a slide with the comparison layout and select, resize, align, and move slide placeholders

**10.5** Use Slide Sorter view and duplicate, move, and delete slides

**10.6** Modify the slide master

**10.7** Add notes and comments to slides

**10.8** Run a presentation in Slide Show view and Presenter view

**10.9** Prepare slides for audience handouts or speaker notes

**Read & Learn**

**Content Online**

The online course includes additional training and assessment resources.

## 10.1 Creating a New Presentation and Inserting Slides

Begin creating a new presentation on the PowerPoint Start screen by choosing a template, theme, and variant on a theme, or by starting with a blank presentation. The first slide in a presentation is a **title slide** with a text **placeholder** for a title and a subtitle. A placeholder is a rectangular container on a slide that can hold text or other content. Each placeholder on a slide can be manipulated independently.

PowerPoint starts a new presentation with a title slide displayed in Normal view. In Normal view, the current slide displays in widescreen format in the **slide pane**. Numbered slide thumbnails display in the **slide thumbnails pane**. A notes pane at the bottom of the slide pane and a Comments pane at the right of the slide pane can be opened as needed.

1. Start PowerPoint and then click the <u>More themes</u> hyperlink near the top right of the PowerPoint Start screen.

2. Click the *Ion* theme in the New backstage area.

The New backstage area has a gallery of design themes. Click to preview a theme along with the theme variants. A **variant** is a different style and color scheme included in the theme family.

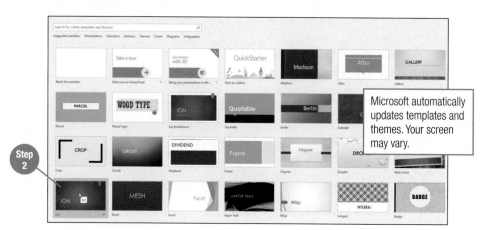

Microsoft automatically updates templates and themes. Your screen may vary.

3. Click the last variant (orange color scheme), and then click the right-pointing arrow below the preview slide next to *More Images*.

Browse through the *More Images* slides to see a variety of content in the color scheme and view the theme or variant style and colors before making a selection.

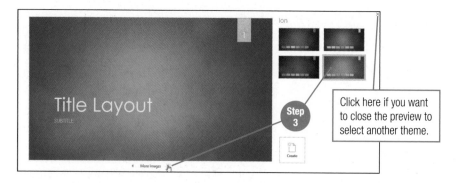

Click here if you want to close the preview to select another theme.

4. Click the second variant (blue color scheme), and then click the right-pointing arrow below the preview slide to view the blue color scheme with a Title and Content layout depicting a chart.

**Watch & Learn**

**Hands On Activity**

**skills**

Create new presentation
Choose theme and theme variant
Insert slide
Edit text on slide

**app tip**

Double-click a theme to start a new presentation using the theme default style and color scheme.

**Tutorial**
Opening a Presentation Based on a Template

⑤ Click the right-pointing arrow below the preview slide two more times to view other types of content with the blue color scheme.

⑥ With the Photo Layout preview displayed, click the Create button.

⑦ Compare your screen with the one shown in Figure 10.1.

**Figure 10.1**

A new PowerPoint presentation with Ion theme and blue color variant in the default Normal view is shown above. See Table 10.1 for a description of screen elements.

**Table 10.1** PowerPoint Features

| Feature | Description |
|---|---|
| Notes button | Button to turn on or turn off the notes pane at the bottom of the slide pane |
| Placeholders | Containers in which you type or edit text, or insert other content such as an image or audio clip |
| Slide pane | Pane that displays the active slide; add or edit content on a slide in this area |
| Slide thumbnails pane | Pane that displays numbered thumbnails of the slides in the presentation; navigate to, insert, delete, or duplicate a slide in this pane |
| Status bar | Bar that displays active slide number with total number of slides in the presentation and displays a message about an action in progress |
| View and Zoom buttons | These buttons change the display of the PowerPoint window. View buttons in order are: Normal, Slide Sorter, Reading View, and Slide Show. Zoom buttons enlarge or shrink the display of the active slide. |

**quick steps**

**Start a New Presentation**
1. Start PowerPoint.
2. Click <u>More themes</u> hyperlink.
3. Click theme.
4. Click variant.
5. Click Create button.

**Insert a Slide**
Click New Slide button in Slides group.
OR
1. Click New Slide button arrow.
2. Click required slide layout.

**Edit Text**
1. Activate slide.
2. Select text or click in placeholder and move insertion point as needed.
3. Type new text or change text as required.

**Tutorial**

Exploring the PowerPoint Screen

**8** Click anywhere in *Click to add title* on the title slide in the slide pane and then type Car Maintenance.

**9** Click anywhere in *CLICK TO ADD SUBTITLE* on the title slide in the slide pane and then type Tips for all seasons.

The subtitle text displays in all capital letters regardless of the case used when you type the text because the Ion theme uses the All Caps font effect for the subtitle text.

Step 8

Step 9

**10** Click in an unused area of the slide to deactivate the subtitle placeholder.

## Inserting New Slides

The **New Slide button** in the Slides group on the Home tab is used to insert a new slide following the active slide. The button has two parts. Clicking the top part of the button adds a new slide with the Title and Content layout, which is the layout used most frequently. The content placeholder in this layout provides options to add text or a table, chart, SmartArt graphic, picture, or video to the slide. Click the New Slide button arrow to choose from a list of slide layouts and other new slide options. A **slide layout** is an arrangement of placeholders that determine the number, position, and type of content placeholders included on a slide.

**11** Click the top part of the New Slide button in the Slides group on the Home tab.

**12** Click anywhere in *Click to add title* in the title placeholder and then type Why maintain a car?.

**13** Click anywhere in *Click to add text* in the content placeholder, type Preserve vehicle value, and then press Enter.

Step 11

Typing text in the content placeholder automatically creates a bulleted list. In the Ion theme, the bullet character is a green, right-pointing arrow.

**14** Type the remaining bulleted list items, pressing Enter after each item except the last one.

Prolong vehicle life

Improve driver safety

Spend less for repairs

Lower operating costs

Improve vehicle appearance

Reduce likelihood of breakdowns

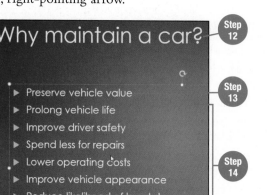

Step 12

Step 13

Step 14

**Tutorial**

Inserting and Deleting Text in Slides

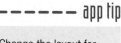
app tip

Change the layout for a slide after a slide has been inserted using the Layout button in the Slides group on the Home tab.

**Tutorials**

Choosing a Slide Layout

Changing a Slide Layout

Inserting a New Slide

15 Click the top part of the New Slide button.

16 Type the text on the third slide as shown in the image below.

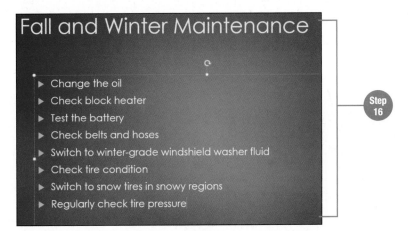

**app tip**

Press Ctrl + Enter to move to the next placeholder on a slide or to create a new slide when pressed in the last placeholder.

## Editing Text on Slides

Activate the slide you want to edit by clicking the slide in the slide thumbnails pane. Select the text you want to change or click inside a placeholder to place an insertion point at the location where you want to edit text and then type new text, change text, or delete text as needed.

17 Click Slide 1 in the slide thumbnails pane.

18 Select *ALL SEASONS* in the subtitle text placeholder and then type car owners so that the subtitle text now reads *TIPS FOR CAR OWNERS.*

**Tutorial**

Checking Spelling

**app tip**

Always carefully spell check and proofread each slide in your presentation. Start with the Spelling button in the Proofing group on the Review tab.

19 Click Slide 2 in the slide thumbnails pane.

20 Click at the beginning of the text in the third bulleted list item, delete *Improve*, and then type Sustain.

21 Click to place an insertion point within the title placeholder and edit the title text so that *m* in *maintain* and *c* in *car* are capital letters. The title should now read "Why Maintain a Car?"

22 Click in an unused area on the slide to deactivate the title placeholder.

23 Save the presentation as **CarMaintenance-YourName** in a new folder named *Ch10* in the CompletedTopics folder on your storage medium. Leave the presentation open for the next topic.

## 10.2 Changing the Theme and Inserting and Modifying a Table

The presentation theme and/or variant can be changed after a presentation has been created. To do this, click the Design tab and browse the themes and theme variants in the Themes and Variants galleries.

**1** With the **CarMaintenance-YourName** presentation open, click Slide 1 in the slide thumbnails pane.

**2** Click the Design tab.

**3** Click the *Facet* theme in the Themes gallery (third option).

When deciding upon a theme, roll the mouse over the various theme options to view the active slide with a live preview of the theme. Changing the theme after slides have been created may cause some changes in capitalization in placeholders. For example, a theme that uses the All Caps font effect in a title or subtitle may mean that you have to do some corrections after the theme is changed if the new theme does not use All Caps.

**4** Click the More button (⬇) at the bottom right of the Variants gallery.

**5** Point to *Colors* and then click *Red Orange* in the Colors gallery.

From the Variant gallery More button are also options to customize the fonts, effects, or background styles used in the theme variant.

**6** Click Slide 3 in the slide thumbnails pane.

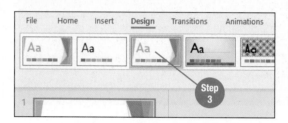
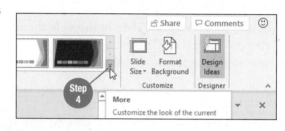
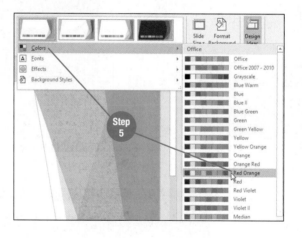

### Inserting a Table on a Slide

PowerPoint includes a Table feature for organizing text on a slide in columns and rows similar to the Table feature in Word. To insert a table on a slide, click the Insert Table button in the content placeholder.

**7** Click the Home tab and then click the top part of the New Slide button to insert a new slide with the Title and Content layout.

**8** Click anywhere in *Click to add title* in the title placeholder and then type Typical Annual Maintenance Costs.

**9** Click the Insert Table button in the content placeholder.

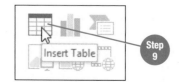

**Watch & Learn**

**Hands On Activity**

Skills

Change theme
Change variant
Insert a table on a slide
Modify table layout

**Tutorials**
Applying a Design Theme
Changing and Modifying Design Themes

Oops!

Design Ideas task pane opens? The Design Ideas task pane may appear when you are making changes to a presentation. The Designer feature works in the background to match content to professionally designed layouts. You can select a design by clicking an option in the task pane. Close the task pane if you don't want the pane taking up screen space.

**Tutorial**
Creating a Table

10  Select 5 in the *Number of columns* text box in the Insert Table dialog box and then type 2.

11  Select 2 in the *Number of rows* text box, type 6, and then click OK.

PowerPoint inserts a table on the slide with the colors in the theme variant.

| Insert Table | ? | × |
|---|---|---|
| Number of columns: | 2 | |
| Number of rows: | 6 | |
| OK | Cancel | |

Step 10

Step 11

12  With the insertion point positioned in the first cell in the table, type Type of Car and then press Tab or click in the second cell.

13  Type Cost.

14  Type the remaining entries in the table by pressing Tab to move to the next cell or by clicking in the next cell and then typing the text as follows:

| Small size | $600 |
|---|---|
| Medium size | $675 |
| Large family sedan | $750 |
| Minivan | $775 |
| SUV | $825 |

## Modifying a Table

The Table Tools Design and Table Tools Layout tabs provide options for modifying a table. These tools are the same table tools you learned in Word. Use the handles to enlarge or shrink the table size, or drag the border of a table to move it on the slide.

15  Drag the right-middle sizing handle to the left until the right border of the table ends approximately below the *c* in *Maintenance* in the title text, as shown in the image at right.

### Typical Annual Maintenance Costs

| Type of Car | Cost |
|---|---|
| Small size | $600 |
| Medium size | $675 |
| Large family sedan | $750 |
| Minivan | $775 |
| SUV | $825 |

Step 15

16  Click in any cell in the second column of the table.

17  Click the Table Tools Layout tab, click the Select button in the Table group, and then click *Select Column*.

18  Click the Center button in the Alignment group.

19  Click in any cell in the first column of the table, select the current entry in the *Width* text box in the Cell Size group, type 3.5, and then press Enter.

20  Drag the top border of the table to move it to the approximate location shown in the image at right.

21  Save the revised presentation using the same name (**CarMaintenance-YourName**). Leave the presentation open for the next topic.

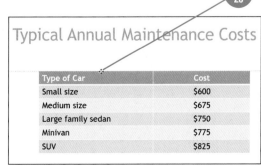

Step 20

### Typical Annual Maintenance Costs

| Type of Car | Cost |
|---|---|
| Small size | $600 |
| Medium size | $675 |
| Large family sedan | $750 |
| Minivan | $775 |
| SUV | $825 |

# 10.3 Formatting Text with Font and Paragraph Options

Font and paragraph formatting options in PowerPoint are the same as those in Word and Excel. Select text within a placeholder and apply a formatting option, or select a placeholder and apply a formatting option to all the text in the placeholder.

A multilevel bulleted list is created using the **Increase List Level button** and the **Decrease List Level button** in the Paragraph group on the Home tab. Each time you click the Increase List Level button, the insertion point or text is indented to the next tab and the bullet character changes to indicate the text is being demoted to the next list level. Use the Decrease List Level button to move the insertion point or text back to the previous tab and promote the text to the previous list level.

1. With the **CarMaintenance-YourName** presentation open, click Slide 3 in the slide thumbnails pane.

2. Insert a new slide with the Title and Content layout.

New slides are inserted after the active slide. The new slide should be positioned between the Fall and Winter Maintenance slide and the Typical Annual Maintenance Costs slide.

3. With the new Slide 4 the active slide, type Spring and Summer Maintenance as the slide title.

4. Type Thoroughly clean vehicle as the first bulleted list item in the content placeholder and then press Enter.

5. With the insertion point positioned at the beginning of the second bulleted list item, click the Increase List Level button in the Paragraph group on the Home tab.

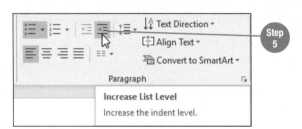

6. Type Prevent rust by removing sand and salt accumulated from winter driving and then press Enter.

7. With the insertion point positioned at the beginning of the third bulleted list item, click the Decrease List Level button in the Paragraph group to move the bullet back to the previous level, type Check cooling system, and then press Enter.

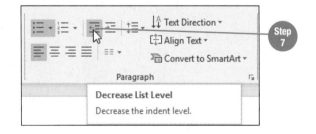

**8** Type the remaining text on the slide as shown in the image below, using the Increase List Level and Decrease List Level buttons as needed.

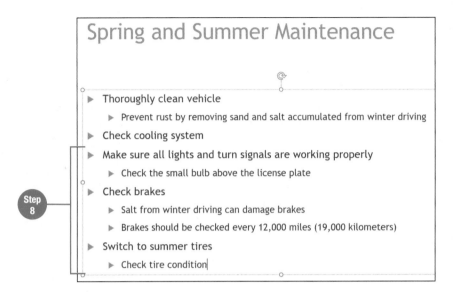

Step 8

**9** Select *12,000* in the content placeholder, click the Font Color button arrow on the Mini toolbar or in the Font group on the Home tab, and then click *Red, Accent 1* (fifth option in first row of *Theme Colors* section).

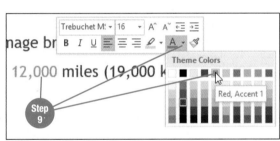

Step 9

**10** Click the Bold button in the Font group.

**11** Click in the title text to activate the title placeholder.

**12** Click the Center button in the Paragraph group.

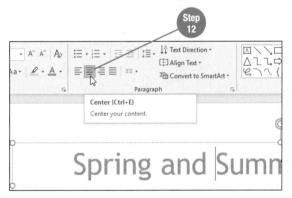

Step 12

**13** Click the Align Left button in the Paragraph group to return the title placeholder to the default paragraph alignment.

**14** Save the revised presentation using the same name (**CarMaintenance -YourName**). Leave the presentation open for the next topic.

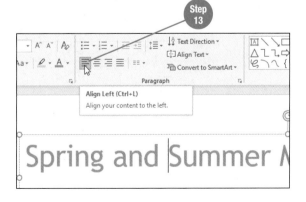

Step 13

# 10.4 Selecting, Resizing, Aligning, and Moving Placeholders

**Watch & Learn**

**Hands On Activity**

The active placeholder displays with a border and selection handles, which are used to resize or move the placeholder. Paragraph or font options apply to the text in which the insertion point is positioned or to selected text. To apply a font or paragraph change to all the text in a placeholder, click the placeholder border to select the placeholder, which also removes the insertion point or deselects text.

**1** With the **CarMaintenance-YourName** presentation open, click Slide 5 in the slide thumbnails pane.

**2** Click the New Slide button arrow and then click *Comparison*.

**3** With the new Slide 6 the active slide, type Top 5 Cars Rated by Maintenance Costs as the slide title.

**4** Type the title and bulleted list text in the left and right content placeholders as shown in the image below.

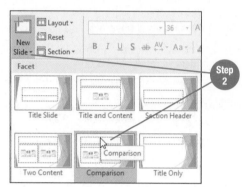

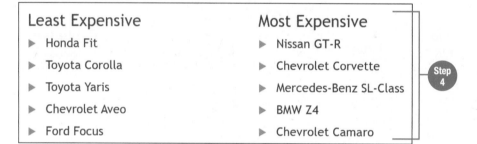

**5** Click anywhere in the bulleted list below the heading *Least Expensive* to activate the placeholder.

**6** Click anywhere along the border of the active placeholder to remove the insertion point, selecting the entire placeholder.

The placeholder border changes to a solid line from a dashed line when the entire placeholder is selected. The next action will affect all the text within the placeholder.

**7** Click the Numbering button in the Paragraph group on the Home tab to change the bulleted list to a numbered list. Do *not* click the down-pointing arrow on the button.

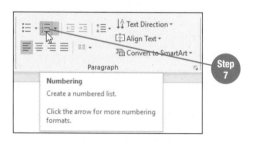

**Skills**

Insert a slide with the comparison layout

Change a bulleted list to a numbered list

Resize a placeholder

Align a placeholder

Move a placeholder

**Tutorial**

Modifying Placeholders

**Oops!**

Only one bullet changed to a number? This occurs if you have an insertion point within the placeholder; only the item in the list at which the insertion point was positioned is changed. Go back to Step 6 and try again.

8  Click anywhere in the bulleted list below *Most Expensive*, click along the border of the active placeholder to select the entire placeholder, and then click the Numbering button.

9  Select the numbered list placeholder below the title *Least Expensive*.

10  Drag the right-middle sizing handle to the left until the right border of the placeholder is at the approximate location shown in the image below.

11  Select the *Least Expensive* title placeholder and then drag the right-middle sizing handle to the left to resize the placeholder until the smart guide appears, indicating the title placeholder is the same width as the content placeholder below it.

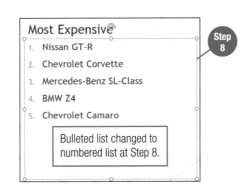

Bulleted list changed to numbered list at Step 8.

**quick steps**

**Resize a Placeholder**
1. Select placeholder.
2. Drag sizing handle as needed.

**Move a Placeholder**
1. Select placeholder.
2. Drag placeholder border as needed.

**app tip**

The AutoFit feature, which is on by default, automatically scales the font size and adjusts spacing between points to fit text within a placeholder.

Smart guides, also called *alignment guides*, appear automatically when moving or resizing objects. A **smart guide** is a colored horizontal or vertical guideline that helps you align, space, or size placeholders or objects evenly.

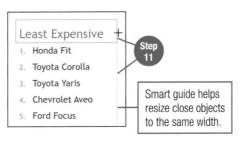

Smart guide helps resize close objects to the same width.

12  With the *Least Expensive* title placeholder still selected, drag the border of the placeholder right until the smart guides appear as shown in the image at right.

13  Select the numbered list placeholder below *Least Expensive* and drag right until left, right, top, and bottom smart guides appear, indicating the placeholder is aligned evenly with the placeholders above and right.

14  Save the revised presentation using the same name (**CarMaintenance-YourName**). Leave the presentation open for the next topic.

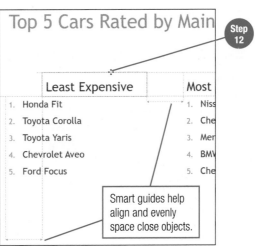

Smart guides help align and evenly space close objects.

## 10.5 Using Slide Sorter View and Moving, Duplicating, and Deleting Slides

**Slide Sorter view** displays all the slides in a presentation as slide thumbnails. Change to Slide Sorter view to perform slide management tasks. For example, you can easily rearrange the order of the slides by dragging a slide thumbnail to a new position within the slide deck. Select a slide in Slide Sorter view or Normal view to duplicate or delete the slide.

1. With the **CarMaintenance-YourName** presentation open, click the View tab and then click the Slide Sorter button in the Presentation Views group.

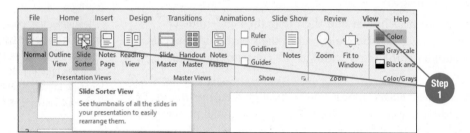

2. Click Slide 5 to select the slide.

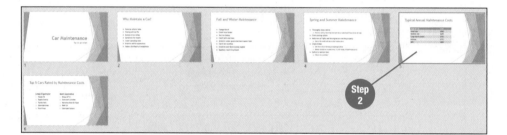

3. Drag Slide 5 to place the slide to the right of Slide 2, and then release the mouse button.

As you drag to move a slide in Slide Sorter view, the existing slides rearrange around the slide and are automatically renumbered.

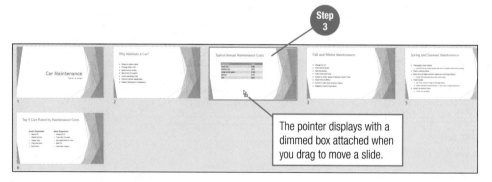

The pointer displays with a dimmed box attached when you drag to move a slide.

### Duplicating a Slide

When you need to create a new slide with the same layout as an existing slide and with the placeholders sized, aligned, and positioned the same, make a duplicate copy of the existing slide. Once the slide is duplicated, all you have to do is change the text inside the placeholders. A duplicated slide is inserted in the presentation immediately after the slide selected to be duplicated.

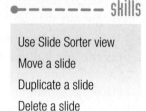

Watch & Learn

Hands On Activity

------- Skills

Use Slide Sorter view
Move a slide
Duplicate a slide
Delete a slide

Tutorial
Changing Views

Tutorial
Rearranging Slides

④ Click to select Slide 6.

⑤ Right-click Slide 6 to display the shortcut menu.

⑥ Click *Duplicate Slide*.

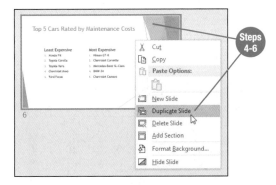

⑦ Double-click Slide 7 to return to Normal view.

## Deleting a Slide

To delete slides in Slide Sorter view or Normal view, select the slide, display the shortcut menu, and then choose *Delete Slide* or press the Delete key. Multiple slides can be deleted all at once. To do this, hold down the Ctrl key while clicking each slide you want to remove. When all slides have been selected, right-click any selected slide and then choose *Delete Slide*.

⑧ Right-click Slide 7 in the slide thumbnails pane to display the shortcut menu.

⑨ Click *Delete Slide*.

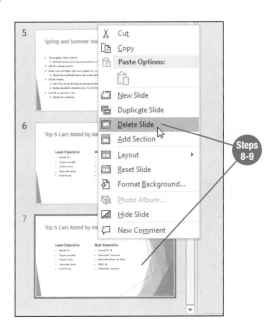

⑩ Save the revised presentation using the same name (**CarMaintenance
-YourName**). Leave the presentation open for the next topic.

## 10.6 Modifying the Slide Master

Each presentation that you create includes a slide master. A **slide master** determines the default formatting and paragraph options for placeholders when you insert new slides. If you want to make a change to a font or paragraph option for the entire presentation, making the change in the slide master will apply the change automatically to all slides in the presentation. For example, if you want a different font color for all the slide titles, change the color on the slide master.

1. With the **CarMaintenance-YourName** presentation open and the View tab active, click the Slide Master button in the Master Views group.

In **Slide Master view**, a slide master at the top of the hierarchy in the slide thumbnails pane controls the font, colors, paragraph options, and background for the entire presentation. Below the slide master is a variety of layouts for the presentation. Changes made to the slide master at the top of the hierarchy affect all the slide layouts below it except the title slide.

2. Scroll up the slide thumbnails pane to the first slide at the top of the hierarchy.

3. Click to select Slide 1.

4. Click the border of the slide master title placeholder on the slide master to select the placeholder.

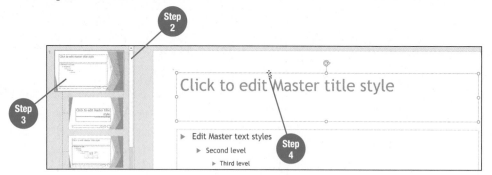

5. Click the Home tab, click the Font Color button arrow, and then click *Dark Red, Accent 6* (last color in first row of *Theme Colors* section).

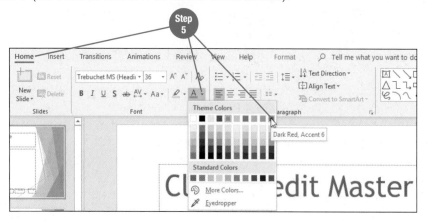

**Watch & Learn**

**Hands On Activity**

— — — — — skills

Display slide master

Format placeholder on slide master

**Tutorial**

Formatting with a Slide Master

— — — — — app tip

Formatting changes made on individual slides override the slide master. Presentations should have a consistent look; therefore, limit individual slide formatting changes to only when necessary, such as to indicate a change in topic or speaker.

6   Click the border of the content placeholder to select the placeholder and then click the Bullets button arrow in the Paragraph group.

7   Click *Bullets and Numbering*.

8   Click the Color button in the Bullets and Numbering dialog box and then click *Dark Red, Accent 6* (last color in first row of *Theme Colors* section).

9   Click the *Hollow Square Bullets* option (first option in second row).

10   Click OK.

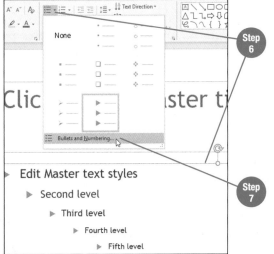

**quick steps**

**Slide Master View**
1. Click View tab.
2. Click Slide Master button.

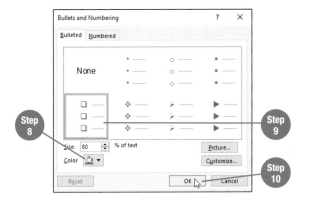

11   Click the Slide Master tab.

12   Click the Close Master View button in the Close group.

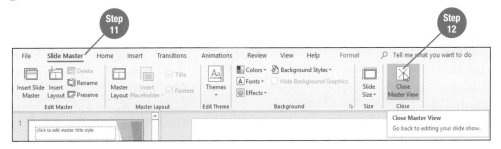

13   Click the Previous Slide or Next Slide buttons at the bottom of the vertical scroll bar or click each slide in the slide thumbnails pane to scroll through and view each slide in the presentation.

Notice that the font color for the title text and the bullet character are changed on each slide *after* the title slide. A title slide has its own slide master and is the first layout below Slide 1 in the slide master hierarchy.

14   Save the revised presentation using the same name (**CarMaintenance -YourName**). Leave the presentation open for the next topic.

**app tip**

Page Up and Page Down also display the previous or next slide in the presentation.

## 10.7 Adding Notes and Comments

Notes, generally referred to as *speaker notes*, are text typed in the **notes pane** below the slide pane in Normal view. Use notes to type reminders for the presenter or to add more details about the slide content for the person giving the presentation. In a presentation designed to be used as a self-study aid, text typed in the notes pane provides more detailed explanations to the learner.

Comments added to slides appear in the **Comments pane** at the right side of the slide pane. If you are creating a shared presentation, use comments to provide feedback or pose questions to others who will be editing the slide deck.

1 With the **CarMaintenance-YourName** presentation open, display Slide 1 in the slide pane.

2 Click the Notes button on the Status bar to turn on the display of the notes pane at the bottom of the slide pane. Skip this step if the notes pane is already visible.

3 Click anywhere in *Click to add notes* in the notes pane and then type Begin this slide with the statistic that approximately 5.2% of motor vehicle accidents are caused by vehicle neglect.

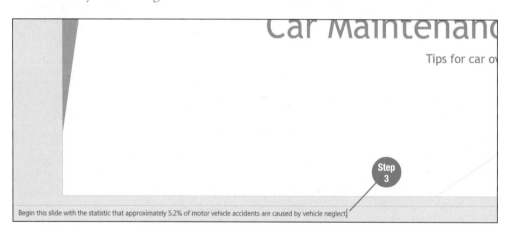

4 Display Slide 3 in the slide pane.

5 Drag the top border of the notes pane upward to increase the height of the pane by approximately a half inch.

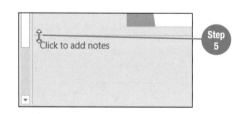

6 Click anywhere in *Click to add notes* in the notes pane and then type Mention that these costs are estimated for a driving distance of 12,000 miles (19,000 kilometers) per year.

7 Press Enter twice and then type Ask the audience if anyone wants to share the total amount paid each year to maintain his or her vehicle.

Mention that these costs are estimated for a driving distance of 12,000 miles (19,000 kilometers) per year.

Ask the audience if anyone wants to share the total amount paid each year to maintain his or her vehicle.

Steps 6-7

8  Click the Notes button to close the notes pane.

9  Click the Review tab and then click the top part of the Show Comments button in the Comments group to display the Comments pane at the right side of the slide pane.

You can also open the Comments pane by clicking the Comments button at the upper right corner of the ribbon. Either button displays or closes the Comments pane.

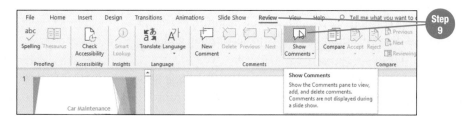

Step 9

10  Click the New button near the top of the Comments pane.

PowerPoint opens a comment box in the Comments pane with your account name associated with the comment.

11  Type Consider adding the source of these statistics to the slide.

12  Click Slide 4 in the slide thumbnails pane, click the New button in the Comments pane, and then type Add more information for any of these points?

13  Click on the slide to close the comment.

A *Reply* text box appears below the comment when the comment box is closed. Reviewers or other editors of the presentation can click in the *Reply* text box to provide feedback.

14  Click the Comments button at the upper right corner of the ribbon to close the Comments pane.

A comment balloon appears on the top left corner of a slide for which a comment has been added.

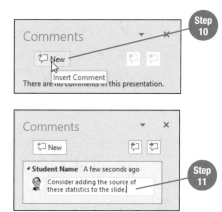

Step 10

Step 11

Use the Previous and Next buttons to navigate to all the comments in a presentation.

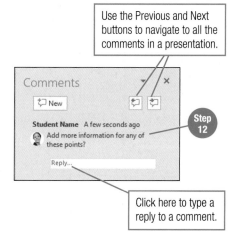

Step 12

Click here to type a reply to a comment.

A comment balloon displays on slides with comments. Click the balloon to open the Comments pane and view the comments and replies.

# Fall and Winter Maintenance

15  Save the revised presentation using the same name (**CarMaintenance –YourName**). Leave the presentation open for the next topic.

## 10.8 Displaying a Slide Show

Display the presentation in **Slide Show view** to preview the slides as they will appear to an audience. Each slide fills the screen with the ribbon and other PowerPoint elements removed; however, tools to navigate and annotate slides are available. Use the **From Beginning button** in the Start Slide Show group on the Slide Show tab to start the slide show at Slide 1.

In **Presenter view**, the slide show displays full screen on one monitor (the monitor the audience will see) and in Presenter view on a second monitor. Presenter view displays a preview of the next slide, notes from the notes pane, a timer, and a slide show toolbar along with other options.

① With the **CarMaintenance–YourName** presentation open, click the Slide Show tab and then click the From Beginning button in the Start Slide Show group.

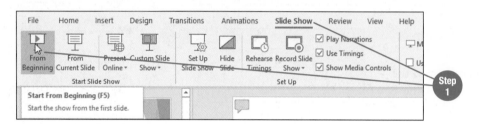

② Click the right-pointing arrow that appears on the Slide Show toolbar near the lower left corner of the screen to move to Slide 2. Move the mouse toward the lower left corner of the screen to display the toolbar if the Slide Show toolbar is not visible.

You can also click anywhere on a slide or press the Page Down key to move to the next slide. The buttons on the Slide Show toolbar are shown in Figure 10.2 and described in Table 10.2 below.

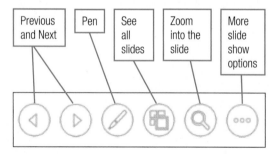

**Figure 10.2**
The Slide Show toolbar. See Table 10.2 for a description of each button.

**Table 10.2** Slide Show Toolbar Buttons

| Button | Description |
|---|---|
| Previous and Next | Displays the previous or next slide in the presentation |
| Pen | Displays a pop-up list of options for using the laser pointer, pen, or highlighter when running the presentation |
| See all slides | View all slides in the presentation similarly to Slide Sorter view. Use this option to jump to a slide out of sequence during a presentation |
| Zoom into the slide | Use this button to click on a portion of a slide that you want to enlarge to temporarily fill the screen for a closer look. Right-click or press the Esc key to restore the slide |
| More slide show options | Displays a pop-up list of options for customizing the presentation; use this button to end the show, access screen options (to show a black screen or a white screen) or display the taskbar |

Watch & Learn

Hands On Activity

━━━━━━ skills

Display a presentation in Slide Show view

Display a presentation in Presenter view

━━━━━━ app tip

Press F5 to start a slide show from Slide 1.

Tutorials

Running a Slide Show

Changing the Display when Running a Slide Show

③ Continue clicking the Next Slide arrow to navigate through the remaining slides in the presentation until the black screen appears.

After the last slide is viewed, a black screen is shown with the message *End of slide show, click to exit.* Many presenters leave the screen black when their presentation is ended until the audience has left because clicking to exit displays the presentation in Normal view on the screen.

④ Click anywhere on the black screen to return to Normal view.

⑤ Display Slide 1 in the slide pane and then click the Slide Show button on the Status bar.

The **Slide Show button** on the Status bar starts the slide show at the active slide.

⑥ Click the More slide show options button (button with three dots) on the Slide Show toolbar and then click *Show Presenter View.*

You can use Presenter view on a system with only one monitor to preview or rehearse a presentation. At a presentation venue, PowerPoint automatically detects the computer setup and chooses the correct monitor on which to show Presenter view.

⑦ Click the Next Slide button on the slide navigator near the bottom of Presenter view until you have navigated to Slide 3 (Figure 10.3).

⑧ Compare your screen with the one shown in Figure 10.3.

⑨ Continue clicking the Next Slide button until you reach Slide 6 and then click END SLIDE SHOW at the top of Presenter view.

⑩ Leave the presentation open for the next topic.

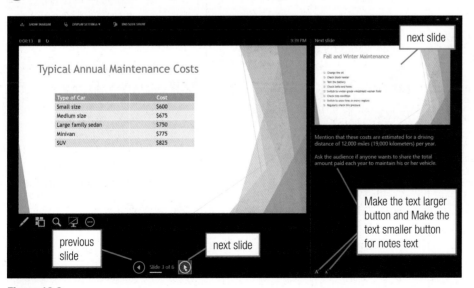

**Figure 10.3**

Shown above is Slide 3 of the **CarMaintenance-YourName** presentation in Presenter view.

## 10.9 Preparing Audience Handouts and Speaker Notes

Some speakers provide audience members with a printout of the slide deck in a format that gives an individual room to handwrite notes during the presentation. PowerPoint provides several options for printing slides as handouts. Speakers who do not use Presenter view during a presentation may also print a copy of the slides with the notes included for reference during the presentation.

1.  With the **CarMaintenance-YourName** presentation open, click the File tab and then click *Print*.

2.  In the Print backstage area, click the *Full Page Slides* option box arrow in the *Settings* category.

3.  Click *3 Slides* in the *Handouts* section.

    The option to print three slides per page provides horizontal lines next to each slide for writing notes, as shown in the Print Preview panel.

4.  Click the *3 Slides* option box arrow and then click *6 Slides Horizontal*.

    Notice that the printout requires two print pages for the six slides in the presentation. By default, comments print with the presentation; the second page is for printing the comments.

5.  Click the *6 Slides Horizontal* option box arrow and then click *Print Comments* to remove the check mark. The printout is now only one page.

6.  Click the *6 Slides Horizontal* option box arrow and then click *Notes Pages* in the *Print Layout* section.

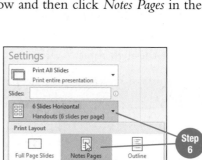

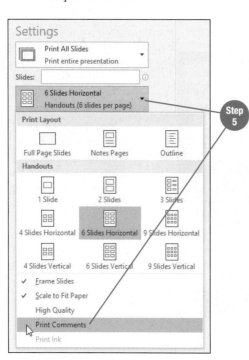

**Notes Pages** prints one slide per page with the slide at the top half of the page and notes or blank space in the bottom half.

**7** Click the Next Page or Previous Page button until Slide 2 is displayed in the Print Preview panel.

**8** Click the Next Page button to display Slide 3.

Notice the notes text is displayed below the slide.

**9** Click the Edit Header & Footer hyperlink at the bottom of the *Settings* section.

The Header and Footer dialog box opens. In this dialog box, add text to print at the top or bottom of each handout. Many speakers put their professional affiliation, company name, or a copyright notice in a header or footer.

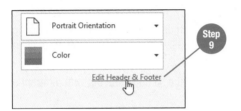

**10** If necessary, click the Notes and Handouts tab.

**11** Click the *Header* check box to insert a check mark, click in the *Header* text box, and then type your first and last names.

**12** Click the *Footer* check box to insert a check mark, click in the *Footer* text box, and then type your school name.

**13** Click the Apply to All button.

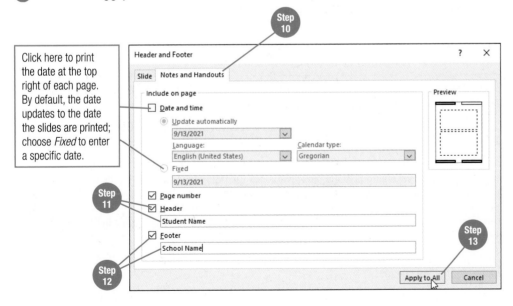

Click here to print the date at the top right of each page. By default, the date updates to the date the slides are printed; choose *Fixed* to enter a specific date.

**14** Preview the header and footer text by scrolling through the remaining slides in the Print Preview panel.

**15** Click the Back button to exit the Print backstage area.

**16** Save the revised presentation using the same name (**CarMaintenance –YourName**) and then close the presentation.

## Topics Review

| Topic | Key Concepts |
|---|---|
| **10.1 Creating a New Presentation and Inserting Slides** | A **slide deck** is a collection of slides in a presentation. |
| | A **presentation application** is software used to create a slide deck. PowerPoint is the presentation application in the Office suite. |
| | A new presentation is created from the PowerPoint Start screen by choosing a template, a theme, or the blank presentation option. |
| | PowerPoint starts a new presentation with a title slide in widescreen format in Normal view. |
| | A **title slide** is the first slide in a presentation. |
| | A **placeholder** is a rectangular container in which you type text or insert other content. |
| | Normal view includes the **slide pane**, which displays the active slide on the right side of the screen, and the **slide thumbnails pane**, which displays numbered thumbnails of all the slides in a single column along the left side of the screen. |
| | A **variant** of a theme family is based upon the same theme but with different colors, styles, and effects. |
| | Add or edit text on a slide by clicking inside the placeholder and then typing or editing text. |
| | Use the **New Slide button** to add a slide to the presentation. |
| | A **slide layout** sets the number, placement, and type of placeholders on a slide. |
| **10.2 Changing the Theme and Inserting and Modifying a Table** | Change the theme and/or variant for a presentation after the presentation has been started using options on the Design tab. |
| | Insert a table on a slide using the Insert Table button within the content placeholder of a new slide. |
| | In the Insert Table dialog box, type the number of columns for the table, type the number of rows for the table, and then click OK. |
| | The Table Tools Design and Table Tools Layout tabs provide options to modify a table. These options are the same tools you learned for modifying a table in Word. |
| **10.3 Formatting Text with Font and Paragraph Options** | Select text within a placeholder or select the entire placeholder to apply formatting changes using options in the Font and Paragraph groups on the Home tab. |
| | Create a multilevel bulleted list using the **Increase List Level button** and the **Decrease List Level button** in the Paragraph group. |
| | Each time you click the Increase List Level button, the insertion point or text is indented to the next tab, and the level changes to the next level. Click the Decrease List Level button to move the insertion point or text back to the previous tab and level. |
| **10.4 Selecting, Resizing, Aligning, and Moving Placeholders** | The active placeholder displays with sizing handles and a border with which you can resize or move the placeholder. |
| | Click the border of a placeholder to remove the insertion point and select the entire placeholder to apply a formatting change. |
| | A **smart guide** is a colored line that appears on the slide as you resize or move a placeholder to assist in aligning the placeholder or evenly spacing the placeholder with other close objects. |

*continued…*

| Topic | Key Concepts |
|---|---|
| **10.5 Using Slide Sorter View and Moving, Duplicating, and Deleting Slides** | **Slide Sorter view** displays all slides as slide thumbnails and is used to rearrange the order of slides or otherwise manage slides in the presentation. |
| | Slide or drag a slide in Slide Sorter view to move the slide to a new position within the presentation. As you move a slide around in Slide Sorter view, existing slides adjust to make room and the slides are automatically renumbered. |
| | Duplicating a slide makes a copy of an existing slide with the placeholders sized, aligned, and positioned the same as the original slide. Use this method to create a new slide based on the same design as an existing slide that has had several placeholders customized. |
| | Delete a slide or group of slides by selecting the slide(s) to be removed and using the *Delete Slide* option from the shortcut menu or by pressing the Delete key. |
| **10.6 Modifying the Slide Master** | Each presentation has a **slide master** that determines the formatting and paragraph options for placeholders on slides. |
| | Display the slide master to make formatting changes that you want to apply to all slides in the presentation. |
| | Change to **Slide Master view** from the View tab to modify the slide master. |
| | In Slide Master view, the slide thumbnails pane displays the slide master at the top of the hierarchy. |
| | Below the slide master, individual slide layouts allow you to format a layout separately from the slide master. |
| **10.7 Adding Notes and Comments** | The **notes pane** appears along the bottom of the slide pane and is used to type speaker notes, reminders for the presenter, or more detailed information about the slide content for a reader. |
| | In a shared presentation, type questions or feedback about a slide for others who will be editing the slide deck in the **Comments pane**. |
| | Reveal or hide the notes pane with the Notes button on the Status bar. |
| | Reveal or hide the Comments pane by clicking the Comments button at the upper right corner of the ribbon or by clicking the Show Comments button in the Comments group on the Review tab. |
| | A comment is added to the active slide by clicking the New button in the Comments pane and then typing the comment text. Click on the slide when finished to close the comment. |
| | A *Reply* text box appears below a comment after the comment is completed for reviewers or editors to type a response to the comment author. |

*continued...*

| Topic | Key Concepts |
|---|---|
| **10.8 Displaying a Slide Show** | **Slide Show view** previews each slide as the audience will see it with a full screen. |
| | Display a slide show starting at Slide 1 by clicking the **From Beginning button** in the Start Slide Show group on the Slide Show tab. |
| | **Presenter view** works with two monitors, where one monitor displays the slide show to the audience, and the second monitor displays the slide show for the presenter with a preview of the next slide, timer controls, and the speaker notes in a larger display. |
| | The Slide Show toolbar provides buttons to navigate slides, annotate slides using the pen tool, see all slides, zoom into a slide, or access other slide show options during the presentation. |
| | After the last slide is shown, a black screen displays, indicating the end of the slide show. |
| | The **Slide Show button** on the Status bar starts the slide show from the active slide in the slide pane. |
| | Start a slide show and use the Show Presenter View option from the More slide show options button on the Slide Show toolbar to switch the view to Presenter view. |
| | PowerPoint automatically detects the correct monitor on which to display Presenter view. |
| **10.9 Preparing Audience Handouts and Speaker Notes** | Preview slides formatted as handouts in the Print backstage area using the *Full Page Slides* option box arrow in the *Settings* category. |
| | The *3 Slides* Handouts option provides lines next to each slide for writing notes. |
| | By default, comments print on a separate page after the slides; to prevent comments from printing, remove the check mark next to *Print Comments*. |
| | Choose the **Notes Pages** option to print one slide per page with the notes from the notes pane. |
| | Add header and/or footer text to a printout using the <u>Edit Header & Footer</u> hyperlink at the bottom of the *Settings* category in the Print backstage area to open the Header and Footer dialog box. |

 **Topics Review**

# Enhancing a Presentation with Multimedia and Animation Effects

## Chapter 11

Presentations are more engaging for audiences when multimedia is used to help speakers communicate their points. Incorporating images, sound, and video into slides in a slide deck can help an audience understand the content and remain focused on the presentation. Animation and transition effects focus attention and add interest when one slide is removed and the next slide appears on the screen or when new content appears on a slide. Several options are available in PowerPoint to add special effects to a presentation that will win over your audience.

Some presentations are not delivered in a venue where a speaker is controlling the slide advancement through the slide show. In these instances, a slide show is customized to run continuously based on a set time for each slide. A presentation delivered this way is called a *self-running presentation*.

In this chapter, you will learn how to add graphic images to slides using images, SmartArt, WordArt, charts, and drawn shapes. You will also learn to add text in a text box, add sound and video, and complete a slide show presentation by adding transitions and animation effects. Lastly, you will learn how to set up a slide show that advances through the slide deck automatically.

### Learning Objectives

**11.1** Insert and resize images on a slide

**11.2** Insert and modify a SmartArt graphic on a slide

**11.3** Convert existing text to a SmartArt graphic and insert and modify a WordArt object on a slide

**11.4** Create and modify a chart on a slide

**11.5** Draw and modify shapes and text boxes on a slide

**11.6** Insert a video clip into a presentation

**11.7** Insert a sound clip into a presentation

**11.8** Add transition and animation effects into a slide show

**11.9** Set up a self-running presentation

 Read & Learn

 Content Online

The online course includes additional training and assessment resources.

# 11.1 Inserting, Resizing, and Aligning Images

Adding an image, illustration, diagram, or chart on a slide emphasizes content, adds visual interest, and helps an audience understand and make connections with the information more easily than with text alone. As you did in Chapter 7, you can insert images from a file on your computer or from an online resource.

## Inserting an Image from a File on Your Computer

To add an image to an existing slide, use the Pictures button in the Images group on the Insert tab to add the image stored in a file on your computer or a computer to which you are connected. Once inserted, move, resize, and/or modify the image using buttons on the Picture Tools Format tab. On a new slide, use the Pictures icon in the content placeholder to add an image to a slide.

1. Start PowerPoint and open the presentation **PaintedBunting** from the Ch11 folder in StudentDataFiles. If necessary, click the Enable Editing button to exit Protected view.

2. Use Save As to save a copy of the presentation as **PaintedBunting-YourName** in a new folder named *Ch11* in the CompletedTopics folder.

3. Browse through the presentation and read the slides.

4. Make Slide 2 the active slide in the slide pane.

5. Click the Insert tab and then click the Pictures button in the Images group.

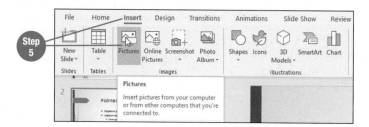

6. Navigate to the Ch11 folder within StudentDataFiles on your storage medium in the Insert Picture dialog box and then double-click **PaintedBunting_NPS.jpg**.

7. Click anywhere on the image to select it. Using one of the four corner selection handles, resize the image to the approximate size shown in the image below.

8. With the mouse pointer positioned anywhere over the selected image, drag the image to the right side of the slide, aligning it with the horizontal and vertical smart guides that appear when the image is even with the top of the text and the right margin on the slide.

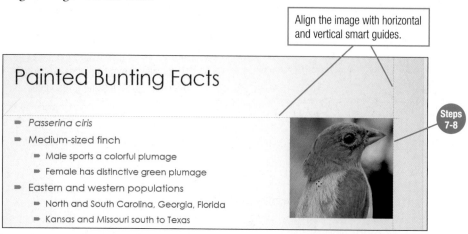

Align the image with horizontal and vertical smart guides.

Steps 7-8

Painted Bunting Facts

- *Passerina ciris*
- Medium-sized finch
    - Male sports a colorful plumage
    - Female has distinctive green plumage
- Eastern and western populations
    - North and South Carolina, Georgia, Florida
    - Kansas and Missouri south to Texas

Watch & Learn

Hands On Activity

Skills

Add an image from a file on your storage medium

Add an image from an online resource

Tutorial

Inserting, Sizing, and Positioning an Image

**9** Insert the image **PaintedBunting_Female.jpg** near the bottom right of the slide (see image below), by completing steps similar to Steps 5 to 8.

**Insert an Image from a File**
1. Activate slide.
2. Click Insert tab.
3. Click Pictures button.
4. Navigate to drive and folder.
5. Double-click desired image.
6. Resize and move as required.

**Insert Image from a Web Resource**
1. Activate slide.
2. Click Insert tab.
3. Click Online Pictures button.
4. Type search keyword.
5. Press Enter.
6. Double-click desired image.
7. Resize and move as required.

## Inserting an Image from a Web Resource

The Online Pictures button in the Images group on the Insert tab or in a content placeholder is used to find an image from a website. Use the search text box in the Online Pictures dialog box to type a word or phrase representing the image you need. As you learned in Chapter 7, search results displayed are from the Bing search engine and apply a copyright filter displaying images tagged with a Creative Commons license.

**10** Make Slide 6 the active slide in the slide pane.

**11** Click the Insert tab and then click the Online Pictures button.

**12** With the insertion point positioned in the search text box, type team and then press Enter.

**13** Double-click the image shown on the slide below. If you cannot locate the same image, close the Online Pictures dialog box and insert the image **Team.png** that is saved in the Ch11 folder in StudentDataFiles.

**14** Resize and move the image to the left side of the bulleted list, aligning it with the horizontal and vertical smart guides that appear when the image is placed evenly with the top and left edge of the bulleted list, as shown on the slide below.

**Oops!**

Close the Design Ideas task pane if the task pane appears after inserting the image on the slide.

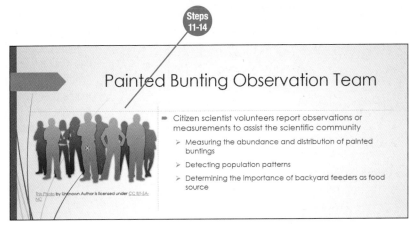

**15** Save the revised presentation using the same name (**PaintedBunting -YourName**). Leave the presentation open for the next topic.

## 11.2 Inserting a SmartArt Graphic

**SmartArt** is a graphic object that visually communicates a relationship in a list, process, cycle, hierarchy, or other diagram. Begin creating a SmartArt graphic by choosing a predesigned layout and then adding text in the Text pane, or by typing text directly in the text placeholders within the shapes. You can add and delete shapes to the graphic as needed and choose from a variety of color schemes and styles. See Table 11.1 for a description of layout category diagrams created using SmartArt.

**Table 11.1** SmartArt Graphic Layout Categories

| Layout Category | Description |
| --- | --- |
| List | Show nonsequential tasks, processes, or other list items |
| Process | Illustrate a sequential series of steps to complete a process or task |
| Cycle | Show a sequence of steps or tasks in a circular or looped process |
| Hierarchy | Show an organizational chart or decision tree |
| Relationship | Show how parts or elements are related to one another |
| Matrix | Depict how individual parts or ideas relate to a whole or central idea |
| Pyramid | Show proportional or hierarchical relationships that build upward |
| Picture | Add pictures inside shapes with small amounts of text to display ideas, a process, or a relationship |

**1** With the **PaintedBunting–YourName** presentation open and Slide 6 the active slide in the slide pane, click the Insert tab.

**2** Click the SmartArt button in the Illustrations group.

**3** In the Choose a SmartArt Graphic dialog box, click *Process* in the Category pane (left pane), click *Basic Chevron Process* in the Layout pane (second option, fifth row in center pane), and then click OK.

PowerPoint places the SmartArt graphic in the center of the slide, overlapping existing content. Three shapes are automatically included in the *Basic Chevron Process* layout.

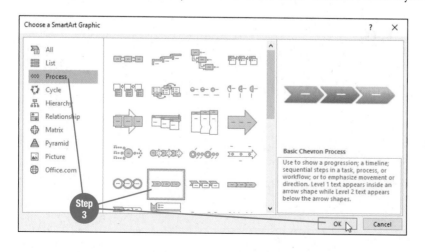

🌥 **Watch & Learn**

🌥 **Hands On Activity**

●———— skills

Add a SmartArt graphic

Modify a SmartArt graphic

🌥 **Tutorial**

Inserting, Sizing, and Positioning SmartArt

●———— app tip

On a new slide with no other content, use the Insert a SmartArt Graphic icon in the content placeholder to create a SmartArt object on a slide.

●———— app tip

If you are unsure which SmartArt graphic to use, click a layout in the center pane so you can read a description of it in the right pane, including suggested usage.

④ With the insertion point in the first bullet Text pane, type Band; click next to the second bullet or press the Down Arrow key and then type Observe; click next to the third bullet or press the Down Arrow key and then type Analyze.

The SmartArt graphic updates as each word is typed in the Text pane to display the text in the shape. You can also add text to the shapes by typing text directly within the text placeholders inside each shape.

⑤ Click the Close button to close the Text pane.

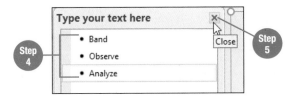

⑥ With the SmartArt Tools Design tab active, click the More button at the bottom of the SmartArt Styles gallery.

⑦ Click *Polished* (first option in *3-D* section).

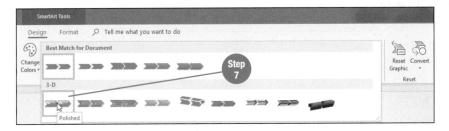

⑧ Click the Change Colors button in the SmartArt Styles group and then click *Colorful – Accent Colors* (first option in the *Colorful* section).

⑨ Drag the border of the SmartArt graphic to reposition the graphic near the bottom center of the slide, and then click on a blank area on the slide to deselect the object (Figure 11.1).

⑩ Save the revised presentation using the same name (**PaintedBunting -YourName**). Leave the presentation open for the next topic.

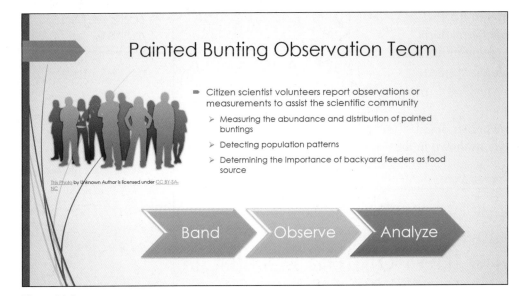

**Figure 11.1**

The completed Slide 6 with SmartArt graphic is shown here.

# 11.3 Converting Text to SmartArt and Inserting WordArt

An existing bulleted list on a slide can be converted to a SmartArt graphic using the **Convert to SmartArt button** in the Paragraph group on the Home tab. **WordArt** is text that is created and formatted as a graphic object. With WordArt, you can create decorative text on a slide with a variety of WordArt Styles and text effects. A WordArt object can also have the text formed around a variety of shapes.

1. With the **PaintedBunting-YourName** presentation open, make Slide 5 the active slide in the slide pane.

2. Click in the bulleted list to activate the placeholder.

3. Click the Convert to SmartArt Graphic button in the Paragraph group on the Home tab.

4. Click *Hierarchy List* (second option in second row).

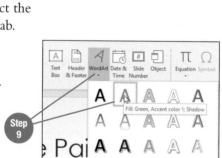

Step 3

Step 4

PowerPoint converts the text in the bulleted list into the selected SmartArt layout. Level 1 text from the bulleted list is placed inside shapes at the top level in the hierarchy diagram, with level 2 text in shapes below the corresponding level 1 box.

5. Close the Text pane if the Text pane is open.

6. Click in the second shape in the top level of the hierarchy, delete *Need to* and capitalize *m* (type M) so that the text inside the shape reads *Manage and preserve natural habitat*.

Step 6

7. Click anywhere along the border of the SmartArt object to select the entire SmartArt object and then change the SmartArt style and color scheme to the same style and color used in the SmartArt graphic on Slide 6.

8. Click on a blank area of the slide to deselect the SmartArt object and then click the Insert tab.

9. Click the WordArt button in the Text group and then click *Fill: Green, Accent color 1; Shadow* (second option in first row).

10. With *Your text here* inside the WordArt object already selected, type Help Save the Painted Bunting!

Step 9

11. Drag the border of the WordArt object to the bottom of the slide, as shown in the image below.

Step 11

Help Save the Painted Bunting!

---

**Watch & Learn**

**Hands On Activity**

— — — — — — Skills

Convert text to a SmartArt graphic

Insert and modify a WordArt object

**Tutorials**

Converting Text and WordArt to a SmartArt Graphic

Inserting and Formatting WordArt

— — — — — — Oops!

Don't remember the style and color used on Slide 6? Refer to page 257, Steps 6 to 8 to see how the SmartArt graphic was formatted on Slide 6.

12 Click the Text Effects button in the WordArt Styles group on the Drawing Tools Format tab, point to *Glow*, and then click *Glow: 5 point; Lime, Accent color 3* (third option in first row of *Glow Variations* section).

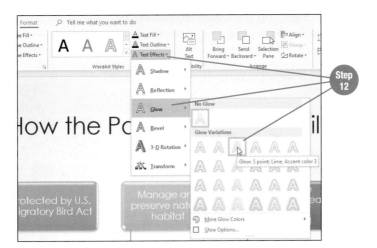

13 Click the Text Fill button and then click *Dark Teal, Accent 4* (eighth option in first row in *Theme Colors* section).

14 Click the Shape Effects button in the Shape Styles group, point to *Shadow*, and then click *Offset: Bottom Right* (first option in *Outer* section).

15 Click a blank area on the slide and compare your slide with the one shown in Figure 11.2. Redo any steps to make corrections as needed.

16 Save the revised presentation using the same name (**PaintedBunting-YourName**). Leave the presentation open for the next topic.

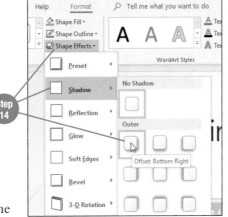

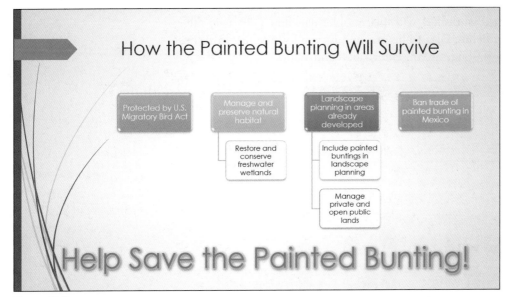

**Figure 11.2**

The completed Slide 5, with a WordArt object and a bulleted list converted to a SmartArt object, is shown.

## 11.4 Creating a Chart on a Slide

Charts similar to the ones you created with Excel in Chapter 9 can be added to a PowerPoint slide. Add a chart using the Insert Chart icon in a content placeholder or with the Chart button in the Illustrations group on the Insert tab. Charts are commonly used in presentations to display dollar figures, targets, budgets, comparisons, patterns, trends, or variations in numerical data.

**1** With the **PaintedBunting-YourName** presentation open, make Slide 3 the active slide in the slide pane.

**2** Click the Insert Chart icon in the content placeholder.

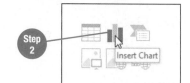

**3** Click OK in the Insert Chart dialog box with *Column* selected in the All Charts category list and *Clustered Column* selected as the chart type.

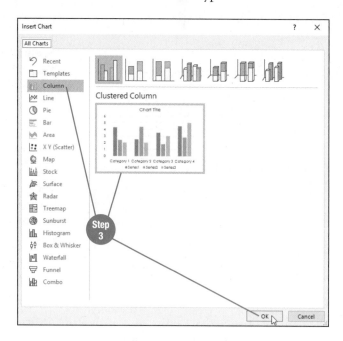

PowerPoint displays a small worksheet with placeholder data and a chart on the slide, as shown in the image below. Data to be graphed is typed inside the worksheet, which updates the chart on the slide as you enter labels and values.

Click this button to open the worksheet in a separate Excel window to edit the chart data.

The chart updates as you replace labels and values in the worksheet. Add or delete rows or columns as needed by right-clicking row numbers or column letters and using options on the shortcut menu.

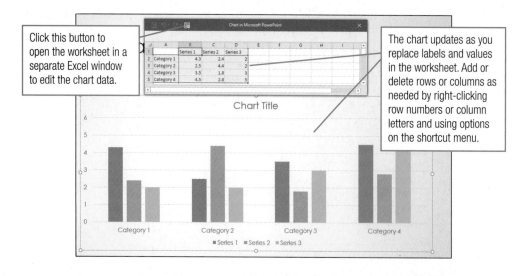

**Watch & Learn**

**Hands On Activity**

● ─ ─ ─ ─ ─ ─ ─ ─ **Skills**

Create a chart on a slide

Modify the chart style and color

**Tutorials**

Creating a Chart

Formatting with Chart Buttons

Changing the Chart Design

④ With B1 in the worksheet the active cell, type 2019.

⑤ Click C1 and then type 2020.

⑥ Type the remaining data in the cells in the worksheet as shown in the image below.

⑦ Close the worksheet.

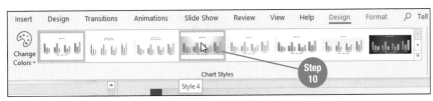

**quick steps**

**Insert a Chart**
1. Activate slide.
2. Click Insert Chart icon in content placeholder.
3. Choose category and chart type.
4. Click OK.
5. Add data in chart data grid.
6. Close worksheet.
7. Format chart as required.

⑧ Click to select the chart title object, select the text *Chart Title*, and then type Monthly Sightings January to April.

⑨ Click the border of the chart to select the entire chart.

⑩ Click *Style 4* (fourth option) in the Chart Styles gallery on the Chart Tools Design tab.

**app tip**

A complex chart or a chart with a lot of data that updates regularly is best created in Excel and copied to a slide in PowerPoint. You will learn how to do this in Chapter 14.

⑪ Click the Change Colors button in the Chart Styles group and then click *Colorful Palette 2* (second row in *Colorful* section).

⑫ Click on the slide outside the chart to deselect the chart object and then compare your slide with the one shown in Figure 11.3.

⑬ Save the revised presentation using the same name (**PaintedBunting –YourName**). Leave the presentation open for the next topic.

**Tutorial**

Changing Chart Formatting

**Oops!**

Your chart data doesn't match Figure 11.3? Reopen the Excel worksheet to edit the entries you typed in Steps 4 to 6. To do this, select the chart object, click the Chart Tools Design tab, and then click the top part of the Edit Data button in the Data group.

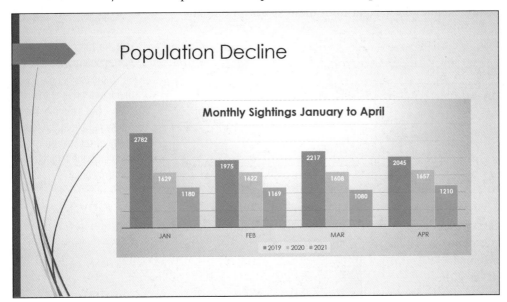

**Figure 11.3**

Shown is the completed Slide 3 with a Clustered Column Chart.

# 11.5 Drawing a Shape and Adding a Text Box

A graphic can be added to a slide by drawing a line, rectangle, circle, arrow, star, banner, or other shape. Once the shape is drawn, text can be added inside the shape, and the shape can be formatted by adding a visual effect or by changing the outline color or fill color. Draw a shape by clicking the **Shapes button** in the Illustrations group on the Insert tab and then selecting the desired shape in the list. To insert the shape at a preset size, click the slide where you want the shape to appear, or drag the crosshairs to create a shape the size that you want.

A text box is text inside a rectangular object that can be manipulated independently from other objects on a slide. To add a text box, click the Text Box button in the Text group on the Insert tab. When the mouse pointer displays as a down-pointing arrow, click on the slide at the position where the text box is to begin and then start typing—the text box width expands to accommodate the amount of text typed.

**1** With the **PaintedBunting-YourName** presentation open and with Slide 3 the active slide in the slide pane, click the Insert tab.

**2** Click the Shapes button in the Illustrations group.

**3** Click the *Arrow: Striped Right* shape (fifth option in second row of the *Block Arrows* section).

A shape can be drawn one of two ways: move the crosshairs to the location where the shape is to appear and then click to insert the shape at the default shape size; or, at the desired position on the slide, drag the crosshairs until the shape is the desired height and width. In the next step, you will create a shape using the first method.

**4** Click inside the chart above the JAN bar for 2020 (the bar with the value *1629*).

**5** With the shape selected, type A 41% decline!

**6** Drag the right-middle sizing handle to the right until the text fits on one line inside the arrow shape.

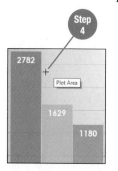

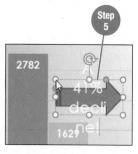

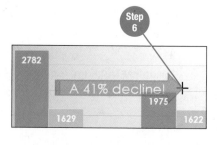

**7** Drag the rotation handle (circled arrow above upper center sizing handle) in an upward diagonal direction toward the left until the shape is at the approximate angle shown in the image at the right.

**Watch & Learn**

**Hands On Activity**

●———————— **SKILLS**

Draw and modify a shape on a slide

Add text inside a shape

Format a shape

Create a text box on a slide

**Tutorials**

Inserting, Sizing, and Positioning Shapes

Formatting Shapes

Copying and Rotating Shapes

Inserting and Formatting Text Boxes

●—————— **app tip**

Use the yellow handles that appear for a selected shape to change the appearance of the shape. For example, the yellow handle at the top of the arrow can be used to change the length of the arrowhead.

8. Drag the border of the arrow shape down to the bottom left of the chart so that it points to the 2020 January bar, as shown in the image at right.

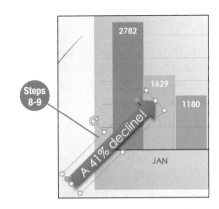

9. With the arrow shape still selected and the Drawing Tools Format tab active, click the More button at the bottom of the Shape Styles gallery and then click *Intense Effect – Turquoise, Accent 6* (last option in last row of *Theme Styles* section).

10. Click the Insert tab and then click the Text Box button in the Text group.

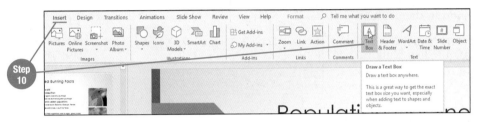

11. With the mouse pointer displayed as a down-pointing arrow, click anywhere below the chart at the left side to insert a text box with an insertion point.

12. Type Source: Painted Bunting Observer Team, University of North Carolina, Wilmington.

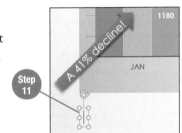

13. Click the border of the text box to remove the insertion point and select the entire placeholder, click the Bold button, and then click the Italic button in the Font group on the Home tab.

14. With the text box placeholder still selected, click the *Font Size* option box arrow and then click *14*.

15. Drag the border of the text box to align the text box at the center below the chart when the smart guides appear, as shown in the image below.

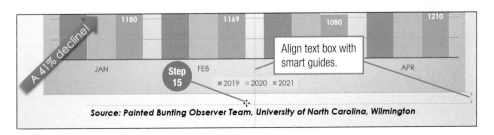

16. Save the revised presentation using the same name (**PaintedBunting -YourName**). Leave the presentation open for the next topic.

---

**quick steps**

**Draw a Shape**
1. Click Insert tab.
2. Click Shapes button.
3. Click required shape.
4. Click on slide to insert default shape size.
5. Type text, if required.
6. Resize, format, and move shape as required.

**Create a Text Box**
1. Click Insert tab.
2. Click Text Box button.
3. Click on slide.
4. Type text.
5. Resize, format, and move text box as required.

**app tip**

Click the View tab and then click the *Gridlines* check box in the Show group to insert a check mark and display evenly spaced horizontal and vertical dotted lines on the slide to assist with placing objects at precise locations.

## 11.6 Adding Video to a Presentation

A high-quality video can demonstrate a process or task that is otherwise difficult to portray using descriptions or pictures. Video is effective for instructional purposes and when used appropriately provides a more enjoyable experience for the audience. Using PowerPoint, you can play a video from a file stored on your PC or link to a video on YouTube or another online source. Use the **Trim Video button** in the Editing group on the Video Tools Playback tab to crop a portion of the video at the beginning or end of the video clip.

1. With the **PaintedBunting-YourName** presentation open, make Slide 6 the active slide in the slide pane.

2. Insert a new slide with the Title and Content layout and then type A Beautiful Bird as the slide title.

3. Click the Insert Video icon in the content placeholder.

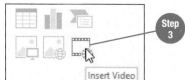

4. Click the <u>Browse From a file</u> hyperlink in the *From a file* section in the Insert Video dialog box.

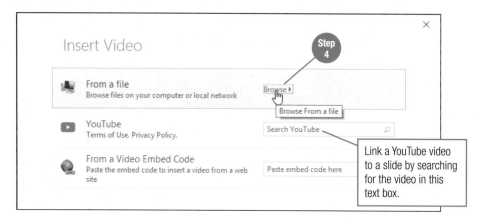

5. Navigate to the Ch11 folder in StudentDataFiles and then double-click the file *PaintedBuntingVideo.mp4*.

6. Click the Play/Pause button below the video to preview the video clip. The video plays for approximately 52 seconds.

**Watch & Learn**

**Hands On Activity**

●- - - - - - - Skills

Insert a video from a file

Trim the video

Set video playback options

**Tutorial**

Inserting and Modifying a Video File

●- - - - - - app tip

PowerPoint recognizes most video file formats, such as QuickTime movies, MP4 videos, MPEG movie files, Windows Media Video (.wmv) files, and Adobe Flash Media.

7  Click the Video Tools Playback tab.

8  Click the Trim Video button in the Editing group.

**quick steps**

**Add Video from a File on a PC**
1. Click Insert Video icon in content placeholder.
2. Click <u>Browse From a file</u>.
3. Navigate to drive and/ or folder.
4. Double-click video file.
5. Edit and/or format video clip object as required.

Trimming a video allows you to display only a portion of a video file if the video is too long or if you want to cut parts at the beginning or end. Drag the green or red slider to start playing at a later starting point and/or end before the video is finished, or enter the start and end times in the Trim Video dialog box.

9  In the Trim Video dialog box, select the current entry in the *Start Time* text box and then type 00:10.

10  Select the current entry in the *End Time* text box and then type 00:30.

11  Click the Play button to preview the shorter video clip.

12  Click OK.

13  Click the *Start* option box arrow (displays *In Click Sequence*) in the Video Options group and then click *Automatically*.

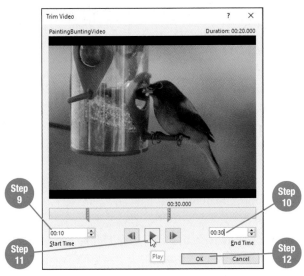

The *Automatically* option means the video will begin playing as soon as the slide is displayed in the slide show.

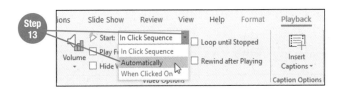

14  Drag the video object left until the smart guide appears at the left, indicating the object is aligned with the slide title.

15  Click the Video Tools Format tab and then click the *Soft Edge Rectangle* option in the Video Styles group (third option).

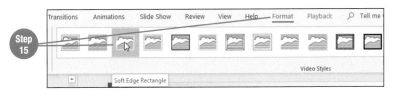

16  Save the revised presentation using the same name (**PaintedBunting –YourName**). Leave the presentation open for the next topic.

## 11.7 Adding Sound to a Presentation

Adding music or other sound during a slide show is another way to interest and entertain your audience. For example, you might time introductory music to play while the title slide displays and your audience gathers, with the end of the music cueing the audience that the presentation is about to begin. Music can also be timed to play during a segment of a presentation to accompany a series of images. To add a sound clip or music to a presentation, activate the slide at which the sound should begin, click the **Audio button** in the Media group on the Insert tab, choose *Audio on My PC*, and then select the sound or music file in the Insert Audio dialog box.

**1** With the **PaintedBunting–YourName** presentation open, make Slide 1 the active slide in the slide pane and then click the Insert tab.

**2** Click the Audio button in the Media group and then click *Audio on My PC*.

**3** Navigate to the Ch11 folder in StudentDataFiles and then double-click the file **Allemande.mp3** in the Insert Audio dialog box.

**4** Drag the sound icon to position the icon and playback tools near the bottom right of the slide.

**5** Click the Play/Pause button to listen to the recording. Click the Pause button after you have heard the beginning of the music clip for about 10 seconds.

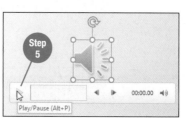

The entire music clip plays for approximately two and a half minutes.

**6** With the Audio Tools Playback tab active, click the *Hide During Show* check box in the Audio Options group to insert a check mark.

**7** Click the *Start* option box arrow (displays *In Click Sequence*) and then click *Automatically*.

This option will start the music as soon as the slide is displayed in the slide show.

**8** Click the *Loop until Stopped* check box to insert a check mark.

The *Loop until Stopped* option replays the music clip continuously until the slide is advanced during the slide show.

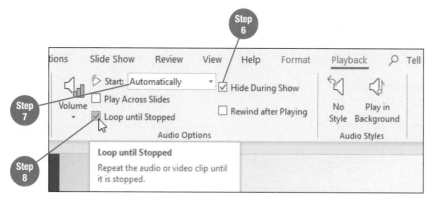

**Watch & Learn**

**Hands On Activity**

————— Skills

Insert sound from a file

Set audio playback options

**Tutorial**

Inserting and Modifying an Audio File

————— app tip

PowerPoint recognizes most audio file formats, including MIDI files, MP3 and MP4 audio files, Windows audio files (.wav), and Windows Media Audio files (.wma).

9  Make Slide 7 the active slide in the slide pane and then insert the audio file **PaintedBunting_Song.mp3** on the slide by completing steps similar to those in Steps 2–8.

The audio recording of the painted bunting bird song is slightly less than two seconds in length.

**quick steps**

**Add Audio from a File**
1.  Click Insert tab.
2.  Click Audio button.
3.  Click *Audio on My PC.*
4.  Navigate to drive and folder.
5.  Double-click audio file.
6.  Edit playback options as required.

10  Insert a new slide after Slide 7 with the Title and Content layout and then type Photo, Video, and Audio Credits as the slide title.

11  Insert a table, type the information shown in Figure 11.4, and then adjust the layout using your best judgment for column widths and position on the slide.

Always credit the source of images, audio, and video used in a presentation if you did not create the multimedia yourself. In this instance, the music from Slide 1 is not credited because the recording is in the public domain and does not require attribution.

12  Save the revised presentation using the same name (**PaintedBunting –YourName**). Leave the presentation open for the next topic.

**in the know**

Many websites offer copyright-free or public domain multimedia. Include the keywords *copyright free* or *public domain* in a search for an image, video, or audio resource. Check the terms of use at a website to credit sources appropriately, because copyright-free allows free usage but usually requires attribution (credit) to the creator.

## Photo, Video, and Audio Credits

| Slide | Item | Credit |
|---|---|---|
| 2 | Photo of male bird | US National Park Service, Amistad National Recreation Area, Del Rio, Texas |
| 2 | Photo of female bird | Dan Pancamo via Wikimedia Commons |
| 6 | Team Image | The "Enea" Comprehensive Institute, Rome via creativecommons.org |
| 7 | Video | Julio C. Rios from Port St Lucie, Florida, via Vimeo |
| 7 | Audio | US Department of the Interior, US Geological Survey, via usgs.gov |

**Figure 11.4**

This table for Slide 8 shows multimedia credits.

# 11.8 Adding Transitions and Animation Effects to Slides for a Slide Show

A **transition** is a special effect that appears as one slide is removed from the screen and the next slide appears. **Animation** adds a special effect to an object on a slide that causes the object to move or change. Animation is a powerful way to enliven a presentation by focusing the viewer's attention on specific text or objects, but don't overuse transition and animation effects, as too much movement can become a distraction.

**1** With the **PaintedBunting-YourName** presentation open, make Slide 1 the active slide in the slide pane.

**2** Click the Transitions tab and then click the More button at the bottom of the Transition to This Slide gallery.

**3** Click the *Blinds* option in the *Exciting* section of the gallery.

PowerPoint previews the effect with the current slide so that you can experiment with various transitions and effects before making your final selection.

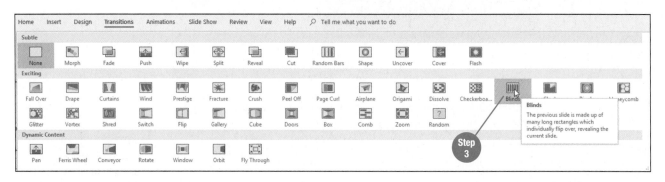

**4** Click the Apply To All button in the Timing group.

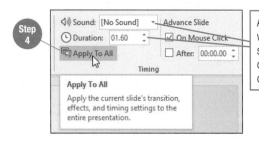

Add a sound effect that plays with the transition and/or speed up or lengthen the time of transition using the two options shown here.

**5** Display the slide show, advance through the first three slides to view the transition effect, and then end the show to return to Normal view.

## Applying Animation Effects Using the Slide Master

To apply the same animation effect to all the titles and/or bulleted lists in a presentation, apply the effect using the slide master. Generally, animation effects for similar objects should be consistent throughout a presentation; a different effect on each slide might become a distraction.

**6** Make Slide 2 the active slide in the slide pane, click the View tab, and then click the Slide Master button.

**7** Click to select the border of the title placeholder on the slide master.

---

**Watch & Learn**

**Hands On Activity**

● ─ ─ ─ ─ ─ ─ SKIllS

Add a transition to all slides

Add an animation effect to an object on the slide master

Add an animation effect to an individual object

**Tutorials**

Adding Slide Transitions

Adding Sound to Slide Transitions

● ─ ─ ─ ─ ─ ─ app tip

Use the Effect Options button in the Transition to This Slide group to choose a variation for the selected transition (such as the direction the blinds move).

8　Click the Animations tab.

9　Click the *Split* option in the Animation gallery.

See Table 11.2 at the bottom of the page for a description of Animation categories.

 **app tip**

The Animation gallery provides the most popular effects. View more effects by category using the options at the bottom of the gallery. For example, *More Entrance Effects* displays all 40 options.

**Tutorials**

Applying and Removing Animations

Modifying Animations

10　Click to select the border of the content placeholder and then click the More button in the Animation gallery.

11　Click the Zoom animation in the *Entrance* section.

12　Click the *Start* option box arrow (displays *On Click*) in the Timing group and then click *After Previous*.

13　Select the current entry in the *Duration* text box, type 1.5, and then press Enter.

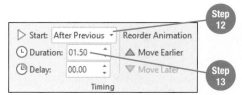

14　Click the Slide Master tab and then click the Close Master View button.

15　Make Slide 1 the active slide in the slide pane, run through the presentation in a slide show to view the transition and animation effects, and then return to Normal view.

**Table 11.2** Animation Effects

| Effect Section | Description |
|---|---|
| Entrance | Animates an object as it appears on the slide; most common animation effect |
| Emphasis | Animates text or object already in place by causing the object to move or to change in appearance; includes effects such as darkening, changing color, bolding, and underlining, to name a few |
| Exit | Animates the text or object after it has been revealed, such as by fading or flying off the slide |
| Motion Paths | Animates an object to move along a linear path, an arc, or some other shape |

## Applying Animation Effects to Individual Objects

As you previewed the slide show, you probably noticed that images and other objects, such as shapes or text boxes, appeared on the slide before the title. You may want these items to remain hidden until the title and text have been revealed. To apply animation to an individual object, display the slide, select the object to be animated, and then apply the desired animation option.

To copy an animation effect from one object to another, use the **Animation Painter button** in the Advanced Animation group on the Animations tab, which operates similarly to the Format Painter button that you learned to use in Chapter 3.

**16** Make Slide 2 the active slide in the slide pane.

**17** Click to select the male bird image at the upper right of the slide.

**18** Click the Animations tab and then click the Wipe animation in the Animation gallery.

**19** Click the *Start* option box arrow and then click *After Previous*.

**20** Click to select the male bird image and then click the Animation Painter button in the Advanced Animation group.

**21** With the mouse pointer displayed with the paint brush icon, click to select the female bird image at the lower right of the slide.

**22** Make Slide 3 the active slide in the slide pane.

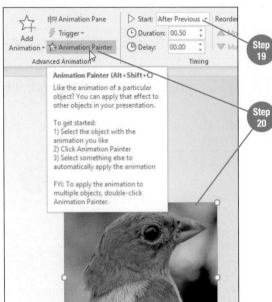

Animation Painter copies animation effects and animation options from one object to another.

• — — — — — — app tip

To adjust the order an object is animated, click the Animation Pane button in the Advanced Animation group, select the object you want to move in the Animation Pane, and then click the Move Earlier or Move Later button in the Reorder Animation group.

**23** Click to select the text box object below the chart and then apply the Fly In animation option.

**24** Click the *Start* option box arrow and then click *After Previous*.

**25** Use the Animation Painter feature to copy the animation options from the text box object to the arrow object on Slide 3.

**26** Make Slide 5 the active slide in the slide pane.

**27** Click to select the WordArt object at the bottom of the slide, apply the Float In animation option, and then change the *Start* option to *After Previous*.

**28** Make Slide 6 the active slide in the slide pane.

**29** Click to select the image at the left side of the slide, apply the Shape animation, and then change the *Start* option to *After Previous*.

**30** Click to select the SmartArt graphic placeholder at the bottom of the slide. Make sure the entire graphic is selected and not an individual shape within the graphic.

**31** Apply the Fly In animation and then change the *Start* option to *After Previous*.

**32** Click the Effect Options button and then click the *One by One* option in the *Sequence* section.

The *One by One* effect option causes each chevron in the graphic to animate on the slide one at a time, starting with the leftmost shape first.

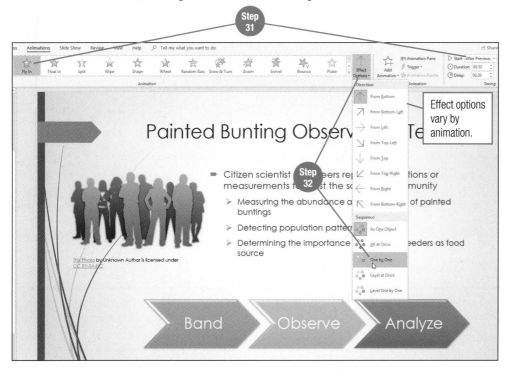

**33** Click to select the text box below the team image that contains copyright information, click the border of the text box to select the entire object, and then press the Delete key. Skip this step if you do not have a copyright text box below the team image.

**34** Run through the presentation in a slide show from the beginning to view the revised animation effects. Return to Normal view when the slide show ends.

**35** Save the revised presentation using the same name (**PaintedBunting -YourName**). Leave the presentation open for the next topic.

## 11.9 Setting Up a Self-Running Presentation

Some presentations are designed to be self-running, meaning that the slides are intended to be displayed continuously at a kiosk or on a PC by an individual instead of within a group setting where a speaker is leading the slide show. To create a presentation that advances through slides automatically, you need to set up a time for each slide to display and ensure that each slide animation is set to start automatically for each object. When the animation options and timing settings are complete, open the Set Up Show dialog box to change the show type to *Browsed at a kiosk* by clicking the **Set Up Slide Show button** in the Set Up group on the Slide Show tab.

**1**  With the **PaintedBunting-YourName** presentation open, use Save As to save a copy of the presentation in the current folder, naming it **PaintedBuntingSelfRunning-YourName**.

**2**  Make Slide 2 the active slide in the slide pane and then display the slide master.

**3**  Select the border of the title placeholder.

**4**  Click the Animations tab and then change the *Start* option in the Timing group to *After Previous*.

**5**  Close Slide Master view.

**6**  Make Slide 1 the active slide in the slide pane and then click to select the sound icon.

**7**  Click the Audio Tools Playback tab and then click the *Play Across Slides* check box in the Audio Options group to insert a check mark.

This option will cause the music that starts at Slide 1 to continue playing through the remaining slides.

**8**  Click the Volume button and then click *Low*.

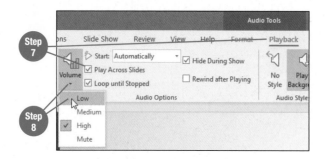

**9**  Make Slide 7 the active slide in the slide pane and then select and delete the sound icon to remove the audio.

**10**  Make Slide 8 the active slide in the slide pane and then select and delete the last row in the table. (Refer to Chapter 7, Topic 7.4, Steps 7–10 if you need assistance with this step.)

**11**  Click the Transitions tab.

**12**  Click the *After* check box in the *Timing* section to insert a check mark, select the current entry in the *After* text box, type 0:25, and then press Enter.

 **Watch & Learn**

 **Hands On Activity**

●━━━━━━ Skills

Add timings to slides

Change the show type to kiosk

 **Tutorials**

Setting Timings for a Slide Show

Looping a Slide Show Continuously

●━━━━━━ app tip

On the Slide Show tab, the Rehearse Timings feature lets you assign a time to each slide as you run through a slide show with a timer active and a Recording toolbar. Use the Next button on the Recording toolbar to advance each slide, and the time for the slide to remain on screen is automatically entered for the slide transition.

**13** Click the Apply To All button.

All the slides will advance automatically after the same 25-second duration. To set individual transition times for slides, activate a slide and enter a different time in the *After* text box.

**14** Make Slide 3 the active slide in the slide pane, select the entry in the *After* text box, type 0:10 and then press Enter.

**15** Change the *After* time for Slide 5 and Slide 6 to 0:15 and for Slide 8 to 0:10.

**16** Click the Slide Show tab and then click the Set Up Slide Show button in the Set Up group.

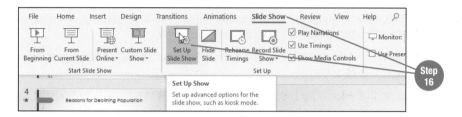

**17** Click the *Browsed at a kiosk (full screen)* option in the *Show type* section in the Set Up Show dialog box.

**18** Click OK.

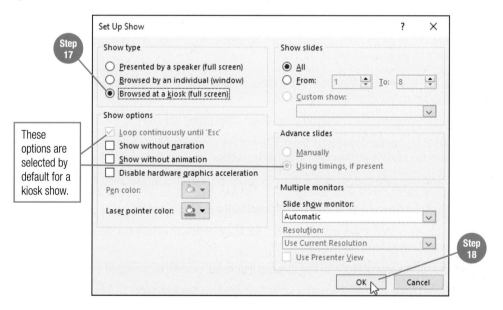

These options are selected by default for a kiosk show.

**19** Start the slide show from the beginning and watch the presentation as it advances through all the slides automatically. End the show when the presentation starts at Slide 1 again by pressing the Esc key.

**20** Save the revised presentation using the same name (**PaintedBuntingSelf Running-YourName**) and then close the presentation.

## Topics Review

| Topic | Key Concepts |
|---|---|
| **11.1 Inserting, Resizing, and Aligning Images** | An image in a file on your PC can be added to a slide using the Pictures button in the Images group on the Insert tab or with the Pictures icon in the content placeholder on a new slide. |
| | Resize, move, or edit an image by selecting the image and using the selection handles and/or buttons on the Picture Tools Format tab. |
| | Find an image from a website using the Online Pictures button in the Images group on the Insert tab. |
| **11.2 Inserting a SmartArt Graphic** | A **SmartArt** graphic is made up of shapes with text to illustrate information in lists, processes, cycles, hierarchies, or other diagrams. |
| | Add a SmartArt graphic on a slide using the SmartArt button in the Illustrations group on the Insert tab or the Insert a SmartArt Graphic icon in a content placeholder. |
| | Choose a SmartArt category and layout in the Choose a SmartArt Graphic dialog box and then click OK to enter the text for each item. |
| | Text can be added to shapes in the Text pane or by typing directly in a shape placeholder. |
| | Modify SmartArt styles or colors or edit the graphic using buttons on the SmartArt Tools Design tab. |
| **11.3 Converting Text to SmartArt and Inserting WordArt** | A bulleted list can be converted into a SmartArt graphic using the **Convert to SmartArt button** in the Paragraph group on the Home tab. |
| | **WordArt** is decorative text inside an independent graphic object on a slide. |
| | Create WordArt using the WordArt button in the Text group on the Insert tab. |
| | Type the WordArt text inside the text box and then add text effects, move, or otherwise edit the object using buttons on the Drawing Tools Format tab. |
| **11.4 Creating a Chart on a Slide** | Insert a chart using the Insert Chart icon in the content placeholder or the Chart button in the Illustrations group on the Insert tab. |
| | Choose the chart category and chart type in the Insert Chart dialog box and then click OK. |
| | Type the data to be graphed in the worksheet on top of the slide placeholders. As you replace the sample data in the worksheet, the chart on the slide updates. |
| | Modify the chart using the buttons on the Chart Tools Design and Chart Tools Format tabs. |
| **11.5 Drawing a Shape and Adding a Text Box** | Draw your own graphics on a slide using the **Shapes button** on the Insert tab. |
| | Type text inside a selected shape and then resize, move, or otherwise modify the shape using buttons on the Drawing Tools Format tab. |
| | A text box is a rectangular object with text that can be moved, resized, or formatted independently. |
| | Create a text box using the Text Box button in the Text group on the Insert tab. |
| **11.6 Adding Video to a Presentation** | Add a video clip to a slide with the Insert Video icon in a content placeholder or with the Video button in the Media group on the Insert tab. |
| | Select a video clip from a file on your PC or by finding a video clip at YouTube. |
| | The **Trim Video button** lets you change the starting and ending position of the video if you do not want to play the entire clip in a slide show. |
| | Change the *Start* option in the Video Options group on the Video Tools Playback tab to *Automatically* to begin playing a video as soon as the slide is displayed in a slide show. |
| | Options on the Video Tools Format tab are used to format the video object. |

*continued...*

| Topic | Key Concepts |
|---|---|
| **11.7 Adding Sound to a Presentation** | Add audio to a slide with the **Audio button** in the Media group on the Insert tab. |
| | Use buttons in the Audio Options group on the Audio Tools Playback tab to hide the sound icon during a slide show, start the audio automatically, play the sound in the background across all slides, or loop the audio continuously until the slide is advanced. |
| | Always credit the sources of all multimedia that you did not create yourself used in a presentation. |
| **11.8 Adding Transitions and Animation Effects to Slides for a Slide Show** | A **transition** is a special effect that appears as one slide is removed from the screen and another is revealed during a slide show. |
| | **Animation** causes an object to move or transform in some way. |
| | Select a transition in the Transition to This Slide gallery on the Transitions tab. |
| | The Apply To All button in the Timing group on the Transitions tab sets the same transition effect to all slides. |
| | Add an animation effect to a placeholder on the slide master to apply the effect to all slides in the presentation. |
| | Animation effects are selected in the Animation gallery on the Animations tab. |
| | Specify how the animation will start and the animation duration using options in the Timing group. |
| | Animation effects are grouped by: Entrance, Emphasis, Exit, and Motion Paths. |
| | Animate an individual object on a slide by selecting the object and then adding an animation effect from the Animation gallery. |
| | The **Animation Painter button** in the Advanced Animation group copies the animation effect and effect options from one object to another. |
| **11.9 Setting Up a Self-Running Presentation** | A self-running presentation is set up to run a slide show continuously. |
| | To create a self-running presentation, each slide needs to have a time entered in the *After* text box in the Timing group on the Transitions tab, and each animated object needs to be set to start automatically. |
| | Open the Set Up Show dialog box from the **Set Up Slide Show button** on the Slide Show tab, and then choose *Browsed at a kiosk (full screen)* to instruct PowerPoint to play the slide show continuously until stopped. |

### Topics Review

## Chapter 12

# Using and Querying an Access Database

Organizations and individuals rely on data to complete transactions, analyze reports, and make decisions. Data that is stored in an organized manner to provide information to meet a variety of needs is called a **database**. A database application is referred to as a **database management system (DBMS)**. Microsoft Access is a DBMS used to organize, store, and maintain data that is used mostly by small- to medium-size businesses.

You interact with a DBMS several times a day as you complete your daily activities. Examples of the types of transactions that involve a DBMS include withdrawing cash from your bank account, completing a purchase, looking up a price at a website, or programming your GPS to find a route to an address.

In this chapter, you will learn database terminology and how to navigate a DBMS, including how to open and close objects, add and maintain records using a datasheet and form, find and replace data, sort and filter data, and use queries to look up information and perform a calculation on a numeric field.

## Learning Objectives

**12.1** Identify a database table, query, report, and form; and define *field*, *field value*, and *record*

**12.2** Add a record using a datasheet

**12.3** Edit and delete records using a datasheet

**12.4** Add, edit, and delete records using a form

**12.5** Find and replace data, and adjust column widths in a datasheet

**12.6** Sort and filter records in a datasheet and form

**12.7** Create a query using a wizard

**12.8** Create a query using Design view

**12.9** Select records in a query using criteria

**12.10** Select records in a query using AND and OR criteria, and sort query results

**12.11** Modify a query, add a calculated field to a query, and preview a database object

**Read & Learn**

**Content Online**
The online course includes additional training and assessment resources.

# 12.1 Understanding Database Objects and Terminology

An Access database is structured and organized to keep track of a large amount of similar data. For example, a library database is organized so that information (such as the title, author, and publisher) is maintained for each book in the library catalog. Making sure that the data is entered and updated in the same manner for each item is important so that information is complete and accurate. For this reason, a database is created with a structure that defines the data that will be collected for each item. Working with objects and data in an existing database will help you understand the terms that are used and how data is organized before you create your own database.

## Identifying Database Objects

An Access database is a collection of related objects in which data is recorded, edited, and viewed. Access opens with a Navigation pane along the left side of the window that is used to open an object. Objects are grouped in the Navigation pane by type. Tables, queries, forms, and reports are the four most common types of database objects and are described in Table 12.1.

**Table 12.1** Four Most Common Access Objects

| Object | Description |
|---|---|
| Table | Data is organized into tables, each of which opens in a datasheet that displays data in columns and rows. A table stores data about one topic or subject only. For example, in the LibraryFines database you will use in this chapter, one table contains data about each student, and another table contains data about each fine. |
| Query | A query is used to extract information from one or more tables in a single datasheet that displays all the data or a subset of data that meets a specific condition. For example, a query could display all library fines that have been issued, or only those fines that remain unpaid. |
| Form | A form provides a user-friendly interface to enter or update data where one record is viewed at a time. The layout of a form can be customized; for example, you can customize a layout to match closely an existing paper form used in a business. |
| Report | Reports are used for viewing or printing data from a table or query. Reports can include summary totals and a customized layout. |

① Start Access and open the database **LibraryFines** from the Ch12 folder in StudentDataFiles.

② Click the Enable Content button in the security warning message bar that appears below the ribbon. Click Yes if you receive a prompt asking if you want to make the database a trusted document.

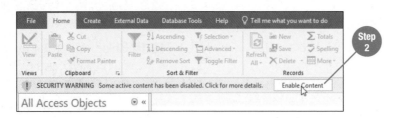

The security warning message bar appears each time you open a database unless the settings for Access on your PC have been changed. Microsoft disables some database content as a way to protect your PC from potentially harmful files that may be embedded in the database without your knowledge (such as a virus). The data files provided with this textbook are safe, so you can click the Enable Content button.

③ Compare your screen with the one shown in Figure 12.1 on the next page.

Watch & Learn

Hands On Activity

● — — — — — — Skills

Open and close a database

Identify objects in a database

Open and close objects

Tutorials

Opening an Existing Database

Closing a Database and Closing Access

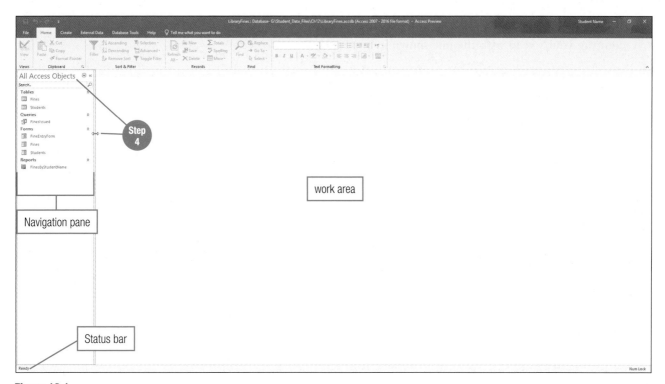

**Figure 12.1**

The **LibraryFines** database is shown opened in the Access window.

4 If necessary, drag the gray border along the right side of the Navigation pane to the right to expand the width of the pane until the title *All Access Objects* is entirely visible.

## Opening and Closing Objects

Data in a database is organized by topic or subject about a person, place, event, item, or other category grouping in an object called a **table**. A database table is the first object that is created. The number of tables varies for each database depending on the information that needs to be stored. Tables are the building blocks for creating other objects, such as a query, form, or report. In other words, you cannot create a query, form, or report without first creating a table.

5 Double-click *Fines* in the Tables group in the Navigation pane.

The table opens in Datasheet view within a tab in the work area. A datasheet resembles a worksheet with data organized in columns and rows. The information about the subject or topic of a table (such as library fines) is divided into columns, each of which is called a **field**. A field should store only one unit of information about a person, place, event, or item. For example, a mailing address is split into at least four fields so the street address, city, state or province, and zip or postal code are separated. This allows the database to be sorted, filtered, or searched by any unit of information.

Each row in the datasheet displays all the fields for one person, place, event, or item and is called a **record**. The data that is stored in one field within a record is called a **field value**.

☁ **Tutorial**

Opening and Closing an Object

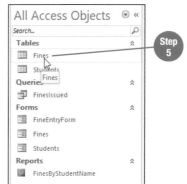

●————— **app tip**

Another way to open any database object is to right-click the object name in the Navigation pane and then click *Open*.

**6** Double-click *Students* in the Tables group in the Navigation pane and then compare your screen with the one shown in Figure 12.2.

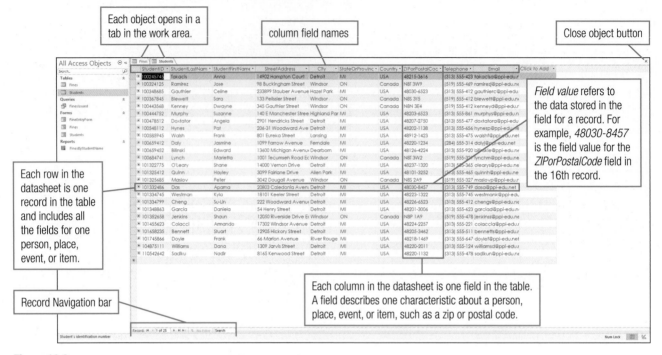

**Figure 12.2**

Shown is the Datasheet view for the Students table in the **LibraryFines** database.

**7** Double-click *FinesByStudentName* in the Reports group in the Navigation pane and review the report content and layout in the work area.

A **report** is designed to view or print data from one or more tables or queries in a customized layout and with summary totals. In this report, library fines are arranged and grouped by student name.

**8** Click the Close button at the upper right of the work area to close the report. Do *not* use the Close button at the upper right of the window, which exits Access.

**9** Double-click *FineEntryForm* in the Forms group in the Navigation pane.

A **form** is used to enter, update, or view one record at a time. Forms can be created that resemble paper-based forms used within an organization.

10  Click the Next record button (right-
pointing arrow) on the Record
Navigation bar located at the bottom of
the form.

Buttons on the Record Navigation bar
are used to move to the first record, previous
record, next record, or last record. Use the
search text box to navigate to a record by
typing a field value.

11  Click the Previous record button (left-
pointing arrow).

12  Click the Last record button (right-
pointing arrow with vertical bar) to
move to the last record in the form.

13  Click the First record button (vertical
bar with left-pointing arrow) to move
to the first record in the form.

14  Click the Close button at the upper
right corner of the work area to close
the form.

15  Double-click *FinesIssued* in the
Queries group in the Navigation
pane.

A **query** opens in a datasheet and displays
data from one or more tables. A query may display
all the records or only a subset of records that meet a specific condition.

**quick steps**

**Open a Database Object**
1. Open database file.
2. Double-click object
   name in Navigation
   pane.

previous record

new (blank) record

first record

last record

search text box

Step 10

| FineID | DateIssued | StudentFirstName | StudentLastName | Telephone | Amount | OverdueBookISBN | BookReturned | FinePaid |
|---|---|---|---|---|---|---|---|---|
| 15-101 | 10-Mar-2021 | Pat | Hynes | (313) 555-6569 | $8.75 | 9780763843011 | ☑ | ☑ |
| 15-102 | 11-Mar-2021 | Angela | Doxtator | (313) 555-4771 | $2.15 | 9780531523541 | ☑ | ☑ |
| 15-103 | 12-Mar-2021 | Pat | Hynes | (313) 555-6569 | $1.25 | 9780412533145 | ☑ | ☐ |
| 15-104 | 12-Mar-2021 | Marietta | Lynch | (519) 555-3214 | $1.40 | 9784123524158 | ☑ | ☐ |
| 15-105 | 13-Mar-2021 | Edward | Bilinski | (313) 555-9200 | $4.25 | 9784125312517 | ☑ | ☑ |
| 15-106 | 13-Mar-2021 | Stuart | Bennett | (313) 555-5112 | $3.25 | 8745125412352 | ☑ | ☐ |
| 15-107 | 15-Mar-2021 | Daniela | Garcia | (313) 555-6235 | $6.50 | 6235841752358 | ☑ | ☑ |
| 15-108 | 15-Mar-2021 | Dwayne | Kenney | (519) 555-4125 | $12.50 | 6498536864758 | ☐ | ☐ |
| 15-109 | 15-Mar-2021 | Daniela | Garcia | (313) 555-6235 | $1.85 | 9774582368541 | ☑ | ☐ |
| 15-110 | 30-Mar-2021 | Jasmine | Daly | (284) 555-3142 | $2.40 | 9774586412352 | ☐ | ☐ |
| 15-111 | 30-Mar-2021 | Pat | Hynes | (313) 555-6569 | $3.25 | 9685412563258 | ☑ | ☑ |
| 15-112 | 04-Apr-2021 | Peter | Maslov | (519) 555-3276 | $7.85 | 3148578685035 | ☑ | ☐ |
| 15-113 | 06-Apr-2021 | Dwayne | Kenney | (519) 555-4125 | $4.60 | 1345685697566 | ☐ | ☐ |
| 15-114 | 08-Apr-2021 | Sara | Blewett | (519) 555-4125 | $3.50 | 8755745862538 | ☑ | ☑ |
| 15-115 | 10-Apr-2021 | Su-Lin | Cheng | (313) 555-4125 | $8.75 | 4574698541255 | ☑ | ☐ |
| 15-116 | 10-Apr-2021 | Anna | Takacis | (313) 555-4235 | $5.60 | 4587456595681 | ☑ | ☑ |
| 15-117 | 12-Apr-2021 | Anna | Takacis | (313) 555-4235 | $2.45 | 6235896485674 | ☐ | ☐ |

A query combines data from more than one table and displays the fields in a specified order.

16  Close the query and the two tables.

Close objects as soon as you are finished viewing or updating data. Some Access
commands will not run if an object is open in the background.

17  Click the File tab and then click *Close*.

Always close a database file using the backstage area before exiting Access so that
all temporary files used by Access while you are viewing and updating records are
properly closed.

## 12.2 Adding Records to a Table Using a Datasheet

To add a new record to a table, open the table and click the **New (blank) record button** on the Record Navigation bar. Type the field values for the new record in each column, pressing Tab or Enter to move to the next column (field) in the datasheet. When you move past the last column in a new row in the datasheet, the record is automatically saved and another new blank record is open.

**1** Open the database **LibraryFines**.

**2** Click the File tab and then click *Save As* to open the Save As backstage area.

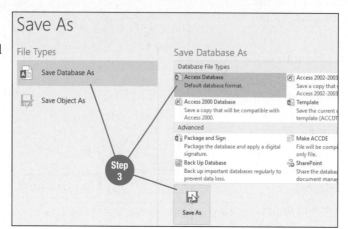

**3** With *Save Database As* already selected in the *File Types* panel and *Access Database* already selected as the *Save Database As* option, click the Save As button.

**4** In the Save As dialog box, navigate to the CompletedTopics folder on your storage medium, create a new folder *Ch12*, double-click the Ch12 folder, change the file name to **LibraryFines-YourName**, and then click the Save button.

**5** Click the Enable Content button on the security warning message bar.

**6** Double-click the Fines table to open the table.

**7** Click the New (blank) record button on the Record Navigation bar.

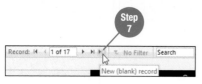

New (blank) record

**8** Type 15-118 in the *FineID* field and then press Tab to move to the next field.

The field *StudentID* looks up names and ID numbers in the Students table using a drop-down list.

**9** Click the down-pointing arrow in the *StudentID* field in the new record field just created, click *100478512 Angela Doxtator*, and then press Tab.

The student ID number becomes the field value, which is connected to Angela Doxtator's record in the Students table.

The Pencil icon indicates the record is being edited. The pencil disappears when Access saves the changes.

| Fines | | | | |
|---|---|---|---|---|
| FineID | StudentID | DateIssue | Amount | Over |
| 15-101 | 100548112 | 10-Mar-2021 | $8.75 | 97807 |
| 15-102 | 100478512 | 11-Mar-2021 | $2.15 | 97805 |
| 15-103 | 100548112 | 12-Mar-2021 | $1.25 | 97804 |
| 15-104 | 100684741 | 12-Mar-2021 | $1.40 | 97841 |
| 15-105 | 100659452 | 13-Mar-2021 | $4.25 | 97841 |
| 15-106 | 101658235 | 13-Mar-2021 | $3.25 | 87451 |
| 15-107 | 101348863 | 15-Mar-2021 | $6.50 | 62358 |
| 15-108 | 100443568 | 15-Mar-2021 | $12.50 | 64985 |
| 15-109 | 101348863 | 15-Mar-2021 | $1.85 | 97745 |
| 15-110 | 100659412 | 30-Mar-2021 | $2.40 | 97745 |
| 15-111 | 100548112 | 30-Mar-2021 | $3.25 | 96854 |
| 15-112 | 101325685 | 04-Apr-2021 | $7.85 | 31485 |
| 15-113 | 100443568 | 06-Apr-2021 | $4.60 | 13456 |
| 15-114 | 100367845 | 08-Apr-2021 | $3.50 | 87557 |
| 15-115 | 101334799 | 10-Apr-2021 | $8.75 | 45746 |
| 15-116 | 100245745 | 10-Apr-2021 | $5.60 | 45874 |
| 15-117 | 100245745 | 12-Apr-2021 | $2.45 | 62358 |
| 15-118 | | | $0.00 | |

| StudentID | | |
|---|---|---|
| 101658235 | Stuart | Bennett |
| 100659452 | Edward | Bilinski |
| 100367845 | Sara | Blewett |
| 101334799 | Su-Lin | Cheng |
| 101455623 | Armando | Colacci |
| 100659412 | Jasmine | Daly |
| 101332486 | Aparna | Das |
| 100478512 | Angela | Doxtator |
| 101745866 | Frank | Doyle |
| 101348863 | Daniela | Garcia |
| 100348685 | Celine | Gauthier |
| 100548112 | Pat | Hynes |
| 101352658 | Shaun | Jenkins |
| 100443568 | Dwayne | Kenney |
| 100684741 | Marietta | Lynch |
| 101325685 | Peter | Maslov |

Step 8

Step 9

**10** Type 12apr2021 in the *DateIssued* field and then press Tab.

The date field has been set up to display underscores and dashes as soon as you begin typing to help you enter the date in the correct format, *dd-mmm-yyyy*. This configuration also ensures that all dates are entered consistently in the database. Access displays the date as *12-Apr-2021*, which is the format set up for date fields in the database.

**11** Type 8.75 in the *Amount* field and then press Tab.

**12** Type 4348973098226 in the *OverdueBookISBN* field and then press Tab.

**quick steps**

**Add New Record**
1. Open table.
2. Click New (blank) record button.
3. Type field values in new row in datasheet.
4. Close table.

| Step 10 | Step 11 | Step 12 | | | |
|---|---|---|---|---|---|
| 15-116 | 100245745 | 10-Apr-2021 | $5.60 | 4587456595681 | ☑ |
| 15-117 | 100245745 | 12-Apr-2021 | $2.45 | 6235896485674 | ☐ |
| 15-118 | 100478512 | 12-Apr-2021 | $8.75 | 4348973098226 | ☐ |
| * | | | $0.00 | | ☐ |

**13** Press the spacebar to insert a check mark in the *BookReturned* field and then press Tab.

*BookReturned* is a field that has been set up to store only one of two possible field values: *Yes* or *No*. Inserting a check mark stores *Yes*, while an empty check box stores *No*.

**14** Click the check box to insert a check mark in the *FinePaid* field and then press Tab.

**15** Type 15apr2021 in the *DatePaid* field and then press Tab.

Moving to the next row in the datasheet automatically saves the record just typed and starts a new record.

**app tip**

Saving in a database is not left to chance! As soon as you complete a new record, Access saves the data to disk.

**16** Add the following field values in the new row in the fields indicated, pressing Tab after typing the data to move to the next field:

| *FineID* | 15-119 | *OverdueBookISBN* | 7349872345760 |
|---|---|---|---|
| *StudentID* | 101348863 Daniela Garcia | *BookReturned* | Yes |
| *DateIssued* | 15apr2021 | *FinePaid* | No (leave blank) |
| *Amount* | 5.25 | *DatePaid* | (leave blank) |

| 15-118 | 100478512 | 12-Apr-2021 | $8.75 | 4348973098226 | ☑ | ☑ | 15-Apr-2021 | |
| 15-119 | 101348863 | 15-Apr-2021 | $5.25 | 7349872345760 | ☑ | ☐ | | Step 16 |
| * | | | $0.00 | | ☐ | ☐ | | |

**17** With the insertion point positioned in the *FineID* field in a new row, close the Fines table. Leave the database open for the next topic.

## 12.3 Editing and Deleting Records Using a Datasheet

Edit a field value in a table using a datasheet by opening the table, selecting the text to be changed, and then typing the new text, or by clicking in the table cell to place an insertion point and then inserting or deleting text as required. Select a record for deletion by clicking in the gray record selector bar along the left edge of the datasheet next to the record and then clicking the Delete button in the Records group on the Home tab. Access requires confirmation before deleting a record.

1. With the **LibraryFines-YourName** database open, double-click *Students* in the Tables group to open the table.

2. Select the text *Murphy* in the *StudentLastName* column in the sixth row in the datasheet and then type Hall as the new last name for Suzanne.

| Students | | |
|---|---|---|
| StudentID ▾ | StudentLastNam ▾ | StudentFirstName ▾ |
| ⊞ 100245745 | Takacis | Anna |
| ⊞ 100324125 | Ramirez | Jose |
| ⊞ 100348685 | Gauthier | Celine |
| ⊞ 100367845 | Blewett | Sara |
| ⊞ 100443568 | Kenney | Dwayne |
| ⊞ 100444752 | Hall | Suzanne |
| ⊞ 100478512 | Doxtator | Angela |

Step 2

3. Press Tab eight times to move to the *Email* field.

4. Press F2 to open the field for editing, move the insertion point as needed, delete *murphy* at the beginning of the email address, and then type hall so that the email address becomes *halls@ppi-edu.net*.

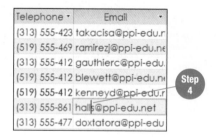

| Telephone ▾ | Email ▾ |
|---|---|
| (313) 555-423 | takacisa@ppi-edu.r |
| (519) 555-469 | ramirezj@ppi-edu.ne |
| (313) 555-412 | gauthierc@ppi-edu. |
| (519) 555-412 | blewett@ppi-edu.ne |
| (519) 555-412 | kenneyd@ppi-edu.r |
| (313) 555-861 | halls@ppi-edu.net |
| (313) 555-477 | doxtatora@ppi-edu |

Step 4

5. Click at the end of *N8T 3W9* in the *ZIPorPostalCode* field in the second row in the datasheet to position the insertion point, press Backspace to remove *3W9*, and then type 2E6.

Changes made to field values in a record are automatically saved as soon as you place an insertion point in another record in the datasheet.

| Country ▾ | ZIPorPostalCoc ▾ | Telephone ▾ |
|---|---|---|
| USA | 48215-3616 | (313) 555-423 |
| Canada | N8T 2E6 | (519) 555-469 |
| USA | 48030-6523 | (313) 555-412 |

Step 5

6. If necessary, scroll left until you can see the student names.

7. Click in the record selector bar (gray bar at left edge of datasheet) next to the record for the student *Das Aparna* when you see the mouse pointer change to a right-pointing black arrow (➜).

The gray bar at the left edge of the datasheet is used to select a record. The mouse pointer displays as a black right-pointing arrow when positioned next to a record in the record selector bar. The top of the record selector bar next to the first field name is called the **Select all button**. Clicking the Select all button selects all field values in all records in the table.

Select all button

Record selector bar

Step 7

8   Click the Delete button in the Records group on the Home tab. Do *not* click the down-pointing arrow on the button.

Step 8

9   Click Yes at the Microsoft Access message box that appears asking whether you are sure you want to delete the record.

Microsoft Access

**You are about to delete 1 record(s).**

If you click Yes, you won't be able to undo this Delete operation. Are you sure you want to delete these records?

Yes     No

Step 9

Always exercise caution when you use the Delete command, because the Undo feature does not work to restore a record. Depending on the purpose of the database, deleting records is generally not done until the records to be deleted are first copied to an archive database (a copy of the database saved for historical record keeping). Alternatively, a database administrator may make an extra backup copy of the database for safekeeping before deleting records. In some workplaces, the delete option is made unavailable to all employees except a few authorized database administrators.

10   Close the Students table. Leave the database open for the next topic.

## 12.4 Adding, Editing, and Deleting Records Using a Form

A form is an Access object that provides a different view for the data stored in a table. Generally only one record at a time is displayed in a columnar layout instead of the spreadsheet-style layout presented in a datasheet. Forms are usually preferred over a datasheet for adding, editing, and deleting records because the user can focus on the data in one record at a time.

1. With the **LibraryFines-YourName** database open, double-click *FineEntryForm* in the Forms group to open the form.

2. Click the New (blank) record button on the Record Navigation bar.

3. Add the field values as shown in the image below, using Tab to move from one field to the next field. After typing the last entry, press Tab.

   A new blank form displays when you press Tab after the last field in a form.

4. Click the First record button on the Record Navigation bar to display the first record in the form.

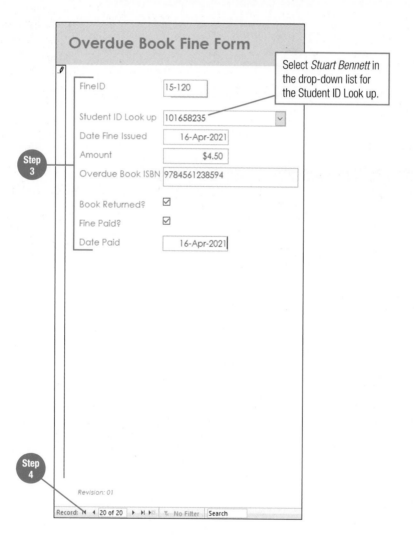

 **Watch & Learn**

**Hands On Activity**

━━━━━━━ skills

Add a record using a form

Edit a record using a form

Delete a record using a form

●━━━━━━ app tip

Print the current record displayed in a form by displaying the Print backstage area, clicking *Print*, and then clicking *Selected Record(s)* in the *Print Range* section of the Print dialog box.

**Tutorials**

Adding and Deleting Records in a Form

Navigating in Objects

**5** Select $8.75 in the *Amount* field and then type 7.25.

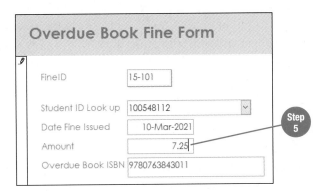

**quick steps**

**Delete a Record Using a Form**
1. Open form.
2. Navigate to desired record.
3. Click Delete button arrow.
4. Click *Delete Record*.
5. Click Yes.

**6** Click the Next record button two times to display record 3 in the form.

**7** Click the Delete button arrow in the Records group on the Home tab and then click *Delete Record*.

**Oops!**

Only the current field value is deleted? This occurs when you do not use the arrow on the button to select the option to delete the entire record. Try Step 7 again, making sure to choose *Delete Record*.

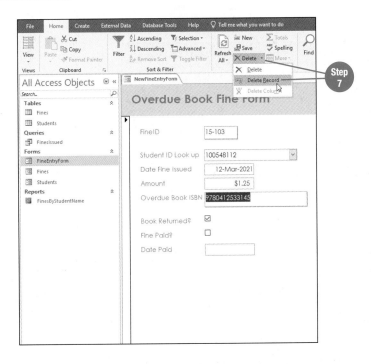

**8** Click Yes at the message box that appears asking whether you are sure you want to delete the record.

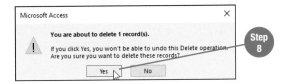

**9** Close the form. Leave the database open for the next topic.

## 12.5 Finding and Replacing Data and Adjusting Column Widths in a Datasheet

The Find feature locates a field value in a datasheet or form and moves the insertion point to each occurrence of the data. Find is helpful to locate a record that you need to edit when the datasheet contains hundreds of records. When a change needs to be made to all occurrences of a field value, use the Replace command to make the change automatically. The column width for a column in a datasheet can be made wider to fully display data for those fields that do not currently display all the field values.

**1** With the **LibraryFines-YourName** database open, double-click *Fines* in the Tables group to open the table.

**2** Click to place an insertion point within the *StudentID* field value in the first record.

**3** Click the Find button in the Find group on the Home tab.

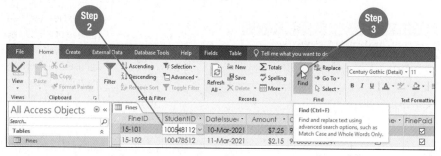

**4** Type 101348863 in the *Find What* text box and then click the Find Next button.

The first record (record 6) that matches the field value is made active.

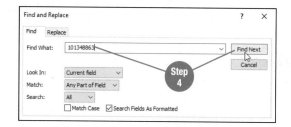

**5** Continue clicking the Find Next button to review all occurrences of the matching field value.

**6** Click OK at the message that Microsoft Access has finished searching records.

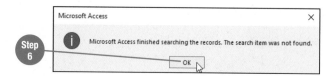

**7** Click the Cancel button to close the Find and Replace dialog box.

**8** Click to place an insertion point within the *FineID* field in the first record.

**Watch & Learn**

**Hands On Activity**

**———— Skills**

Find and replace data using a datasheet

Adjust column width in a datasheet

**Tutorials**

Finding Data

Finding and Replacing Data

Adjusting Field Column Width

**—————— Oops!**

No records found? Click OK and then check that you typed the ID number without errors.

⑨ Click the Replace button in the Find group.

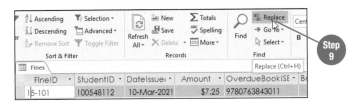

⑩ Type 15- in the *Find What* text box and then press Tab.

⑪ Type MCL- in the *Replace With* text box.

⑫ Click the *Match* option box arrow and then click *Any Part of Field*.

⑬ Click the Replace All button.

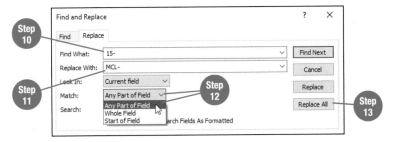

⑭ Click Yes at the message box asking whether you want to continue and informing you that the Replace operation cannot be undone.

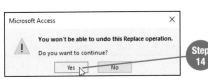

⑮ Click the Close button in the top right corner of the Find and Replace dialog box.

⑯ Click to place the insertion point in any field value within the *DateIssued* field.

⑰ Click the More button in the Records group on the Home tab and then click *Field Width*.

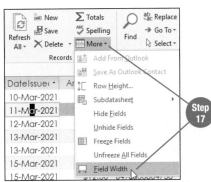

⑱ Click Best Fit in the Column Width dialog box.

**Best Fit** adjusts the width of the column to accommodate the length needed to display the longest field value.

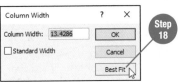

⑲ Close the Fines table. Click Yes when prompted to save the changes to the layout of the table. Leave the database open for the next topic.

Saving changes to the layout of the table means that Access will retain the new column width for the *DateIssued* field when the table is reopened.

**quick steps**

**Find a Record**
1. Open table or form.
2. Click in field to be searched.
3. Click Find button.
4. Type field value in *Find What* text box.
5. Click Find Next until done.
6. Click OK.
7. Close dialog box.

**Replace a Field Value**
1. Open table or form.
2. Click in field to be searched.
3. Click Replace button.
4. Type field value in *Find What* text box.
5. Type new field value in *Replace With* text box.
6. Click Replace or Replace All, as needed.
7. Click Yes.
8. Close dialog box.

**Adjust Column Width**
1. Open table.
2. Click in any record in column.
3. Click More button.
4. Click *Field Width*.
5. Type value for column width or click Best Fit.

**app tip**

A column width can also be widened or narrowed by dragging the right column boundary in the field names row right or left. Double-click the column boundary to best fit the column width.

# 12.6 Sorting and Filtering Records

Records are initially arranged in the datasheet alphanumerically by the field in the table that has been defined as the primary key. A **primary key** is a field that contains the data that uniquely identifies each record in the table. Generally, the primary key is an identification number, such as *StudentID* in the Students table. To change the order of the records, click in the column to sort and then use the Ascending or Descending buttons in the Sort & Filter group on the Home tab.

**1** With the **LibraryFines-YourName** database open, double-click *Students* in the Tables group.

The primary key field in the Students table is the field named *StudentID*. Notice the records in the datasheet are arranged in order of the StudentID field values.

**2** Click to place an insertion point within any field value in the *StudentLastName* column.

**3** Click the Ascending button in the Sort & Filter group on the Home tab.

Notice the records in the table are now arranged in order by the student last name field values.

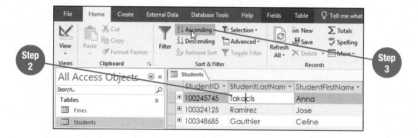

**4** Close the Students table. Click Yes when prompted to save the changes to the design of the table.

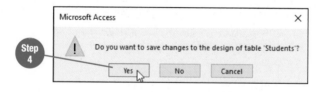

Selecting Yes to save changes to the design of the table means that the table will remain sorted by the *StudentLastName* field when you reopen the datasheet. Each object based upon the Students table can have its own sort option saved.

**5** Open the Students form.

**6** Click the Next record button a few times to view the first few records. Notice the records are arranged by *StudentID*.

**7** Click the First record button to return the display to the first record.

**8** Click to place an insertion point in the *Student Last Name* field.

**9** Click the Ascending button in the Sort & Filter group.

---

**Watch & Learn**

**Hands On Activity**

● – – – – – – Skills

Sort records

Filter records

● – – – – – app tip

When a datasheet is sorted by a field other than the primary key, an up-pointing arrow (ascending order indicator) or down-pointing arrow (descending order indicator) displays next to the field name used to sort.

**Tutorials**

Sorting Records in a Table

Filtering Records

**10** Scroll through the first 10 records in the form to view the sorted order and then close the form.

**11** Open the Students table. Notice the up-pointing arrow next to *StudentLastName* in the field names row, indicating the records are arranged alphabetically in ascending order by the student last names.

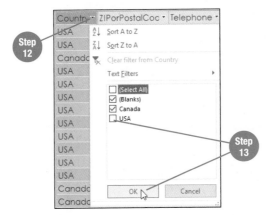

You can filter a datasheet in Access using the same techniques you learned for filtering a table in Excel in Chapter 9. Recall that a filter temporarily hides the rows that you do not want to view.

**12** Click the filter arrow (down-pointing arrow) next to *Country*.

**13** Click the check box next to *USA* to clear the check mark from the box and then click OK.

The datasheet is filtered to display records for students who reside in Canada only. The indicators that a datasheet is filtered include a funnel icon next to the filtered field name, the message *Filtered* (shaded with an orange background) on the Record Navigation bar, and also at the right end of the Status bar along the bottom of the window.

**14** Click the Toggle Filter button in the Sort & Filter group to clear the filter, redisplaying all records in the datasheet.

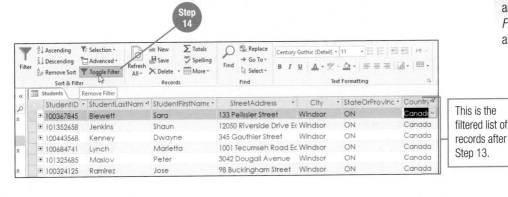

This is the filtered list of records after Step 13.

**15** Close the Students table. Click No when prompted to save changes to the design of the table. Leave the database open for the next topic.

Note that a sort order can be saved with a datasheet, but a filter option is not saved when the table is closed.

# 12.7 Creating a Query Using the Simple Query Wizard

Queries extract information from one or more tables in the database and display the results in a datasheet. Some queries display fields from more than one table in the same datasheet. For example, in the LibraryFines database, the student names are in one table and the fines are in another table; a query can combine the names and fines in one datasheet. Other queries are designed to answer a question about the data; for example, *Which library fines are unpaid?* The **Simple Query Wizard** assists a user with creating a query by prompting the user to make selections in a series of dialog boxes.

**1** With the **LibraryFines-YourName** database open, click the Create tab.

**2** Click the Query Wizard button in the Queries group.

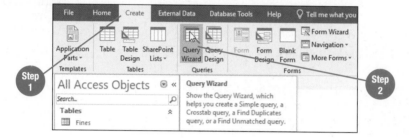

**3** Click OK in the New Query dialog box with *Simple Query Wizard* already selected.

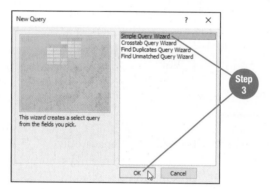

**4** In the first Simple Query Wizard dialog box, click the *Tables/Queries* option box arrow and then click *Table: Students*. Skip this step if *Table: Students* is already displayed in the *Tables/Queries* option box.

The first step in creating a query is to choose the tables or queries and the fields from each table or query that you want to display in a datasheet.

**5** With *StudentID* already selected in the *Available Fields* list box, click the Add Field button (displays as a right-pointing arrow) to move *StudentID* to the *Selected Fields* list box.

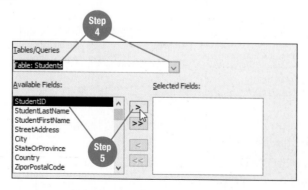

---

**Watch & Learn**

**Hands On Activity**

— — — — — — **Skills**

Create a query using a wizard

**Tutorial**
Creating a Query Using the Simple Query Wizard

— — — — — — **Oops!**

Added the wrong field or the order of fields is wrong? Move a field back to the *Available Fields* list box by selecting the field and using the Remove Field button (left-pointing arrow).

**6** Double-click *StudentFirstName* in the *Available Fields* list box to move the field to the *Selected Fields* list box.

Add fields to the *Selected Fields* list box in the order you want the fields displayed in the query results datasheet.

**7** Double-click the following fields in the *Available Fields* list box to move each field to the *Selected Fields* list box.

> *StudentLastName*
> *Telephone*
> *Email*

**8** Click the *Tables/Queries* option box arrow and then click *Table: Fines*.

**9** Double-click the following fields in the *Available Fields* list box to move each field to the *Selected Fields* list box.

> *DateIssued*
> *Amount*
> *FinePaid*

**10** Click Next.

**11** Click Next in the second Simple Query Wizard dialog box to accept *Detail (shows every field of every record)* for the query results.

**12** Select the current text in the *What title do you want for your query?* text box in the third Simple Query Wizard dialog box, type StudentsWithFines, and then click Finish.

**13** Review the query results datasheet. The first five rows are shown in the image below. Notice the fields are displayed in the order selected in the first Simple Query Wizard dialog box.

**14** Close the StudentsWithFines query. Leave the database open for the next topic.

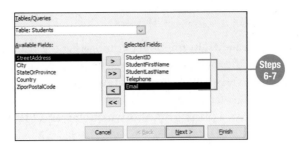

Steps 6-7

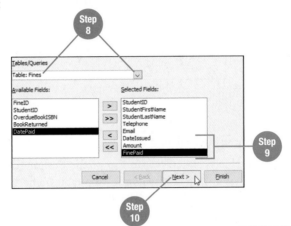

Step 8

Step 9

Step 10

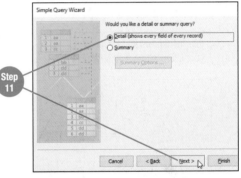

Step 11

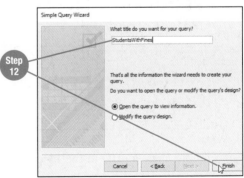

Step 12

**quick steps**

**Create a Query Using Simple Query Wizard**
1. Click Create tab.
2. Click Query Wizard button.
3. Click OK.
4. Choose each table and/or query and fields in required order.
5. Click Next.
6. Click Next.
7. Type title for query.
8. Click Finish.

first five rows of query results datasheet

| StudentID | StudentFirstName | StudentLastName | Telephone | Email | DateIssued | Amount | FinePaid |
|---|---|---|---|---|---|---|---|
| 100245745 | Anna | Takacis | (313) 555-4235 | takacisa@ppi-edu.net | 10-Apr-2021 | $5.60 | ☑ |
| 100245745 | Anna | Takacis | (313) 555-4235 | takacisa@ppi-edu.net | 12-Apr-2021 | $2.45 | ☐ |
| 100367845 | Sara | Blewett | (519) 555-4125 | blewett@ppi-edu.net | 08-Apr-2021 | $3.50 | ☑ |
| 100443568 | Dwayne | Kenney | (519) 555-4125 | kenneyd@ppi-edu.net | 15-Mar-2021 | $12.50 | ☐ |
| 100443568 | Dwayne | Kenney | (519) 555-4125 | kenneyd@ppi-edu.net | 06-Apr-2021 | $4.60 | ☐ |

# 12.8 Creating a Query Using Design View

Every Access object has at least two views. In one view, you browse the data in the table, query, form, or report. This is the view that is active when you open the object from the Navigation pane. Another view, called **Design view**, is used to set up or define the structure and/or layout of a table, query, form, or report. A query can be created in Design view, which displays a blank grid into which you add the fields you want to display in the query results.

1. With the **LibraryFines –YourName** database open and with the Create tab active, click the Query Design button in the Queries group.

2. In the Show Table dialog box with the *Fines* table selected, click the Add button.

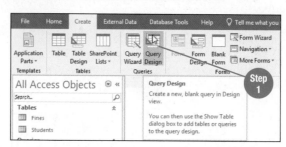

A field list box for the Fines table is added to the top of the *Query1* design grid in the work area.

3. Double-click *Students* in the Show Table dialog box.

A field list box for the Students table is added to the top of the design grid beside the Fines table field list box. A black join line connects the two tables. The black line displays 1 and an infinity symbol (∞), which indicates the type of relationship for the two tables. You will learn about relationships in the next chapter.

4. Click the Close button in the Show Table dialog box.

join line (shows tables are connected with a relationship)

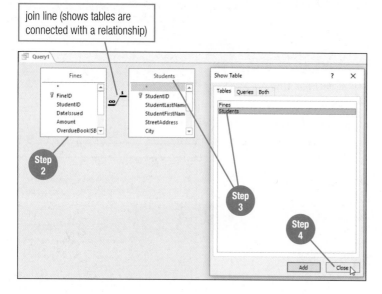

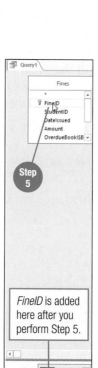

*FineID* is added here after you perform Step 5.

5. Double-click *FineID* in the Fines table field list box.

*FineID* is added to the *Field* text box in the first column of the design grid. The blank columns in the bottom of the window represent the query results datasheet. Build a query by adding fields to the blank columns in the order you want them to appear in the datasheet. Double-clicking a field name in a table field list box in the top half of the window adds a field to the next available column, as you saw when you completed Step 5.

Watch & Learn

Hands On Activity

━ ━ ━ ━ ━ ━ Skills

Create a query using Design view

Tutorials

Creating a Query in Design View

Creating a Query in Design View Using Multiple Tables

●━━━━━ Oops!

Closed the Show Table dialog box by mistake? Reopen the dialog box using the Show Table button in the Query Setup group on the Query Tools Design tab.

6  Double-click the following fields in the Fines table field list box to add each field to the next available column in the query design grid.

   *DateIssued*
   *Amount*
   *BookReturned* (scroll down the table field list box to the field)

7  Double-click the following fields in the Students table field list box to add each field to the next available column in the query design grid.

   *StudentFirstName*
   *StudentLastName*

**quick steps**

**Create a Query Using Design View**
1. Click Create tab.
2. Click Query Design button.
3. Add tables to design grid.
4. Close Show Table dialog box.
5. Double-click field names in table field list boxes in desired order for query results datasheet.
6. Click Run button.
7. Click Save button.
8. Type query name.
9. Click OK.

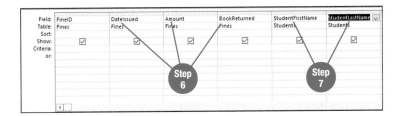

8  Click the Run button in the Results group on the Query Tools Design tab to view the query results datasheet.

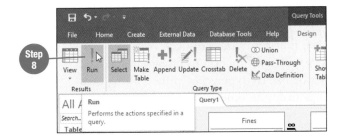

The **Run button** instructs Access to produce the query results datasheet by assembling the data from the tables according to the query instructions. A query is a set of instructions with table names and field names to display in a datasheet. The query results datasheet is not a duplicate copy of the data—each time a query is opened or run, the data is generated by extracting the field values from the tables. The first five rows in the query results datasheet are shown in the image below.

| FineID | DateIssued | Amount | BookReturned | StudentFirstName | StudentLastName |
|---|---|---|---|---|---|
| MCL-116 | 10-Apr-2021 | $5.60 | ☑ | Anna | Takacis |
| MCL-117 | 12-Apr-2021 | $2.45 | ☐ | Anna | Takacis |
| MCL-114 | 08-Apr-2021 | $3.50 | ☑ | Sara | Blewett |
| MCL-108 | 15-Mar-2021 | $12.50 | ☐ | Dwayne | Kenney |
| MCL-113 | 06-Apr-2021 | $4.60 | ☐ | Dwayne | Kenney |

first five rows of query results datasheet

9  Click the Save button on the Quick Access Toolbar.

10 Type FinesWithBookReturnList in the Save As dialog box and then press Enter or click OK.

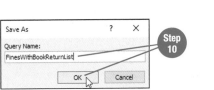

11 Close the FinesWithBookReturnList query. Leave the database open for the next topic.

## 12.9 Entering Criteria to Select Records in a Query

Both query results datasheets for the queries you created using the Simple Query Wizard and using Design view displayed all records in the tables. Queries are often created to select records from tables that meet one or more conditions. For example, in this topic, you will add a criterion to display only those records in which the fines are unpaid. A query that extracts records based on criteria is called a **select query**.

1  With the **LibraryFines-YourName** database open, double-click the *StudentsWithFines* query.

2  Click the top part of the View button in the Views group on the Home tab.

The View button is used to switch between the query results datasheet and Design view.

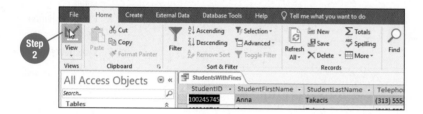

3  Click in the *Criteria* box in the *FinePaid* column in the design grid, type No, press the spacebar, and then press Enter.

Access displays functions in a drop-down list as you type text that matches the letters in a function name. As you type *No*, the function wizard displays *Now* in a drop-down list. Typing a space after *No* causes the *Now* function to disappear. *FinePaid* is a field in which the field value is either *Yes* or *No*. By typing *No* in the *Criteria* box, you are instructing Access to select the records from the Fines table in which *No* is the field value for *FinePaid*.

| Email | DateIssued | Amount | FinePaid |
| Students | Fines | Fines | Fines |
| ☑ | ☑ | ☑ | ☑ |
| | | | No |

Step 3

4  Click the Run button.

Notice that 10 records are selected in the query results datasheet and that the check box in the *FinePaid* column for each record is empty.

| StudentID | StudentFirstName | StudentLastName | Telephone | Email | DateIssued | Amount | FinePaid |
|---|---|---|---|---|---|---|---|
| 100245745 | Anna | Takacis | (313) 555-4235 | takacisa@ppi-edu.net | 12-Apr-2021 | $2.45 | ☐ |
| 100443568 | Dwayne | Kenney | (519) 555-4125 | kenneyd@ppi-edu.net | 15-Mar-2021 | $12.50 | ☐ |
| 100443568 | Dwayne | Kenney | (519) 555-4125 | kenneyd@ppi-edu.net | 06-Apr-2021 | $4.60 | ☐ |
| 100659412 | Jasmine | Daly | (284) 555-3142 | dalyj@ppi-edu.net | 30-Mar-2021 | $2.40 | ☐ |
| 100684741 | Marietta | Lynch | (519) 555-3214 | lynchm@ppi-edu.net | 12-Mar-2021 | $1.40 | ☐ |
| 101325685 | Peter | Maslov | (519) 555-3276 | maslovp@ppi-edu.net | 04-Apr-2021 | $7.85 | ☐ |
| 101334799 | Su-Lin | Cheng | (313) 555-4125 | chengs@ppi-edu.net | 10-Apr-2021 | $8.75 | ☐ |
| 101348863 | Daniela | Garcia | (313) 555-6235 | garciad@ppi-edu.net | 15-Mar-2021 | $1.85 | ☐ |
| 101348863 | Daniela | Garcia | (313) 555-6235 | garciad@ppi-edu.net | 15-Apr-2021 | $5.25 | ☐ |
| 101658235 | Stuart | Bennett | (313) 555-5112 | bennetts@ppi-edu.net | 13-Mar-2021 | $3.25 | ☐ |

This is the query results datasheet displaying unpaid fines only.

**Watch & Learn**

**Hands On Activity**

————— Skills

Select records in a query using criteria

**Tutorial**
Adding a Criteria Statement to a Query

————— Oops!

Empty datasheet? Use the View button to return to Design view and check that *No* is in the *Criteria* box of the *FinePaid* column. A spelling mistake will cause an empty datasheet.

5. Click the File tab and then click *Save As*.

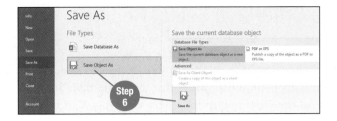

6. Click *Save Object As* in the *File Types* panel and then click the Save As button in the *Save the current database object* panel.

7. Type UnpaidFines in the *Save 'StudentsWithFines' to* text box in the Save As dialog box and then press Enter or click OK.

8. Close the UnpaidFines query.

9. Double-click the *StudentsWithFines* query.

10. Click the View button to switch to Design View.

11. Click in the *Criteria* box in the *DateIssued* column, type Between March 1, 2021 and March 31, 2021 and then press Enter.

12. Position the mouse pointer on the *DateIssued* right column boundary line in the gray bar above the columns and drag to the right when the mouse pointer displays with a black vertical bar and a left- and right-pointing arrow until the entire criteria statement is visible.

A criteria statement added to a date field is automatically converted to the date format *m/d/yyyy*, and a pound symbol (#) is added to the beginning and end of each date. Other ways to select records by date use the greater than (>) or less than (<) operator. For example, >*March 1, 2021* displays as >*#3/1/2021#* and selects all records in the table with a date *after* March 1, 2021. Table 12.2 provides more criteria statement examples.

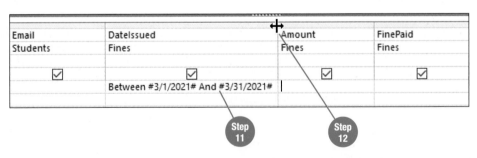

13. Click the Run button. Notice only records with a fine issued in March 2021 are displayed.

14. Use *Save Object As* to save the revised query as *March2021Fines*.

15. Close the March 2021Fines query. Leave the database open for the next topic.

**quick steps**

**Select Records in a Query**
1. Open query.
2. Click View button.
3. Type criterion in *Criteria* box of field by which to select records.
4. Run query.
5. Save query or use Save As to save revised query using new name.

**Oops!**

Syntax error message appears? A typing mistake in keywords *Between* or *And* or a typing error in the date is called a *syntax error*. Click OK to close the message and correct typing errors in the criteria statement.

**Table 12.2** Criteria Examples

| Field | Entry Typed in *Criteria* Box | Records Selected |
|---|---|---|
| *Amount* | <=5 | Fines issued that were $5.00 or less |
| *Amount* | >5 | Fines issued that were more than $5.00 |
| *DateIssued* | March 15, 2021 (entry converts automatically to *#3/15/2021#*) | Fines issued on March 15, 2021 |
| *StudentLastName* | Kenney (entry converts automatically to "Kenney") | Fines issued to student with the last name *Kenney* |

## 12.10 Entering Multiple Criteria to Select Records and Sorting a Query

Watch & Learn

Hands On Activity

●――――――――― Skills

Select records using *AND* criteria

Select records using *OR* criteria

Sort query results

Tutorials

Designing a Query with an AND Criteria Statement

Designing a Query with an OR Criteria Statement

Sorting Data and Hiding Fields in Query Results

More than one criterion can be entered in the query design grid to select records. For example, you may want a list of all unpaid fines that are more than $5.00. When more than one criterion is on the same row in the query design grid, it is referred to as an *AND* statement, meaning that each criterion must be met for a record to be selected. When more than one criterion is on different rows in the query design grid, it is referred to as an *OR* statement, meaning that any criterion can be met for a record to be selected.

1 With the **LibraryFines-YourName** database open, double-click the *UnpaidFines* query to open the query.

2 Click the top part of the View button in the Views group on the Home tab to switch to Design view.

3 Click in the *Criteria* box in the *Amount* column, type >5, and then press Enter.

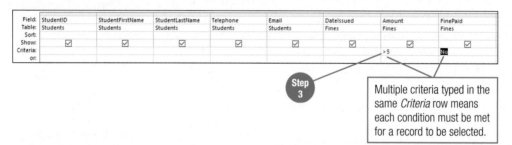

Step 3

Multiple criteria typed in the same *Criteria* row means each condition must be met for a record to be selected.

4 Click the Run button. Compare your results with the datasheet shown in the image below.

| StudentID | StudentFirstName | StudentLastName | Telephone | Email | DateIssued | Amount | FinePaid |
|---|---|---|---|---|---|---|---|
| 100443568 | Dwayne | Kenney | (519) 555-4125 | kenneyd@ppi-edu.net | 15-Mar-2021 | $12.50 | ☐ |
| 101325685 | Peter | Maslov | (519) 555-3276 | maslovp@ppi-edu.net | 04-Apr-2021 | $7.85 | ☐ |
| 101334799 | Su-Lin | Cheng | (313) 555-4125 | chengs@ppi-edu.net | 10-Apr-2021 | $8.75 | ☐ |
| 101348863 | Daniela | Garcia | (313) 555-6235 | garciad@ppi-edu.net | 15-Apr-2021 | $5.25 | ☐ |

This is the query results datasheet displaying unpaid fines over $5.00.

5 Use *Save Object As* to save the revised query as *UnpaidFinesOver$5*.

6 Close the UnpaidFinesOver$5 query.

7 Click the Create tab and then click the Query Design button.

8 Double-click *Students* in the Show Table dialog box and then click the Close button.

9 Double-click the following fields in the Students table field list box to add each field to the next available column in the query design grid.

> *City*
> *StudentID*
> *StudentFirstName*
> *StudentLastName*
> *Telephone* (scroll down the table field list box to the field)

| Field: | City | StudentID | StudentFirstName | StudentLastName | Telephone |
|---|---|---|---|---|---|
| Table: | Students | Students | Students | Students | Students |
| Sort: | | | | | |
| Show: | ☑ | ☑ | ☑ | ☑ | ☑ |

Step 9

10  Click in the *Criteria* box in the *City* column, type Detroit, click in the row below
*Detroit* (next to *or*), type Windsor, and then press Enter.

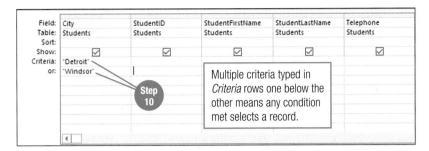

Step 10

Multiple criteria typed in
*Criteria* rows one below the
other means any condition
met selects a record.

●---- quick steps

**Select Records Using
AND**
1.  Open query in Design
    view.
2.  Type criterion in
    *Criteria* box of first field
    by which to select.
3.  Type criterion in
    *Criteria* box of second
    field by which to select.
4.  Run query.
**Select Records Using OR**
1.  Open query in Design
    view.
2.  Type criterion in
    *Criteria* box of first field
    by which to select.
3.  Type criterion in *or*
    box of second field by
    which to select.
4.  Run query.
**Sort a Query**
1.  Open query in Design
    view.
2.  Click in *Sort* box of col-
    umn by which to sort.
3.  Click *Sort* option box
    arrow.
4.  Click *Ascending* or
    *Descending*.

Access inserts double quotation marks at the beginning and end of a criterion for a
field that contains text, such as a city, name, or other field not used for calculating values.

11  Click the Run button.

Students who reside in Detroit *or* Windsor are displayed in the query results datasheet.

12  Click the View button to return to Design view.

A query is sorted by choosing *Ascending* or *Descending* in the *Sort* box of the
column by which you want to sort. At Step 9, *City* was placed first in the design
grid because Access sorts query results by column left to right. To arrange the records
alphabetically by student last name grouped by cities, the *City* field needed to be
positioned left of the *StudentLastName* field.

13  Click in the *Sort* box in the *City* column to place an insertion point and display
the option box arrow, click the *Sort* option box arrow, and then click *Ascending*.

14  Click in the *Sort* box in
the *StudentLastName*
column, click the
option box arrow that
appears, and then click
*Ascending*.

Step 13                                Step 14

| Field: | City | StudentID | StudentFirstName | StudentLastName |
| Table: | Students | Students | Students | Students |
| Sort: | Ascending | | | |
| Show: | ☑ | ☑ | ☑ | Ascending |
| Criteria: | "Detroit" | | | Descending |
| or: | "Windsor" | | | (not sorted) |

15  Click the Run button.

The query results datasheet is sorted alphabetically by city and then by the student
last name within each city.

DetroitAndWindsorStudents

| City | StudentID | StudentFirstName | StudentLastNam | Telephone |
|------|-----------|------------------|----------------|-----------|
| Detroit | 101658235 | Stuart | Bennett | (313) 555-511 |
| Detroit | 101334799 | Su-Lin | Cheng | (313) 555-412 |
| Detroit | 101455623 | Armando | Colacci | (313) 555-221 |
| Detroit | 100478512 | Angela | Doxtator | (313) 555-477 |
| Detroit | 101348863 | Daniela | Garcia | (313) 555-623 |
| Detroit | 100548112 | Pat | Hynes | (313) 555-656 |
| Detroit | 101322775 | Shane | O'Leary | (313) 555-365 |
| Detroit | 110542642 | Nadir | Sadiku | (313) 555-478 |
| Detroit | 100245745 | Anna | Takacis | (313) 555-423 |
| Detroit | 101334745 | Kyla | Westman | (313) 555-745 |
| Detroit | 104875111 | Dana | Williams | (313) 555-124 |
| Windsor | 100367845 | Sara | Blewett | (519) 555-412 |
| Windsor | 101352658 | Shaun | Jenkins | (519) 555-478 |
| Windsor | 100443568 | Dwayne | Kenney | (519) 555-412 |
| Windsor | 100684741 | Marietta | Lynch | (519) 555-321 |
| Windsor | 101325685 | Peter | Maslov | (519) 555-327 |
| Windsor | 100324125 | Jose | Ramirez | (519) 555-469 |

This query results
datasheet displays
students who reside
in either Detroit or
Windsor and is sorted
first by *City* and then by
*StudentLastName*.

16  Save the query and name it **DetroitAndWindsorStudents**.

17  Close the DetroitAndWindsorStudents query. Leave the database open for the
next topic.

# 12.11 Modifying a Query, Calculating in a Query, and Previewing a Datasheet

☁ **Watch & Learn**

☁ **Hands On Activity**

Modify a query by adding and removing columns in the query design grid using buttons in the Query Setup group on the Query Tools Design tab. A calculated field can be created in a query that performs a mathematical operation on a numeric field. A database design best practice is to avoid including a field in a table for storing data that can be generated by performing a calculation on another field. For example, assume that in the LibraryFines database, each fine is assessed a $2.50 administrative fee. Because the fee is a constant value, adding a field in the table to store the fee is not necessary. In this topic, you will use a query to calculate the total fine, including the administrative fee.

① With the **LibraryFines-YourName** database open, double-click the *FinesIssued* query to open the query.

② Switch to Design view.

③ Click in any cell in the *Telephone* column in the query design grid.

④ Click the Delete Columns button in the Query Setup group on the Query Tools Design tab.

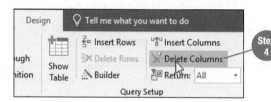

Buttons in the Query Setup group are used to modify a query by deleting columns or inserting new columns between existing fields.

⑤ Delete the *OverdueBookISBN* and *BookReturned* fields by completing steps similar to Steps 3 and 4.

⑥ With *FinePaid* the active field, click the Insert Columns button in the Query Setup group.

⑦ With an insertion point positioned in the *Field* box in the new column between *Amount* and *FinePaid*, type Fine with Admin Fee: [Amount]+2.50 and then press Enter.

⑧ Drag the right column boundary line in the gray bar above the calculated field until you can see the entire formula as shown in the image below. Note that Access drops the zero at the end of the formula.

The text before the colon is the column heading for the new field *Fine with Admin Fee*. After the colon, the mathematical expression *[Amount]+2.5* is stored. A field name used in a formula is typed within square brackets.

⑨ Click the Run button.

Notice that the calculated field is not formatted the same as the *Amount* field, and that the column needs to be widened to display the entire column heading.

⑩ Click in any cell in the *Fine with Admin Fee* column, click the More button in the Records group, click *Field Width*, and then click the Best Fit button in the Column Width dialog box.

● ─ ─ ─ ─ ─ Skills

Delete and insert columns in a query

Create a calculated field

Format a field

Print Preview a datasheet

☁ **Tutorials**

Performing Calculations in a Query

Previewing and Printing a Table

● ─ ─ ─ ─ ─ Oops!

Error message appears? Check your typing to make sure you typed a colon after *Fine with Admin Fee*, used square brackets, and that the entry has no other spelling errors.

● ─ ─ ─ ─ ─ app tip

Use the same mathematical operators in Access that you would use in a formula in Excel: + to add, - to subtract, * to multiply, and / to divide.

11 Switch to Design view.

12 Click to place an insertion point in the *Fine with Admin Fee* column heading in the query design grid and then click the Property Sheet button in the Show/Hide group on the Query Tools Design tab.

13 Click in the *Format* box in the Property Sheet task pane, click the option box arrow that appears, and then click *Currency*.

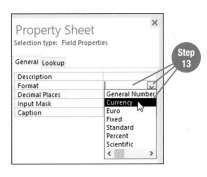

14 Close the Property Sheet task pane.

15 Click the Run button.

16 Use *Save Object As* to save the revised query as *FinesWithAdminFee*.

| FineID | DateIssued | StudentFirstName | StudentLastName | Amount | Fine with Admin Fee | FinePaid |
|--------|-----------|------------------|-----------------|--------|---------------------|----------|
| MCL-116 | 10-Apr-2021 | Anna | Takacis | $5.60 | $8.10 | ☑ |
| MCL-117 | 12-Apr-2021 | Anna | Takacis | $2.45 | $4.95 | ☐ |
| MCL-114 | 08-Apr-2021 | Sara | Blewett | $3.50 | $6.00 | ☑ |
| MCL-108 | 15-Mar-2021 | Dwayne | Kenney | $12.50 | $15.00 | ☐ |
| MCL-113 | 06-Apr-2021 | Dwayne | Kenney | $4.60 | $7.10 | ☐ |

These are the first five records in the FinesWithAdminFee query displaying the *Fine with Admin Fee* calculated column.

A table datasheet or query results datasheet should be viewed in Print Preview before printing to make adjustments as necessary to the orientation and margins.

17 Click the File tab, click *Print*, and then click *Print Preview* in the Print panel of the backstage area.

The entire datasheet does not fit on one page in the default portrait orientation.

18 Click the Landscape button in the Page Layout group on the Print Preview tab.

Notice that Access prints the query name and the current date at the top of the page and the page number at the bottom of the page.

19 Click the Close Print Preview button in the Close Preview group.

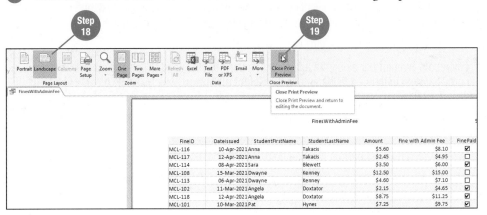

20 Close the FinesWithAdminFee query.

21 Click the File tab and then click *Close* to close the LibraryFines database.

## Topics Review

| Topic | Key Concepts |
|---|---|
| **12.1 Understanding Database Objects and Terminology** | A **database** stores data in an organized manner to provide information for a variety of purposes. |
| | A **database management system (DBMS)** is application software to organize, store, and maintain a database. |
| | An Access database is a collection of objects used to enter, maintain, and view data. |
| | Access opens with a Navigation pane along the left side of the window used to select an object to open and view data. |
| | A **table** stores data about a single topic or subject such as people, places, events, or items. |
| | Each characteristic about the subject or topic of a table is called a **field**. |
| | The data stored in a field is called a **field value**. |
| | A set of fields for one person, place, event, item, or other subject of a table is called a **record**. |
| | A table opens in a datasheet where the columns are fields and the rows are records. |
| | A **report** is an object used to view or print data from tables in a customized layout and with summary totals. |
| | A **form** is another interface to view, enter, or edit data and displays only one record at a time in a customized layout. |
| | Use buttons on the Record Navigation bar to scroll records in a form. |
| | A **query** is used to combine fields from one or more tables in a single datasheet and may display all records or only some records that meet a condition. |
| **12.2 Adding Records to a Table Using a Datasheet** | New records are added to a table by opening the table datasheet and then clicking the **New (blank) record button** on the Record Navigation bar. |
| | Type field values in each column in the table datasheet, pressing Tab to move from one field to the next field. |
| | A field that displays with a down-pointing arrow means that you can enter the field value by selecting an entry from a drop-down list. |
| | Date fields can be set up to display underscores and hyphens to make sure dates are entered consistently in the correct format. |
| | A field that displays with a check box stores *Yes* if the box is checked and *No* if the box is left empty. |
| | Access automatically saves a new record as soon as you press Tab to move past the last field. |
| **12.3 Editing and Deleting Records Using a Datasheet** | Edit a field value in a datasheet by selecting text to be changed and typing new text, or by clicking to place an insertion point within a field (cell) and inserting or deleting text as required. |
| | Function key F2 opens a field for editing. |
| | The gray record selector bar along the left edge of a datasheet is used to select a record. At the top of the gray bar next to the first field name is the **Select all button**, which selects all field values in all records in the table. |
| | Click the Delete button in the Records group on the Home tab to delete the selected record from the table. |
| | Access requires that you confirm a deletion before the record is removed. |
| | Generally, records are not deleted until data has been copied to an archive database and/or a backup copy of the database has been made. |

*continued…*

| Topic | Key Concepts |
|---|---|
| **12.4 Adding, Editing, and Deleting Records Using a Form** | A form is the preferred object for adding, editing, and deleting records because one record at a time is displayed in a columnar layout in the work area. |
| | Add new records and edit field values in records using the same techniques that you used for adding and editing records using a datasheet. |
| | To delete a record using a form, display the record, click the Delete button arrow in the Records group on the Home tab, and then click *Delete Record*. |
| **12.5 Finding and Replacing Data and Adjusting Column Widths in a Datasheet** | Click in the column in a datasheet that contains the field value you want to locate and use the Find command to move to all occurrences of the text entered in the *Find What* text box. |
| | Use the Replace command to find all occurrences of an entry and replace the field value with new text. |
| | Activate any cell in a column for which the width needs to be adjusted, click the More button in the Records group on the Home tab, click *Field Width*, and then enter the desired width or click the Best Fit button. |
| | The **Best Fit** option adjusts the width of the column to accommodate the longest entry in the field. |
| **12.6 Sorting and Filtering Records** | Initially, a table is arranged alphanumerically by the **primary key** field values. A primary key is a field in the table that uniquely identifies each record, such as *StudentID*. |
| | To sort by a field other than the primary key, click in the field by which to sort and then click the Ascending button or Descending button in the Sort & Filter group on the Home tab. |
| | Filter a datasheet by clearing check boxes for items you do not want to view in the *Sort & Filter* list box accessed from the filter arrow next to the field name. |
| | Use the Toggle Filter button in the Sort & Filter group to redisplay all records in a filtered datasheet. |
| **12.7 Creating a Query Using the Simple Query Wizard** | Queries extract information from one or more tables in a single datasheet. |
| | The **Simple Query Wizard** helps you build a query by making selections in three dialog boxes. |
| | In the first Simple Query Wizard dialog box, choose each table or query and the fields in the order that you want them in the query results datasheet. |
| | In the second Simple Query Wizard dialog box, choose a detail or summary query. |
| | Assign a name to the query in the third Simple Query Wizard dialog box. |
| **12.8 Creating a Query Using Design View** | Every object in Access has at least two views. Opening an object from the Navigation pane opens the table, query, form, or report in the view that displays the object with table data. |
| | **Design view** is used to set up or define the structure or layout of an object. |
| | Design view for a query presents a blank grid of columns in which you add the fields in the order you want them in the query results datasheet. |
| | Add table field list boxes to the query from the Show Table dialog box by selecting the table names and clicking the Add button. Close the Show Table dialog box when all tables have been added. |
| | Double-click field names in the table field list boxes in the order you want the columns in the query results datasheet. Each field is added to the next available column in the query design grid. |
| | The **Run button** is used after building a query in Design view to instruct Access to generate the query and display the query results datasheet. |

*continued...*

| Topic | Key Concepts |
|---|---|
| **12.9 Entering Criteria to Select Records in a Query** | A query that displays only those records that meet one or more conditions is called a **select query**.<br><br>Use the View button in the Views group on the Home tab to switch between Datasheet view and Design view in a query.<br><br>Type the criterion by which you want records selected in the *Criteria* box of the column by which records are to be selected.<br><br>In a field that displays check boxes, the criterion is either *Yes* or *No*.<br><br>A date added as a criterion is displayed in the format *#m/d/yyyy#*. Use date keywords *Between* and *And* to select records based on a range of dates. |
| **12.10 Entering Multiple Criteria to Select Records and Sorting a Query** | More than one criterion entered in the same *Criteria* row in the query design grid is an *AND* statement, which means each criterion must be met for the record to be selected.<br><br>More than one criterion entered in *Criteria* rows one below the other is an *OR* statement, which means that any condition can be met for the record to be selected.<br><br>Choose *Ascending* or *Descending* from the *Sort* option box arrow for the column by which to sort query results in Design view.<br><br>Access sorts a query by column left to right. If necessary, position the field to be sorted first to the left of another field that is to be sorted. |
| **12.11 Modifying a Query, Calculating in a Query, and Previewing a Datasheet** | A calculated field can be created in a query that generates values using a mathematical expression.<br><br>Use the Delete Columns and Insert Columns buttons in the Query Setup group on the Query Tools Design tab to remove or add new columns in a query.<br><br>A calculated column is created by typing in the *Field* box a column heading, a colon (:), and then the mathematical expression.<br><br>Type a field name in a mathematical expression within square brackets.<br><br>Open the Property Sheet task pane to change the format of a calculated field.<br><br>A table or query results datasheet should be previewed before printing to make adjustments to page orientation and/or margins. |

## Topics Review

## Chapter 13

# Creating a Table, Form, and Report in Access

Planning a new database involves understanding the purpose of the database and the information each database user will need. Prior to starting Access and creating a new database file, the database designer carefully analyzes the data to be collected and decides the best way to define and group the data elements into logical units. Tables are created first because they are the basis for all other objects. Next, tables that need to be connected for queries, forms, or reports are joined in a relationship. Objects such as queries, forms, and reports are created after the tables and relationships are defined.

In Chapter 12, you examined an existing database and added and edited data in a table and form. You also created queries to select records for a variety of purposes. Now that you have seen how Access data interacts with objects, you are ready to build a new database on your own. In this chapter, you will learn to create a new database, create a new table, assign a primary key, modify field properties, edit relationships, create a form, create a report, compact and repair a database, and create a backup copy of a database.

## Learning Objectives

**13.1** Create a new database and describe guidelines for designing tables

**13.2** Create a new table using Datasheet view and assign a caption to a field

**13.3** Create a new table using Design view and assign a primary key

**13.4** Add a field to a table

**13.5** Change the field size and add a default value for a field using Design view

**13.6** Create a lookup list for a field

**13.7** Identify a one-to-one relationship and a one-to-many relationship, and edit a relationship

**13.8** Create and edit a form

**13.9** Create, edit, and view a report

**13.10** Compact and repair, and back up a database

**Read & Learn**

**Content Online**

The online course includes additional training and assessment resources.

# 13.1 Creating a New Database File and Understanding Table Design Guidelines

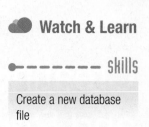

Watch & Learn

● ------- Skills

Create a new database file

The first step in creating a database is to assign a name and storage location for the new database file. Because Access saves records automatically as data is added to a table, the file name and storage location are required in advance. Once the file is created, Access displays a blank table. Before you create a new table, you must carefully plan the fields and field names and identify a primary key. The tables you will create in this chapter have already been planned. The guidelines in Table 13.1 provide you with an overview of the table design process to understand how the tables you create were structured.

**Table 13.1** Guidelines for Planning a New Table

| Guideline | Description |
|---|---|
| Divide data into the smallest possible units | A field should be segmented into the smallest units of information to facilitate sorting and filtering. For example, a person's name could be split into three fields: first name, middle name, and last name. |
| Assign each field a name | Up to 64 characters can be used in a field name with a combination of letters, numbers, spaces, and some symbols. Database programmers prefer short field names with no spaces. A field to store a person's last name could be assigned the name *LName*, *Last*, *LastName*, or *Last_Name*. Short names are preferred because Access provides the ability to enter a longer descriptive title for column headings in datasheets, forms, and reports that is separate from the field name. |
| Assign each field a data type | *Data type* refers to the type of information that will be entered as field values. Look at examples of data to help you determine the data type. By assigning the most appropriate data type, Access can verify data as it is being entered for the correct format or type of characters. For example, a field defined as a Number field will cause Access to reject alphabetic letters typed into the field. The most common data types are Short Text, Number, Currency, Date/Time, and Yes/No. Data types are described in Table 13.2 in the next topic. |
| Decide the field to be used as a primary key | Each table should have one field that uniquely identifies a record, such as a student number, receipt number, or email address. Access creates an ID field automatically in a blank datasheet that can be used if the table data does not have a unique identifier. In some cases, a combination of two or more fields is used as a primary key. |
| Include a common identifier field in a table that will be joined to another table | Data should not be duplicated in a database. For example, a book title would not be stored in both the Books table and the Sales table. Instead, the book title is stored in the Books table only, and a book ID field in the Sales table is used to join the two tables in a relationship. Relationships will be explained in a later topic. |

**1** Start Access.

**2** Click *Blank database* on the Access Start screen.

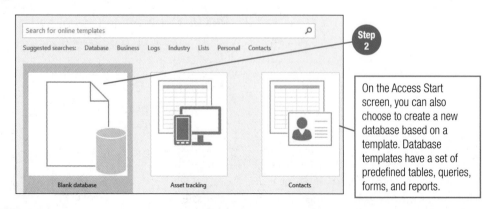

**Tutorial**

Opening a Blank Database

On the Access Start screen, you can also choose to create a new database based on a template. Database templates have a set of predefined tables, queries, forms, and reports.

**3** With the current text in the *File Name* text box already selected, type UsedBooks -YourName and then click the Browse button (file folder icon).

**4** In the File New Database dialog box, navigate to the CompletedTopics folder on your storage medium, create a new folder named *Ch13*, double-click to open the Ch13 folder, and then click OK.

**5** Click the Create button.

Access creates the database file and opens a new table datasheet named *Table1* in the work area, as shown in Figure 13.1.

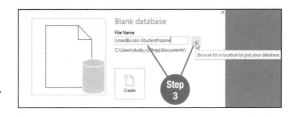
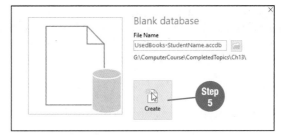

**quick steps**

**Create a New Database**
1. Start Access.
2. Click *Blank database*.
3. Type file name.
4. Click Browse button.
5. Navigate to drive and/ or folder location.
6. Click OK.
7. Click Create.

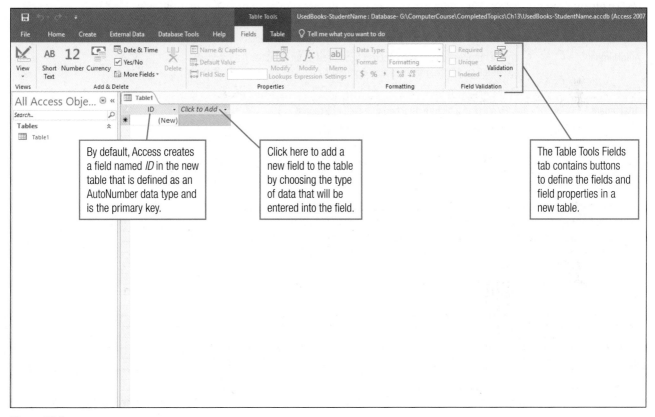

**Figure 13.1**
Access creates a blank table datasheet, like the one shown here, in a new database file.

**6** Leave the blank table datasheet open for the next topic.

Each table in a database should contain information about one subject only. In this chapter, you will create tables for a used-textbook database that a student organization may use to keep track of students, textbooks, and sales. The three tables you will create in this chapter are described as follows:

**Books:** shows the title, author, condition, and asking price for each book

**Sales:** tracks each sale with the date, amount, and payment method

**Students:** shows information about each student who has textbooks for sale

## 13.2 Creating a New Table Using a Datasheet

Create a new table in a blank datasheet in Datasheet view by adding a column for each field. Begin by specifying the data type for a field and then typing the field name. Data types are described in Table 13.2. Once the fields have been defined, use the Save button on the Quick Access Toolbar to assign the table a name.

**Table 13.2** Field Data Types

| Data Type | Use for This Type of Field Value |
| --- | --- |
| Short Text | Alphanumeric text up to 255 characters for names, identification numbers, telephone numbers, or other similar data |
| Number | Numeric data other than monetary values |
| Currency | Monetary values such as sales, costs, or wages |
| Date & Time or Date/Time | Dates or times that you want to verify, sort, select, or calculate |
| Yes/No | Data that can only be Yes or No, or True or False |
| Lookup & Relationship or Lookup Wizard | Option list with field values from another table, or from a predefined list of options (items) |
| Long Text or Rich Text | Alphanumeric text of more than 255 characters; Rich Text enables formatting options such as font, font color, bold, and italic in the field values |
| AutoNumber | Unique number generated by Access to be used as an identifier field; Access generates sequential numbers starting from 1 |
| Hyperlink | Stores web addresses |
| Attachment | A file such as a picture attached to a field in a record |
| Calculated Field | Formula that calculates the field value using data in other fields |

**1** With the **UsedBooks –YourName** database open and with the blank datasheet for Table1 open, click the Date & Time button in the Add & Delete group on the Table Tools Fields tab.

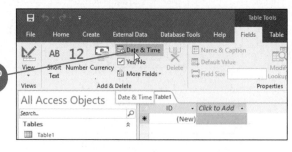

Choosing the most appropriate data type for a field is important for sorting, calculating, and verifying data. Access expects dates to be entered in the format *m/d/y* unless the region setting in the Control Panel is changed to another format.

**2** With *Field1* selected in the field name box for the new column, type SaleDate and then press Enter.

The *Click to Add* column opens the Data Type option list for the next new field. Add a new field using either the *Click to Add* option list or the buttons in the Add & Delete group on the Table Tools Fields tab.

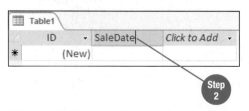

**3** Click *Short Text* in the *Click to Add* option list.

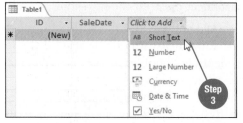

④ Type BookID as the field name and then press Enter.

⑤ Click *Currency* in the *Click to Add* option list, type Amount, and then press Enter.

⑥ Click *Short Text*, type PayMethod, and then press Enter.

⑦ Double-click the *SaleDate* field name to select the field.

⑧ Click the Name & Caption button in the Properties group on the Table Tools Fields tab.

The **Caption property** is used to type a descriptive title for a field that includes spaces between words or the full text of an abbreviated field name.

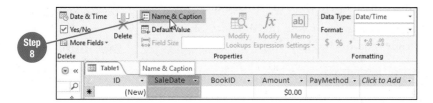

⑨ Click in the *Caption* text box, type Sale Date, and then press Enter or click OK in the Enter Field Properties dialog box.

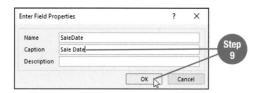

⑩ Click to select the *Amount* field, click the Name & Caption button, click in the *Caption* text box, type Sale Amount, and then press Enter or click OK.

⑪ Drag the right column boundary of the *Sale Amount* column until the entire column heading is visible, if necessary.

⑫ Click to select the *PayMethod* field, click the Name & Caption button, click in the *Caption* text box, type Payment Method, and then press Enter or click OK.

⑬ Double-click the right column boundary of the *Payment Method* column.

Double-clicking the right column boundary of a field sets the column width to the Best Fit option—the length of the longest entry in the column.

⑭ Click the Save button on the Quick Access Toolbar.

⑮ Type Sales in the *Table Name* text box in the Save As dialog box and then press Enter or click OK.

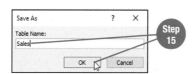

completed Sales table datasheet

⑯ Close the Sales table. Leave the database open for the next topic.

## 13.3 Creating a New Table Using Design View and Assigning a Primary Key

A new table can be created in Design view in which fields are defined in rows in the top half of the work area. In the previous topic, the Sales table, created in a blank datasheet, had an ID field automatically created and designated as the primary key. In Design view, an ID field is not created for you. After creating the fields in the Design view window, assign the primary key to the field that will uniquely identify each record and then save the table.

1. With the **UsedBooks-YourName** database open, click the Create tab.

2. Click the Table Design button in the Tables group. Close the Property Sheet task pane if the task pane opens at the right side of the window.

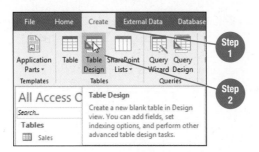

3. With the insertion point positioned in the first row in the *Field Name* column, type StudentID and then press Enter.

4. With *Short Text* in the *Data Type* column, press Enter to accept the default data type.

5. Press Enter to move past the *Description* column and move down to the next row to start a new field.

Descriptions are optional entries. A description is used to add information about a field or to enter instructions to end users who see the description in the Status bar of a datasheet when the field is active.

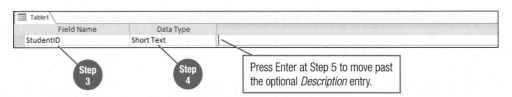

Press Enter at Step 5 to move past the optional *Description* entry.

6. Type LName in the *Field Name* column in the second row and then press Enter three times to move to the *Field Name* column in the third row.

7. Enter the remaining fields as shown in the image below.

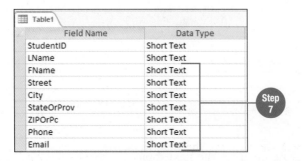

---

**Watch & Learn**

**Hands On Activity**

● – – – – – – **Skills**

Create a new table using Design view

Assign a primary key

**Tutorials**

Creating a Table in Design View

Setting the Primary Key Field

Managing Fields in Design View

● – – – – – – **app tip**

You can also use the Tab key to move to the next column in Design view.

**8** Click to place an insertion point within the *StudentID* field name.

**9** Click the Primary Key button in the Tools group on the Table Tools Design tab.

A key icon in the field selector bar (gray bar along left edge of *Field Name* column) indicates the field is designated as the primary key for the table.

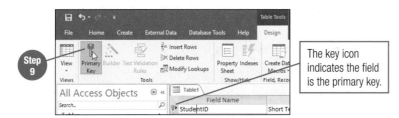

The key icon indicates the field is the primary key.

**10** Click the Save button on the Quick Access Toolbar, type Students in the *Table Name* text box in the Save As dialog box, and then press Enter or click OK.

**11** Close the Students table.

**12** Click the Create tab and then click the Table Design button.

**13** Create the first five fields in the new table using the default Short Text data type and without descriptions as follows:

> *BookID*
> *StudentID*
> *Title*
> *Author*
> *Condition*

**14** Type AskPrice in the *Field Name* column in the sixth row and then press Enter.

**15** Click the *Data Type* option box arrow and then click *Currency*.

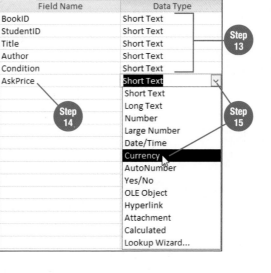

**16** Click to place an insertion point within the *BookID* field name and then click the Primary Key button.

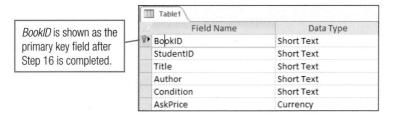

*BookID* is shown as the primary key field after Step 16 is completed.

**17** Click the Save button on the Quick Access Toolbar, type Books, and then press Enter or click OK.

**18** Close the Books table. Leave the database open for the next topic.

## 13.4 Adding Fields to an Existing Table

Open a table in Datasheet view and use the *Click to Add* column to add a new field, or make active a field in the datasheet and use the buttons on the Table Tools Fields tab to add a new field after the active field.

**1** With the **UsedBooks-YourName** database open, double-click the Books table to open the table.

**2** Click the *Click to Add* column heading and then click *Currency*.

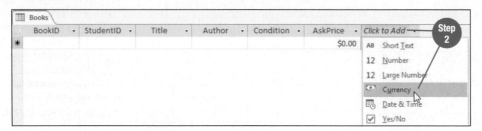

**3** Type StopPrice as the field name and then press Enter.

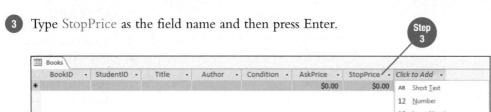

**4** Click the top part of the View button in the Views group on the Table Tools Fields tab to switch to Design view.

**5** Click in the *Description* column in the *StopPrice* field row and then type Do not sell for lower than the student's stop price value.

| Field Name | Data Type | |
| --- | --- | --- |
| BookID | Short Text | |
| StudentID | Short Text | |
| Title | Short Text | |
| Author | Short Text | |
| Condition | Short Text | |
| AskPrice | Currency | |
| StopPrice | Currency | Do not sell for lower than the student's stop price value |

**6** Save and then close the Books table.

**7** Double-click to open the Students table.

**8** Click the *Phone* field name to select the column.

**9** Click the Table Tools Fields tab and then click the Yes/No button in the Add & Delete group to add the new field in the column to the right of the *Phone* field.

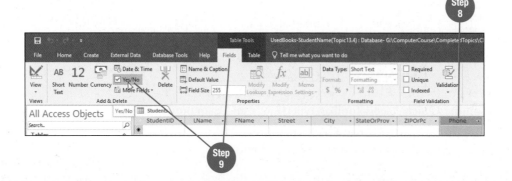

**Watch & Learn**

**Hands On Activity**

------- Skills

Add a field to a table

**Tutorials**

Managing Fields in Datasheet View

Formatting Table Data

10 Type DirectDeposit as the field name and then press Enter.

11 Double-click the right column boundary of the *DirectDeposit* field to best fit the column width.

**Add a Field**
1. Open table.
2. Click *Click to Add* column.
3. Click required data type.
4. Type field name.
5. Press Enter.
6. Save table.
OR
1. Open table.
2. Click field name to select column left of where you want the new field located.
3. Click the required data type button.
4. Type field name.
5. Press Enter.
6. Save table.

12 Save and then close the Students table.

13 Open the Books table.

14 Click the *StopPrice* field name to select the column.

15 Look at the message *Do not sell for lower than the student's stop price value* that displays in the Status bar.

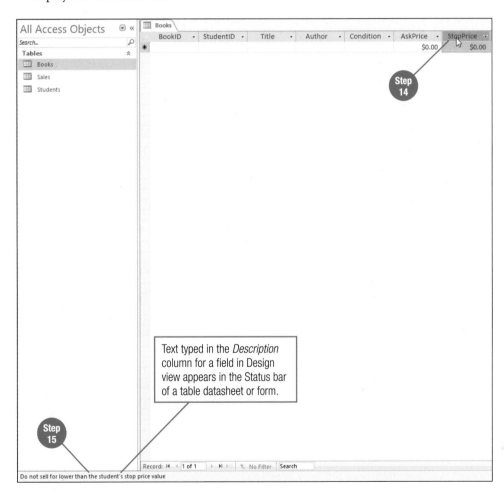

Text typed in the *Description* column for a field in Design view appears in the Status bar of a table datasheet or form.

In this topic, you added a new field using the table datasheet. A field can also be added to a table using Design view. Open a table, switch to Design view, and then type the new field name and select the data type in the next available row. To add a new field positioned elsewhere, click in the field name row *below* the position you want the new field to be placed, and then click the Insert Rows button in the Tools group on the Table Tools Design tab.

16 Close the Books table. Leave the database open for the next topic.

## 13.5 Modifying Field Properties Using Design View

Each field in a table has a set of field properties associated with the field. A **field property** is a single characteristic or attribute of a field. For example, the field name is a field property, and the data type is another field property. The properties in each field can be modified to customize, format, or otherwise change the behavior of a field. The lower half of the work area of a table in Design view contains the **Field Properties pane** that is used to modify properties other than the field name, data type, and description.

**1** With the **UsedBooks-YourName** database open, right-click the *Students* table name in the Navigation pane and then click *Design View*.

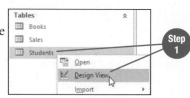

**2** Click in the *StateOrProv* field name to select the field and display the associated field properties in the Field Properties pane.

Options displayed in the Field Properties pane vary by the field data type. For example, a Date & Time field does not have a Field Size property.

**3** Double-click *255* or drag to select *255* in the *Field Size* property box in the Field Properties pane, type *2*, and then press Enter.

Setting a field size for a state or province field ensures that all new field values use the two-character abbreviation for addressing letters or creating labels from the database.

**4** Click in the *Default Value* property box, type MI, and then press Enter.

Access automatically adds quotation marks to the text in a *Default Value* property box. The text entered into a *Default Value* property box is automatically entered as the field value in new records; the end user presses Enter to accept the value or types an alternative entry. Adding a default value entry not only saves time when a new record is being added to the table, but it also makes sure that the field value is entered consistently and with correct spelling. Use the property whenever a field is likely to have several records with the same entry.

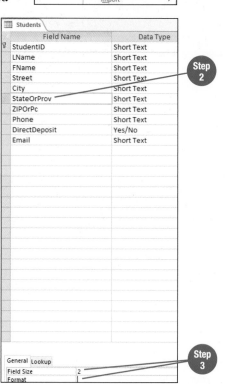

**5** Click in the *Caption* property box and then type State or Province.

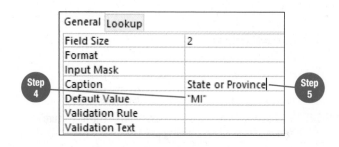

**Watch & Learn**

**Hands On Activity**

●-------  skills

Change the field size

Add a default value

Add a caption in Design view

**Tutorials**

Modifying Field Properties in Design View

Applying a Validation Rule in Design View

●-------  app tip

Buttons on the Table Tools Fields tab are available for changing the most common field properties in Datasheet view.

⑥ Click in the *StudentID* field name to select the field, click in the *Caption* property box in the Field Properties pane, type Student ID, and then press Enter.

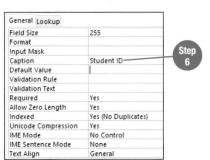

●---- **quick steps**

**Modify a Field Property**
1. Open table in Design view.
2. Click in field name in top half of work area.
3. Click in property box or select current value in property box.
4. Type entry or select option from option list.
5. Save table.

⑦ Add the following caption properties like you did for Step 6.

| Field Name | Caption |
|---|---|
| *LName* | Last Name |
| *FName* | First Name |
| *Street* | Street Address |
| *ZIPOrPC* | ZIP or Postal Code |
| *Phone* | Telephone |
| *DirectDeposit* | Direct Deposit |
| *Email* | Email Address |

⑧ Save the table.

⑨ Click the top part of the View button in the Views group to switch to Datasheet view.

⑩ Double-click the right column boundary of column headings that are not entirely visible to best fit the column widths.

Notice that *MI* appears in the *State or Province* column by default.

⑪ Click in the *Student ID* column to place an insertion point, type 999, and then press Tab.

⑫ Type your name and a fictitious address in the remaining fields similar to the data shown in the image below. Adjust column widths as needed to view all data.

⑬ Close the table. Click Yes to save changes to the table layout. Leave the database open for the next topic.

See Table 13.3 for a description of additional field properties that are often used.

**Table 13.3** Additional Field Properties and Examples

| Field Property | Description | Example |
|---|---|---|
| Format | Modify this property to change the display of data entered into the field from the default format. | A field for storing dollar values can be formatted to show commas in thousands and a dollar symbol. |
| Validation Rule and Validation Text | An expression in the Validation Rule property is checked as data is entered; invalid data is not saved. The Validation Text entry displays a message to the user when a rule is in effect. | The expression *>=5* in a numeric field ensures values less than 5 are rejected. A *Validation Text* entry such as *Please enter a value over 5* displays on the screen when a field value is rejected. |
| Required | A *Yes* in this property ensures that the field is not left blank in a new record. | This is used often for a *ZIP or PostalCode* field to make sure each record has a valid mailing code. |

## 13.6 Creating a Lookup List

A **lookup list** presents a list of field values when a lookup field is made active while new records are added in a datasheet or form. The list entries can be a fixed list of options created for the field, or the list can present field values that exist in a field in another table that is linked to the lookup field. A lookup list has many advantages, including consistency, accuracy, and efficiency when adding data in new records. Access provides the **Lookup Wizard** to assist with creating field properties in which the user chooses options and types values in dialog boxes.

1. With the **UsedBooks-YourName** database open, right-click the *Books* table name in the Navigation pane and then click *Design View*.

2. Click in the *Condition* field name to select the field.

3. Click in the *Data Type* box, click the option box arrow that appears, and then click *Lookup Wizard*.

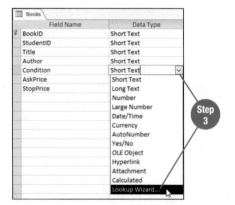

4. Click *I will type in the values that I want.* in the first Lookup Wizard dialog box and then click Next.

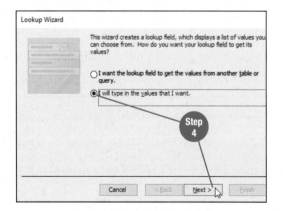

5. Click in the first blank row below *Col1* in the second Lookup Wizard dialog box and then type Excellent – No wear or markings.

6. Drag the *Col1* right column boundary to the right to increase the column width as shown in the image below.

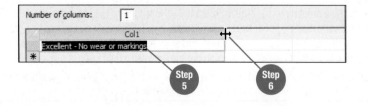

7. Click in the second row and then type Very Good – Minor wear on cover.
8. Type the remaining entries in the list as shown in the image below.

You can also use the Down Arrow key to move to the next row in the column.

9. Click Next.

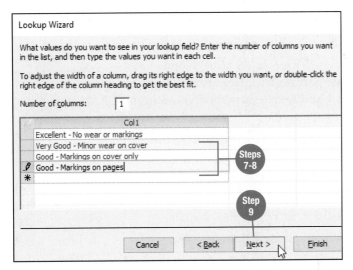

**quick steps**

**Create a Lookup List**
1. Open table in Design view.
2. Click *Data Type* option box arrow for field.
3. Click *Lookup Wizard*.
4. Click *I will type in the values that I want*.
5. Click Next.
6. Type list entries in *Col1* column and adjust column width.
7. Click Next.
8. Click Finish.
9. Save table.

**Oops!**

Pressed Enter by mistake while typing a list entry in the second Lookup Wizard dialog box? Click the Back button to return to the list if you did not finish typing all the entries in Step 8.

10. Click Finish in the last Lookup Wizard dialog box.
11. Click the Lookup tab in the Field Properties pane and review the entries made to each lookup property by the Lookup Wizard.

The entries typed in Steps 5 to 8 appear in the Row Source property. If you made a typing mistake while typing the list items, you can correct the error by clicking in the *Row Source* property box and inserting or deleting text as needed. Be cautious not to modify or delete syntax (quotations or semicolons) while correcting typing errors. If desired, you can limit field values in new records to only those items in the list by changing the Limit To List property to *Yes*.

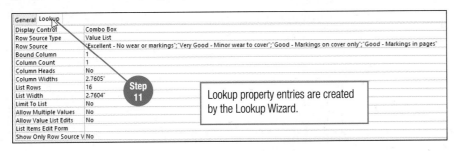

Lookup property entries are created by the Lookup Wizard.

12. Save the table and then click the top part of the View button to switch to Datasheet view.
13. Type the data in a new record as shown in the image below. In the *Condition* field, click the option box arrow that appears and click *Good - Markings on pages*.
14. Adjust column widths as needed to show all data in each column.

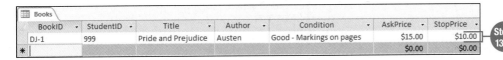

15. Save the table layout changes and then close the table.
16. Close the database.

# 13.7 Displaying and Editing a Relationship

A relationship allows you to create queries, forms, or reports with fields from two tables by connecting the two tables on a common field. Joining two tables in a relationship allows data to be reused, which prevents duplication and ensures consistency in records. For example, an ID, name, or title of a book can be looked up in one table rather than repeating the information in another table.

**1** Open the database **UsedTextbooks** from the Ch13 folder in StudentDataFiles.

**2** Click the File tab and then click *Save As*. With *Save Database As* and *Access Database (*.accdb)* already selected in the Save As backstage area, click the Save As button. Save a copy of the database as **UsedTextbooks-YourName** in the Ch13 folder in CompletedTopics on your storage medium.

**3** Click the Enable Content button in the security warning message bar.

This file is similar to the database you have been working on in this chapter but with the Books table modified, additional lookup lists, and with 10 records added to each table.

**4** Open the Books table, review the datasheet, and then close the table.

**5** Open the Sales table, review the datasheet, and then close the table.

**6** Open the Students table and then change *Doe* in the last record in the *Last Name* column to your last name.

**7** Change *Jane* in the last record in the *First Name* column to your first name and then close the table.

**8** Click the Database Tools tab and then click the Relationships button in the Relationships group.

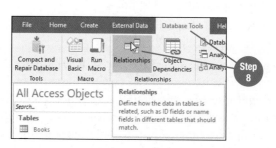

A field list box for each table displays in the Relationships window. A black join line connecting two table field list boxes indicates a relationship has been created between the two tables. Observe that each join line connects the tables on the same field name in each table field list box.

**9** Click to select the black join line that connects the Books table field list box to the Sales table field list box, and then click the Edit Relationships button in the Tools group on the Relationship Tools Design tab.

In the Edit Relationships dialog box that appears, *One-To-One* is shown in the *Relationship Type* section. A **one-to-one relationship** means that the two tables are joined on the primary key in each table. (*BookID* displays a key icon next to the field

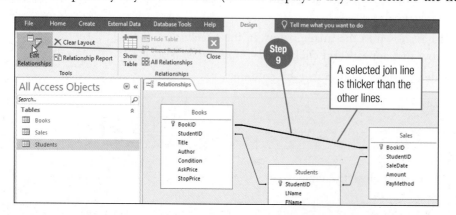

A selected join line is thicker than the other lines.

## Watch & Learn

## Hands On Activity

— — — — — — SKILLS

Identify a one-to-one relationship

Enforce referential integrity

Identify a one-to-many relationship

## Tutorials

Creating a One-to-One Relationship

Creating a One-to-Many Relationship

Editing and Deleting a Relationship

— — — — — — APP TIP

When you create a lookup list that looks up field values from a field in another table, Access automatically creates a relationship between the two tables if one does not exist.

name in each table field list box.) In this type of relationship, only one record can exist for a *BookID* in each table.

10 Click to insert a check mark in the *Enforce Referential Integrity* check box and then click OK.

Turning on **Enforce Referential Integrity** means that a record in Books is entered first before a record with a matching *BookID* is entered in Sales. Notice in the screen image at the right, Books is the left table name below *Table/Query*, and Sales is the right table name below *Related Table/Query*. The table identified below *Table/Query* (in this case the Books table) is the one for which referential integrity is applied—the table in which new records are entered first. The table shown at the left is also referred to as the **primary table**.

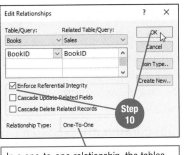

In a one-to-one relationship, the tables are joined on the primary key field in each table.

11 Click to select the black join line that connects the Books table field list box to the Students table field list box and then click the Edit Relationships button.

12 Click to insert a check mark in the *Enforce Referential Integrity* check box and then click OK.

A **one-to-many relationship** occurs when the common field used to join the two tables is the primary key in only one table (the primary table). *One* student can have *many* textbooks for sale. In this instance, a record must first be entered into Students (primary table) before a record with a matching student ID can be entered into Books (related table). A field added to a related table that is not a primary key and is included for the purpose of creating a relationship is called a **foreign key**. A one-to-many relationship is the most common type of relationship in a database.

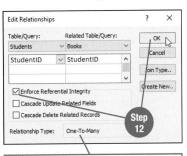

In a one-to-many relationship, the field joining the two tables is the primary key in one table and the foreign key in the other table.

13 Click the Close button in the Relationships group. Click Yes if prompted to save changes to the layout of the relationships, and leave the database open for the next topic.

**quick steps**

**Display Relationships**
1. Click Database Tools tab.
2. Click Relationships button.

**Enforce Referential Integrity**
1. Display Relationships.
2. Click to select black join line.
3. Click Edit Relationships button.
4. Click *Enforce Referential Integrity* check box.
5. Click OK.

**app tip**

To create a new relationship, drag the common field name in the primary table field list box to the same field in the related table field list box and then release the mouse. Change options as necessary and then click the Create button in the Edit Relationships dialog box that opens when the mouse is released.

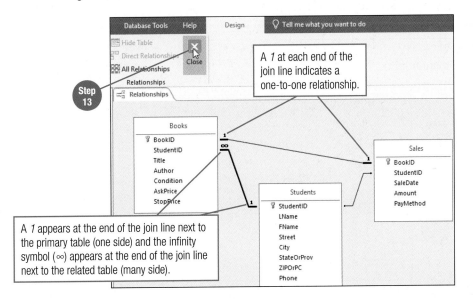

A *1* at each end of the join line indicates a one-to-one relationship.

A *1* appears at the end of the join line next to the primary table (one side) and the infinity symbol (∞) appears at the end of the join line next to the related table (many side).

# 13.8 Creating and Editing a Form

The Forms group on the Create tab includes buttons to create a variety of forms. A form can be as simple as all fields in a selected table arranged in a single column, to a complex form with fields from multiple tables and with some arranged in a subform within a form. Once created, a form can be modified using buttons on the Form Layout Tools Design, Arrange, and Format tabs.

**1** With the **UsedTextbooks–YourName** database open, click to select the *Books* table name in the Navigation pane.

**2** Click the Create tab.

**3** Click the Form button in the Forms group.

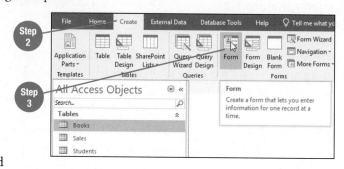

A form is created with all the fields in the selected table arranged in a vertical layout and displayed in Layout view, as shown in Figure 13.2. **Layout view** is the view in which you edit a report or form structure and appearance. **Form view** is the view in which data is viewed, entered, and updated and is the view in which a form is opened when you double-click the form name in the Navigation pane.

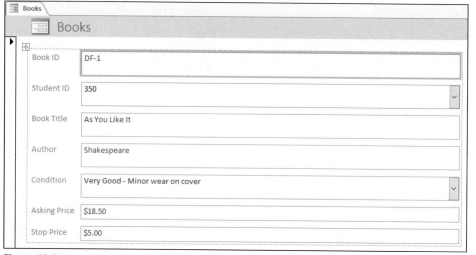

**Figure 13.2**

Shown is the Books form created using the Form button in Step 3.

**4** Click the Themes button in the Themes group on the Form Layout Tools Design tab.

A theme sets the color scheme and fonts used on a form or report.

**5** Click *Slice* in the *Office* section.

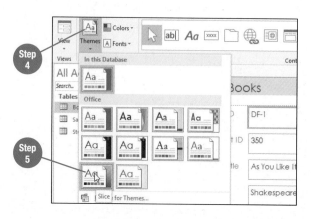

**Watch & Learn**

**Hands On Activity**

●————— skills

Create a form

Apply a theme

Add and format an image on a form

Format the form title

●————— app tip

Buttons to switch between views are at the right end of the Status bar.

**Tutorials**

Creating a Form Using the Form Button

Managing Control Objects in a Form

**6** Click the Logo button in the Header/Footer group.

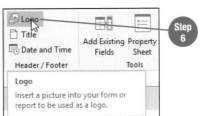

**7** In the Insert Picture dialog box, navigate to the Ch13 folder in StudentDataFiles and then double-click the file **Textbooks.jpg**.

The image is inserted into a selected logo **control object** near the top left of the form. A control object is a rectangular content placeholder in a form or report used to place images, text, field names, and field values. Each item on a form is placed in a control object. Each control object can be selected and edited to modify the appearance or other properties for the content.

**8** With the logo control object still selected, click the Property Sheet button in the Tools group.

The Property Sheet task pane opens at the right side of the window and displays all the properties available for the selected control object.

**9** Click the Format tab in the Property Sheet task pane, click in the *Size Mode* property box, click the option box arrow that appears, and then click *Zoom*.

The *Zoom* option for the Size Mode property fits the image to the control object placeholder size, maintaining the image proportions.

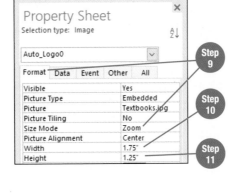

**10** Select the current value in the *Width* property box and then type 1.75.

**11** Select the current value in the *Height* property box, type 1.25, and then press Enter.

**12** Close the Property Sheet task pane.

**13** Click on the *Books* title to select the form title control object.

An orange border around the control object indicates the object is selected.

**14** Click the Form Layout Tools Format tab.

**15** Click the *Font Size* option box arrow and then click *48*.

**16** Click the Save button on the Quick Access Toolbar and then click OK in the Save As dialog box to accept the default form name *Books*.

**17** Close the Books form.

**18** Double-click the *Books* form name in the Navigation pane to reopen the form, scroll through a few records, and then close the form. Leave the database open for the next topic.

# 13.9 Creating, Editing, and Viewing a Report

Create a report using techniques similar to those used to create a form. The Reports group on the Create tab has a Report tool similar to the Form tool. Other buttons in the Reports group include options to design a report from a blank page, create a report using the Report Wizard, or generate mailing labels using the Label Wizard. Modify a report with buttons on the Report Layout Tools tabs. Change the page layout options for printing purposes with buttons on the Report Layout Tools Page Setup tab.

1. With the **UsedTextbooks–YourName** database open, click to select the *Sales* table name in the Navigation pane.

2. Click the Create tab and then click the Report button in the Reports group.

A report is created with all the fields in the Sales table arranged in a tabular layout. By default, Access includes the current date and time, page numbering, and totals for numeric fields in all reports. Reports use the same theme as forms so that all objects have a consistent appearance.

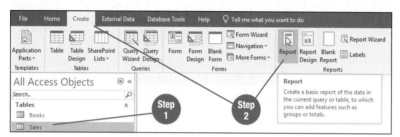

3. Click the Logo button in the Header/Footer group on the Report Layout Tools Design tab and then double-click the file *Textbooks.jpg*.

4. With the logo image control object still selected, open the Property Sheet task pane; change the Size Mode, Width, and Height properties to the same settings that you applied to the image in the previous topic; and then close the Property Sheet task pane. If necessary, refer to Steps 8 to 12 in Topic 13.8.

5. Click the *Sales* title to select the report title control object, click the Report Layout Tools Format tab, click the *Font Size* option box arrow, and then click *48*.

6. Click to select the current date control object near the top right of the report.

7. Drag the right border of the control object to the left until the control ends just left of the vertical dashed line that extends the height of the report. If pound symbols display after resizing the control, drag the left border of the control object towards the left margin until the date displays again.

The vertical dashed line indicates a page break. Resize control objects so that all objects are to the left of the vertical dashed line to fit on one page.

8. Click to select the *Book ID* column heading control object.

9. Drag the right border of the control object left about one-half inch to resize it to the approximate width shown in the image at right.

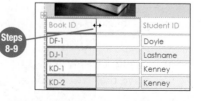

**Watch & Learn**

**Hands On Activity**

------- Skills

Create a report
Resize control objects

**Tutorials**
Creating a Report
Formatting a Report

------- Oops!

Textbooks image not shown? Access defaults to the last folder used in the Insert Picture dialog box. If necessary, navigate to the Ch13 folder in StudentDataFiles.

------- app tip

A control object is filled with pound symbols (#) if the object is made too narrow for the content to be fully displayed. In that case, increase the width of the control object until the content redisplays.

Notice that resizing a column heading control object resizes the entire column.

**10** Click to select the *Page 1 of 1* control object and resize the control until the right border is just left of the vertical dashed line.

**11** Click to select the control object with the total at the bottom of the *Sale Amount* column and then drag the bottom border of the control down until the value is entirely visible within the object.

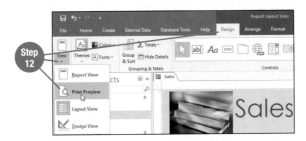

page number control resized to end left of the page break in Step 10

**12** Click the Report Layout Tools Design tab, click the View button arrow in the Views group, and then click *Print Preview*. Compare your report with the one shown in Figure 13.3. If necessary, switch to Layout view, resize control objects, and then switch back to Print Preview.

Buttons to switch views are also available at the right end of the Status bar.

**Figure 13.3**

Shown is the Sales report for Topic 13.9 displayed in Print Preview.

**13** Click the Close Print Preview button in the Close Preview group.

**14** Click the View button arrow and then click *Report View*.

**Report view** is the view in which a report opens from the Navigation pane. Report view is used to view data on the screen instead of printing a hard copy. A report cannot be edited in Report view.

**15** Close the Sales report, saving changes to the report design and accepting the default report name of *Sales*. Leave the database open for the next topic.

**quick steps**

**Create a Report**
1. Click to select table or query name in Navigation pane.
2. Click Create tab.
3. Click Report button.
4. Modify report as required.
5. Click Save button.
6. Type report name.
7. Click OK.

**Tutorials**

Managing Control Objects in a Report

Customizing a Report in Print Preview

## 13.10 Compacting, Repairing, and Backing Up a Database

A database file becomes larger and fragmented over time as new records are added, edited, and deleted. The file size for the database may become larger than is necessary if the space previously used by records that have since been deleted is not compacted. Access provides a **Compact & Repair Database button** that eliminates unused space and reduces the file size.

Backing up a database file should be done regularly for record keeping and data-loss prevention purposes. In most businesses, backing up of a database is a scheduled activity that occurs automatically, with some parts of the database backed up hourly. A database administrator is responsible for scheduling the backup routines. In this topic, you will perform a manual backup procedure, which can be done at any time a backup is deemed necessary.

1. With the **UsedTextbooks-YourName** database open, click the File tab.

2. Click the Compact & Repair Database button in the Info backstage area.

Access closes all objects and the Navigation pane during a compact and repair routine. The Navigation pane redisplays when the compacting and repairing is complete. For a large database file, compacting and repairing the database file may take a few moments to process.

If a database is shared, make sure no one else is using the database before starting a compact and repair operation.

3. Click the File tab and then click *Options*.

In the Access Options dialog box, a database file can be set to compact and repair each time the file is closed.

4. Click *Current Database* in the left pane of the Access Options dialog box.

5. Click to insert a check mark in the *Compact on Close* check box in the *Application Options* section.

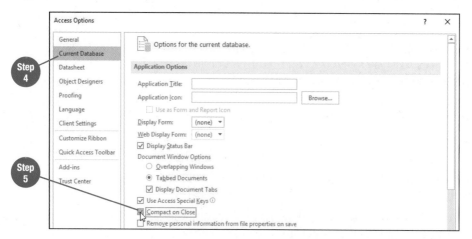

**Watch & Learn**

**Hands On Activity**

●  —  —  —  —  —  —  Skills

Compact and repair a database

Create a backup copy of a database

**Tutorial**

Compacting and Repairing a Database

⑥ Click OK to close the Access Options dialog box.

⑦ Click OK in the message box that says the database must be closed and reopened for the option to take effect.

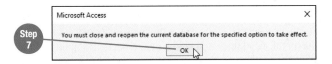

⑧ Click the File tab and then click *Save As*.

⑨ In the Save As backstage area, click *Back Up Database* in the *Advanced* section in the Save Database As panel.

⑩ Click the Save As button.

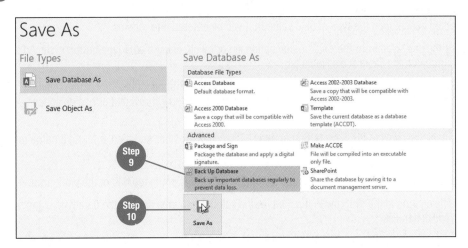

⑪ Click the Save button in the Save As dialog box.

By default, the backup copy of the database is saved in the same folder as the current database and the file name is the same database file name with the current date added after an underscore at the end of the name, for example, UsedTextbooks -YourName_*currentdate*.

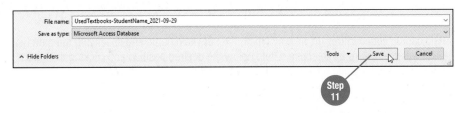

A backup copy of a database is normally saved on another storage medium, such as a server connected to the PC housed in another building, or on a storage device at a cloud service provider, such as OneDrive. By saving a backup copy of a database to an offsite location, the data can be restored in the event of a fire, theft, or other loss of the PC on which the original database is stored.

⑫ Close the database.

● - - - - **quick steps**

**Compact and Repair a Database**
1. Click File tab.
2. Click Compact & Repair Database button.

**Turn on Compact on Close**
1. Click File tab.
2. Click *Options*.
3. Click *Current Database*.
4. Click *Compact on Close*.
5. Click OK twice.

**Back Up a Database**
1. Click File tab.
2. Click *Save As*.
3. Click *Back Up Database*.
4. Click Save As button.
5. Click Save.

## Topics Review

| Topic | Key Concepts |
|---|---|
| **13.1 Creating a New Database File and Understanding Table Design Guidelines** | Access requires the name and file storage location before creating a new database because Access saves changes automatically as you work with data. |
| | Planning a new table involves several steps, some of which include dividing the data into fields, assigning each field a name, assigning each field a data type, deciding the field that will be the primary key, and including a common field to join a table to another table if necessary. |
| | To create a new database file, choose *Blank database* on the Access Start screen, type the file name, browse to the desired drive and/or folder, and then click the Create button. |
| | When a new database is created, Access displays a blank table datasheet. |
| **13.2 Creating a New Table Using a Datasheet** | Begin a new field in a table datasheet by first selecting the data type and then typing the field name. |
| | Several data types are available, such as Short Text, Number, Currency, and Date & Time. A data type is selected based upon the type of field value that will be entered into records. |
| | Choose the data type from the *Click to Add* option list or by choosing a data type button in the Add & Delete group on the Table Tools Fields tab. |
| | An entry in the **Caption property** is used as a descriptive title that becomes the column heading for a field in a datasheet. Use a caption if the field name has been abbreviated or has no spaces between words. |
| | Save a table by clicking the Save button on the Quick Access Toolbar and then entering a name for the table. |
| **13.3 Creating a New Table Using Design View and Assigning a Primary Key** | In Design view, a new table is created by defining fields in rows in the top half of the work area. |
| | Type a field name in the first row in the *Field Name* column and then specify the data type using the *Data Type* option list. |
| | An optional description can be added for a field, with additional information about the purpose of the field or with instructions on what to type into the field. |
| | Once the fields are defined, assign the primary key by placing an insertion point anywhere within the field name and then clicking the Primary Key button in the Tools group on the Table Tools Design tab. |
| **13.4 Adding Fields to an Existing Table** | Open a table and use the *Click to Add* column to add a new field to the end of an existing table datasheet. |
| | Select a column in a datasheet and use the buttons in the Add & Delete group on the Table Tools Fields tab to add a new field at the right of the selected field. |
| | A new field can be added in Design view by typing the new field name and specifying the data type in the next available row, or by selecting a field below the desired position and clicking the Insert Rows button in the Tools group on the Table Tools Design tab. |
| **13.5 Modifying Field Properties Using Design View** | A **field property** is a single characteristic or attribute of a field that customizes, formats, or changes the behavior of the field. Each field in a table has a set of associated field properties. |
| | The lower half of the Design view window is the **Field Properties pane** in which properties for a selected field are modified. |
| | The Field Size property is used to limit the number of characters that can be entered into a field. |
| | The Default Value property is used to specify a field value that is automatically entered in the field in new records. |
| | Other field properties that are often modified are the Format, Validation Rule, Validation Text, and Required field properties. |

*continued...*

| Topic | Key Concepts |
|---|---|
| **13.6 Creating a Lookup List** | A **lookup list** is a list of field values that displays when a field is made active in a datasheet or form. |
| | Items in the lookup list can be predefined or extracted from one or more fields in another table. |
| | The **Lookup Wizard** presents a series of dialog boxes to help create the field properties for a lookup list field. |
| | Create a custom list of predefined entries by choosing *I will type in the values that I want.* in the first Lookup Wizard dialog box. |
| | Type the list entries in the *Col1* column and adjust the column width in the second Lookup Wizard dialog box. Click Finish in the third dialog box. |
| | List entries and list options can be edited using the Lookup tab in the Field Properties pane. |
| **13.7 Displaying and Editing a Relationship** | A relationship is when two tables are joined together on a common field. |
| | Black join lines connecting a common field between two table field list boxes indicates a relationship has been created. |
| | A **one-to-one relationship** means that the two tables are joined on the primary key in each table. |
| | **Enforce Referential Integrity** causes Access to check that new records are entered into the primary table first before records with a matching field value are entered into the related table. |
| | A **primary table** is the table in which the common field is the primary key and into which new records are entered first. The primary table name is the table name at the left below *Table/Query* in the Edit Relationships dialog box. |
| | In a **one-to-many relationship**, the common field used to join the tables is the primary key in only one table (the primary table). |
| | A field added to a table that is not a primary key and is added for the purpose of creating a relationship is called a **foreign key**. |
| **13.8 Creating and Editing a Form** | The Form button in the Forms group on the Create tab creates a new form with all fields in the selected table or query arranged in a columnar layout. |
| | **Layout view** is the view in which you modify a report or form structure and appearance using buttons on the three Form Layout Tools tabs: Design, Arrange, and Format. |
| | **Form view** is the view in which a form is displayed when opened from the Navigation pane and is the view used to add, edit, and delete data. |
| | The Themes button on the Form Layout Tools Design tab is used to change the color scheme and fonts for a form. |
| | Use the Logo button to choose an image to display near the upper left of a form. |
| | A **control object** is a rectangular placeholder for content placed in a form or report. Each item on a form or report is placed in a control object. |
| | Each control object can be selected and modified to change the appearance or properties of the content. |
| | Open the Property Sheet task pane to make changes to a control object properties. |
| | Change the Size Mode property of an image to *Zoom* to fit the content to the object size with the height and width proportions maintained. |
| | A control object can be resized by changing the values for the *Width* and *Height* in the Property Sheet task pane. |

*continued...*

| Topic | Key Concepts |
|---|---|
| **13.9 Creating, Editing, and Viewing a Report** | The Report tool in the Reports group on the Create tab creates a report with all the fields in the selected table or query in a tabular arrangement. |
| | Access creates a current date and time control, a page number control, and a total control for each numeric column in a new report. |
| | The vertical dashed line in a report indicates a page break. |
| | Resizing a column heading control object resizes the entire column. |
| | Buttons on the Report Layout Tools Page Setup tab are used to change page layout options for printing purposes. |
| | **Report view** is the view in which a report is displayed when opened from the Navigation pane and is the view that displays the data in the report. |
| **13.10 Compacting, Repairing, and Backing Up a Database** | The compact and repair process eliminates unused space in the database file. |
| | Use the **Compact & Repair Database button** in the Info backstage area to perform a compact and repair operation. |
| | During the compact and repair routine, Access closes all objects and the Navigation pane. |
| | Turn on the *Compact on Close* option in the Access Options dialog box with the Current Database pane active to instruct Access to perform a compact and repair operation each time the database is closed. |
| | Display the Save As backstage area and choose *Back Up Database* in the *Advanced* section to create a backup copy of the current database. |
| | Access adds the current date after an underscore character to the end of the current database file name when a backup is created. |
| | A backup copy of a database is normally saved to an offsite storage location, such as a server in another building or a cloud provider, so data can be restored in the event of fire, theft, or other loss. |

## Topics Review

# Integrating Word, Excel, PowerPoint, and Access Content

## Chapter 14

The Office suite is designed to easily share and integrate data or objects among the applications or to other resources, such as a web page. For some tasks, you may have portions of a project distributed across more than one application. For example, you may have a chart in Excel and a list in Access that you want to add into a report in Word. The ability to integrate data from one application to another allows you to use the application that best fits each task and to assemble a comprehensive product without duplicating the of project contributors. In other cases, you may want to distribute content created in a format not native to the source application.

In Chapter 3, you used the Copy and Paste buttons in the Clipboard group to copy text, an image, and a chart between Word, Excel, and PowerPoint. Copy and paste is the best method for situations when the data to be shared is not large and is not likely to need updating. You also learned in Chapter 3 how to publish a document as a PDF, which makes a copy of the document in a format that can be viewed in a browser or PDF reader app. In this chapter, you will learn other methods for integrating and distributing data and objects that include importing, exporting, embedding, and linking. These methods allow greater automation and collaboration to aid in content management. For example, if you update a native chart file it automatically updates in other linked files. This can save a lot of content management time—something employers value.

### Read & Learn

## Learning Objectives

**14.1** Import Excel worksheet data into a table in Access

**14.2** Export an Access query to Excel

**14.3** Embed an Excel chart into a Word document

**14.4** Embed Excel data into a PowerPoint presentation and edit the embedded data

**14.5** Link an Excel chart with a PowerPoint presentation, edit the chart, and update the link

**14.6** Create slides from a Word outline

**14.7** Export a PowerPoint presentation as a video

### Content Online

The online course includes additional training and assessment resources.

# 14.1 Importing Excel Worksheet Data into Access

**Watch & Learn**

A new Access table can be created from data in an Excel worksheet, or Excel data can be appended to the bottom of an existing Access table. Because an Excel worksheet and an Access datasheet use the same column-and-row structure, the two applications are often used to interchange data. To facilitate the import, the Excel worksheet should be set up like an Access datasheet, with the field names in the first row and with no blank rows or columns within the data. The **Import Spreadsheet Wizard** in Access is used to facilitate the import operation by providing a series of dialog boxes that prompt the user through the steps to create the table or add data to the bottom of an existing datasheet.

**— — — — — Skills**

Create a table by importing Excel data

Modify imported table design

1. Start Access and open the **Parking** database from the Ch14 folder in StudentDataFiles.

2. Use Save As to save a copy of the database as **Parking-YourName** in a new folder, *Ch14*, within CompletedTopics on your storage medium. Accept the default options *Save Database As* and *Access Database* in the Save As backstage area.

3. Click the Enable Content button in the security warning message bar.

4. Click the External Data tab.

5. Click the New Data Source button in the Import & Link group, point to *From File*, and then click *Excel*.

**Tutorial**

Importing Data to a New Table

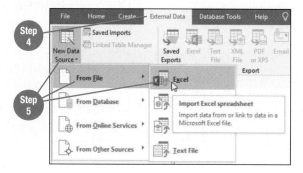

**— — — — — — app tip**

In Excel, click the Data tab and use buttons in the Get External Data group to import data from other sources into a worksheet. You can import into Excel from an Access table, from a web page, and from a text file.

6. Click the Browse button in the Get External Data – Excel Spreadsheet dialog box.

7. Navigate to the Ch14 folder in StudentDataFiles in the File Open dialog box and then double-click *ParkingRecords*.

8. Click OK to accept the selected option *Import the source data into a new table in the current database*. Click Open if a Microsoft Access Security Notice message box appears warning you that a potential security concern has been identified.

**Tutorial**

Linking Data to a New Table

Use the *Append* option if the table already exists in the database and you want to add new records from an Excel worksheet to the end of the existing datasheet table.

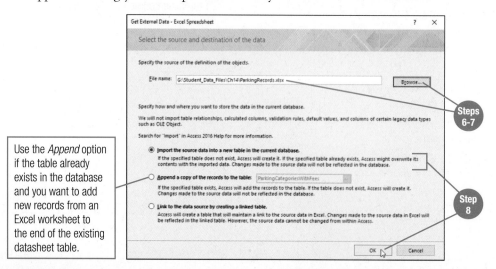

**9** In the first Import Spreadsheet Wizard dialog box, Click Next to accept the worksheet *Student Parking Records*.

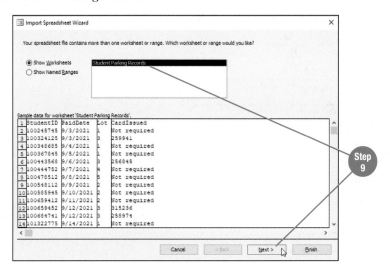

**10** In the second Import Spreadsheet Wizard dialog box, click Next with a check mark already inserted in the *First Row Contains Column Headings* check box.

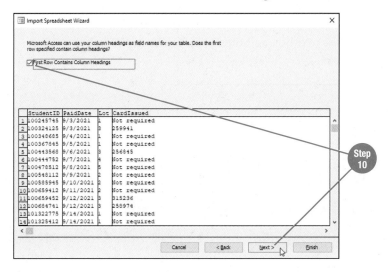

**11** In the third Import Spreadsheet Wizard dialog box, click the *PaidDate* column heading and look at the option selected in the *Data Type* option box.

Notice that Access has correctly identified the data as a Date field. In this dialog box, you can review each column and modify the options in the *Field Options* section as needed, or you can elect to make changes in Design view after the import is completed. If a column exists in the Excel worksheet that you do not wish to import into the table, select the column and insert a check mark in the *Do not import field (Skip)* check box.

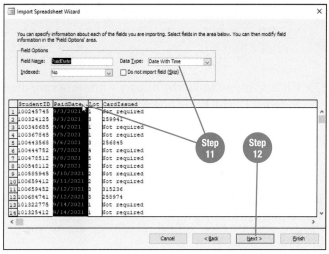

**12** Click Next.

**13** In the fourth Import Spreadsheet Wizard dialog box, click *Choose my own primary key*.

Access inserts the *StudentID* field name in the primary key option box (the first column in the worksheet).

**14** Click Next.

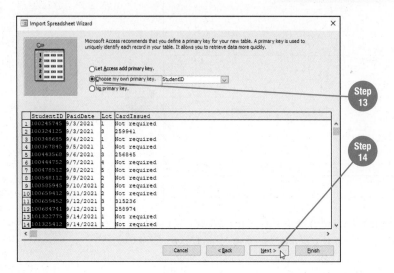

**15** Type ParkingSales in the *Import to Table* text box and then click Finish in the last Import Spreadsheet Wizard dialog box.

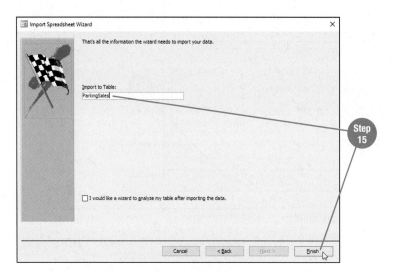

**16** Click Close to finish the import without saving the import steps in the Get External Data - Excel Spreadsheet dialog box.

For situations in which you frequently import from Excel to Access, you can save the import specifications and repeat the import later using the same settings.

**app tip**

You can import text into a Word document from another source by using the Open dialog box. Word will attempt to convert text from another file type into a new Word document. Word successfully imports text from text files, rich text format files, OpenDocument text files, and WordPerfect files. You can even open a PDF document in Word, and Word converts the PDF file into an editable Word document.

**17** Open the ParkingSales table from the Navigation pane and review the datasheet.

| StudentID | PaidDate | Lot | CardIssued | Click to Add |
|---|---|---|---|---|
| 100245745 | 9/3/2021 | 1 | Not required | |
| 100324125 | 9/3/2021 | 3 | 259941 | |
| 100348685 | 9/4/2021 | 1 | Not required | |
| 100367845 | 9/5/2021 | 1 | Not required | |
| 100443568 | 9/6/2021 | 3 | 256845 | |
| 100444752 | 9/7/2021 | 4 | Not required | |
| 100478512 | 9/8/2021 | 5 | Not required | |
| 100548112 | 9/9/2021 | 2 | Not required | |
| 100585945 | 9/10/2021 | 2 | Not required | |
| 100659412 | 9/11/2021 | 2 | Not required | |
| 100659452 | 9/12/2021 | 3 | 315236 | |
| 100684741 | 9/12/2021 | 3 | 258974 | |
| 101322775 | 9/14/2021 | 1 | Not required | |
| 101325412 | 9/14/2021 | 1 | Not required | |
| 101325685 | 9/14/2021 | 4 | Not required | |
| 101332486 | 9/17/2021 | 3 | 315235 | |
| 101334745 | 9/17/2021 | 2 | Not required | |
| 101334799 | 9/19/2021 | 1 | Not required | |
| 101348863 | 9/21/2021 | 5 | Not required | |
| 101352658 | 9/21/2021 | 3 | 248975 | |
| 101455623 | 9/21/2021 | 2 | Not required | |
| 101658235 | 9/21/2021 | 5 | Not required | |
| 101745866 | 9/22/2021 | 1 | Not required | |
| 104875111 | 9/22/2021 | 2 | Not required | |
| 110542642 | 9/22/2021 | 3 | 289456 | |

Datasheet view for table created from imported Excel worksheet data

**quick steps**

**Import Worksheet Data into Access**
1. Open database.
2. Click External Data tab.
3. Click New Data Source button in Import & Link group, point to *From File*, and click *Excel*.
4. Click Browse.
5. Navigate to and double-click Excel file.
6. Click OK.
7. Click Next with worksheet selected.
8. Click Next with *First Row Contains Column Headings* selected.
9. Change *Field Options* for columns, if desired, and click Next.
10. Choose primary key field and click Next.
11. Type table name and click Finish.
12. Click Close.

**18** Click the Home tab, click the top part of the View button in the Views group to switch to Design view, and then review the field names and data types for the new table.

**19** Click the *Data Type* option box arrow for the *Lot* field name and then click *Short Text*.

In the Parking database, the *Lot* field should be defined as Short Text, because lot numbers are not field values that you would add or subtract in a calculation. When Access imported the Excel worksheet data, the *Lot* column contained all numeric values, so Access assigned the field the Number data type option. In a database, a numeric data type option is used for those fields that will be calculated or totaled in queries, forms, and reports.

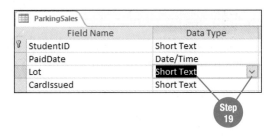

| ParkingSales | |
|---|---|
| Field Name | Data Type |
| StudentID | Short Text |
| PaidDate | Date/Time |
| Lot | Short Text |
| CardIssued | Short Text |

Step 19

**20** Save and then close the table. Leave the database open for the next topic.

## 14.2 Exporting an Access Query to Excel

Access table data can be exported to Excel for situations where you want to use the mathematical analysis tools in Excel. Access creates a copy of the selected table or query data in an Excel worksheet file in a drive and folder that you specify. Buttons in the Export group on the External Data tab provide options to send a copy of Access data in a variety of file formats.

**1** With the **Parking–YourName** database open, click the Create tab and then click the Query Design button.

**2** In the Show Table dialog box, double-click each of the four table names to add all four table field list boxes to the query, and then click the Close button.

When a table is created by importing, a relationship will not exist between the new table and other tables in the database. All tables in a query should be joined so that records are not duplicated in the query results datasheet. When a relationship is necessary (such as in a query), you can join the tables by dragging the common field name from one table field list box to the same field in the other table field list box.

**3** Drag the *StudentID* field name in the ParkingSales field list box to *StudentID* in the Students field list box.

A black join line appears next to the *StudentID* field name between the table field list boxes when you release the mouse, as shown in the image at the right.

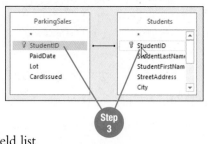

**4** Drag the *Lot* field name in the ParkingSales field list box to *LotNo* in the ParkingLots field list box.

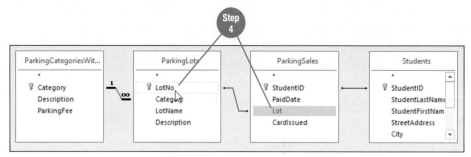

**5** Double-click the following fields to add the fields to the query design grid. (Note that you are selecting fields in all four table field list boxes.)

| Field Name | Table Name |
|---|---|
| *StudentID* | ParkingSales |
| *StudentFirstName* | Students |
| *StudentLastName* | Students |
| *PaidDate* | ParkingSales |
| *Lot* | Parking Sales |
| *Description* | ParkingLots |
| *ParkingFee* | ParkingCategoriesWithFees |

| StudentID | StudentFirstName | StudentLastName | PaidDate | Lot | Description | ParkingFee |
|---|---|---|---|---|---|---|
| ParkingSales | Students | Students | ParkingSales | ParkingSales | ParkingLots | ParkingCategoriesWit |
| ☑ | ☑ | ☑ | ☑ | ☑ | ☑ | ☑ |

Step 5

**6** Click the Run button in the Results group on the Query Tools Design tab.

**Watch & Learn**

------- skills

Export an Access query to Excel

**Tutorials**

Exporting Access Data to Excel

Exporting Access Data to Word

------- app tip

Create relationships in the Relationships window if the tables to be joined will be used together in other objects, such as a form or report.

7. Save the query as **ParkingSales2021** and then close the query.

8. Click to select the *ParkingSales2021* query name in the Navigation pane.

9. Click the External Data tab and then click the Excel button in the Export group.

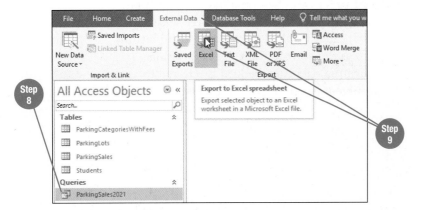

● - - - - quick steps

**Export a Query to Excel**
1. Open database.
2. Select query name.
3. Click External Data tab.
4. Click Excel button in Export group.
5. Click Browse button.
6. Type file name and navigate to destination folder.
7. Click Save.
8. Specify export options.
9. Click OK.
10. If Excel opened, review worksheet and close Excel.
11. Click Close.

10. In the Export – Excel Spreadsheet dialog box, click the Browse button, type ParkingSales-YourName in the *File name* text box, navigate to the Ch14 folder in CompletedTopics on your storage medium, and then click Save.

11. Click to insert a check mark in the *Export data with formatting and layout.* check box and in the *Open the destination file after the export operation is complete* check box.

12. Click OK.

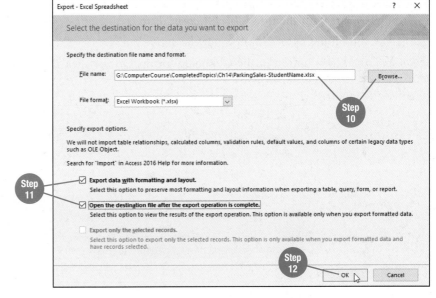

Excel is started automatically with the data from the query results datasheet shown in a worksheet. Notice the first row contains the field names from the query, and the worksheet tab is renamed to the query name.

| | A | B | C | D | E | F | G |
|---|---|---|---|---|---|---|---|
| 1 | StudentID | StudentFirstName | StudentLastName | PaidDate | Lot | Description | Parking Fee |
| 2 | 100245745 | Anna | Takacis | 9/3/2021 | 1 | Main campus north of Administration building | $250.00 |
| 3 | 100324125 | Jose | Ramirez | 9/3/2021 | 3 | Main campus south of Technology wing | $200.00 |
| 4 | 100348685 | Celine | Gauthier | 9/4/2021 | 1 | Main campus north of Administration building | $250.00 |
| 5 | 100367845 | Sara | Blewett | 9/5/2021 | 1 | Main campus north of Administration building | $250.00 |
| 6 | 100443568 | Dwayne | Kenney | 9/6/2021 | 3 | Main campus south of Technology wing | $200.00 |

first six rows in Excel after query is exported

Green triangles display next to data in the *StudentID* and *Lot* columns because the data is numeric but was exported from Access as text. Green triangles in Excel flag data that is a potential error. You can ignore the error flags here.

13. Close Excel to return to Access.

14. Click Close in the Export - Excel Spreadsheet dialog box to finish the export without saving the export steps.

15. Close the database and then close Access.

## 14.3 Embedding an Excel Chart into a Word Document

In Chapter 3, you used Copy and Paste features to duplicate text and a chart between applications. Another way to duplicate content between applications is to paste copied data as an **embedded object**. Embedding, like copying and pasting, inserts a duplicate of the selected text or object at the desired location. The application in which the data originally resides is called the **source application**, and the data that is copied is referred to as the **source data**. Using a similar naming scheme, the application in which the data is embedded is referred to as the **destination application**, and the document, worksheet, or presentation into which the embedded object is placed is referred to as the **destination document**.

**Watch & Learn**

------- Skills

Embed an Excel chart into a document

**Tutorial**
Copying and Pasting Data between Programs

1. Start Excel and then open **SocialMediaStats** from the Ch14 folder in StudentDataFiles. If necessary, click the Enable Editing button.

2. Start Word and then open **SocialMediaProject** from the Ch14 folder in StudentDataFiles. If necessary, click the Enable Editing button.

3. Use Save As to save a copy of the Word document as **SocialMediaProject –YourName** in the Ch14 folder within CompletedTopics.

4. Click the Excel button on the taskbar to switch to Excel and then click to select the pie chart with the title *Global Market Share*.

5. Click the Home tab if Home is not the active tab.

6. Click the Copy button in the Clipboard group. Do *not* click the arrow on the button.

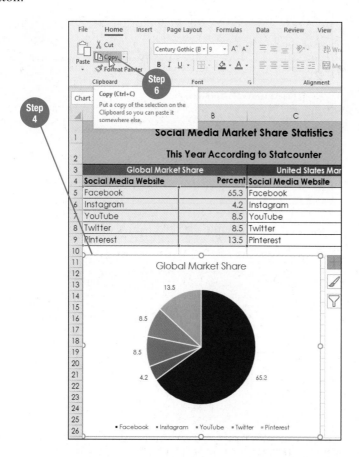

7. Click the Word button on the taskbar to switch to Word and then click to position the insertion point at the left margin on the blank line two line spaces below the first table.

8  Click the Paste button arrow and then click *Use Destination Theme & Embed Workbook* (first button in *Paste Options* section).

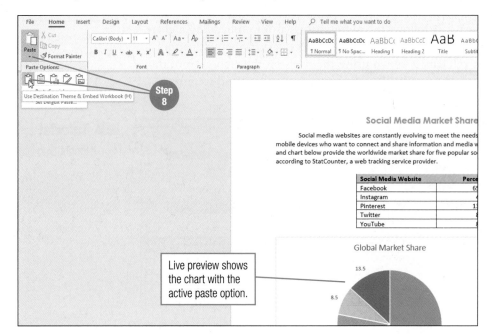

●━━━━ quick steps

**Embed an Excel Chart into Word**
1. Open worksheet in Excel.
2. Open document in Word.
3. Make Excel active, select and copy chart.
4. Switch to Word.
5. Position insertion point.
6. Click Paste button arrow.
7. Click *Use Destination Theme & Embed Workbook*.

☁ **Tutorial**
   Embedding an Object

9  Click to select the chart object in the document and then click the Center button in the Paragraph group on the Home tab.

The Chart feature is standardized in Word, Excel, and PowerPoint. A chart embedded within any of the three applications offers the Chart Tools tabs and three chart-editing buttons with which the chart can be modified after being embedded.

10  Click the Chart Elements button (plus symbol), point to *Legend* (right-pointing arrow appears), click the right-pointing arrow, and then click *Right*. Click in the document away from the chart to deselect the chart object.

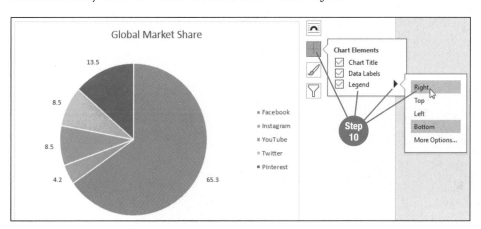

11  Click the Excel button on the taskbar, click the pie chart with the title *United States Market Share*, and then click the Copy button.

12  Click the Word button on the taskbar. Click to position the insertion point at the bottom of the document, and then embed, center, and format the pie chart by completing steps similar to Steps 8 through 10.

13  Save the revised document using the same name (**SocialMediaProject –YourName**) and then close Word. Leave Excel and the **SocialMediaStats** workbook open for the next topic.

## 14.4 Embedding Excel Data into PowerPoint and Editing the Embedded Data

Embedding text or worksheet data follows the same process as for embedding a chart. Double-click an embedded object to edit text or worksheet data in the destination location. Embedded text or cell data is edited using the tools on the ribbon from the source application. Click outside an embedded object to end editing and to restore the ribbon of the destination application.

**Watch & Learn**

─────── Skills

Embed Excel data into a presentation

Edit an embedded table

**Tutorial**

Embedding Objects

1. With Excel active and the **SocialMediaStats** workbook open, select A3:B9 and then click the Copy button.

2. Start PowerPoint and open **SocialMediaPres** from the Ch14 folder in StudentDataFiles.

3. Use Save As to save a copy of the presentation as **SocialMediaPres –YourName** in the Ch14 folder within CompletedTopics.

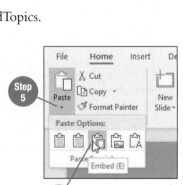

4. Make Slide 3 the active slide.

5. Click the Paste button arrow.

6. Click *Embed* (third option in *Paste Options* section).

7. Click the Drawing Tools Format tab, click the Shape Fill button in the Shape Styles group, and then click *White, Text 1* (second option in *Theme Colors* section).

8. Resize and position the embedded object to the approximate size and position shown in the image below.

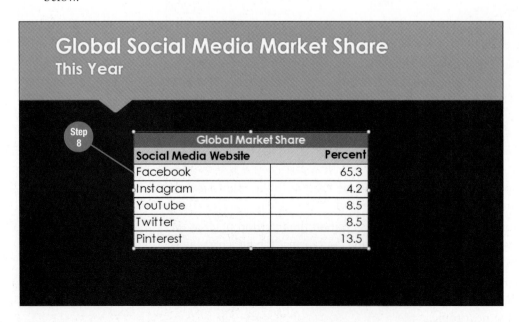

⑨ Double-click anywhere over the cells to open the embedded object for editing.

Notice the embedded cells open in an Excel worksheet and the ribbon changes to Excel's ribbon.

⑩ Select B5:B9 and then click the Decrease Decimal button in the Number group on the Home tab.

**quick steps**

**Edit an Embedded Object**
1. Double-click embed-
   ded object.
2. Make desired changes.
3. Click outside object.

⑪ Click on the slide away from the embedded object to end editing and restore the PowerPoint ribbon.

⑫ Click the Excel button on the taskbar, select C3:D9, and then click the Copy button.

⑬ Click the PowerPoint button on the taskbar. Make Slide 4 the active slide, and embed, format, resize, and position the copied cells by completing steps similar to those in Steps 5 through 11.

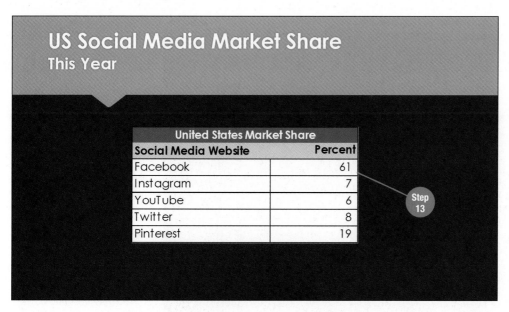

⑭ Save the revised presentation using the same name (**SocialMediaPres –YourName**) and then close PowerPoint. Leave Excel and the **SocialMediaStats** workbook open for the next topic.

# 14.5 Linking an Excel Chart with a Presentation and Updating the Link

If the data that you want to integrate between two applications is continuously updated, copy and link the data instead of copying and pasting or embedding. When copied data is linked, changes made to the source data can be automatically updated in any other document, worksheet, or presentation to which the data is linked. Linking avoids duplicating work and ensures that errors are not made when the same data is changed in more than one location. Linked objects are managed using the **Links dialog box**, where a link can be set to update automatically, a link can be broken, or the source location for a linked object can be changed.

1. With Excel active and the **SocialMediaStats** workbook open, use Save As to save a copy of the workbook as **LinkedSocialMediaStats-YourName** in the Ch14 folder within CompletedTopics.

2. Start PowerPoint and then open **SocialMediaPres**.

3. Use Save As to save a copy of the presentation as **LinkedSocialMediaPres -YourName** in the Ch14 folder within CompletedTopics.

4. Click the Excel button on the taskbar, click to select the *Global Market Share* pie chart, and then click the Copy button.

5. Click the PowerPoint button on the taskbar and then make Slide 3 the active slide.

6. Click the Paste button arrow.

7. Click *Use Destination Theme & Link Data* (third option in *Paste Options* section). If the Design Ideas task pane appears at the right side of the window, close the task pane.

8. Resize and move the chart to the approximate size and position shown in the image below.

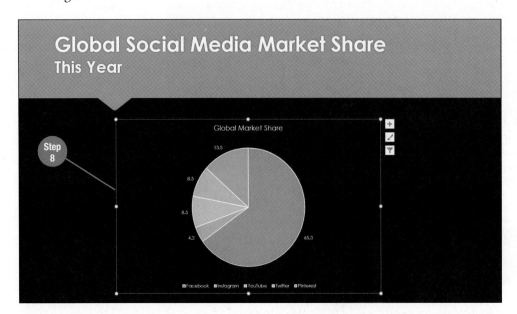

9. Click the Excel button on the taskbar, select the *United States Market Share* pie chart, and then click the Copy button.

**Watch & Learn**

------- Skills

Link an Excel chart with a presentation

Turn on automatic link updates

Edit a linked chart

Update links

**Tutorial**

Linking Objects

----- in the know

Linking is not just for integrating data between two different applications; you can link two documents in Word, two worksheets in Excel, or two tables in Access.

**10** Click the PowerPoint button on the taskbar. Make Slide 4 the active slide and then link, resize, and move the chart by completing steps similar to those in Steps 6 through 8.

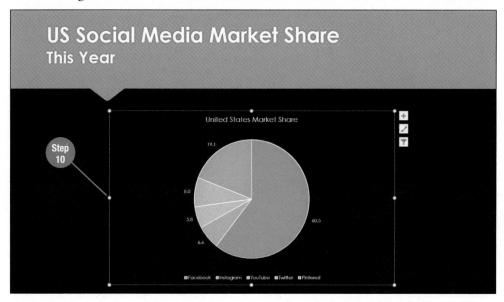

**11** Click the File tab and then click *Edit Links to Files* in the *Related Documents* section at the bottom of the Properties panel (right panel) in the Info backstage area.

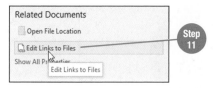

**12** In the Links dialog box, click to select the first link in the *Links* list box and then click to insert a check mark in the *Automatic Update* check box.

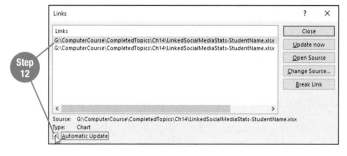

**13** Click to select the second link and then click to insert a check mark in the *Automatic Update* check box.

**14** Click the Close button in the Links dialog box.

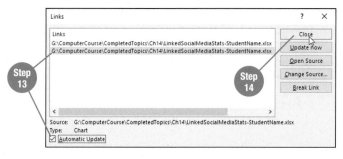

**app tip**

Open the Links dialog box to update the location of the source data if the source data file has been moved to another drive or folder or to break a link. Click the link and then click the Change Source button to browse to the new file location or click the Break Link button to remove the link to the content.

15  Click the Back button to exit the Info backstage area and return to the presentation.

16  Save the revised presentation using the same name (**LinkedSocialMediaPres -YourName**) and then close the presentation. Leave PowerPoint open for a task at Step 22.

17  Click the Excel button on the taskbar.

18  Change the value in D9 from *19.1* to *26.1*.

19  Change the value in D6 from *6.6* to *1.6*.

20  Change the value in D5 from *60.5* to *58.5*.

Notice the pie chart updated after each change in value. The revised chart is noticeably different from the original pie chart.

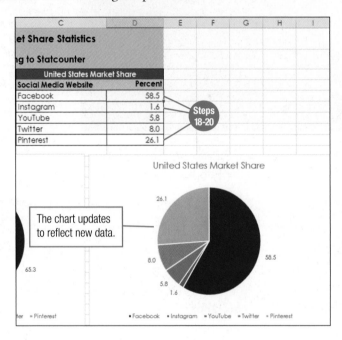

21  Save the revised worksheet using the same name (**LinkedSocialMediaStats -YourName**) and then close Excel.

22  Switch to PowerPoint, if PowerPoint is not already active, and then open **LinkedSocialMediaPres-YourName**.

Because the presentation contains linked data that is set to automatically update, you are prompted to update links. A dialog box with a security notice and the **Update Links button** appears when you open a file that has links to objects outside the document, worksheet, or presentation.

23  Click the Update Links button in the Microsoft PowerPoint Security Notice dialog box.

●──────  app tip

If the source and destination files are both open at the same time, changes made to the source reflect in the destination file immediately.

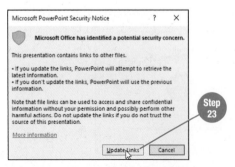

In this instance, clicking the Update Links button is safe from a security standpoint because you are working with files that you have opened and edited. However, if you receive a dialog box like the one you saw in Step 23 for a file that you did not create, exercise caution before updating links. Take a moment to consider where the linked object originated and whether you are sure the source application and data are from a trusted source. Malware is sometimes circulated in Office documents, and updating a link to an infected file could spread malware to your device. If in doubt, the best option is to click the Cancel button and contact the source data creator for more information.

24  Make Slide 4 the active slide. Notice that the chart is updated to reflect the same data as the revised Excel chart.

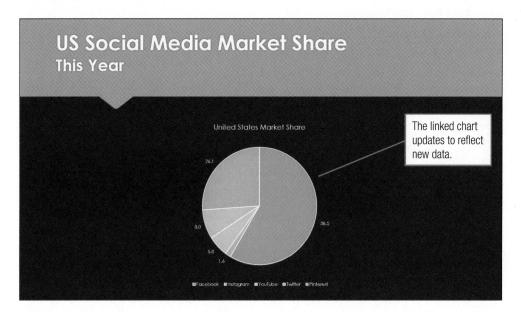

The linked chart updates to reflect new data.

25  Make Slide 3 the active slide and then delete the title inside the chart above the pie.

26  Make Slide 4 the active slide and then delete the title inside the chart above the pie.

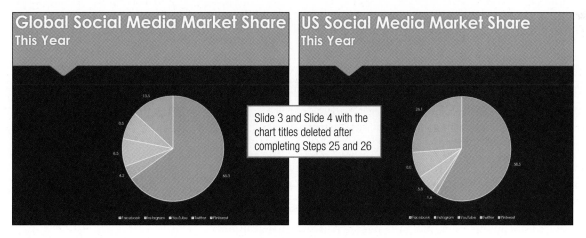

Slide 3 and Slide 4 with the chart titles deleted after completing Steps 25 and 26

27  Save the revised presentation using the same name (**LinkedSocialMediaPres -YourName**) and then close PowerPoint.

# 14.6 Creating Slides from a Word Outline

The text for a presentation can be typed in a Word document as an outline and then imported into PowerPoint to create new slides. In the Word document, apply heading styles to control how the text will be formatted on a slide. A paragraph formatted with the *Heading 1* style becomes the slide title for a new slide. Similarly, a paragraph formatted with the *Heading 2* style becomes a level one bulleted list entry. Each subsequent heading style moves the text to the next bulleted list level.

1 Start Word and then open **AutonomousObjects** from the Ch14 folder in StudentDataFiles.

2 Use Save As to save a copy of the document as **AutonomousObjects -YourName** in the Ch14 folder in CompletedTopics.

3 Examine the document, noting that all text is formatted as a *Heading 1* style, a *Heading 2* style, or a *Heading 3* style.

4 Click to position the insertion point at the end of the sentence *Amazon and others are developing delivery drones for the US market* and press Enter to start a new line.

5 Type Drone taxis are being developed by several companies and then press Enter.

6 Type Uber, in partnership with NASA, is planning to test drone taxi service in two US cities in 2020.

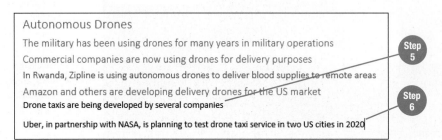

7 Click to position the insertion point in the sentence *Drone taxis are being developed by several companies* and then click the *Heading 2* style option in the Styles group on the Home tab.

8 Click to position the insertion point in the sentence *Uber, in partnership with NASA, is planning to test drone taxi service in two US cities in 2020* and then click the *Heading 3* style option.

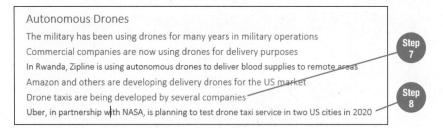

9 Save the revised document using the same name (**AutonomousObjects -YourName**) and then close Word.

10 Start PowerPoint and then open the presentation **AutonomousObjectsPres** from the Ch14 folder in StudentDataFiles.

11 Use Save As to save the presentation as **AutonomousObjectsPres-YourName** in the Ch14 folder within CompletedTopics.

**Watch & Learn**

- - - - - - - **Skills**

Create PowerPoint slides from a Word document

**Tutorial**

Importing a Word Outline

12 Click the New Slide button arrow in the Slides group on the Home tab and then click *Slides from Outline*.

13 In the Insert Outline dialog box, if necessary, navigate to the Ch14 folder in CompletedTopics and then double-click ***AutonomousObjects–YourName***.

PowerPoint converts the Word document text into slides. A progress message displays on the Status bar, and the slides appear when the conversion is done. A document formatted with a theme different from the theme for the presentation may need to have the presentation slides reset after importing the text. Resetting the imported slides reformats all slides to the presentation theme.

14 Click each slide in the presentation to view how the document was converted.

15 Click Slide 2 in the slide thumbnails pane. Hold down the Shift key and then click Slide 5 in the slide thumbnails pane.

This action selects Slide 2 through Slide 5.

16 Click the Reset button in the Slides group on the Home tab.

17 Click each slide in the presentation to view the changes.

Resetting the slides that were imported from Word caused the new slides to format according to the presentation theme (Circuit) instead of the Word document theme (Office).

18 Save the revised presentation using the same name (**AutonomousObjectsPres –YourName**). Leave the presentation open for the next topic.

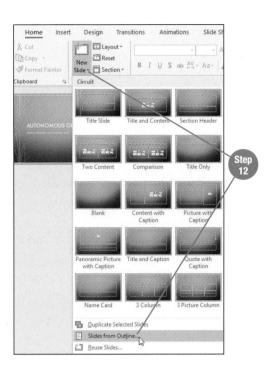

Step 12

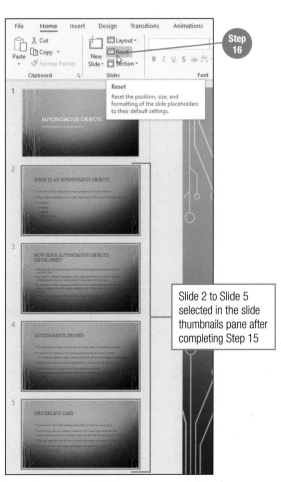

Step 16

Slide 2 to Slide 5 selected in the slide thumbnails pane after completing Step 15

**quick steps**

**Create Slides from a Word Document**
1. Open presentation file in PowerPoint.
2. Click New Slide button arrow in Slides group on Home tab.
3. Click *Slides from Outline*.
4. Navigate to and double-click Word document.

**Reset Slides to Presentation Theme**
1. Select slides to reset.
2. Click Reset button in Slides group on Home tab.

## 14.7 Exporting a Presentation as a Video

After you have completed a presentation, you can export the presentation in a video format to share the content with others. PowerPoint includes options to create a video from the slides at various file sizes and qualities. You can export the presentation with preprogrammed transitions and animations in effect, or you can export the video using a set number of seconds for each slide. Once the video is created, the file can be shared by linking to the video file on a web page or by publishing the video to YouTube or other media-sharing websites.

**Watch & Learn**

━━━━━━ **Skills**

Export a presentation as a video

**Tutorials**

Exporting Presentations

Saving a Presentation in a Different Format

1. With the **AutonomousObjectsPres–YourName** presentation open, click Slide 1 in the slide thumbnails pane if Slide 1 is not the active slide, and then click the Transitions tab.

2. Click the *Wipe* option in the Transition to This Slide gallery and then click the Apply To All button in the Timing group.

3. Select the current entry in the *After* text box in the Timing group, type 0:10, and then press Enter.

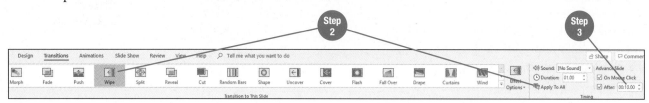

4. Make Slide 2 the active slide, select the entry in the *After* text box, type 0:15 and then press Enter.

5. Change the *After* time for Slide 3, Slide 4, and Slide 5 to 0:25 by completing a step similar to Step 4.

6. Make Slide 2 the active slide, click the View tab, and then click the Slide Master button.

7. Click to select the border of the title placeholder on the slide master.

8. Click the Animations tab.

9. Click the *Float In* option in the Animation gallery.

10. Click to select the border of the bulleted list content placeholder and then click the *Random Bars* option in the Animation gallery.

11. Click the *Start* option box arrow (displays *On Click*) in the Timing group and then click *After Previous*.

12. Select the current entry in the *Duration* text box, type 3.0, and then press Enter.

13. Click the Slide Master tab and then click the Close Master View button.

14. Make Slide 1 the active slide, run through the presentation in a slide show to view the transition and animation effects, and then return to Normal view.

15. Save the revised presentation using the same name (**AutonomousObjectsPres –YourName**).

**16** Click the File tab and then click *Export*.

**17** Click *Create a Video* in the Export backstage area.

**18** With the *Full HD (1080p)* and *Use Recorded Timings and Narrations* options already selected in the Create a Video panel, click the Create Video button.

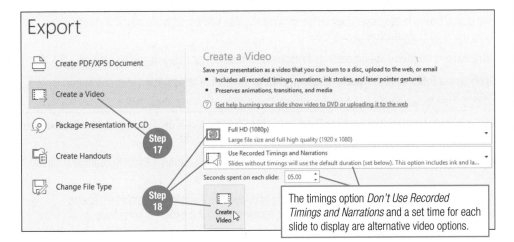

**19** Click the Save button in the Save As dialog box with the file name *AutonomousObjectsPres-YourName* in the *File name* text box and with the *Save as type* option set to MPEG-4 Video (*.mp4).

PowerPoint displays a progress message on the Status bar as the video file is created. The video creation process can take from a few minutes to a few hours depending on the number of slides in the presentation and the complexity of the content, transition, and animation effects.

**20** When the video creation is complete (watch progress message on Status bar), close PowerPoint.

**21** Click the File Explorer button on the taskbar.

**22** Navigate to the Ch14 folder in CompletedTopics on your storage medium and then double-click the video file *AutonomousObjectsPres-YourName*.

**23** Watch the video in the default media player app on your PC (or other video app that you prefer) and then close the window when the video is complete.

**24** Close the File Explorer window.

## Topics Review

| Topic | Key Concepts |
|---|---|
| **14.1 Importing Excel Worksheet Data into Access** | In Access, the **Import Spreadsheet Wizard** starts when you choose an Excel workbook from the New Data Source option in the Import & Link group on the External Data tab. |
| | Five dialog boxes in the Import Spreadsheet Wizard guide you through the steps to create a new table using data in an Excel worksheet. |
| | You can save the Excel import settings to repeat the import later using the same settings. |
| | Open an imported table in Access in Design view to modify the table design after the import is complete. |
| **14.2 Exporting an Access Query to Excel** | Export Access table or query data to Excel using the Excel button in the Export group on the External Data tab. |
| | Specify the file name, drive and folder, and export options in the Export – Excel Spreadsheet dialog box. |
| | You can elect to export the data with formatting and layout options in the datasheet, and to automatically open Excel with the worksheet displayed when the export is complete. |
| | Export specifications can be saved to repeat the export later. |
| **14.3 Embedding an Excel Chart into a Word Document** | An **embedded object** is data that has been copied from a document, worksheet, or presentation and pasted as an object in another application. |
| | The application from which data is copied is called the **source application**. The data that is copied is referred to as the **source data**. |
| | The **destination application** is the application that receives the copied data. The **destination document** refers to the document, worksheet, or presentation into which copied data is pasted as an object. |
| | Click the Paste button arrow and then choose the desired embed option to embed copied data as an object in the destination document. |
| **14.4 Embedding Excel Data into PowerPoint and Editing the Embedded Data** | Double-click an embedded object to open the object data for editing using the source application ribbon and tools. |
| | Click outside the embedded object to end editing and restore the destination application ribbon. |
| **14.5 Linking an Excel Chart with a Presentation and Updating the Link** | Source data that is continuously updated should be linked (instead of copied and pasted or copied and embedded) to avoid duplication of work and to reduce errors made when data is entered more than once. |
| | Click the Paste button arrow in the destination document and then choose the desired link option to link copied data. |
| | Click *Edit Links to Files* in the Properties panel in the Info backstage area to open the Links dialog box in which you manage links to source data. |
| | Select a link in the **Links dialog box** and insert a check mark in the *Automatic Update* check box to turn on automatic updates for the link. |
| | Click the **Update Links button** in the security warning dialog box that appears when you open a file with a linked object to update data from the source. |
| | Exercise caution when updating links to source data that you did not create yourself. |

*continued…*

| Topic | Key Concepts |
|---|---|
| **14.6 Creating Slides from a Word Outline** | Text typed in a Word document as an outline can be imported into PowerPoint to create new slides. |
| | Text formatted in the document with the *Heading 1* style becomes a new slide title and text formatted with the *Heading 2* style becomes a first level bulleted list entry. |
| | Click the New Slide button arrow in the Slides group on the Home tab and then click *Slides from Outline* to import text in a Word document into PowerPoint. |
| | The imported new slides may need to be reset in PowerPoint to apply the presentation theme formatting if the document theme is different than the presentation theme. |
| | Select the slides to be reset and then click the Reset button in the Slides group on the Home tab. |
| **14.7 Exporting a Presentation as a Video** | Export a presentation as a video to share the content with others by linking to the video file on a web page or by publishing the video to a media sharing website, such as YouTube. |
| | Apply the desired transition, animation, and timing options to the slides before exporting. |
| | When the presentation is complete, click the File tab, click *Export*, and then click the *Create a Video* option. |
| | Change the video quality, file size, and timings options, if necessary, and then click the Create Video button in the Create a Video panel. Navigate to the drive and folder where you want to save the video, change the file name if desired, and then click the Save button. |
| | When the video is completed, open a File Explorer window, navigate to the drive and folder, and then double-click the video file name to watch the video in the default media player app on your device. |

**Topics Review**

## Chapter 15

# Using Office Online and OneDrive

Office Online includes the web-based version of Word, Excel, and PowerPoint, which can be accessed from OneDrive. Computing, software, and storage services accessed entirely from the web is called **cloud computing**. With cloud computing, all you need is a computer with a web browser to create and edit a document, worksheet, or presentation. With cloud computing technology, you do not need to install software on your PC or mobile device, because all software and storage of documents is online. Microsoft offers its web-based productivity apps free to Microsoft account holders.

In Chapter 3, you learned how to save a presentation to OneDrive within PowerPoint. In this chapter, you will learn to create and edit files using Office Online from OneDrive and upload, download, and share files in OneDrive.

## Learning Objectives

**15.1** Create a document in Word Online

**15.2** Create a worksheet in Excel Online

**15.3** Create a presentation in PowerPoint Online

**15.4** Edit a presentation in PowerPoint Online

**15.5** Upload and download files to and from OneDrive

**15.6** Share a document in OneDrive

### Read & Learn

### Content Online

The online course includes additional training and assessment resources.

## 15.1 Creating a Document Using Word Online

**Word Online** is similar to the desktop version of Word that you used in Chapters 6 and 7; however, Word Online has fewer features than the desktop version. The program looks similar, but you will notice that for some features, functionality within a browser environment is different than the desktop version. As you work in Word Online, the document is saved automatically as soon as you make changes. A document created in Word Online is saved in the same file format as the desktop version of Word, which means documents are transferable between editions, desktops, and online.

1. Open a browser window.

2. With the insertion point positioned in a blank Address bar, type https:// onedrive.live.com and then press Enter.

3. If necessary, click *Sign in*, type the email address or phone number for your Microsoft account and click Next, type your password and click the Sign in button on the second sign-in screen. Skip this step if you are already signed in to OneDrive, which occurs with the Microsoft Edge browser when you are already signed in to your Microsoft Account for Windows.

Once signed in, the OneDrive window appears similar to the one shown in Figure 15.1.

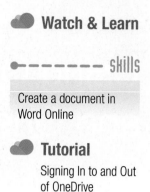

**Watch & Learn**

— — — — — — SKILLS

Create a document in Word Online

**Tutorial**

Signing In to and Out of OneDrive

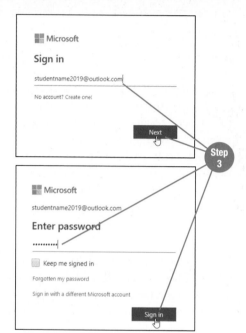

— — — — — — Oops!

Don't know your Microsoft account? If you have a hotmail.com, live.com, or outlook .com email address, your email login is your Microsoft account; otherwise, click the link to create a new account.

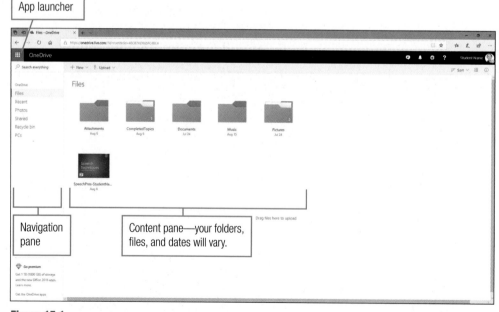

**Figure 15.1**

The OneDrive window initially shows the *Files* list in the Content pane for the signed-in user.

**4** Click the App launcher in the upper left of the screen next to OneDrive (displays as a waffle icon and is sometimes called *the waffle button*) and then click the Word tile.

**5** Click *Got it!* if a screen displays with information about new features in Office Online; this message appears the first time you use Office Online and when changes have been made to the software.

**6** Click New blank document on the Word Online Start screen.

Word Online launches, as shown in Figure 15.2. By default, the **Simplified Ribbon** displays, with one row of buttons and without the group names.

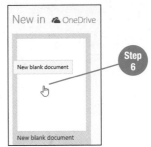

------- go online

**https://CA3 .ParadigmEducation .com/Office**

Go here to start Word Online without first starting OneDrive. You can access all the same web-based apps from https://www.office.com/. Click the desired tile and then sign in with a Microsoft account or a school account.

**Tutorial**

Creating a New Document in an Office Online App

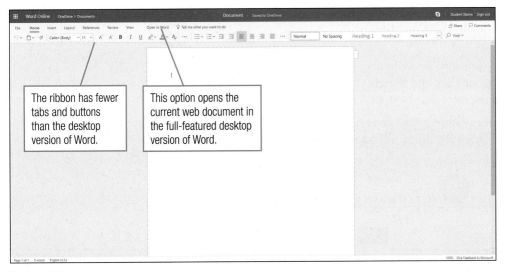

The ribbon has fewer tabs and buttons than the desktop version of Word.

This option opens the current web document in the full-featured desktop version of Word.

**Figure 15.2**
The Word Online new document window with the Simplified Ribbon is shown.

**7** Type the following text in the document window, using all the default settings:

What Is Green Computing?

Green computing refers to the use of computers and other electronic devices in an environmentally responsible manner. Green computing can encompass new or modified computing practices, policies, and procedures. This trend is growing with more individuals and businesses adopting green computing strategies every year.

Strategies include the reduction of energy consumption by computers and other devices; reduction in use of paper, ink, and toner; and reuse, recycling, or proper disposal of electronic waste.

**8** Proofread carefully and correct any typing errors that you find. If necessary, use the Spelling & Grammar button on the Review tab to spell check the document.

9 Click to place the insertion point within the title *What Is Green Computing?* and then click the Align Center button on the Home tab.

10 Select all the text in the document and then change the font size to 12 using the *Font Size* option box arrow.

11 Select the two paragraphs of text below the title, click the More Paragraph Options button (displays as three dots), point to *Line Spacing*, and then click *1.5*.

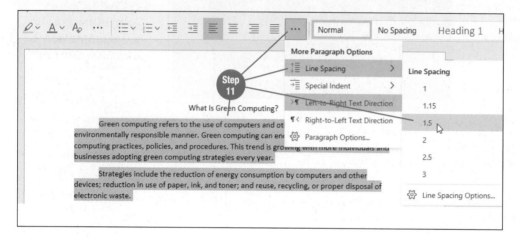

12 Click at the end of the last paragraph to position the insertion point and then press Enter to create a new line.

13 Click the Insert tab, click the Picture button, and then click *Bing*.

14 Type recycling in the *Search Bing Images* text box and then press Enter or click the Search button (magnifying glass).

15 Click the green recycling symbol image shown below and then click Insert. If the image shown is not available, insert the image using the student data file **recyclinglogo.jpg**.

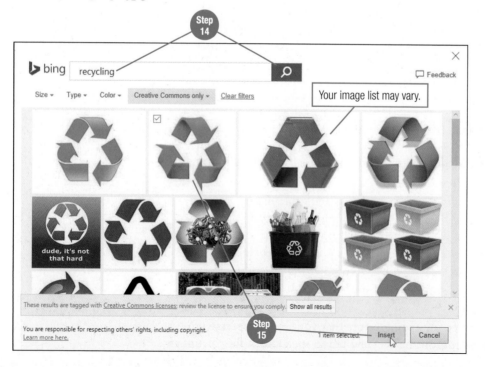

16 With the image selected, drag the lower right corner selection handle upward and left until the image is the approximate size shown in the image below.

17 With the image still selected, click the Home tab and then click the Align Center button.

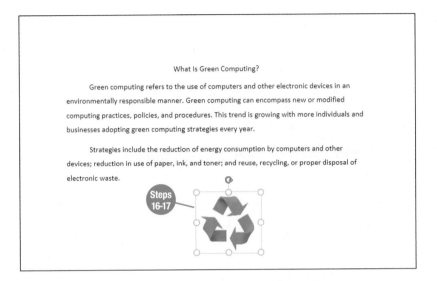

18 Click within the document text to deselect the image.

19 Click the File tab and then click *Save As*.

20 Click *Rename* in the Save As panel in the backstage area.

21 Type GreenComputing-YourName in the *Enter a name for this file* text box in the Rename dialog box and then press Enter or click OK.

22 Close the document browser tab.

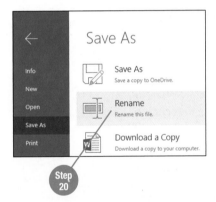

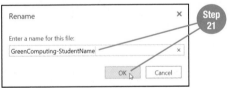

23 Click the Documents folder in the <u>Files</u> list in the OneDrive Content pane. Notice the Word document thumbnail for the **GreenComputing-YourName** document. Notice also the file management options on the bar below the OneDrive banner. Leave OneDrive open for the next topic.

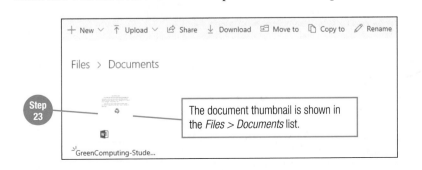

## 15.2 Creating a Worksheet Using Excel Online

**Excel Online** looks similar to the full-featured desktop edition of Excel; however, like Word Online, the ribbon contains fewer options, and functionality for some features varies from the desktop edition. Use Excel Online to create a basic worksheet or basic collaborative worksheet, but for a worksheet that requires advanced formulas or editing, the desktop version of Excel is the better choice.

**1** With OneDrive open, click the App launcher (waffle icon), click the Excel tile, and then click <u>New blank workbook</u> on the Excel Online Start screen.

Excel Online opens a window similar to the window shown in Figure 15.3. Like Word Online, Excel Online workbooks are saved in the same file format as the desktop version of Excel and are transferable between software editions.

**Watch & Learn**

●━━━━━━━ Skills

Create a worksheet in Excel Online

**Tutorial**

Creating a Worksheet Using Excel Online

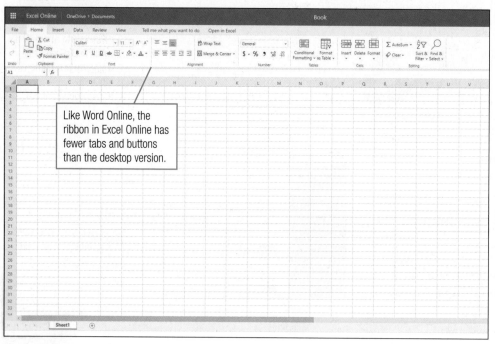

Like Word Online, the ribbon in Excel Online has fewer tabs and buttons than the desktop version.

**Figure 15.3**

The Excel Online new blank workbook window is shown.

**2** Type the labels and values in the cells as shown in the image below. In A1 type the entire worksheet title and change *Your Name* to your first and last names.

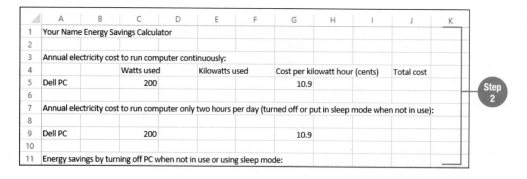

| | A | B | C | D | E | F | G | H | I | J | K |
|---|---|---|---|---|---|---|---|---|---|---|---|
| 1 | Your Name Energy Savings Calculator | | | | | | | | | | |
| 2 | | | | | | | | | | | |
| 3 | Annual electricity cost to run computer continuously: | | | | | | | | | | |
| 4 | | | Watts used | | Kilowatts used | | Cost per kilowatt hour (cents) | | | Total cost | |
| 5 | Dell PC | | 200 | | | | 10.9 | | | | |
| 6 | | | | | | | | | | | |
| 7 | Annual electricity cost to run computer only two hours per day (turned off or put in sleep mode when not in use): | | | | | | | | | | |
| 8 | | | | | | | | | | | |
| 9 | Dell PC | | 200 | | | | 10.9 | | | | |
| 10 | | | | | | | | | | | |
| 11 | Energy savings by turning off PC when not in use or using sleep mode: | | | | | | | | | | |

Step 2

**3** Click cell E5 to make E5 the active cell and then type the formula =(c5*24*365)/1000.

The formula multiplies the watts used by a PC running continuously 24 hours per day and 365 days per year, and then divides the result by 1000 to convert watts to kilowatts.

**4** Click cell J5 to make J5 the active cell and then type the formula: =e5*(g5/100).

The cost per kilowatt hour in G5 is divided by 100 to convert 10.9 to a decimal value representing cents.

**5** Type the remaining formulas in the cells indicated:

E9    =(c9*2*365)/1000
J9    =e9*(g9/100)
J11    =j5–j9

| E | F | G | H | I | J |
|---|---|---|---|---|---|
| nuously: | | | | | |
| Kilowatts used | | Cost per kilowatt hour (cents) | | | Total cost |
| 1752 | | 10.9 | | | 190.968 |
| two hours per day (turned off or put in sleep mode when not in use): | | | | | |
| 146 | | 10.9 | | | 15.914 |
| use or using sleep mode: | | | | | 175.054 |

Steps 3-5

**6** Select A1:J1 and then click the Merge & Center button in the Alignment group on the Home tab.

**7** With A1:J1 still selected, apply bold formatting and change the font size to 12.

**8** Click J5 to make J5 the active cell, click the Accounting Number Format button in the Number group on the Home tab, and then click $ English (United States).

**9** Apply the Accounting Number Format to J9 and J11.

**10** With J11 the active cell, click the Borders button arrow in the Font group on the Home tab and then click Outside Borders.

**11** With J11 still the active cell, apply bold formatting and the Green font color option (sixth color in Standard Colors section).

**12** Click in any blank cell to view the border in J11 and then proofread carefully and correct any typing errors that you find.

| | A | B | C | D | E | F | G | H | I | J |
|---|---|---|---|---|---|---|---|---|---|---|
| 1 | | | | Your Name Energy Savings Calculator | | | | | | |
| 2 | | | | | | | | | | |
| 3 | Annual electricity cost to run computer continuously: | | | | | | | | | |
| 4 | | | Watts used | | Kilowatts used | | Cost per kilowatt hour (cents) | | | Total cost |
| 5 | Dell PC | | 200 | | 1752 | | 10.9 | | | $ 190.97 |
| 6 | | | | | | | | | | |
| 7 | Annual electricity cost to run computer only two hours per day (turned off or put in sleep mode when not in use): | | | | | | | | | |
| 8 | | | | | | | | | | |
| 9 | Dell PC | | 200 | | 146 | | 10.9 | | | $  15.91 |
| 10 | | | | | | | | | | |
| 11 | Energy savings by turning off PC when not in use or using sleep mode: | | | | | | | | | $ 175.05 |

Steps 6-11

**13** Click the File tab, click Save As, and then click Rename. Type EnergySavings –YourName in the Rename dialog box and then press Enter or click OK.

**14** Close the workbook tab. Leave OneDrive open for the next topic.

An Excel workbook thumbnail is added to the Files > Documents list in OneDrive.

## 15.3 Creating a Presentation Using PowerPoint Online

A basic presentation that does not need to incorporate a chart or audio can be created using **PowerPoint Online**. Other PowerPoint Online features that vary from the full-featured desktop version include fewer animation and transition options, the inability to customize a slide show with timings, less advanced setup slide show options (see page 374), and fewer views.

(see page 374)

**Watch & Learn**

------ skills

Create a presentation in PowerPoint Online

**Tutorial**

Creating a Presentation Using PowerPoint Online

1. With OneDrive open and with the Documents folder displayed, click the App launcher, click the PowerPoint tile, and then click <u>New blank presentation</u> on the PowerPoint Online Start screen.

2. Click the DESIGN tab, click the More Themes option arrow at the right end of the Themes gallery, and then click *Banded*.

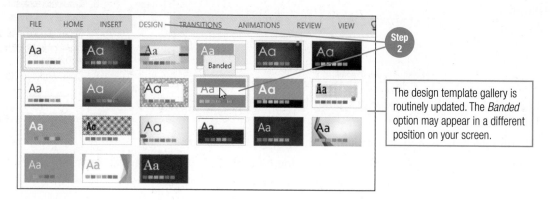

The design template gallery is routinely updated. The *Banded* option may appear in a different position on your screen.

3. Click *Variant 4* in the Variants gallery.

A new presentation with the selected theme is started in the PowerPoint Online window, as shown in Figure 15.4. Similar to Word and Excel, presentations are saved in the same file format and are transferable between PowerPoint Online and the desktop version of PowerPoint.

**Figure 15.4**

The PowerPoint Online window is shown. PowerPoint Online does not have the Slide Show tab found in the desktop version of PowerPoint.

4. Click anywhere in *CLICK TO ADD TITLE* on the title slide and then type Green Computing. Click anywhere in *Click to add subtitle* and then type your name. (The theme converts the title text to all uppercase text.)

5. Click the HOME tab and then click the New Slide button in the Slides group.

6. In the New Slide dialog box, click the *Title and Content* layout if it is not already selected, and then click Add Slide.

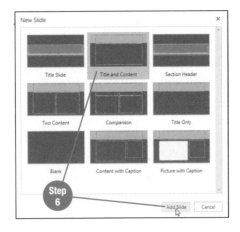

7. Type the following text on Slide 2. If Autocorrect capitalizes the first character in the last three lines, edit *C* to a lowercase character.

*Slide title*    What Is Green Computing?
*Bulleted list* Use of computers and other electronic devices in an environmentally responsible manner including new or modified:
       computing practices
       computing policies
       computing procedures

8. Add another new slide with the Title and Content layout and then type the following text on Slide 3:

*Slide title*    Green Computing Strategies
*Bulleted list* Reduction in energy consumption
       Reduction in use of paper, ink, and toner
       Reuse or recycling of devices
       Proper disposal of ewaste

9. Make Slide 2 the active slide, click the INSERT tab, and then click the Online Pictures button in the Images group.

10. Type computer in the *Search Bing Images* text box and then press Enter or click the Search button (magnifying glass).

11. Scroll down, click to select the image shown at the right, and then click Insert. Insert the student data file **computer.png** if the image shown is not available.

12. Move the image to the approximate position as shown in the slide at the right.

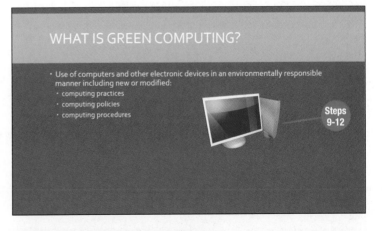

13. Make Slide 3 the active slide. Insert and position the image as shown in the slide at the right. Search for the image by typing recycling in the *Search Bing Images* text box. If necessary, insert the student data file **recycling.gif**.

14. Click the File tab, click *Rename*, and then click *Rename*. Type GreenComputingPres –YourName and then press Enter or click OK.

15. Close the presentation tab. Leave OneDrive open for the next topic.

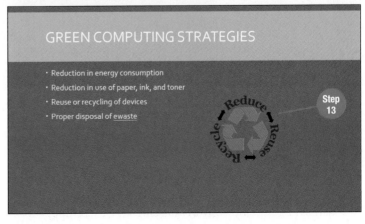

●----- **quick steps**

**Create a Presentation in PowerPoint Online**
1. Open a browser window.
2. Navigate to https://onedrive.live.com.
3. If necessary, sign in with Microsoft account.
4. Click the App launcher.
5. Click PowerPoint tile and then click New blank presentation.
6. Enter and format slides.
7. Rename presentation.
8. Close presentation tab.

## 15.4 Editing a Presentation in PowerPoint Online

Opening a presentation from OneDrive in PowerPoint Online displays the slides in Editing view so you can make changes to the presentation by adding and deleting content on existing slides or by adding new slides.

1. With OneDrive open and with the Documents folder displayed, click to insert a check mark in the check circle at the top right corner of the **GreenComputingPres–YourName** thumbnail if the check circle is empty.

   A check mark in the check circle indicates the presentation is selected. Additional options display along the top of the OneDrive window when a presentation is selected.

2. Click Open and then click Open in PowerPoint Online.

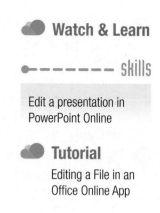

**Watch & Learn**

— — — — — — **Skills**

Edit a presentation in PowerPoint Online

**Tutorial**
Editing a File in an Office Online App

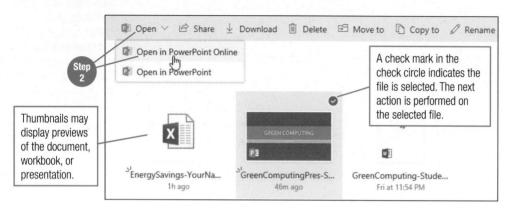

The presentation opens in Editing view, as shown in Figure 15.5.

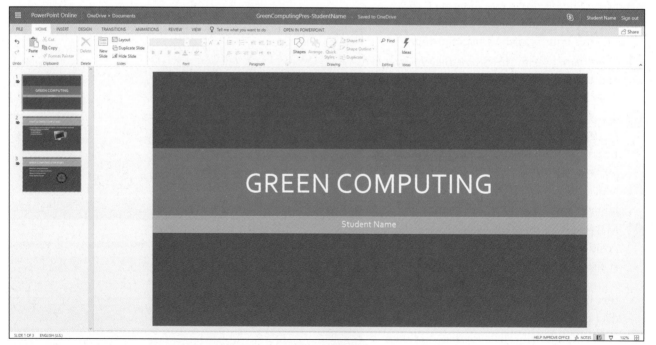

**Figure 15.5**

A presentation displays in Editing view in PowerPoint Online when opened from OneDrive.

③ Make Slide 3 the active slide.

④ Add a new slide with the Title and Content layout and then type the following text on Slide 4:

*Slide title*    Green Computing Example

*Bulleted list*    A desktop PC can use up to 1700 kilowatt hours per year if left on continuously

Turning off or putting the PC in sleep mode when not in use can save over 1600 kilowatt hours per year for average use of 2 hours per day

This strategy can save $175 per year for electricity cost at 10.9 cents per kilowatt hour

⑤ Search for the image shown on the slide below using the key phrase *power on*. Insert and move the image to the approximate position as shown. If the image shown is not available, insert the student data file **poweron.png**.

● — — — — **quick steps**

**Edit a Presentation in PowerPoint Online**
1. Open a browser window.
2. Navigate to https:// onedrive.live.com.
3. If necessary, sign in with Microsoft account.
4. Select presentation thumbnail.
5. Click <u>Open</u>.
6. Click <u>Open in PowerPoint Online</u>.
7. Edit presentation as required.
8. Close presentation tab.

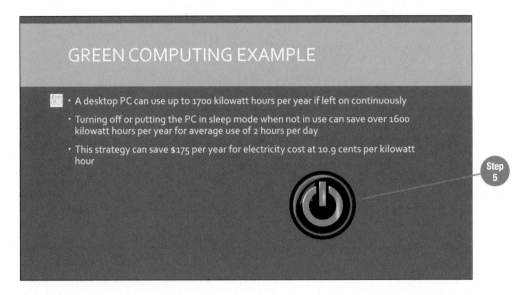

⑥ Click the TRANSITIONS tab.

⑦ Click the *Wipe* option in the Transition to This Slide gallery and then click the Apply To All button in the Timing group.

⑧ Make Slide 1 the active slide and then click the Slide Show button at the right end of the Status bar. Click Okay if a message pops up that the slide show will run full screen.

⑨ Click through each slide to the end of the slide show and then exit the slide show to return to the PowerPoint Online editing screen.

⑩ Close the presentation tab. Leave OneDrive open for the next topic.

## 15.5 Downloading Files from and Uploading Files to OneDrive

☁ **Watch & Learn**

●------- Skills

Download files from OneDrive

Upload files to OneDrive

Move files to a folder on OneDrive

Although you can work in Office applications with OneDrive as the default storage location, many people prefer to store files on a storage medium (such as their hard disk drive or a USB flash drive). In that case, OneDrive is used as a backup storage receptacle or to make a file accessible on multiple devices without having to copy the file to each PC. A file is also copied to OneDrive to share the file with other people. Similarly, you can download a file from OneDrive to your local PC or mobile device to view or edit the file offline.

**1** With OneDrive open, click to clear the check mark in the check circle for the **GreenComputingPres-YourName** thumbnail and then click to insert a check mark in the check circle for the **GreenComputing-YourName** Word document.

**2** Click <u>Download</u> in the bar along the top of the OneDrive window.

**3** Click the Save button in the pop-up message at the bottom of Microsoft Edge with the options to Open

●------- Oops!

No check circle visible on thumbnail? Move the mouse over the thumbnail and a selection circle will appear.

or Save downloads and then click the *Dismiss* option (displays as ✕) when the message displays that the file finished downloading. Skip these steps if you are using a different browser and the messages shown do not display.

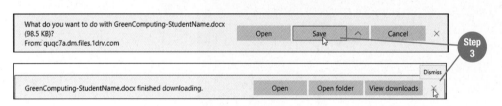

●------- app tip

You can copy and paste files to and from OneDrive from a File Explorer window. OneDrive is shown in the File Explorer Navigation pane above This PC.

By default, files downloaded from OneDrive are saved in the Downloads folder for the signed in user when you are using OneDrive in the Microsoft Edge browser. Another browser may display a Save As dialog box when you click the Download option. In that case you can select to save the file in the Downloads folder on your computer.

**4** Click to clear the check mark in the check circle for the **GreenComputing -YourName** thumbnail.

**5** Select and download the **EnergySavings-YourName** workbook to the Downloads folder by completing steps similar to those in Steps 1 through 3.

**6** Clear the check mark for the **EnergySavings-YourName** thumbnail and then select and download the **GreenComputingPres-YourName** presentation to the Downloads folder by completing steps similar to those in Steps 1 through 3.

☁ **Tutorial**

Downloading Files from OneDrive

You can select and download multiple files in one operation instead of doing each file individually. OneDrive creates a zipped folder when more than one file is selected to download. Files in a zipped folder should be extracted after downloading.

**7** Click the File Explorer button on the taskbar to open File Explorer.

**8** Click *Downloads* in the Navigation pane.

The three files downloaded from OneDrive are shown in the Content pane, as seen in the image at the right.

**9**   Select and copy the three downloaded files to a new folder named *Ch15* in the CompletedTopics folder on your storage medium. Refer to Chapter 1, Topic 1.9 if you need assistance with this step.

**10**   Close the File Explorer window to return to the browser window with OneDrive active. Clear the check mark in the check circle for the **GreenComputingPres–YourName** thumbnail if the file is still selected.

In the next steps you will copy three images from your storage medium to your account storage at OneDrive.

**11**   Click Upload in the bar along the top of the OneDrive window, and then click Files.

**12**   In the Open dialog box, navigate to the Ch15 folder in StudentDataFiles on your storage medium, select the three files in the folder whose names begin with **PaintedBunting**, and then click Open.

The three files are uploaded to the current folder in your account storage in OneDrive. Thumbnails appear for uploaded files when the upload is complete.

**13**   Click to insert a check mark in the check circle for each of the three painted bunting image thumbnails and then click Move to in the bar along the top of the window.

**14**   Click *Pictures* in the Move items to panel at the right side of the window and then click Move at the top of the panel. Leave OneDrive open for the next topic.

Files are uploaded to the active folder in OneDrive. An alternative way to complete this topic would be to first display the Pictures folder in the Content pane and then upload the three image files instead of uploading and then moving the files as you did in Steps 11 to 14.

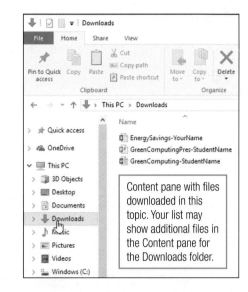

Content pane with files downloaded in this topic. Your list may show additional files in the Content pane for the Downloads folder.

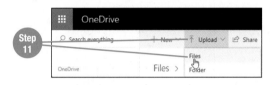

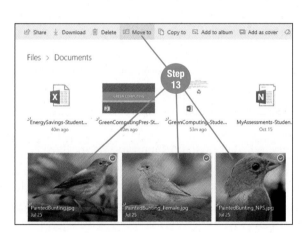

**quick steps**

**Download a File from OneDrive**
1. Sign in to OneDrive.
2. Select file.
3. Click Download.
4. Click *Save*, then *Dismiss* option.
5. Open File Explorer.
6. Display Downloads folder in Content pane.
7. Select, copy, and paste downloaded file to desired location.

**Upload a File to OneDrive**
1. Sign in to OneDrive.
2. Click Upload, then Files.
3. Navigate to drive and folder.
4. Select file(s) to upload.
5. Click Open.
6. If necessary, move files to desired destination folder.

**Tutorial**

Uploading Files

## 15.6 Sharing a File on OneDrive

OneDrive is an excellent tool for collaborating on documents when working with a team. A team leader can create or upload documents to OneDrive and then share the files with the team members who need them. An individual with shared access to a document receives an email with a link to the file. Changes to the file are made to the copy in OneDrive so that only one document, worksheet, or presentation has to be managed. Collaborating by sharing a file on OneDrive is less cumbersome than sending a file as an email attachment and then trying to manage multiple versions of the same document.

*Note: In this topic, you will share a Word Online document with a classmate. Check with your instructor for instructions on with whom you should share the Word document. If necessary, share the document with yourself by using an email address other than your Microsoft account.*

1 With OneDrive open, select the **GreenComputing-YourName** document.

2 Click <u>Share</u> in the bar along the top of the window.

3 If a message box pops up with an *Expiring links* message and a Get Premium button, click the close button (displays as ×).

4 Click <u>Email</u> in the Share *GreenComputing-YourName* option box.

5 Click in the *Enter a name or email address* text box and type the email address for a classmate.

More than one email address can be entered in the text box by typing a semicolon to separate email addresses.

6 Click in the *Add a message here* text box and then type Please make your changes to the file accessed from this link.

7 Click <u>Share</u>.

A message displays in the notifications area when the file is successfully shared.

**Watch & Learn**

●  ━ ━ ━ ━ ━ ━  **Skills**

Share a file on OneDrive

**Tutorials**

Sharing a File on OneDrive

Sharing a Folder from OneDrive

●  ━ ━ ━ ━ ━ ━  **app tip**

You can share files with anyone with a valid email address—the recipient does not have to have a Microsoft account.

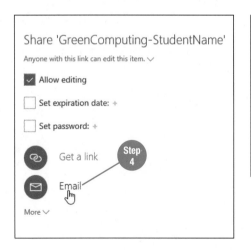

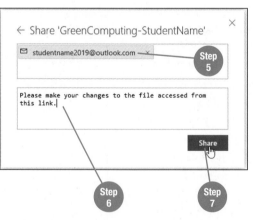

8 Click the App launcher and then click the Mail tile. Click *Got it!* if a message displays with information about changes made to the Mail application in OneDrive.

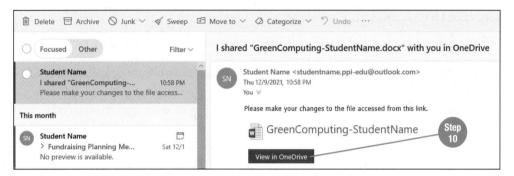

9 With *Inbox* the active mail folder, click the message header for the message received from a classmate with the subject line informing you the **GreenComputing-StudentName.docx** file has been shared with you in OneDrive to view the message in the message pane.

10 Click the View in OneDrive button in the message window to open the document in Word Online. Notice the *Edit Document* option in the bar along the top of the window.

**quick steps**

**Share a File on OneDrive**
1. Sign in to OneDrive.
2. Select file.
3. Click Share, then Email.
4. Click in email address text box and type recipient email address(es).
5. Click in message text box and type message.
6. Click Share.

**Oops!**

No message? Check the email address that the classmate used to ensure the correct address was typed. If an address other than Outlook, Hotmail, or Live was used, you need to go to another mail program to find the message with the link. In that case, sign out of OneDrive, launch your other mail program, and complete Steps 9 and 10. Note also that some mail programs may flag the message as junk mail. Check your Junk Mail folder if the message is not in your Inbox.

11 Close the document tab and close the Mail tab to return to the OneDrive window.

12 Click Shared in the Navigation pane to view the file details of files shared by you and with you in the Content pane.

You will have two entries in the Shared panel if you shared the document with a classmate and a classmate shared his or her document with you.

13 Click the *Account* option that displays your account picture or the generic user silhouette at the right end of the OneDrive bar to view the My accounts pane.

14 Click Sign out.

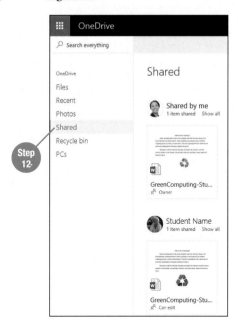
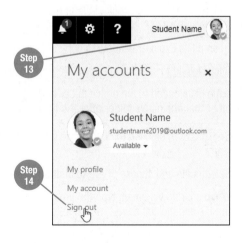

15 Close the browser window.

## Topics Review

| Topic | Key Concepts |
|---|---|
| **15.1 Creating a Document Using Word Online** | **Cloud computing** is a service in which computing resources, software, and storage are provided online, accessed via a web browser. |
| | **Word Online** is the web-based version of Word accessed from OneDrive. The application has fewer ribbon tabs and options than the desktop version of Word, and the functions of some features may vary. |
| | Documents created in Word Online are saved automatically in the same file format as the desktop version of Word, meaning files can be transferred between editions. |
| | Sign in to OneDrive with a Microsoft account, click the App launcher, click the Word tile, and then click New blank document to start a new document. |
| | Word Online starts with the **Simplified Ribbon** displayed by default that shows only one row of buttons and without group names. |
| | Use the Save As backstage area to rename the document saved automatically by Word Online. |
| **15.2 Creating a Worksheet Using Excel Online** | **Excel Online** is best suited for basic worksheets; use the desktop version of Excel for worksheets that need advanced formulas or editing. |
| | Worksheets created in Excel Online are saved in the same file format as the desktop version of Excel, meaning files can be transferred between editions. |
| | Like Word Online, Excel Online has fewer features than the desktop version, and some functionality may vary. |
| **15.3 Creating a Presentation Using PowerPoint Online** | Use **PowerPoint Online** to create a presentation that does not need charts, audio, advanced animation or transition effects, or advanced slide show setup options. |
| | Presentations created in PowerPoint Online are saved in the same file format as the desktop version of PowerPoint, meaning files can be transferred between editions. |
| **15.4 Editing a Presentation in PowerPoint Online** | Select a presentation file thumbnail, click Open and then click the Open in PowerPoint Online option in OneDrive to open a presentation in Editing view. |
| | View the slide show for a presentation in PowerPoint Online by clicking the Slide Show button at the right end of the Status bar. |
| **15.5 Downloading Files from and Uploading Files to OneDrive** | A check mark in the check circle at the top right of a file thumbnail indicates the file is selected. |
| | Click Download to copy a selected file from OneDrive to the Downloads folder on your computer. Open a File Explorer window to copy and paste the file to the desired location. |
| | Click Upload to select and copy a file from your computer to your OneDrive storage. |
| | A file is uploaded to the active folder in OneDrive. Use the Move to option to move a file to a folder other than the active folder after the file is uploaded. |
| **15.6 Sharing a File on OneDrive** | OneDrive can be used to collaborate with team members by sharing one copy of a file among several users. |
| | Select a file and choose Share to distribute a shared file via email or by a link to the shared file. |
| | When sharing via email, type the email address of individual(s) who will be allowed to view and edit the file and include a brief message to the recipient(s). |
| | Click Shared in the OneDrive Navigation pane to view files that you have shared and files that have been shared with you by others. |

Topics Review

# Glossary/Index

## Image Credits